CONTEMPORARY MATHEMATICS

Titles in this Series

Volume

1 **Markov random fields and their applications,** Ross Kindermann and J. Laurie Snell

2 **Proceedings of the conference on integration, topology, and geometry in linear spaces,** William H. Graves, Editor

3 **The closed graph and P-closed graph properties in general topology,** T. R. Hamlett and L. L. Herrington

4 **Problems of elastic stability and vibrations,** Vadim Komkov, Editor

5 **Rational constructions of modules for simple Lie algebras,** George B. Seligman

6 **Umbral calculus and Hopf algebras,** Robert Morris, Editor

7 **Complex contour integral representation of cardinal spline functions,** Walter Schempp

8 **Ordered fields and real algebraic geometry,** D. W. Dubois and T. Recio, Editors

9 **Papers in algebra, analysis and statistics,** R. Lidl, Editor

10 **Operator algebras and K-theory,** Ronald G. Douglas and Claude Schochet, Editors

11 **Plane ellipticity and related problems,** Robert P. Gilbert, Editor

12 **Symposium on algebraic topology in honor of José Adem,** Samuel Gitler, Editor

13 **Algebraists' homage: Papers in ring theory and related topics,** S. A. Amitsur, D. J. Saltman and G. B. Seligman, Editors

14 **Lectures on Nielsen fixed point theory,** Boju Jiang

15 **Advanced analytic number theory. Part I: Ramification theoretic methods,** Carlos J. Moreno

16 **Complex representations of GL(2, K) for finite fields K,** Ilya Piatetski-Shapiro

17 **Nonlinear partial differential equations,** Joel A. Smoller, Editor

18 **Fixed points and nonexpansive mappings,** Robert C. Sine, Editor

19 **Proceedings of the Northwestern homotopy theory conference,** Haynes R. Miller and Stewart B. Priddy, Editors

20 **Low dimensional topology,** Samuel J. Lomonaco, Jr., Editor

Titles in this series

Volume

21 **Topological methods in nonlinear functional analysis,** S. P. Singh, S. Thomeier, and B. Watson, Editors

22 **Factorizations of $b^n \pm 1$, $b = 2, 3, 5, 6, 7, 10, 11, 12$ up to high powers,** John Brillhart, D. H. Lehmer, J. L. Selfridge, Bryant Tuckerman, and S. S. Wagstaff, Jr.

23 **Chapter 9 of Ramanujan's second notebook—Infinite series identities, transformations, and evaluations,** Bruce C. Berndt and Padmini T. Joshi

24 **Central extensions, Galois groups, and ideal class groups of number fields,** A. Fröhlich

25 **Value distribution theory and its applications,** Chung-Chun Yang, Editor

26 **Conference in modern analysis and probability,** Richard Beals, Anatole Beck, Alexandra Bellow and Arshag Hajian, Editors

27 **Microlocal analysis,** M. Salah Baouendi, Richard Beals and Linda Preiss Rothschild, Editors

Conference in Modern Analysis and Probability

CONFERENCE IN MODERN ANALYSIS AND PROBABILITY

HELD AT YALE UNIVERSITY,

NEW HAVEN, CONNECTICUT

JUNE 8–11, 1982

1980 *Mathematics Subject Classification*. Primary 46Bxx, 60Gxx, 28Dxx.

Library of Congress Cataloging in Publication Data

Conference in Modern Analysis and Probability (1982: Yale University)
Conference in Modern Analysis and Probability.
(Contemporary mathematics, ISSN 0271-4132; v. 26)
Bibliography: p.
1. Mathematical analysis—Congresses. 2. Probabilities—Congresses. I. Beales, Richard, 1938– . II. Title. III. Series.
QA299.6.C66 1984 515 84-484
ISBN 0-8218-5030-X

CONTEMPORARY MATHEMATICS

Volume 26

Conference in Modern Analysis and Probability

Richard Beals, Anatole Beck,
Alexandra Bellow and Arshag Hajian,
Editors

AMERICAN MATHEMATICAL SOCIETY
Providence · Rhode Island

This volume is dedicated to

PROFESSOR SHIZUO KAKUTANI

with the affection, admiration and
respect of his students and colleagues.

TABLE OF CONTENTS

Foreword........ 1

Program........ 3

List of Participants........ 5

1. Roy L. Adler and Leopold Flatto: Cross section map for the geodesic flow on the modular surface 9

2. M. A. Akcoglu and L. Sucheston: On identification of superadditive ergodic limits 25

3. J. R. Baxter and R. V. Chacon: The equivalence of diffusions on networks to Brownian motion 33

4. A. Bellow and V. Losert: On sequences of density zero in ergodic theory 49

5. J. van den Berg and M. Keane: On the continuity of the percolation probability function 61

6. Felix E. Browder: Coincidence theorems, minimax theorems, and variational inequalities 67

7. J. R. Choksi: Recent developments arising out of Kakutani's work on completion regularity of measure 81

8. J. R. Choksi and S. J. Eigen: An automorphism of a homogeneous measure algebra which does not factorize into a direct product 95

9. Daniel I. A. Cohen: Another generalization of the Brouwer fixed point theorem 101

10. Yael Naim Dowker: An ergodic theorem 105

11. Nathaniel A. Friedman: Higher order partial mixing 111

12. Hillel Furstenberg: IP-systems in ergodic theory 131

13. Arshag Hajian and Yuji Ito: Induced transformations on a section ... 149

14. Edwin Hewitt: Conjugate Fourier series on the character group of the additive rationals 159

15. Kiyosi Itô: A stochastic differential equation in infinite dimensions 163

16. Kinrad Jacobs: Ergodic theory and combinatorics 171

17. William B. Johnson and Joram Lindenstrauss: Extensions of Lipschitz mappings into a Hilbert space 189

18. Robert R. Kallman: A uniqueness result for a class of compact connected groups 207

19. L. A. Karlovitz: Two extremal properties of functions 213

20. Robert Kaufman: On Bernoulli convolutions 217

21. Harvey B. Keynes and Mahesh G. Nerurkar: Generic theorems for lifting dynamical properties by continuous affine cocycles 223

22. Bruce Kitchens: Linear algebra and subshifts of finite type 231

23. Anthony Lo Bello: The etymology of the word ergodic 249

24. Peter A. Loeb: A functional approach to nonstandard measure theory 251

25. Dorothy Maharam: On positive operators 263

26. Brian Marcus, Karl Petersen and Susan Williams: Transmission rates and factors of Markov chains 279

27. I. Namioka: Ellis groups and compact right topological groups 295

28. William Parry and Klaus Schmidt: Invariants of finitary isomorphisms with finite expected code-lengths 301

29. Marina Ratner: Ergodic theory in hyperbolic space 309

30. Haskell P. Rosenthal: Embedding of L^1 in L^1 335

31. Daniel J. Rudolph: Inner and barely linear time changes of ergodic R^k-actions 351

32. M. J. Sharpe: Processes evolving from the indefinite past 373

33. Erik G. F. Thomas: An infinitesimal characterization of Gelfand pairs 379

34. Nils Tongring: Multiple points of Brownian motion 387

35. Benjamin Weiss: Measurable dynamics 395

36. Kôsaku Yosida and Shigetake Matsuura: A note on Mikusiński's proof of the Titchmarsh convolution theorem 423

37. Robert J. Zimmer: Ergodic actions of arithmetic groups and the Kakutani-Markov fixed point theorem 427

INTRODUCTION

The Conference in Modern Analysis and Probability in honor of Professor Shizuo Kakutani was held on June 8-11, 1982, at Yale University on the occasion of his retirement. In these Proceedings we present the papers that were submitted for this Conference. Some of the invited authors were unable to attend, and due to time constraints not all the papers submitted were included in the program of the Conference.

The three major areas of mathematics on which the Conference focused were functional analysis, probability theory, and ergodic theory. Most of the articles presented were works by the respective authors on problems that were pioneered by Professor Kakutani in the past. Questions in Brownian motion, induced transformations, representation of M-spaces, and fixed point theorems were discussed.

Members of the organizing committee were Richard Beals, Anatole Beck, Alexandra Bellow, and Arshag Hajian. Due to geographical restrictions, however, the weight of the preparatory work fell on Richard Beals. The assistance given to the Committee by Phyllis Stevens of the Mathematics Department at Yale was invaluable; her experience and competence insured the smoothness of the Conference. The painstaking work of retyping the manuscripts for the Proceedings was done by the secretarial staff of the Mathematics Department at Yale under the careful guidance of Regina Hoffman; we owe much to her in the preparation of these Proceedings.

We are indebted to the National Science Foundation[1] for the initial funding of this Conference, without which it would have been difficult for this Conference to have taken place.

[1] NSF Grant - MCS-8118339

FOREWORD

As indicated by this conference honoring Shizuo Kakutani on his formal retirement from his Professorship at Yale, his talents and achievements in his chosen field are well known and appreciated throughout the world of mathematics.

I would like to take this opportunity to make a few remarks about his achievements as a humanitarian, - a person devoted to promoting the welfare of humanity, in this case, particularly those who showed an interest in mathematics, though this was not a necessary condition.

I first met Shizuo in the spring of 1941 at the Institute for Advanced Study. I would not remember this if it were not for the fact that he wrote me a letter dated May 8, 1941, in which he expressed his pleasure at having met me and sent a solution to a problem concerning the transitivity properties of a dynamical system which I had posed to him. As always, he was careful about accreditations and said that John Oxtoby had suggested the method for constructing the desired example.

I came to Yale in 1948. The presiding chairman of the department was W. R. Longley, who was scheduled to retire in 1949, and though it was generally assumed that I would inherit the august post, the administration did not make it official until well along in the spring of 1949. Early in 1949, Jake informed me that our mutual friend, Shizuo Kakutani, who was then visiting the Institute for Advanced Study, was seeking a position in the United States. We informed Longley and persuaded him to go forward with the necessary steps leading to an appointment for Shizuo. I think there was some concern as to how the undergraduates would react to being taught by a Japanese whose interests might be mainly concentrated in mathematical research. But the appointment went through and, to our good fortune, Shizuo joined us.

As it turned out, there was no need of any concern. Shizuo took great interest in the undergraduate teaching and in the students themselves, an action which is often conducive to more effort on the part of the students. His courses were never routine and he must have put much time, effort and ingenuity into preparing his lectures. He never paraded his superior knowledge or unusual talents in mathematics. He set high standards of achievements but was always compassionate for those who could only reach lesser heights. The excellence of his teaching did not go unrecognized and in 1968 he was presented with the William Clyde DeVane Award at a meeting of the undergraduate membership of Phi Beta Kappa.

Let me turn to Shizuo, the library builder. When I came to Yale, the Department of Mathematics had its own library. It was housed in a fairly large room on the top floor of Leet Oliver and was by no means insignificant. There were many volumes and even journals such as Acta Mathematica, which, like many of the books, was duplicated in the Main Library. There was no librarian,-it was run on the honor system, - and it was very useful. One of the members of the department was in charge of the library and there was little in the way of duties connected with the position.

A few years after I came, Ed Begle, who had been in charge of the library, had more important things to do, and I asked Shizuo to take on the task, assuming that he would give about as much time to it as the others had.

Quite unintentionally, I think I made some of his early days on the job the unhappiest of his life. There were some fine old volumes, beautifully bound, in the departmental library, but these were duplicated in the Main Library, and I didn't think that we needed two copies of the same book. Since there was little support for our library, I decided to sell the duplicates. One of the New York book dealers came to New Haven, offered what seemed like a reasonable sum and took them away. I have never seen anyone looking more woebegone than Shizuo was, as the books were carried off. He looked as if he were losing some of his best friends. I didn't realize until long afterward that those books were among his best friends.

But he recovered and soon made it clear that he wasn't going to take just a casual interest in the departmental library, - he aimed at making it rank with the best mathematical libraries in the country. And that is just what he did, with the backing of Charles Rickart, particularly, during whose chairmanship our library finally acquired a librarian.

I hardly need to say anything about the remarkable effectiveness of Shizuo's teaching at the graduate level. Many of his doctoral students were at this conference and each has his own tale of the profound influence that Shizuo has had on his life. His concern for the welfare of his students didn't end with their attainment of the doctoral degree. I think he always knew where all his students were, what they were doing and, when needed, he gave them support. When he didn't have enough to worry about as far as his own students were concerned, he used to worry about mine.

But Shizuo's teaching has reached far beyond the confines of Yale University and its students. In a sense he has been the effective teacher of a host of us, a stimulating and gracious giver, sharing his knowledge, ideas, ingenuity and the precious gift of time, with one and all. For this, we extend our thanks to our good friend Shizuo and hope that we can continue to learn from him for a long time to come.

G. A. Hedlund

Program of the

Conference on Modern Analysis and Probability

Tuesday, June 8, 1982

9:30-10:20	J. Lindenstrauss: "Some recent results on Banach lattices"
11:00-11:50	K. Itô: "Stochastic differential equations in infinite dimensions"
2:00-2:40	F. Browder: "Degree theories, old and new"
2:50-3:30	E. Hewitt: "Conjugate Fourier series on the character group of the rationals"
4:00-4:40	W. Johnson: "Extensions of Lipschitz mappings"
5:00-5:40	L. Karlovitz: "Chebyshev and related extremal polynomials"

Wednesday, June 9, 1982

9:30-10:20	D. Sullivan: "Riemann surfaces and dynamics"
11:00-11:50	A. Dvoretzky: "The Fatou inequality and cluster points of random variables"
2:00-2:40	P. Loeb: "A Daniell integral and extensions of nonstandard measure theory"
2:50-3:30	R. Anderson: "Some applications of nonstandard analysis in probability theory"
4:00-4:40	R. Kaufman: "Bernoulli convolutions and differentiable functions"
4:50-5:30	M. Sharpe: "Processes evolving from the indefinite past"

Thursday, June 10, 1982

9:00-9:45 K. Jacobs: "Ergodic theory and combinatorics, a survey"

10:00-10:45 H. Furstenberg: "IP sets and ergodic theory"

11:15-12:00 M. Ratner: "Rigidities of horocycle flow"

2:00-2:40 W. Parry: "Invariants of finitary isomorphisms with finite expected code-lengths"

2:50-3:30 D. Maharam Stone: "Some problems related to measure algebras"

4:00-4:40 M. Keane: "The unicity of infinite clusters and the continuity of the percolation probability"

4:50-5:30 B. Weiss: "Equivalence of transformations and flows"

Friday, June 11, 1982

9:00-9:45 A. Katok: "Special representations for group actions and Kakutani equivalence"

10:00-11:15 J. Feldman and D. Ornstein: "Kakutani equivalence"

11:30-12:15 R. Zimmer: "Actions of arithmetic groups"

2:00-2:40 J. Choksi: "Some recent developments arising out of Kakutani's work on completion regularity of measures"

2:50-3:30 J. Oxtoby: "Probability limit identification functions"

4:00-4:40 K. Petersen: "Efficient encoding and metrically sofic systems"

LIST OF PARTICIPANTS

ADLER, Roy L.	IBM T.J. Watson Research Center
ANAGNOSTAKIS, Christopher	Albertus Magnus College
ANDERSON, Robert	Princeton University
ARNOLD, Leslie K.	Daniel H. Wagner Associates
AUSLANDER, Joseph	University of Maryland, College Park
AUSLANDER, Louis	CUNY, Graduate Center
BADE, William G.	University of California, Berkeley
BARTLE, Robert G.	University of Illinois, Urbana-Champaign
BEALS, Richard W.	Yale University
BECK, Anatole	University of Wisconsin, Madison
BELLOW, Alexandra	Northwestern University
BLAKLEY, G.R.	Texas A & M University
BLEI, Ron C.	University of Connecticut, Storrs
BROWDER, Felix E.	University of Chicago
BROWN, Leon	Wayne State University
BURKEL, Robert B.	Kansas State University
CANAVATI, Jose A.	Universidad Autonoma Metropolitana, Mexico
CHIANG, G. Carina	Princeton University
CHOKSI, Jal R.	McGill University, Canada
COHEN, Daniel A.	Rockefeller University
COLE, Brian	Brown University
COMFORT, W. Wistar	Wesleyan University
CONLON, Lawrence W.	Washington University
CORWIN, Lawrence	Rutgers University
COVEN, Ethan M.	Wesleyan University
CURTIS, Philip C. Jr.	University of California, Los Angeles
del JUNCO, Andreas	Ohio State University
DJAFARI-ROUHANI, Behzad	California State University, Long Beach
DVORETZKY, Aryeh	Hebrew University, Israel
DYNKIN, Eugene B.	Cornell University
EDELMAN, Alan	Yale University
EIGEN, Stanley	McGill University, Canada
ELSON, Constance	Ithaca College
ELTON, John H.	University of Texas, Austin
ENGEL, David D.	Daniel H. Wagner Associates
ENGEL, Frank P.	Daniel H. Wagner Associates
FEIT, Sidnie	Yale University
FELDMAN, Jacob	University of California, Berkeley

FELDMAN, Paul	Yale University
FELL, Harriet	Northeastern University
FIELDSTEEL, Adam	Wesleyan University
FLATTO, Leopold	Bell Telephone Laboratories
FRENKEL, Igor	Yale University
FURSTENBERG, Hillel	Hebrew University, Israel
GAMELIN, Theodore	University of California, Los Angeles
GLASS, Michael	CUNY, Bronx Community College
GLICKSBERG, Irving	University of Washington, Seattle
GODDARD, John	Universidad Autonoma Metropolitana, Mexico
GORDON, Yehoram	Texas A & M University
GREEN, Leon W.	University of Minnesota, Minneapolis
HAJIAN, Arshag B.	Northeastern University
HAWKINS, Jane	SUNY, Stony Brook
HEDLUND, Gustav A.	Yale University
HELLERMAN, Leo	IBM, Poughkeepsie
HEWITT, Edwin	University of Washington, Seattle
HOWE, Roger	Yale University
ITÔ, Kiyosi	Gakushuin University, Japan
ITO, Yuji	Rikkyo University, Japan
JACOBS, Konrad	University of Erlangen, West Germany
JACOBSON, Florie	Albertus Magnus College
JACOBSON, Nathan	Yale University
JOHNSON, William B.	Ohio State University
KAKUTANI, Shizuo	Yale University
KARLOVITZ, Les A.	Georgia Institute of Technology
KATOK, Anatole	University of Maryland, College Park
KATZNELSON, Yitzhak	Hebrew University, Israel
KAUFMAN, Robert	University of Illinois, Urbana-Champaign
KEANE, Michael	Delft University of Technology, Netherlands
KENNY, Patrick	McGill University, Canada
KEY, Eric	Cornell University
KEYNES, Harvey B.	University of Minnesota, Minneapolis
KNOWLES, Robert	University of Connecticut, Waterbury
KOLCHIN, Ellis	Columbia University
KRIEGER, Wolfgang	University of Heidelberg, West Germany
KRIKELES, Basil C.	Florida International University
LACEY, H. Elton	Texas A & M University
LAWNICZAK, Anna T.	Louisiana State University
LIND, Douglas A.	University of Washington, Seattle
LINDENSTRAUSS, Joram	Hebrew University, Israel

LO BELLO, Anthony J.	Allegheny College
LOEB, Peter	University of Illinois, Urbana-Champaign
LYONS, Terence J.	University of London, England
MA, Lawrence	Yale University
MACKEY, George W.	Harvard University
MAGYAR, Peter	Yale University
MARCUS, Brian	University of North Carolina, Chapel Hill
MASSEY, William S.	Yale University
MILNE, Stephen	Texas A & M University
MOSTOW, George D.	Yale University
MULVEY, Irene	Wesleyan University
NAMIOKA, Isaac	University of Washington, Seattle
NEYMAN, Abraham	Harvard University
ODELL, Edward	University of Texas, Austin
ONOYAMA, Takuji	Keio University, Japan
ORNSTEIN, Donald S.	Stanford University
OXTOBY, John C.	Bryn Mawr College
PARRY, William	Warwick University, England
PETERSEN, Karl	University of North Carolina, Chapel Hill
PRASAD, V.S.	University of Sherbrooke, Canada
PUCKETTE, Stephen E.	University of the South
RANDOL, Burton	CUNY, Graduate Center
RATNER, Marina	University of California, Berkeley
RHODES, Frank	University of Southampton, England
RICKART, Charles E.	Yale University
RIGELHOF, R.	McGill University, Canada
ROSEN, William G.	National Science Foundation
ROSS, Kenneth	University of Oregon, Eugene
ROTA, Gian-Carlo	Massachusetts Institute of Technology
ROTHMAN, Neal J.	National Science Foundation
ROVNYAK, James	University of Virginia, Charlottesville
SAITO, Tosiya	Keio University, Japan
SALTMAN, David	Yale University
SCHREIBER, Bertram M.	Wayne State University
SCHWARTZMAN, Sol	University of Rhode Island, Kingston
SCOVILLE, Richard	Duke University
SEDRANSK, Nell	SUNY, Center at Albany
SELIGMAN, George	Yale University
SHARPE, Michael	University of California, San Diego
SHIELDS, Paul	Stanford University
SIDNEY, Stuart	University of Connecticut, Storrs

SINE, Robert	University of Rhode Island, Kingston
SMITH, Mi-Soo	SUNY, Old Westbury
SNELL, J. Laurie	Dartmouth College
STARR, Norton	Amherst College
STEWART, Mark	Northeastern University
STONE, Arthur H.	University of Rochester
STONE, Dorothy Maharam	University of Rochester
STONE, Marshall H.	University of Massachusetts, Amherst
SULLIVAN, Dennis	CUNY, Graduate Center
SZCZARBA, Robert	Yale University
TAMAGAWA, Tsuneo	Yale University
TANAKA, Junichi	Tsuru University, Japan
THOMAS, Erik	University of Groningen, Netherlands
THOUVENOT, Jean Paul	University of Paris VI, France
TONGRING, Nils	Yale University
VOAS, Charles H.	Yale University
WADA, Junzo	Waseda University, Japan
WEISS, Benjamin	Hebrew University, Israel
WEISSLER, Fred	University of Texas, Austin
WERMER, John	Brown University
WHITE, Brian	Courant Institute of Mathematical Sciences
WILLIAMS, Susan	University of North Carolina, Chapel Hill
WOOD, Carol	Wesleyan University
YU, Kai F.	Yale University
ZIMMER, Robert	University of Chicago
ZUCKERMAN, Gregg	Yale University

Contemporary Mathematics
Volume 26, 1984

CROSS SECTION MAP FOR THE GEODESIC FLOW ON THE MODULAR SURFACE

Roy L. Adler and Leopold Flatto

1. INTRODUCTION

In this paper we investigate the relationship between two topics which at first sight seem unrelated. The first deals with the ergodic properties of geodesic flows on two-dimensional surfaces of constant negative curvature, a rather active area in the thirties studied by many well known mathematicians. For a detailed survey of the work of that period see [H]. The second one deals with ergodic properties of non-invertible maps of the unit interval, a current popular subject and one also with an interesting history going back to Gauss (see [B]). We shall show how each of these subjects sheds light on the other. Ergodic properties of interval maps can be used to derive ergodic properties of flows and conversely. Furthermore explicit formulas for invariant measures of interval maps can be gotten from the invariance of the hyperbolic measure for the flows. This fact is particularly interesting as there is a paucity of formulas for invariant measures of interval maps and we have here a method of deriving a class of these. In particular, we show how Gauss's formula for the invariant measure associated with continued fractions, which seems to have been produced ad hoc, can be derived anew.

The connection between geodesic flows and non-invertible maps of the unit interval is achieved by a series of reductions. The first one, attributed to Poincaré, reduces the study of a flow to that of a cross section map. For the geodesic flows which we are considering, the motion takes place on a 3-dimensional space, namely the unit tangent bundle to a 2-dimensional surface. The first reduction yields a cross section map T_C on a 2-dimensional region C. If the measure of C is infinite, an intermediate reduction, called inducing, can be made in order to replace C by a subset of finite measure. A suitable coordinatization of C shows that a second reduction is possible - namely the existence of a one-dimensional factor map whose natural extension is T_C.

In this paper we restrict our discussion to one particular cross section for the geodesic flow on the modular surface. We have chosen one of the simplest examples to illustrate our ideas. The geometrical constructions associated

0271-4132/84 $1.00 + $.25 per page

with other cross sections and other Fuchsian groups get more complex and we hope to present these in subsequent papers. The example that we have chosen is also interesting because it deals with a cross section map which is the natural extension of the so called continued fraction transformation. We shall see how the invariant measure guessed by Gauss can be derived from our recipe.

Our arguments are based on simple geometric ideas. Unfortunately, to make these rigorous, niggling details are encountered involving certain sets of measure zero. We remove these sets from discussion, in order to facilitate the ensuing arguments so that the reader does not loose track of the main ideas. The arising disadvantage is that results become incomplete as certain geodesics are not accounted for. In [AF], we give a complete discription of the flow, and an accompanying symbolic description accounting for all geodesics.

2. GEODESIC FLOW ON THE MODULAR SURFACE

We give in this section a description of the geodesic flow on the modular surface. We first describe the corresponding flow on the hyperbolic plane.

Let $H = \{x+iy : y > 0\}$ be the upper half plane endowed with the metric $ds^2 = \frac{dx^2+dy^2}{y^2}$. H is called the hyperbolic plane and has constant negative curvature -1. The geodesics of H are half-circles or straight lines orthogonal to the x-axis. Let U be the unit tangent bundle consisting of the unit tangent vectors on H. U is coordinatized by $u = u(x,y,\theta)$, where (x,y) is the base point of $u \in U$ and θ is the angle measured in the counterclockwise direction between the positive x-axis and u. The hyperbolic measures on H, U are defined respectively by $dA = \frac{dxdy}{y^2}$, $dm = dA d\theta$ The geodesic flow G_t, $-\infty < t < \infty$, is the class of homeomorphisms of U defined by $u \to u_t$, where u and u_t are unit tangent vectors to the initial and terminal points of a geodesic segment of hyperbolic length t. G_t has a simple description in the following coordinatization of U. To each $u \in U$, assign the triple (ξ,η,s) where ξ,η are the points of intersection of the geodesic determined by u and the x-axis, ξ being in the forward direction, and s is the hyperbolic arc length parameter along the geodesic. The origin of s on a half circle is chosen to be its mid-point, but for vertical lines some other convention must be adopted. In these coordinates, the flow G_t becomes

$$G_t : (\xi,\eta,s) \to (\xi,\eta,s+t) \qquad 2.1)$$

A Jacobian computation, which we omit, yields the formula

$$dm = \frac{dx\,dy\,d\theta}{y^2} = 2\frac{d\xi\,d\eta\,ds}{(\xi-\eta)^2} \tag{2.2}$$

2.2) puts into evidence the fact that dm is invariant under G_t.

We carry over the above concepts and formulas to the modular surface. Let Γ be the modular group, i.e. $\Gamma = \{\tau(z) = \frac{az+b}{cz+d}$: $ad - bc = 1$ and $a,b,c,d \in Z\}$. Abstractly Γ is the group generated by $\alpha(z) = z + 1$, $\beta(z) = -\frac{1}{z}$ satisfying the relations:

$$\beta^2 = (\beta\alpha)^3 = \text{identity} \tag{2.3}$$

Γ acts both on $\mathbf{H}$ and $\mathbf{U}$, the action on the latter given by $\bar{\tau}(z,\theta) = (\tau(z),\ \theta + \arg \tau'(z))$. We refer to Γ as $\bar{\Gamma}$, when acting on $\mathbf{U}$. Let M and $\mathbf{M}$ denote respectively the spaces of Γ-and $\bar{\Gamma}$-orbits of $\mathbf{H}$ and $\mathbf{U}$, i.e.

$$M = \{\Gamma z : z \in H\}, \quad \mathbf{M} = \{\bar{\Gamma}u : u \in \mathbf{U}\}$$

where

$$\Gamma z = \{\tau z : \tau \in \Gamma\}, \quad \bar{\Gamma}u = \{\tau u : \tau \in \bar{\Gamma}\}$$

One obtains concrete realizations of M and $\mathbf{M}$ in the following manner. Let

$$F = \{z=x+iy: |z| > 1, -\tfrac{1}{2} \le x < \tfrac{1}{2}\} \cup \{z=x+iy: |z| = 1, -\tfrac{1}{2} \le x < 0\}.$$

F is a fundamental domain for Γ (see [L]) and H is tessellated with the images of F under Γ. Opposite vertical boundary lines of F are identified under α, and the left half of the bottom boundary with the right under β. Consequently M can be thought of as the closure of F under the above identifications. Similarly unit vectors with base points on the boundary of F can also be identified under $\bar{\alpha}$ and $\bar{\beta}$, and so $\mathbf{M}$ can be thought of as unit vectors emanating from points of $\bar{F}$ modulo these identifications.

To topologize M and $\mathbf{M}$ we introduce the projection maps π, $\bar{\pi}$ from $\mathbf{H}$ to M and $\mathbf{U}$ to $\mathbf{M}$ resp. i.e.

$$\pi(z) = \Gamma z,\ z \in \mathbf{H}, \quad \bar{\pi}(u) = \bar{\Gamma}u,\quad u \in \mathbf{U}$$

Projections by π and $\bar{\pi}$ of neighborhoods of $\mathbf{H}$ and $\mathbf{U}$ form a basis of neighborhoods for M and $\mathbf{M}$. Topologically M is a punctured sphere (open disk) and M an open solid torus. With the exception of the Γ-orbits of the points i, $\frac{1}{2} + \frac{\sqrt{3}}{2}i$ and the $\bar{\Gamma}$-orbits of the unit vectors based at i, $\frac{1}{2} + \frac{\sqrt{3}}{2}i$, the projection maps are locally 1-1, so that x,y or ξ,η provide local coordinates for the points of $M - \{\pi(i),\ \pi(\frac{1}{2} + \frac{\sqrt{3}}{2}i)\}$, and x,y,θ or ξ,η,s provide local coordinates for $\mathbf{M} - \{\bar{\pi}(0,1,\theta),\ \bar{\pi}(\frac{1}{2} + \frac{\sqrt{3}}{2},\theta)\},\ 0 \le \theta < 2\pi\}$. If we avoid the exceptional points, then the formulas for ds, dA, dm carry over to M and $\mathbf{M}$. Because G_t commutes with the action of $\bar{\Gamma}$ on $\mathbf{U}$, $\mathbf{M}$ inherits a geodesic flow $\bar{G}_t = \bar{\pi}G_t$ from $\mathbf{U}$. The inherited measure for $\mathbf{M}$, denoted by $\bar{m} = m\bar{\pi}^{-1}$, is invariant under the inherited flow. We have

$$\bar{m}(\mathbf{M}) = \int_F dm = \int_0^{2\pi}\int_{-1/2}^{1/2}\int_{\sqrt{1-x^2}}^{\infty} \frac{dydxd\theta}{y^2} = \frac{2\pi^2}{3}$$

In order to normalize the measure so that $\bar{m}(\mathbf{M}) = 1$, we set for the rest of the paper

$$d\bar{m} = \frac{3}{2\pi^2}\frac{dxdyd\theta}{y^2} = \frac{3}{\pi^2}\frac{d\xi d\eta d\theta}{(\xi-\eta)^2}$$

3. CROSS SECTION MAPS

To study the flow on $\mathbf{M}$, we introduce the notion of a cross section map. Roughly speaking, a cross section is a subset which every orbit hits again and again. The correspondence between successive points of return to the cross section serves to define a cross section map. Furthermore, a measure on a space which is invariant under a flow induces a measure on a cross section which is invariant under the cross section map. The cross section for the geodesic flow on the modular surface which we select consists essentially of projections into $\mathbf{M}$ of elements of $\mathbf{U}$ emanating from the boundary of F and pointing into its exterior. We say essentially because we are confronted with the difficulty that there exist orbits which do not visit the proposed cross section infinitely often both past and future - these are the $\bar{\pi}$-projections of G_t orbits starting or terminating at cusp points. Analytically, they are described by $\bar{\pi}(\xi,\eta,s)$, $-\infty < s < \infty$, where either ξ or η is rational or ∞. Thus for some elements of our proposed cross section, the cross section map would remain undefined.

To remove this difficulty we can either i) augment the cross section so that all orbits intersect it infinitely often both past and future or ii) remove the above mentioned orbits from $\mathbf{M}$ - and their points of intersection from the proposed cross section. Of the two approaches, (i) has the advantage of accounting for all orbits. Unfortunately, the details encountered in carrying out i) prove rather cumbersome and we pursue instead ii) (i) is presented in [AF]). Furthermore, in order to give a convenient description of the cross section map in terms of the ξ,η coordinates introduced below, we shall also remove from $\mathbf{M}$ the orbits passing through $\pi(\frac{1}{2}, \frac{\sqrt{3}}{2})$. The totality of orbits removed from $\mathbf{M}$ forms a set of measure zero, the same holding for the removed points from the cross section.

We call the resulting cross section C, and decompose it into four parts as $C = \bigcup_{i-1}^{4} C_i$ where: C_1 and C_4 are the $\bar{\pi}$-projections of vectors emanating respectively from the right and left boundaries of F and pointing to its exterior, C_2 and C_3 are the $\bar{\pi}$-projections of vectors emanating from the lower boundary of F and pointing to its exterior, to the right in case of C_2 and to the left in case of C_3. The sets C_i are depicted in Figure 1.

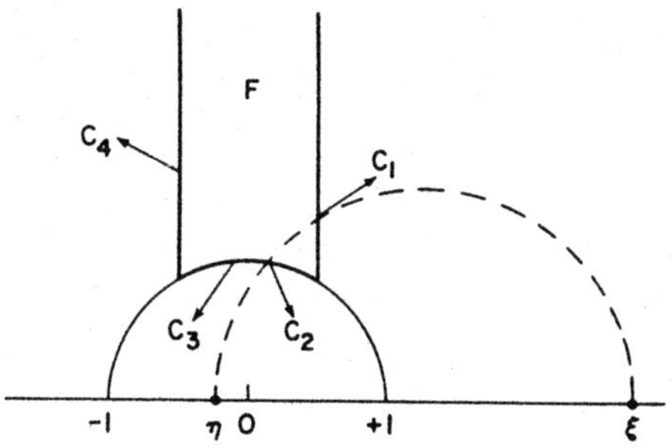

Figure 1

To each element $v \in C$, we assign the coordinates $(\xi.\eta)$ where $u(\xi,\eta,s)$ is the unique vector in $\mathcal{U}$ emanating from ∂F and pointing out of F such that $\bar{\pi}(u) = v$. This coordinate assignment is 1-1 on C. It is an exericse in analytic geometry to express the C_i's in these coordinates. We designate the set in the (ξ,η) plane representing C_i again by C_i. We have:

$$C_1 = \{(\xi,\eta):(\xi,\eta) \notin D,\ (\xi - \tfrac{1}{2})(\eta - \tfrac{1}{2}) < -\tfrac{3}{4},\ \xi > \eta\}$$

$$C_2 = \{(\xi,\eta):(\xi,\eta) \notin D,\ (\xi + \tfrac{1}{2})(\eta + \tfrac{1}{2}) < -\tfrac{3}{4} < (\xi - \tfrac{1}{2})(\eta - \tfrac{1}{2}), \xi > \eta\}$$

$$C_3 = -C_2,\ C_4 = -C_1 \qquad 3.1)$$

where

$$D = \{(\xi,\eta):\tau\xi = \infty \text{ or } \tau\eta = \infty \text{ or } (\tau\xi,\tau\eta) = [\tfrac{1}{2}, \tfrac{\sqrt{3}}{2}] \text{ for some } \tau \in \Gamma\} \qquad 3.2)$$

The condition $(\xi,\eta) \notin D$, tacitly assumed from now on, arises from the fact that we have removed from the flow those orbits which are $\bar{\pi}$-projections of G orbits having at least one end point at a cusp, and those passing through $\pi[\frac{1}{2}, \frac{\sqrt{3}}{2}]$.

In the ξ,η coordinates, the cross section map T_C is given by

$$T_C(\xi,\eta) = \begin{cases} (\xi-1,\eta-1), & \text{on } C_1 \\ (-\frac{1}{\xi}, -\frac{1}{\eta}) & \text{on } C_2 \cup C_3 \\ (\xi+1,\eta+1) & \text{on } C_4 \end{cases} \qquad 3.3)$$

The sets C_i and their images $C_i' = T_C(C_i)$ are depicted in Figure 2, where we have decomposed C_i, C_i' as $C_{i+} \cup C_{i-}$, $C_{i+}' \cup C_{i-}'$, $i = 2,3$, and have replaced the symbols C_i, etc. by i, etc.

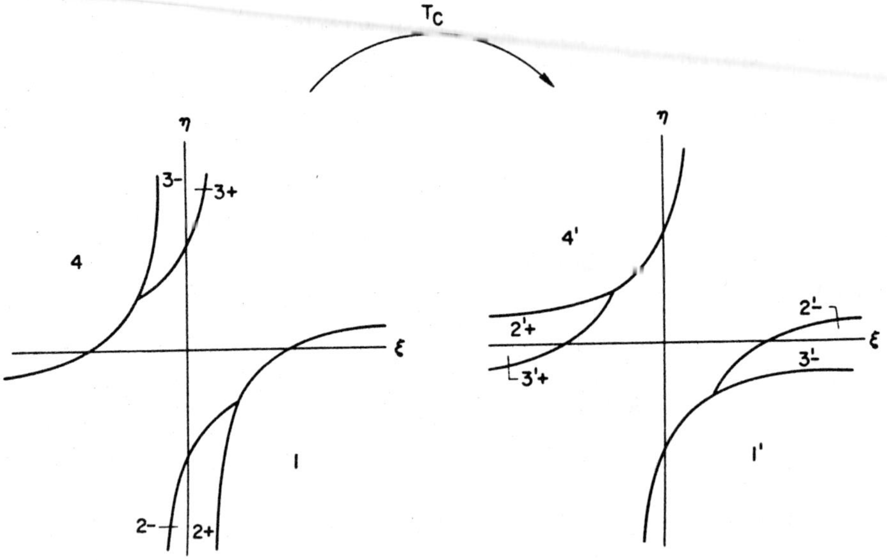

Figure 2

We verify 3.3) for $(\xi,\eta) \in C_1$, a similar verification holding for the other cases. A geodesic leaving F from the right vertical wall at (ξ,η,s) is identified under α^{-1} with a geodesic entering F on the left vertical wall at $(\xi-1,\eta-1,s)$. This geodesic leaves F at some point $(\xi-1,\eta-1,s')$. Thus the next return to C from an element $(\xi,\eta) \in C_1$ is the element $T_C(\xi,\eta) = (\xi-1,\eta-1)$, be it either in C_1 or C_3.

The $\bar{G}_t$-invariant measure $\bar{m}$ induces a T_C-invariant measure denoted by m_C, on the cross section C. m_C is defined by

$$m_C(B) = \lim_{\Delta s\to 0} \frac{1}{\Delta s}\bar{m}\{G_t u : u \in B, 0 \le t \le \Delta s\}$$

for any measurable subset B of C. In the ξ,η coordinates dm_C is obtained from dm by dropping s, i.e.

$$dm_C = \frac{3}{2\pi^2}\frac{d\xi d\eta}{(\xi-\eta)^2} \qquad 3.4)$$

The T_C-invariance of m_C can either be derived directly from its definition [AF] or from a Jacobian computation which shows that $\frac{d\xi d\eta}{(\xi-\eta)^2}$ is preserved when ξ,η are subjected to the same fractional linear transformation. Integrating 3.4), we obtain $m_C(C) = \infty$.

4. CONJUGACY OF CURVILINEAR AND RECTILINEAR MAPS

It seems rather difficult, if not impossible, to obtain the ergodic properties of $T_C(\xi,\eta)$ directly from formula 3.3). To do so, we introduce an auxiliary map $T_R(\bar{\xi},\bar{\eta})$ which will be shown conjugate to $T_C(\xi,\eta)$. $T_R(\bar{\xi},\bar{\eta})$ is defined as follows

Let $R = \bigcup_{i=1}^{4} R_i$, where

$$\begin{aligned} R_1 &= \{(\bar{\xi},\bar{\eta}):(\bar{\xi},\bar{\eta}) \notin D,\ 1 < \bar{\xi} < \infty,\ -\infty < \bar{\eta} < 0\} \\ R_2 &= \{(\bar{\xi},\bar{\eta}):(\bar{\xi},\bar{\eta}) \notin D,\ 0 < \bar{\xi} < 1,\ -\infty < \bar{\eta} < -1\} \\ R_3 &= -R_2,\ R_4 = -R_1 \end{aligned} \tag{4.1}$$

D is defined by 3.2) with $\bar{\xi},\bar{\eta}$ replacing ξ,η. $T_R(\bar{\xi},\bar{\eta})$ is the 1-1 map from R onto itself given by:

$$T_R(\bar{\xi},\bar{\eta}) = \begin{cases} (\bar{\xi}-1,\bar{\eta}-1) & \text{on } R_1 \\ (-\frac{1}{\bar{\xi}},-\frac{1}{\bar{\eta}}) & \text{on } R_2 \cup R_3 \\ (\bar{\xi}+1,\bar{\eta}+1) & \text{on } R_4 \end{cases} \tag{4.2}$$

The sets R_i and their images $R_i' = T_R(R_i)$ are depicted in Figure 3, where we have replaced R_i, R_i' resp. by $\bar{i}$, $\bar{i}'$. The R_i's are straightened out versions of the C_i's. For this reason we call C and R respectively the curvilinear and rectilinear domains. Similarly we call T_C, (ξ,η), the curvilinear transformation and coordinates, and T_R, $(\bar{\xi},\bar{\eta})$, the rectilinear transformations and coordinates.

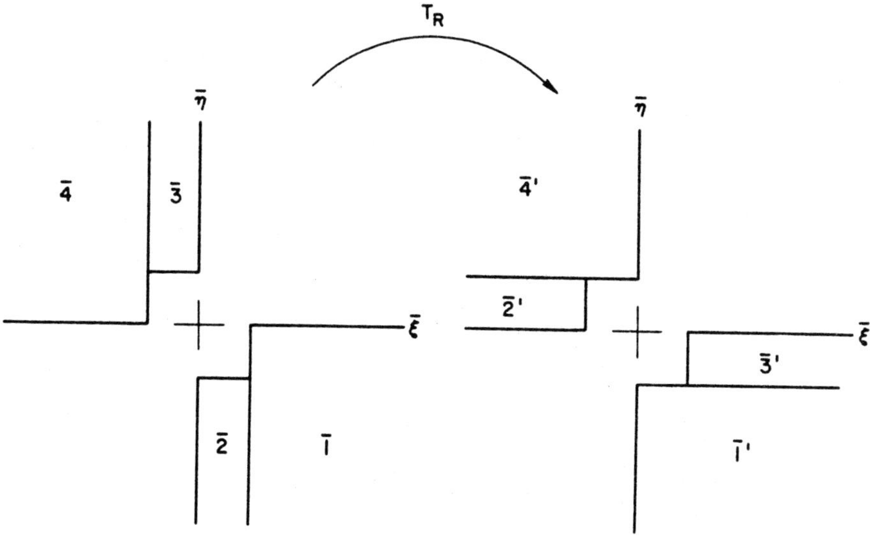

Figure 3

Since $T_R(\bar{\xi},\bar{\eta})$ is defined by piecewise transforming the variables $\bar{\xi},\bar{\eta}$ by the same fractional linear transformation, it also preserves the measure $dm_C = \frac{3}{2\pi^2}\frac{d\bar{\xi}d\bar{\eta}}{(\bar{\xi}-\bar{\eta})^2}$. An integration yields $m_C(R) = \infty$.

The reason for introducing T_R stems from the fact that it has a feature which T_C does not have - namely T_R maps vertical lines in R into vertical lines and T_R^{-1} maps horizontal lines into horizontal ones. Thus we have

$$T_R(\bar{\xi},\bar{\eta}) = (f_E(\bar{\xi}),\cdot) \qquad 4.3)$$

where $f_E(\bar{\xi})$ is given by

$$f_E(\bar{\xi}) = \begin{cases} \bar{\xi}-1 & , \quad 1 < \bar{\xi} < \infty \\ -\frac{1}{\bar{\xi}} & , \quad 0 < |\bar{\xi}| < 1 \\ \bar{\xi}+1 & , \quad -\infty < \bar{\xi} < 1 \end{cases} \qquad 4.4)$$

$f_E(\bar{\xi})$ maps the set $E = (-\infty,\infty)$-rationals onto itself. We call f_E a factor of T_R, and T_R an extension of f_E. Remarkably, we also have

$$T_R^{-1}(\bar{\xi},\bar{\eta}) = (\cdot,f_E(\bar{\eta})) \qquad 4.5)$$

with the same function f_E. In section 6, we use formulas 4.3), 4.5), to derive the ergodic properties of T_R.

We set up a conjugacy between T_C and T_R by constructing a 1-1 map Φ from C onto R satisfying

$$T_R \circ \Phi = \Phi \circ T_C \qquad 4.6)$$

$$\Phi = \text{identity on } C \cap R \qquad 4.7)$$

We show that conditions 4.6), 4.7) serve to determine Φ. Indeed, they force the manner in which Φ maps the four pieces of $C - R = U_1 \cup U_2 \cup V_1 \cup V_2$ respectively onto the four pieces of $R - C = \bar{U}_1 \cup \bar{U}_2 \cup \bar{V}_1 \cup \bar{V}_2$ defined below and depicted in Figure 4.

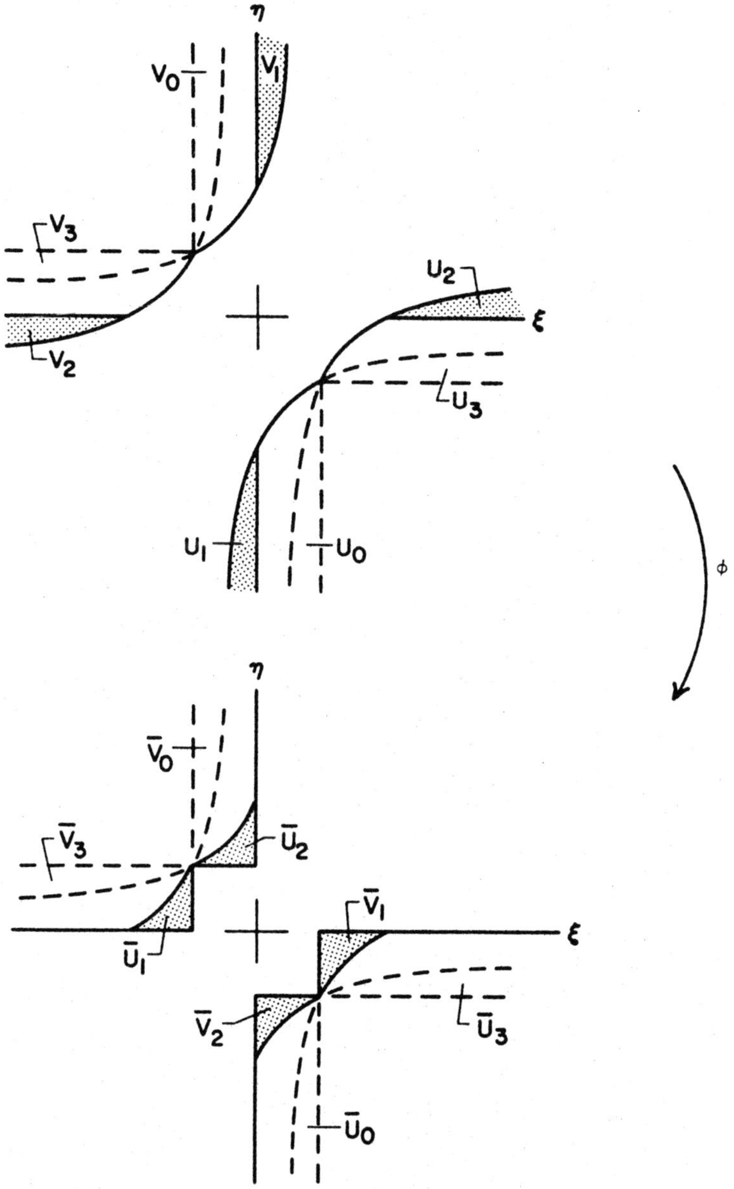

"BULGES INTO CORNERS"

Figure 4

$$U_1 = C_2 - R \cap C$$
$$U_2 = C_1 - R \cap C$$
$$V_1 = C_3 - R \cap C$$
$$V_2 = C_2 - R \cap C$$

$$\bar{U}_1 = R_4 - R \cap C$$
$$\bar{U}_2 = R_3 - R \cap C$$
$$\bar{V}_1 = R_1 - R \cap C$$
$$\bar{V}_2 = R_2 - R \cap C$$

In Figure 4, the first four sets are referred to as "bulges" and the last four as "corners".

To show that Φ is fixed by 4.6), 4.7) we introduce the following four subsets of $R \cap C$

$$U_0 = \bar{U}_0 = T_C^{-1} U_1$$
$$U_3 = \bar{U}_3 = T_C U_2$$
$$V_0 = \bar{V}_0 = T_C^{-1} V_1$$
$$V_3 = \bar{V}_3 = T_C V_2$$

Recalling that $\alpha(z) = z+1$, $\beta(z) = -\frac{1}{z}$, we define, with a slight abuse of notation, $\alpha(\xi,\eta) = (\alpha\xi,\alpha\eta)$ and $\beta(\xi,\eta) = (\beta\xi,\beta\eta)$. In these terms

$$T_C(\xi,\eta) = \begin{cases} \alpha^{-1}(\xi,\eta) & \text{on } C_1 \\ \beta(\xi,\eta) & \text{on } C_2 \cup C_3 \\ \alpha(\xi,\eta) & \text{on } C_4 \end{cases} \qquad 4.8)$$

with a similar formula for $T_R(\bar{\xi},\bar{\eta})$, replacing the C_i's by R_i's. Verifying 4.6) amounts to checking the commutativity of the following diagrams

$$\begin{array}{ccccccc}
U_0 & \xrightarrow{T_C = \alpha^{-1}} & U_1 & \xrightarrow{T_C = \beta} & U_2 & \xrightarrow{T_C = \alpha^{-1}} & U_3 \\
\downarrow \Phi=I & & \downarrow \Phi & & \downarrow \Phi & & \downarrow \Phi=I \\
\bar{U}_0 & \xrightarrow{T_R = \beta} & \bar{U}_1 & \xrightarrow{T_R = \alpha} & \bar{U}_2 & \xrightarrow{T_R = \beta} & \bar{U}_3
\end{array}$$

$$\begin{array}{ccccccc}
V_0 & \xrightarrow{T_C = \alpha} & V_1 & \xrightarrow{T_C = \beta} & V_2 & \xrightarrow{T_C = \alpha} & V_3 \\
\downarrow \Phi=I & & \downarrow \Phi & & \downarrow \Phi & & \downarrow \Phi=I \\
\bar{V}_0 & \xrightarrow{T_R = \beta} & \bar{V}_1 & \xrightarrow{T_R = \alpha^{-1}} & \bar{V}_2 & \xrightarrow{T_R = \beta} & \bar{V}_3
\end{array}$$

The commutativity forces Φ to satisfy

$$\Phi = \begin{cases} \beta\alpha & \text{on } U_1 \\ (\alpha\beta)^2 & \text{on } U_2 \\ \beta\alpha^{-1} & \text{on } V_1 \\ (\alpha^{-1}\beta)^2 & \text{on } V_2 \\ (\beta\alpha)^3 & \text{on } U_3 \\ (\beta\alpha^{-1})^3 & \text{on } V_3 \end{cases} \qquad 4.9)$$

Observe that the group relations yield $(\beta\alpha)^3 = (\beta\alpha^{-1})^3 =$ identity, which is consistent with condition 4.7), as $U_3, V_3 \subseteq R \cap C$.

We abbreviate 4.9) as

$$\Phi = \begin{cases} (\xi,\eta) & \text{on } C \cup R \\ \left(-\frac{1}{\xi+1}, -\frac{1}{\eta+1}\right) & \text{on } U_1 \cup V_2 \\ \left(-\frac{1}{\xi-1}, -\frac{1}{\eta-1}\right) & \text{on } U_2 \cup V_1 \end{cases} \qquad 4.10)$$

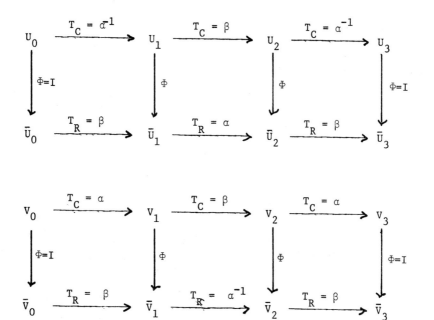

5. INDUCING

Ergodic theory of factors and extensions was developed by Rohlin [R] for transformations on finite measure spaces. In Section 6 we use this theory to prove the ergodicity of T_R, and hence that of $\bar{G}_t$, but we must first overcome the difficulty that T_R acts on the infinite measure space R. This is achieved by introducing a related map T_Q, called the induced rectilinear map, which acts on $Q = R_1 \cup R_2$.

We have

$$m_C(Q) = \frac{3}{\pi^2} \int_Q\int \frac{d\bar{\xi}d\bar{\eta}}{(\bar{\xi}-\bar{\eta})^2} = \frac{6 \log 2}{\pi^2} < \infty \tag{5.1}$$

and

$$R = \bigcup_{n=1}^{\infty} T_R^{-n}Q = \bigcup_{n=1}^{\infty} T_R^{n}Q \tag{5.2}$$

Define

$$T_Q(\bar{\xi},\bar{\eta}) = T_R^{n(\bar{\xi},\bar{\eta})}(\bar{\xi},\bar{\eta}) \tag{5.3}$$

where

$$n(\bar{\xi},\bar{\eta}) = \inf(n:n \geq 1, T_R^n(\bar{\xi},\bar{\eta}) \in Q)$$

From 5.2), we conclude that T_Q is a 1-1 map from Q onto itself. T_Q preserves m_C on measurable subsets of Q. This fact follows from 5.2) and standard set manipulations. T_Q is given by

$$T_Q(\bar{\xi},\bar{\eta}) = \begin{cases} (-(1/\bar{\xi}), - 1/\bar{n} + [1/\bar{\xi}]) & \text{on } R_2 \\ - T_Q(-\bar{\xi},-\bar{\eta}) & \text{on } R_3 \end{cases} \tag{5.4}$$

where $(\bar{\xi})$, $[\bar{\xi}]$ denote respectively the fractional and integral part of $\bar{\xi}$.

It is convenient to introduce the change of variables $(\tilde{\xi},\tilde{\eta}) = \psi(\bar{\xi},\bar{\eta}) = (\bar{\xi}-1/\bar{n})$. Let $P_i = \psi R_i$, $P = P_2 \cup P_3$, $T_P = \psi T_Q \psi^{-1}$. 5.4) transforms to

$$[\tilde{\xi}',\tilde{\eta}'] = T_P(\tilde{\xi},\tilde{\eta}) = \begin{cases} - \left((1/\tilde{\xi}), \dfrac{1}{[1/\tilde{\xi}]+\tilde{\eta}}\right) & \text{on } P_2 \\ - T_P(\tilde{\xi},\tilde{\eta}) & \text{on } P_3 \end{cases} \tag{5.5}$$

T_P has the normalized invariant measure $m_P = \dfrac{m_C\psi^{-1}}{m_C(Q)}$ given by

$$dm_P = \frac{1}{2 \log 2} \frac{d\tilde{\xi}d\tilde{\eta}}{(1+\tilde{\xi}\tilde{\eta})^2} \tag{5.6}$$

We remark that T_P, like T_R is conjugate to a cross section map - namely, if we set $B = \Phi^{-1}\psi^{-1}P$ and define $T_B = (\psi\phi)^{-1}T_P\psi\phi$, then T_B is the cross section map induced by $\bar{G}_t$ on the cross section B.

6. FACTOR MAPS

Let $f(\xi) = (\frac{1}{\xi})$, $0 < \xi < 1$, $\xi \in I^+ = (0,1) \cap E$. $f(\xi)$ is called the continued fraction map, the reason for the name stemming from the fact that $f[a_0 \dots a_n \dots] = [a_1 \dots a_n \dots]$, where $[a_0 \dots a_n \dots]$ is the continued fraction expansion of ξ. $f(\xi)$ is known to be ergodic w.r.t. Lebesgue measure [B], and Gauss observed that $f(\xi)$ has the invariant measure $\frac{d\xi}{1+\xi}$. These facts together with the ergodic theorem yield various interesting metrical theorems about the digits of continued fraction expansions [B].

We show in the present section that $f(\xi)$ is a factor map of T_P which was shown in Section 4 to be conjugate to the cross section map T_B of the flow $\bar{G}_t$. From this fact, we shall derive: i) the invariant measure of $f(\xi)$ from that of $\bar{G}_t$, ii) ergodic properties of $\bar{G}_t$ from those of $f(\xi)$ and vice versa.

We drop tildas from the variables $\tilde{\xi},\tilde{\eta}$ and augment P of Section 5 to $P = S \cup -S$, where $S = \{(\xi,\eta) : 0 < \xi,\eta < 1,\ \xi \text{ and } \eta \text{ irrational}\}$.

For the augmented P, T_P is defined as in 5.5), replacing P_2, P_3 resp. by S, $-S$. Since we have only added a zero set to P, the new T_P and the old one have identical ergodic properties. To take advantage of the symmetry of the action of T_P on P, we define

$$T_S(\xi,\eta) = \left((\frac{1}{\xi}),\ \frac{1}{[\frac{1}{\xi}]+\eta} \right), \quad (\xi,\eta) \in S \tag{6.1)}$$

It is readily verified that T_S is a 1-1 map from S onto itself, with

$$T_S^{-1}(\xi,\eta) = \left(\frac{1}{[\frac{1}{\eta}]+\xi},\ (\frac{1}{\eta}) \right) \tag{6.2)}$$

If we identify $(\dots a_{-2}, a_{-1}; a_0, a_1 \dots)$ with $(\xi,\eta) = ([a_0, a_1, \dots], [a_{-1}, a_{-2}, \dots])$, then $T_S(\dots a_{-2}, a_{-1}; a_0, a_1, \dots) = (\dots a_{-1}, a_0; a_1, a_2, \dots)$ and $T_S^{-1}(\dots a_{-2}, a_{-1}; a_0, a_1 \dots) = (\dots a_{-3}, a_{-2}; a_{-1}, a_0, \dots)$.

Let $q(\xi,\eta) = (|\xi|, |\eta|)$, which maps P onto S. On S we obtain the T_S-invariant measure $m_S = m_P q^{-1}$. In ξ,η coordinates

$$dm_S = \frac{1}{\log 2} \frac{d\xi d\eta}{(1+\xi\eta)^2} \tag{6.3)}$$

Let $S_2 = \{1,-1\}$, $m_2(1) = m_2(-1) = \frac{1}{2}$, and endow $S \times S_2$ with the measure $m_S \times m_2$. Let $T_2(\varepsilon) = -\varepsilon$, $\varepsilon = \pm 1$. $(\xi,\eta) \to (q(\xi,\eta), \operatorname{sgn} \xi)$ is an isomorphism between P and $S \times S_2$, T_P being conjugate to $T_S \times T_2$.

Let $P_\xi(\xi,\eta) = \xi$, $(\xi,\eta) \in S$. On I^+, define the measure $\mu = m_S P_\xi^{-1}$.

THEOREM 1:

i) μ is f-invariant, i.e. $\mu(f^{-1}A) = \mu(A)$ for any measurable set $A \subset I^+$.

ii) $d\mu = \frac{1}{\log 2}\frac{d\xi}{1+\xi}$, i.e. μ is the Gauss measure.

PROOF:

i) $T_S(\xi,\eta) = (f(\xi),\cdot)$ so that

$$p_\xi \circ T_S = f \circ p_\xi \tag{6.4}$$

Hence

$$\mu(f^{-1}A) = m_S(p_\xi^{-1}f^{-1}A) = m_S(T_S^{-1}p_\xi^{-1}A) = m\ (p_\xi^{-1}A) = \mu(A) \tag{6.5}$$

ii) For $A \in I^+$, $p_\xi^{-1}(A) = A \times I^+$. Hence

$$\mu(A) = m_S(A \times I^+) = \frac{1}{\log 2}\int_A d\xi \int_0^1 \frac{d\eta}{(1+\xi\eta)^2} = \frac{1}{\log 2}\int_A \frac{d\xi}{1+\xi} \tag{6.6}$$

We derived the ergodic properties of $\bar{G}_t$ from those of T_S, and those of T_S from those of f. Concerning the latter we have

THEOREM 2: f is exact, in other words it has trivial tail field - i.e. if $A \in \bigcap_{n=1}^{\infty} f^{-n}\mathbf{B}_{I^+}$, where $\mathbf{B}_{I^+}$ denotes the field of Borel subsets of I^+, then $\mu(A) = 0$ or 1. A short proof of this can be found in [MPV, lemma 4, p.114]. We prove that T_S is the natural extension of f.

THEOREM 3: T_S is the natural extension of f, which means

i) T_S is 1-1 onto a.e.

and

ii) $\bigcup_{n=0}^{\infty} T_S^r\mathbf{A} = \mathbf{B}$ where $\mathbf{B}$ = field of Borel subsets of S and $\mathbf{A} = p_\xi^{-1}\mathbf{B}_{I^+}$.

PROOF: As mentioned earlier i) follows from the definition of T_S. To prove ii) consider the partitions $\{I_n = I^+ \cap (\frac{1}{n+1},\frac{1}{n}),\ n = 1,2,\ldots)$ and $\mathbf{P} = \{p_\xi^{-1}I_n,\ n = 1,2,\ldots\}$. It suffices to show that the common refinement of the partitions $T_S^k\mathbf{P}$, $-n \le k \le n$, consists of rectangles, the largest of whose diameters decreases to 0 as $n \to \infty$. From formulas 6.1), 6.2), it is readily deduced that T_S contracts vertical distances and T_S^{-1} horizontal ones. Under iteration, the rate of convergence can be shown to be exponential.

We now apply the results of [R] about endomorphisms and automorphisms of a Lebesgue space. These are the objects we are investigating, the above terms referring respectively to almost everywhere many-to-one and one-to-one onto measure preserving transformations acting on measure spaces which are measure theoretically the same as the unit interval.

DEFINITION: An automorphism T of a Lebesgue space is a Kolmogoroff automorphism iff there exista a sub σ-algebra of the family $\mathbf{B}$ of Borel sets such that

i) $\mathbf{A} \subsetneq T\,\mathbf{A}$

ii) $\bigcup_{n=-\infty}^{\infty} T^n \mathbf{A}$ generates $\mathbf{B}$

iii) $\bigcap_{n=-\infty}^{\infty} T^n \mathbf{A}$ is trivial.

THEOREM 4: An endomorphism is exact iff its natural extension is Kolmogoroff.
From Theorems 2-4, we obtain the

COROLLARY: T_S is a Kolmogoroff automorphism. In particular all its powers are ergodic, especially T_S^2.

The ergodicity of $\bar{G}_t$ follows from the following equivalences.

THEOREM 5: The following are equivalent:

i) $\bar{G}_t$ is ergodic

ii) T_C is ergodic

iii) T_R is ergodic

iv) T_P is ergodic

v) $T_S \times T_2$ is ergodic

vi) T_S^2 is ergodic

PROOF: The equivalence of ii)-iii) and iv)-v) follows from conjugacy. The equivalence of i)-ii), iii)-iv), v)-vi) follows from elementary set considerations.

Finally, the ergodicity of f may be obtained from that of $\bar{G}_t$. For, by Theorem 5, the ergodicity of $\bar{G}_t$ implies that of T_S^2. Since f^2 is a factor of T_S^2, f^2 is ergodic, which implies the same for f.

REFERENCES

[AF] R. L. Adler and L. Flatto, Cross Section Maps for Geodesic Flows, I (The Modular Surface), Ergodic Theory and Dynamical Systems, vol. 2, Proc. Special Year Md. 1979-1980, Progress in Math., Birkhauser, Boston Basel & Stuttgart, p. 103-161.

[B] P. Billingsley, Ergodic Theory and Information, John Wiley Sons, Inc., New York (1965).

[H] G. A. Hedlund, The dynamics of geodesic flows. Bull.Amer. Math. Soc. 45 (1939), 241-261.

[L] J. Lehner, Discontinuous Groups and Automorphic Functions, A.M.S., Providence, R.I. (1964).

[MPV] J. Moser, E. Phillips, and S. Varadhan, Ergodic Theory, A Seminar, Lecture Notes, New York Univ., N.Y. (1975).

[R] V. A. Rohlin, Exact endomorphisms of a Lebesgue space. Izv. Akad. Nauk. SSSR Ser. Mat. 25, (1961), 499-530, Russian Amer. Math. Soc. Transl., Series 2, vol. 39, (1964), 1-37, English.

I.B.M. Thomas J. Watson Research Center
Yorktown Heights, New York 10598

Bell Laboratories
Murray Hill, New Jersey 07974

Contemporary Mathematics
Volume 26, 1984

ON IDENTIFICATION OF SUPERADDITIVE ERGODIC LIMITS

M. A. Akcoglu* and L. Sucheston**

1. INTRODUCTION

Let (X,F,μ) be a σ-finite measure space. The relations below are often defined only modulo sets of measure zero; the words a.e. may or may not be omitted. Let L_1 be the usual Banach Space of integrable functions on X and let L be the class of functions on X that are bounded below by an integrable function. We will assume that functions in L take values in $(-\infty,\infty]$. Any positive linear bounded operator on L_1 has a natural extension to L; we will not distinguish between the original operator on L_1 and its extension to L.

Let T be a conservative positive linear contraction on L_1. _Conservative_ means that for some (equivalently, for all) strictly positive function f in L_1, $\Sigma_{i=0}^{\infty} T^i f = \infty$ a.e.

Let $(s_n)_{n\geq 0}$ be a sequence of function in L such that

(1.1) $$s_n + T^n s_m \leq s_{n+m} \quad \text{for all} \quad n,m\geq 0.$$

Such a sequence is called a superadditive process [1],[10]. Let g be a fixed integrable function on X, strictly positive a.e. As we will discuss in the next section in more detail, it is then known that both of the following limits exist in $(-\infty,\infty]$:

(1.2) $$\gamma = \lim_n \frac{1}{n} \int s_n \, d\mu,$$

(1.3) $$\rho = \text{a.e.} \lim_n \frac{s_n}{\Sigma_{i=0}^{n-1} T^i g} .$$

The main purpose of Section 2 is to show (Theorem (2.9)) that

(1.4) $$\gamma = \int \rho \, g \, d\mu .$$

---

*Research partially supported by NSERC Grant A3974

**Research partially supported by NSF Grant MCS 8005395

In section 3 this relation is then applied to a particular case to prove Theorem (3.1). Here we state two consequences of this theorem in a self-contained form.

(1.5) THEOREM. Let τ be a conservative measure preserving transformation of a σ-finite measure space (X,F,μ). Let p be a number, $0 < p < 1$, and let f be a non negative function on X such that f^p is integrable. Then, for any strictly positive function h on X,

$$\lim_n \frac{(\Sigma_{i=0}^{n-1} f \cdot \tau^i)^p}{(\Sigma_{i=0}^{n-1} h \cdot \tau^i} = 0 \text{ a.e. and} \tag{1.6}$$

$$\lim_n \frac{1}{n} (\Sigma_{i=0}^{n-1} f \cdot \tau^i)^p = 0 \qquad \text{a.e.} \tag{1.7}$$

(1.8) THEOREM. Let τ be as in the previous theorem. Let $\varepsilon > 0$ and let f be a non-negative function on X, with a strictly positive integral. Then

$$\lim_n \frac{1}{n} \int (\Sigma_{i=0}^{n-1} f \cdot \tau^i)^{1+\varepsilon} d\mu = \infty. \tag{1.9}$$

Theorem (1.5) is of main interest when $\int f\, d\mu = \infty$ or when τ is an ergodic measure preserving transformation of an infinite measure space. We note that first example of such a transformation was constructed by Kakutani [9]. If the measure is finite, (then τ is necessarily conservative) and the sequence $(f \cdot \tau^i)_{i\geq 0}$ is an independent sequence, then the conclusion (1.7) is due to Marcinkiewicz (see e.g. [12], p. 153); see also [5] and [2].

In connection with Theorem (1.8), note that Hajian and Kakutani [6],[7] considered ergodic measure preserving transformations τ such that $\lim_n \mu(A \cap \tau^{-n} A) = 0$ for all $A \in F$ with finite measure. For such a τ the conclusion (1.9) may be a little unexpected.

2. PRELIMINARIES

Let $(\alpha_n)_{n\geq 0}$ be a sequence of numbers in $(-\infty,\infty]$ such that

$$\alpha_n + \alpha_m \leq \alpha_{n+m} \qquad \text{for all } n,m \geq 0. \tag{2.1}$$

Then $\lim_n \frac{1}{n} \alpha_n$ exists in $(-\infty,\infty]$ and is equal to $\sup_{n>0} \frac{1}{n} \alpha_n$. (see e.g. [8], p. 244).

(2.2) LEMMA. Let $(\Psi_n)_{n\geq 0}$ be a sequence of functions in L such that

$$(2.3)\qquad \Psi_n + \Psi_m \leq \Psi_{n+m} \qquad \text{for all } n,m \geq 0.$$

Then

$$(2.4)\qquad -\infty < \lim_n \frac{1}{n}\int \Psi_n \, d\mu = \int \lim_n \frac{1}{n}\Psi_n \, d\mu \leq \infty .$$

PROOF. By induction on (2.3) we obtain $\Psi_1 \leq \frac{1}{n}\Psi_n$. Hence the sequence $\frac{1}{n}\Psi_n$ is bounded from below by a single integrable function. Then we have, by Fatou's lemma and by the observation preceding Lemma (2.2),

$$-\infty < \int \Psi_1 \, d\mu \leq \int \lim_n \frac{1}{n}\Psi_n \, d\mu \leq \lim_n \int \frac{1}{n}\Psi_n \, d\mu$$

$$\leq \int \sup_{n>0} \frac{1}{n}\Psi_n \, d\mu = \int \lim_n \frac{1}{n}\Psi_n \, d\mu \leq \infty . \qquad \square$$

Let $g \in L_1^+$ be a fixed function, strictly positive a.e. Let $\mathcal{I}$ be the σ-algebra of invariant sets of T. Let M be the conditional expectation operator with respect to $\mathcal{I}$, relative to the finite measure $g d\mu$. Hence, if $gf \in L_1$ when Mf is an $\mathcal{I}$-measurable function such that $\int_I fg d\mu = \int_I (Mf) g d\mu$ for all $I \in \mathcal{I}$. We then let $Rf = M(\frac{1}{g} f)$ for each $f \in L_1$ and observe that the positive linear operator R satisfies

$$(2.5)\qquad RTf = Rf, \quad f \in L_1,$$

$$(2.6)\qquad \int f d\mu = \int gRf \, d\mu, \quad f \in L_1.$$

Hence gR is a positive linear contraction of L_1 and it can be applied to functions in L. Therefore R can also be applied to functions in L.

The Chacon-Ornstein ergodic theorem [4] and the identification of the ratio ergodic limits [3] give that

$$(2.7)\qquad Rf = \text{a.e.} \lim_n \frac{\Sigma_{i=0}^{n-1} T^i f}{\Sigma_{i=0}^{n-1} T^i g} .$$

The relation (2.7) is usually stated for $f \in L_1$, but it is clear that it holds for $f \in L$.

(2.8) LEMMA. Let $(s_n)_{n\geq 0}$ be a superadditive sequence. Then

$$-\infty < \lim_n \int \frac{1}{n} s_n \, d\mu = \int \lim \frac{1}{n} gRs_n \, d\mu \leq \infty .$$

PROOF. By definition, (s_n) is a sequence of functions in L satisfying

$$s_n + T^n s_m \leq S_{n+m} \quad \text{for all} \quad n,m \geq 0 .$$

Applying gR to this inequality, we obtain, because of (2.5) and (2.6) that (gRs_n) is a sequence of functions in L satisfying

$$gRs_n + gRs_m \leq gRs_{n+m} \quad \text{for all} \quad n,m \geq 0 .$$

Then we apply Lemma (2.2) to complete the proof, observing that $\int \frac{1}{n} s_n \, d\mu = \int \frac{1}{n} gRs_n \, d\mu$. □

If $(s_n)_{n\geq 0}$ is a superadditive sequence then Theorem (3.3) in [1] gives that a.e. $\lim_n \frac{s_n}{\Sigma_{i=0}^{n-1} T^i g}$ exists and is equal to a.e. $\lim_n \frac{1}{n} Rs_n$. (This result is stated only for superadditive sequences consisting of functions in L_1, but the extension to functions in L is trivial.) Hence we have the following theorem.

(2.9) THEOREM. <u>If (s_n) is a superadditive sequence then</u>

$$-\infty < \lim_n \frac{1}{n} \int s_n \, d\mu = \int \lim_n \frac{s_n}{\Sigma_{i=0}^{n-1} T^i g} \cdot g d\mu \leq \infty .$$

3. POINT TRANSFORMATIONS

We will now apply the preceding result to a particular case. Assume that the operator T is defined by $Tf = f \cdot \tau$, where τ is a measure preserving transformation of X. Such a transformation is called conservative if the operator T defined above is conservative.

(3.1) THEOREM. <u>Let τ be a conservative measure preserving transformation of a σ-finite measure space (X,F,μ). Let ϕ be a function defined on the positive real line $[0,\infty)$, taking values in $(-\infty,\infty)$, and assume that</u>

$$\phi(\alpha) + \phi(\beta) \leq \phi(\alpha+\beta) \quad \text{for all} \quad \alpha,\beta \geq 0 . \tag{3.2}$$

<u>Let f be a measurable function on X, taking values in $[0,\infty)$, such that $\phi(f)$ is an integrable function. Then both of the limits $c = \lim_{\alpha\to\infty} \frac{1}{\alpha} \phi(\alpha)$ and $\gamma = \lim_n \frac{1}{n} \int \phi(\Sigma_{i+0}^{n-1} f \cdot \tau^i) \, d\mu$ exist in $(-\infty,\infty]$ and we have that</u>

$$\gamma = c. \int f \, d\mu, \tag{3.3}$$

with the convention that $0.\infty = \infty . 0 = 0$.

Furthermore, if g is a strictly positive integrable function, then

$$\lim_n \frac{\phi(\Sigma_{i=0}^{n-1} f \cdot \tau^i)}{\Sigma_{i=0}^{n-1} g \cdot \tau^i} = c\ Rf \quad \text{a.e.} \tag{3.4}$$

again with the convention that $0 . \infty = \infty . 0 = 0$.

PROOF. The existence of $c = \lim_{\alpha\to\infty} \frac{\phi(\alpha)}{\alpha} \leq \infty$ follows as in the discrete case; also, as before, $c = \sup_{\alpha>0} \frac{\phi(\alpha)}{\alpha}$.

Define a sequence $(s_n)_{n\geq 0}$ as follows

$$s_n = \begin{cases} 0 & \text{if } n = 0 \\ \phi(\Sigma_{i=0}^{n-1} f \cdot \tau^i) & \text{if } n > 0 . \end{cases} \tag{3.5}$$

Then (3.2) implies that

$$s_n + s_m \cdot \tau^n \leq s_{n+m} . \tag{3.6}$$

Hence $ns_1 \leq s_n$ for all $n \geq 1$. Since $s_1 = \phi(f)$ is an integrable function, we see that $(s_n)_{n\geq 0}$ is a sequence in L. Hence (s_n) is a superadditive sequence. Therefore we already know that $\gamma = \lim_n \frac{1}{n} \int s_n \, d\mu$ exists.

Our plan for the rest of the proof is as follows. We first obtain (3.4) under the assumption that $c \neq 0$ or $\int f \, d\mu < \infty$. Because of Theorem (2.9), this also implies (3.3) under the same assumption. If $c = 0$ and $\int f \, d\mu = \infty$ we then show that (3.3) holds, i.e. that $\gamma = 0$. This, in turn, implies (3.4) in this case, again, because of Theorem (2.9).

Let $A = \{x | \Sigma_{i=0}^{\infty} f(\tau^i x) > 0\}$. Then $\Sigma_{i=0}^{n-1} f(\tau^i x) = 0$ for all n, on A^c, and consequently, $s_n = \phi(0)$ on A^c. If $\mu(A^c) > 0$ then $\phi(0) < \infty$ (since $s_1 \in L_1$) and therefore the limit in (3.4) is zero a.e. on A^c. Since $Rf = 0$ on A^c, the equation (3.4) is correct on A^c.

For each $x \in A$ there is an integer $n_0 = n_0(x)$ such that $\Sigma_{i=0}^{n-1} f(\tau^i x) > 0$ if $n \geq n_0$.

Hence we have, on A,

$$\lim_n \frac{s_n}{\Sigma_{i=0}^{n-1} g \cdot \tau^i} = \lim_n \frac{\phi(\Sigma_{i=0}^{n-1} f \cdot \tau^i)}{\Sigma_{i=0}^{n-1} f \cdot \tau^i} = \frac{\Sigma_{i=0}^{n-1} f \cdot \tau^i}{\Sigma_{i=0}^{n-1} g \cdot \tau^i}$$

$$= cRf$$

as we wanted to prove. Note that this proof is valid only if $c \neq 0$ or $c = 0$ but $Rf < \infty$ a.e. Since $\int gRf\, d\mu = \int f\, d\mu$ the condition $Rf < \infty$ a.e. is satisfied if $\int f\, d\mu < \infty$. If, however, $c = 0$ and $\int f\, d\mu = \infty$ we can not apply the above argument.

To complete the proof of the theorem, we first make a few general observations.

Let $\phi'(\alpha) = \phi(\alpha)$ if $\alpha > 0$ and $\phi'(0) = 0$. Then ϕ' still satisfies $\phi'(\alpha) + \phi'(\beta) \leq \phi'(\alpha+\beta)$ for all $\alpha,\beta \geq 0$ and we have $c' = \lim_{\alpha\to\infty} \frac{1}{\alpha}\phi'(\alpha) = c$. If $\phi(f)$ is in L_1, then $\phi'(f)$ is also in L_1 and consequently $s'_n = \phi'(\Sigma_{i=0}^{n-1} f \cdot \tau^i)$ is still a superadditive process. Also.

$$\lim_n \frac{s_n}{\Sigma_{i=0}^{n-1} g \cdot \tau^i} = \lim_n \frac{s'_n}{\Sigma_{i=0}^{n-1} g \cdot \tau^i} \quad \text{a.e.,}$$

and, consequently, $\gamma' = \lim_n \frac{1}{n}\int s'_n\, d\mu = \gamma$, because of (2.9). Therefore there will be no loss of generality in assuming

$$\phi(0) = 0 \tag{3.7}$$

and we now make this assumption.

We claim that if f is a non-negative function such that $\phi(f) \in L_1$ and if $\varepsilon > 0$ then there are two non-negative functions f' and f'' such that $f = f' + f''$, $f' \in L_1$ and $\|\phi(f'')\| < \varepsilon$, where the norm is the L_1-norm.

To see this let (X_k) be an increasing sequence of sets of finite measure with $X = \bigcup_k X_k$. Let

$$B_k = X_k \cap \{x | 0 \leq f(x) \leq k\}$$

and let $f_k = f\chi_k$, where χ_k is the characteristic function of B_k. Then $f_k \in L_1$ and $\phi(f-f_k) = (1-\chi_k)\,\phi(f)$, because of (3.7), and, consequently, $\|\phi(f-f_k)\|$ converges to 0 as $k \to \infty$, since $\lim_k \chi_k = 1$ a.e. Therefore we can take $f' = f_k$, $f'' = f-f_k$ with a sufficiently large k.

Finally we observe that if $c = 0 = \sup_{\alpha>0} \frac{\phi(\alpha)}{\alpha}$, then $\phi(\alpha) \leq 0$ for all $\alpha \geq 0$, and therefore,

$$|\alpha(\alpha+\beta)| \leq |\phi(\alpha)| + |\phi(\beta)| , \quad \alpha,\beta \geq 0. \tag{3.8}$$

Now it is easy to complete the proof of the theorem. If $c = 0$ and $\int f\, d\mu = \infty$ then we take an arbitrary $\varepsilon > 0$ and find non-negative functions f' and f'' such that $f = f' + f''$, $f' \in L_1$, $\|\phi(f'')\| < \varepsilon$. Hence, because of (3.8),

$$|s_n| \le |\phi(\Sigma_{i=0}^{n-1} f' \cdot \tau^i)| + \sum_{i=0}^{n-1} |\phi(f'' \cdot \tau^i)| .$$

But

$$\lim_n \frac{1}{n} \int |\phi(\Sigma_{i=0}^{n-1} f' \cdot \tau^i| d\mu$$

$$= - \lim_n \frac{1}{n} \int \phi(\Sigma_{i=0}^{n-1} f' \cdot \tau^i) d\mu = 0,$$

by the first part of the proof, since $c = 0$ and $\int f' \, d\mu < \infty$. Also, since τ is measure preserving,

$$\frac{1}{n} \int \sum_{i=0}^{n-1} |\phi(f'' \cdot \tau^i)| \, d\mu = \int |\phi(f'')| \, d\mu < \varepsilon .$$

Hence we see that $|\gamma| \le \overline{\lim}_n \frac{1}{n} \int |s_n| \, d\mu < \varepsilon$. □

Note that Theorems (1.5) and (1.8) stated in the Introduction are just special cases of Theorem (3.1), corresponding to $\phi(\alpha) = -\alpha^p$ and to $\phi(\alpha) = \alpha^{1+\varepsilon}$, respectively.

We also observe that the assumption that τ is conservative is not needed in the proof of (1.7). As noted above, if $\mu(X) < \infty$, then τ is conservative. In general, X is a disjoint union of two <u>invariant</u> sets, C and D, such that τ restricted to C is conservative, and τ restricted to D is dissipative (see [11], Proposition 2.1). In [11] C and D are defined in terms of wandering sets, but it is easy to see that this decomposition coincides with the usual Hopf decomposition (see e.g. [12], p. 196) where D is characterized by the property that $\Sigma_{i=0}^{\infty} g \cdot \tau^i < \infty$ a.e. on D for each g in L_1^+. Choosing $g = f^p$ and using the subadditivity of the function α^p for $0 < p < 1$, we obtain that (1.7) holds also on D. Since there is no communication between C and D, the restrictions of τ to each set can be studied separately. Finally, we note that in this argument the pth power for $0 < p < 1$ can be replaced by any subadditive function from $[0,\infty)$ to $[0,\infty)$.

Professor Ulrich Krengel has recently brought to our attention that the decomposition of X into invariant sets C and D for point-transformations was first proved by J. Feldman [13]. Denote by $F(A)$ the number $\limsup \mu(A \cap \tau^{-1}A)$. Call a measure-preserving point transformation τ of type zero (of positive type) if for every non-null set A of finite measure, $F(A) = 0$ $(F(A) > 0)$. It is shown in [11] that X decomposes into two invariant sets X_0 and X_+ such that τ restricted to X_0 is on zero type, and τ restricted to X_+ is of positive type. Thus the part X_0 can be studied separately, and it is shown in [11] that on X_0 one has

$$\lim_n \frac{1}{n^{1+\epsilon}} \int \left(\sum_{i=0}^{n-1} f \cdot \tau^{k_i} \right)^{1+\epsilon} d\mu = 0$$

for every function f in $L_{1+\epsilon}$, uniformly in the set of all strictly increasing sequences of integers (k_i). This renders (1.9) above somewhat more surprising. Finally, it may be of interest to determine whether (1.9) holds for a class of subsequences (k_i).

REFERENCES

[1] M. A. Akcoglu and L. Sucheston: A Ratio Ergodic Theorem for Superadditive Processes. Z. Wahrscheinlichkeitstheorie verw. Geb. 44, 269-278, 1978.

[2] B. Bru, H. Heinich and J. C. Lootgieter: Lois des grands nombres pour les variables echangeables. C. R. Acad. Sc. Paris, 293, 485-488, 1981.

[3] R. V. Chacon: Convergence of Operator Averages. Ergodic Theory, pp. 89-120, New York, Academic Press, 1963.

[4] R. V. Chacon and D. Ornstein: A general Ergodic Theorem. Illinois J. Math. 4, 153-160, 1960.

[5] G. A. Edgar and L. Sucheston: Demonstrationes de lois des grands nombres par les sousmartingales descendantes. C. R. Acad. Sc. Paris, 292, 967-970, 1981.

[6] A. Hajian and S. Kakutani: Weakly wandering sets and invariant measures. Trans. Amer. Math. Soc. 110, 136-151, 1964.

[7] A. Hajian and Y. Ito: Transformations that do not accept a finite invariant measure. Bull. Amer. Math. Soc. 84, 417-427, 1978.

[8] E. Hille and R. S. Phillips: Functional Analysis and Semi-Groups. 1957.

[9] S. Kakutani: Induced measure preserving transformations. Proc. Imp. Acad. Sci. Tokyo, 19, 635-641, 1943.

[10] J. F. C. Kingman: Subadditive ergodic theory. Ann. Prob. 1, 883-905, 1973.

[11] U. Krengel and L. Sucheston: On mixing in infinite measure spaces. Z. Wahrscheinlichkeitstheorie verw. Geb. 13, 150-164 (1969).

[12] J. Neveu: Mathematical Foundations of the Calculus of Probability. 1965.

[13] J. Feldman: Subinvariant measures for Markoff operators. Duke Math. J., 29, 71-98, 1962.

M. A. Akcoglu
Department of Mathematics
University of Toronto
Toronto, Ontario, M5S 1A1
CANADA

L. Sucheston
Department of Mathematics
The Ohio State University
Columbus, OH 43210
U.S.A.

Contemporary Mathematics
Volume 26, 1984

THE EQUIVALENCE OF DIFFUSIONS ON NETWORKS TO BROWNIAN MOTION

J. R. Baxter and R. V. Chacon

§1. INTRODUCTION. A time change in a Markov process does not alter the hitting probabilities. On the other hand, given the hitting probabilities, the process is determined up to a time change. Thus, for many purposes, one may regard Markov processes which differ from each other only by a time change as essentially equivalent. Given a particular Markov process, it is natural to look for that time change which makes the resulting process as simple as possible.

A familiar example of these ideas is the theorem, due to Feller and others ([1], [3]) characterizing diffusions on the line. Feller's theorem says that, given a regular diffusion (X_t) on the real line, it is possible to find a time change σ, and an increasing continuous function F such that $X_t = F(B_{\sigma(t)})$, where (B_t) is Brownian motion. If $G = F^{-1}$, then

$$\frac{1}{2}a\, G'' + vG' = 0, \quad \text{where } \frac{1}{2}a \frac{d^2}{dx^2} + v \frac{d}{dx} \tag{1.1}$$

is the infinitesimal generator L of the process (X_t).

The time change σ and its inverse τ are defined by either of the following equations:

$$\sigma(t) = \int_0^t \left(\frac{(F')^2 \circ G}{a}\right)^{-1} (X_s)\, ds, \tag{1.2}$$

$$\tau(t) = \int_0^t \left(\frac{(F')^2}{a \circ F}\right) (X_s)\, ds \tag{1.3}$$

In the present paper we wish to consider the problem of extending this representation for diffusions on the line to two general situations: first, to diffusions on smooth manifolds of higher dimension, and second, to diffusions on networks (which are one-dimensional but not smooth or simply-connected). The results for networks seem to be more difficult than those for smooth manifolds. This is perhaps not surprising, since we conjecture that very general Markov processes are approximable by the network diffusions we consider here. Before stating the results, we will reformulate the theorem just stated for the

0271-4132/84 $1.00 + $.25 per page

line. The process $F(B_t)$ can be characterized by its infinitesimal generator L, which by Ito's formula is $L = \frac{1}{2}(F' \circ F^{-1})^2 \frac{d^2}{dx^2} + \frac{1}{2}(F'' \circ F^{-1}) \frac{d}{dx}$. This operator is easily seen to be equal to the Laplace-Beltrami operator associated with that metric on the line whose metric tensor B consists of the element $2(G')^2$ in the standard coordinate system. Thus the B-length of an arc α is, up to a constant factor, the ordinary length of $F^{-1}(\alpha)$. For any Riemannian manifold, we shall refer to the diffusion whose infinitesimal generator is the Laplace-Beltrami operator on the manifold as the Brownian motion associated with the Riemannian metric. Feller's result quoted above can now be expressed by saying that there is a time change τ such that the process (X_τ) is the Brownian motion associated with a certain Riemannian metric on the line. We shall show in section 2 that this form of Feller's result holds for any diffusion on a smooth manifold, provided that the dimension is not equal to 2, and provided that the drift satisfies an exactness condition (Theorem 1). For a general drift, the result is false, even in dimensions other than 2. However, we show in dimension other than 2, that there is a time change τ such that X_τ has infinitesimal generator equal to the sum of two terms, the first being the Laplace-Beltrami operator with respect to a certain Riemannian metric, and the second a drift term in which the velocity field has zero divergence with respect to the same metric. Thus, the diffusion X_τ and the drift both leave invariant the volume measure associated with the metric (Theorem 2).

In section 3 we define what we mean by a diffusion on a network, and establish some basic properties. Then, in section 4, we prove results for network diffusions analogous to those of section 2. (Theorems 1, 2, and 3).

§2. Let M be a C^∞-manifold of dimension n. Let $(\Omega, \mathfrak{F}, \mathfrak{F}_t, X_t, Pr^x)$ be a diffusion on M. We shall assume that there is a second order differential operator L on M such that for every C^∞-function f with compact support on M,

$$E^x[f \circ X_\gamma] = E^x[f \circ X_0] + \int_0^\gamma (Lf) \circ X_s \, ds \tag{2.1}$$

for every stopping time $\gamma \leq$ the exit time of some compact set. We will call L the generator for the process (X_t) and generally not distinguish processes with the same generator, even if they live on different sample spaces. We assume the coefficients in L are C^2 on M, and that the matrix of second order coefficients is strictly positive definite everywhere.

We will consider time changes τ such that the inverse σ of τ is given by

$$\sigma(t) = \int_0^t h^{-1} \circ X_s \, ds, \tag{2.2}$$

where h is a positive C^2-function on M. In this case the new process $(\Omega, \mathcal{F}, \mathcal{F}_\tau, X_\tau, Pr^x)$ is easily seen to have generator hL. In general, for a given L, the corresponding process may have a finite lifetime ζ. This may be true for X_τ even if not for X_t. However, it is easy to see that the lifetime will always be greater than the exit time of any compact set. Thus all local statements are well defined whether the lifetime is finite or not. For this reason we will not explicitly refer to the lifetime of the process again.

The matrix T of second order coefficients in L depends on the coordinates chosen. It is easy to see that T is a symmetric, positive definite contravariant tensor of order 2. Then $A \equiv T^{-1}$ is a Riemannian metric tensor on M. Let Δ_A denote the corresponding Laplace-Beltrami operator. $L - \Delta_A$ is a first order differential operator, independent of coordinates because L and Δ_A are independent of coordinates. Thus there exists a C^1 vector field v on M such that $L - \Delta_A = D_v$, where $D_v f \equiv \langle v, df \rangle$. If we let ∇_A denote the gradient operator associated with A, then

$$L = \Delta_A + D_v = \Delta_A + v \cdot \nabla_A. \tag{2.3}$$

Let $B = h^{-1}A$. The process (X_τ) defined above has generator

$$hL = \Delta_B + D_w = \Delta_B + w \cdot \nabla_B \tag{2.4}$$

for some vector field w.

In order to generalize Feller's theorem on the real line, we seek an h such that $w = 0$.

It is easy to show that $\nabla_B f = h\nabla_A f$, for any f, that $\nabla_B \cdot u = h^{n/2}\nabla_A \cdot h^{-n/2}u$, for any u, and $\Delta_B = h\Delta_A + (1 - \frac{n}{2})(\nabla_A h) \cdot \nabla_A$. Hence

$$w = hv + (\tfrac{n}{2} - 1)\nabla_A h. \tag{2.5}$$

THEOREM 1. <u>If</u> $n = 2$ <u>and</u> $v \neq 0$ <u>then</u> $w \neq 0$, <u>no matter what</u> h <u>is chosen. For</u> $n \neq 2$, $w = 0$ <u>can be achieved if and only if</u> $v = \nabla_A g$ <u>for some</u> g, <u>or, equivalently, when the form obtained by lowering</u> v <u>with</u> A <u>is exact. In this case,</u> $w = 0$ <u>if and only if</u> $h = c \exp((1 - \frac{n}{2})^{-1} g)$.

Theorem 1 is obvious from 2.5. When v is not of the form $\nabla_A g$, then we may try to obtain what seems to be the next best condition on w instead of $w = 0$, namely $\nabla_B \cdot w = 0$. This condition is easily seen to be equivalent to

$$\Delta_A \varphi - \nabla_A \cdot (\varphi v) = 0, \quad \text{where } \varphi = h^{1-(n/2)} \tag{2.6}$$

or equivalently

$$L^{*}\varphi = 0, \tag{2.7}$$

where L^{*} is the adjoint of L with respect to A-volume.

We have at once

THEOREM 2. *If* $n = 2$ *and* $\nabla_A \cdot v \neq 0$ *then* $\nabla_B \cdot w \neq 0$, *no matter what* h *is chosen. For* $n \neq 2$, $\nabla_B \cdot w = 0$ *can be achieved if and only if* $h^{1-(n/2)}$ *is an invariant density of* (X_t) *with respect to A-volume.*

§3. We now wish to study diffusions on a network Λ. The network will consist of a finite or countable collection of bounded intervals called *branches*, joined together at their endpoints in an arbitrary way, at *junction points*. We assume that only finitely many branches are attached to each junction point, and we assume that Λ is connected in the obvious topology. Both ends of a branch may be connected to the same endpoint. Λ is not considered to be embedded in Euclidean space.

A function f defined on a branch I will be called branchwise C^k on I if when we imbed the interior of I in the real line there is a C^k-function defined on the real line which agrees with f on the interior of I. This definition presupposes a choice of coordinate each branch.

We now define the class $\mathfrak{D}$ of diffusions $(\Omega, \mathcal{F}, \mathcal{F}_t, X_t, Pr^x)$ which we will study on Λ. Each process in $\mathfrak{D}$ is assumed to be continuous and strong Markov. For each process (X_t) in $\mathfrak{D}$, a second-order differential operator L is defined on each branch of Λ, such that the coefficients of L are branchwise C^2 on each branch, and such that the second order coefficient is positive and bounded away from 0 on each branch, in appropriate coordinates. We assume that for each branch I, each x in the interior of I, each stopping time γ which is less than or equal to the first exist time of I, and each f which is piecewise C^2 on I, that (2.1) holds. This condition is described by saying that (X_t) has *generator* L on each branch. It is also assumed that for each junction point x,

$$\lim E^x \gamma(U) = 0, \tag{3.1}$$

where $\gamma(U)$ denotes the first exit time of an open set containing x, and the limit is taken for U sufficiently small shrinking to $\{x\}$. This condition is described by saying that (X_t) has *no holding points*.

We will consider time changes τ defined using a function h, as in (2.2), where h is assumed to be positive, branchwise $-C^2$, and bounded away from 0 on each branch. Then, if $(\Omega, \mathcal{F}, \mathcal{F}_t, \Pr^x)$ is in $\mathcal{D}$, so is $(\Omega, \mathcal{F}, \mathcal{F}_\tau, X_\tau, \Pr^x)$.

For every generator L on the branches of Λ, there is, as described in section 2, a metric A and a vector field v on each branch such that $L = \Delta_A + D_v$. Let $\mathcal{D}_0$ consist of those processes (X_t) in $\mathcal{D}$ such that the corresponding v is 0 on each branch.

When convenient, we will specify a direction for a branch, or any subinterval of a branch, so that the interval can be traversed in a positive or negative sense by (X_t). A directed branch has an origin and a terminus, which may of course be the same point. If I is a directed branch, $-I$ will denote the same branch given opposite sense.

LEMMA 1. Let (X_t) be a process in $\mathcal{D}_0$, A the corresponding metric. Let x be a junction point. Let $I_1,\ldots,I_k$ be the branches at x, directed so that x is the origin of each I_i. If some branch at x both begins and ends at x we count it twice, corresponding to the two possible senses. Let α_i, β_i be given, with $0 < \alpha_i$, $\beta_i \leq A$- length I_i. Let y_i, z_i be points in I_i, such that the subinterval J_i of I_i from x to y_i with the same sense as I_i has A- length α_i, and the subinterval K_i of I_i from x to z_i with the same sense as I_i has A- length β_i. Let γ be the first time one of the intervals J_i is traversed, starting from x, with correct sense. Let η be the first time one of the intervals K_i is traversed, starting from x, with correct sense. Let p_i denote the probability that, starting from x, (X_t) traverses J_i with correct sense before any J_j, $j \neq i$. Let q_i denote the probability that, starting from x, (X_t) traverses K_i with correct sense before any K_j, $j \neq i$.

Then, with $c^{-1} = \sum_{i=1}^{k} q_i\beta_i/\alpha_i$,

$$p_i\alpha_i = c\, q_i\beta_i, \quad i = 1,\ldots,k \quad \text{and} \tag{3.2}$$

$$E^x\gamma = c\, E^x\eta + \frac{1}{2} \sum_{i=1}^{k} p_i\alpha_i(\alpha_i - \beta_i). \tag{3.3}$$

PROOF. First, assume $\beta_i \leq \alpha_i$, and β_i is so small that $K_i \cap K_j = \{x\}$ $\forall i \neq j$, and $E^x\eta < \infty$, which is possible by (3.1). Consider starting (X_t) at x and stopping at $\{z_1,\ldots,z_k\}$. Clearly the mass on z_i is q_i, and the stopping time is η. Now release (X_t) and stop when $\{x,y_1,\ldots,y_k\}$ is reached. Since (X_t) is the Brownian motion associated with A on each branch, if ζ is the final stopping time, at time ζ a fraction $q_i\beta_i/\alpha_i$ of the paths will be at y_i, having traversed J_i before any J_j, $i \neq j$, and the rest of the paths will have returned to x, having not yet traversed any

J_i. Furthermore, since the variance per unit time of (X_t) with respect to the A-metric is 2, $E^x\zeta = E^x\eta + \frac{1}{2}\sum_{i=1}^{k} q_i\beta_i(\alpha_i - \beta_i)$. The fraction of paths at x, that have not yet traversed any J_i, is $m = 1 - \sum_{i=1}^{k} q_i\beta_i/\alpha_i$. Repeating this whole procedure over and over, we find

$p_i = (\sum_{n=0}^{\infty} m^n)\, q_i\beta_i/\alpha_i = c\, q_i\beta_i/\alpha_i$, and

$E^x\gamma = (\sum_{n=0}^{\infty} m^n)\{E^x\eta + \frac{1}{2}\sum_{i=1}^{k} q_i\beta_i(\alpha_i - \beta_i)\} = c\,E^x\eta + \frac{1}{2}\sum_{i=1}^{k} p_i\alpha_i(\alpha_i - \beta_i)$.

Thus lemma 1 is proved in this case. The general case follows, because we can apply the special case twice to express both p_i and q_i, $E^x\gamma$ and $E^x\eta$ in terms of a third stopping time.

We have not used all of condition (3.1) in proving lemma 1. Letting $\beta_i \to 0$, (3.1) applied to (3.3) gives

$$E^x\gamma = \frac{1}{2}\sum_{i=1}^{k} p_i\alpha_i^2. \tag{3.4}$$

DEFINITION 1. The <u>branch probability</u> $p(I_i)$ is defined to be p_i, where the numbers α_i in lemma 1 are all chosen equal.

Lemma 1 shows that this definition does not depend on the choice of α_i. Definition 1 assigns a nonnegative number $p(I)$ to every <u>directed branch</u> I in Λ, such that $\Sigma\, p(I) = 1$ for every sum over the directed branches originating at a junction point x. Any such function $p(I)$ on the directed branches of I will be called a branch probability function p.

DEFINITION 2. Let W denote the set of continuous functions f on Λ with compact support, such that f is piecewise C^2 on each branch, and such that at each junction point x,

$$\sum_{i=1}^{k} p_i \langle u_i, df\rangle = 0, \tag{3.5}$$

where $I_1,\ldots,I_k$ are the directed branches originating at x, $p_i = p(I_i)$, and u_i is a vector at x of unit A-length, pointing into I_i.

W depends on p and A. We may write $W(p,A)$ or $W(X_t)$ where necessary.

In definition 3 below we define the branch probabilities p for any (X_t) in $\mathcal{D}$. $W(X_t)$ is then defined by definition 2.

By a standard approximation argument we can show that, for f in W, (2.1) holds. It follows that the generator L and the branch probabilities p for a process (X_t) in $\mathcal{D}_0$ uniquely determine the process.

Now let (X_t) be any process in $\mathcal{D}$. Let x be a junction point, I a directed branch at X. If necessary, subdivide I, so that the terminus of I is different from x. By theorem 1 of section 2, there is a unique function h on I such that $\lim_{y\to x} h(y) = 1$, y in I, and such that if the time change τ is determined by (2.1) then (X_t) has zero drift on I. Choosing h in this way on each branch at x, and similarly on the remaining branches of Λ, we obtain a time change τ such that $(X_\tau) \in \mathcal{D}_0$, and $\lim_{y\to x} h(y) = 1$.

DEFINITION 3. We define the *branch probabilities* $p(I)$ for I at x, for (X_t), to be equal to those of the process (X_τ) just constructed.

Since $p(I)$ for (X_τ) only depends on the behaviour of (X_t) near x, this definition of $p(I)$ does not depend on the manner in which any subdivisions of I were carried out.

LEMMA 2. *Let* h *be any allowable function defining a time change* τ *for a process* (X_t) *in* $\mathcal{D}$ *by* (2.2). *Let* x *be a junction point. Let* $I_1,\dots,I_k$ *be the branches at* x, $\tilde p_1,\dots,\tilde p_k$ *the branch probabilities for* (X_τ). *Let* $\lim_{y\to x} h(y) = h_i$, *when* $y \to x$ in I_i. *Then*

$$\tilde p_i = c p_i/\sqrt{h_i}, \quad \text{where} \quad c^{-1} = \sum_{i=1}^{k} p_i/\sqrt{h_i} . \tag{3.6}$$

PROOF. By subdividing branches if necessary we assume x is not the terminus of any I_i. By making an initial time change on (X_t), and a final time change on (X_τ), both of the sort described before definition 3, we may assume that both (X_t) and (X_τ) are in $\mathcal{D}_0$, while p_i, $\tilde p_i$, and h_i are unchanged. In this case h must be constant on each branch, so $h = h_i$ on I_i. Let A be the metric associated with (X_t), $\tilde A$ with (X_τ). Then $\tilde A = h_i^{-1} A$ on I_i. A point y_i on I_i at an A-distance of δ from x will lie at a A-distance of $\sqrt{h_i}\,\delta$ from x. (3.6) then follows from (3.2).

COROLLARY. $W(X_\tau) = W(X_t)$, *and* (2.1) *holds for every* f *in* $W(X_t)$.

As a consequence of lemma 2, we note that the generator L and branch probabilities p for a process (X_t) uniquely determine the generator and branch probabilities for any time-changed process (X_τ) in $\mathcal{D}_0$, hence determine $(X\ )$, as noted above, and thus determine the original process (X_t).

We now give an explicit construction for the process (X_t) in $\mathcal{D}$ corresponding to a given generator L on each branch of Λ and given branch probabilities p.

First, fix a particular junction point x. Let the branches at x be $I_1,\dots,I_k$. Let $J_1,\dots,J_k$ be distinct copies of the positive half of the real line. Extend L in any smooth bounded fashion from each I_i to J_i. Let S be the network consisting of $J_1,\dots,J_k$ joined at the single junction point x. We will define a process (ξ_t) on S.

Let $L = a\frac{d^2}{du^2} + w\frac{d}{du}$ on J_i, where u is any coordinate on J_i. a and w depend on i. a is second order contravariant, so $A = a^{-1}$ is second order convariant. Let $v = w - (\frac{da}{du} / 2)$. Then
$L = \sqrt{a}\,\frac{d}{du}\,\sqrt{a}\,\frac{d}{du} + v\frac{d}{du} = \Delta_A + D_v$. Define g on J_i by

$$(3.7)\qquad g(x) = 0,\quad dg = Av.\quad (\text{so that } v = \nabla_A g)$$

This defines g on S. Let $h = e^{2g}$, $B = h^{-1}A$. Define the coordinate r_i on J_i by

$$(3.8)\qquad \sqrt{2}\, r_i(y) = B\text{-distance from } x \text{ to } y, \text{ for every } y \text{ in } J_i.$$

We note that

$$(3.9)\qquad hL = \frac{1}{2}\Delta_B = \frac{1}{2}\frac{d^2}{dr_i^2}.$$

Let $n_t(r) = \frac{1}{\sqrt{2\pi t}} e^{-r^2/2t}$, the usual Brownian transition density. Let $q_t(r,\rho)$ denote the transition density for standard Brownian motion to go from r to ρ at time t without hitting 0, that is,
$q_t(r,\rho) = n_t(\rho - r) - n_t(\rho + r)$. Define

$$(3.10)\qquad p_t(y,z) = 2p_j[n_t(r_j(z) - r_i(y)) - q_t(r_i(y), r_j(z))] + {} + \delta_{ij}\, q_t(r_i(y), r_j(z)),$$

where y is in J_i, z is in J_j, and $p_j = p(I_j)$.

It is a straightforward matter to check that $(p_t(y,z))$ is a well-behaved semigroup. Let (ξ_t) be the associated Markov process. Let K denote the generator for the (ξ_t)-process. It is easy to check that
$K = \frac{1}{2}\frac{d^2}{dr_i^2} = hL$ on each J_i, and that (ξ_t) has branch probability $p(I_i)$ for each branch J_i. (Strictly speaking (ξ_t) is not in $\mathcal{D}$ since the branches of S are unbounded, but the definition of branch probabilities clearly still makes sense.)

Let

(3.11) $$\tau(t) = \int_0^t h(\xi_s)\, ds, \quad \sigma = \tau^{-1}$$

Let I_i' be the closed subinterval of I_i extending from x to the midpoint of I_i. Let $S'(x)$ consist of the intervals I_i' joined at the single junction point x.

Define the process (U_t^x) on $S'(x)$ to be the process (ξ_σ) stopped at the ends of the intervals I_i'. Clearly (U_t^x) has branch probabilities $p(I_i)$ into each I_i' and has generator $h^{-1}K = L$ on each I_i'. (U_t^x) is thus part of the desired (X_t) process. To finish the construction of (X_t) we define the process $(V_t(I))$ on any branch I to be the diffusion on I with generator L and stopping at both ends of I. (X_t) is now defined as follows: a random particle starting at y begins in some $S'(x)$. It moves under (U_t^x) until reaching an endpoint of $S'(x)$, which is the midpoint of some branch I. The particle then moves under $(V_t(I))$ until reaching an endpoint of I, which is some junction point z. The particle then moves in $S'(z)$ under (U_t^z), etc.

For any f in W, (2.1) holds. This is true because the corresponding equation holds with (X_t) replaced by (ξ_σ), (U_t^x), or $(V_t(I))$. Since (2.1) holds, (X_t) is easily seen to be in $\mathcal{D}$ with generator L and branch probabilities p, as desired.

§4. We have already defined the branch probability p for a process in $\mathcal{D}$. We now define the <u>hitting probability</u> Q has follows:

DEFINITION 1. If I is a directed branch at x, $Q(I)$ is the probability that I is the first directed branch at x traversed with correct sense starting from x.

LEMMA 1. *Let τ be any time change such that (X_τ) is in $\mathcal{D}_0$. Let B be the metric associated with (X_τ). Let x be a junction point with branches $I_1,\ldots,I_k$. Let ℓ_i be the B-length of I_i, and $\tilde{p}_i$ the branch probability of I_i from x, for the (X_τ)-process. Then*

(4.1) $$Q(I_i) = (\tilde{p}_i/\ell_i) / \sum_{j=1}^{k} (\tilde{p}_j/\ell_j).$$

PROOF. Follows at once from (3.2).

$Q(I_i)$ can be written explicitly in terms of A and v, where $L = \Delta_A + D_v$ is the generator of (X_t), if we express τ using (2.2), and express h in terms of A and v, as in theorem 1 of section 2. We let g

be defined by $g(x) = 0$, $\nabla_A g = v$. Let $h = e^{2g}$, τ as in (2.2). Then $B = h^{-1}A$, and $\tilde{p}_i = p_i$, the branch probability for the original (X_t)-process. Also $\ell_i = \int_{I(i)} e^{-g} d(A\text{-length})$. These quantities can then be substituted into (4.1).

Another approach to computing $Q(I_i)$ is to recall that hitting probabilities are harmonic functions of the starting point (cf. [2]). This method gives (4.1) again, since $\ell_i = f(y_i) - f(x)$ for any function f in W which is (X_τ)-harmonic and satisfies $\langle u, df\rangle = 1$, when u is a vector of unit B-length at x pointing into I_i. f is then also harmonic for (X_t).

THEOREM 1. *Let* (X_t) *and* (Y_t) *be processes in* $\mathcal{D}_0$ *with the same hitting probabilities* Q, *such that the associated metrics for the two processes give the same length to every branch of* Λ. *Then the two processes have the same branch probabilities* p, *and there is a homeomorphism* F *of* Λ, *such that* $F(x) = x$ *for every junction point* x, *and such that* X_t *and* $F(Y_t)$ *are the same process.*

PROOF. (X_t) and (Y_t) have the same branch probabilities by (4.2). Let A, B be the metrics associated with (X_t), (Y_t). Let F be the homeomorphism of each branch that is isometric from B-length to A-length. It is easy to see that (X_t) and $F(Y_t)$ have the same L and p on Λ, so the theorem is proved.

COROLLARY. *Let* (X_t) *and* (Y_t) *be processes in* $\mathcal{D}$, *with the same hitting probabilities* Q. *Then there is a time change* τ *such that* (X_τ) *and* (Y_t) *have the same branch probabilities* p, *and there is a homeomorphism* F *of* Λ, *such that* $F(x) = x$ *for every junction point* x *in* Λ, *and* (X_τ) *and* $F(Y_t)$ *are the same process.*

PROOF. We may assume (Y_t) is in $\mathcal{D}_0$ by making a preliminary time change. Let τ be that time change such that (X_τ) is in $\mathcal{D}_0$ and such that the metrics for (X_τ) and (Y_t) give the same length to every branch.

To see how Feller's theorem should generalize to a network, let us consider the network Λ obtained by decomposing the real line into a sequence of intervals. Feller's theorem says that, for any (X_t) in $\mathcal{D}$, there is a time change τ such that (X_τ) is in $\mathcal{D}_0$, and such that (X_τ) has branch probabilities $\frac{1}{2}$, $\frac{1}{2}$ at each junction point. This suggests that we might be able to prescribe the branch probabilities for the time changed process.

THEOREM 2. *Let hitting probabilities* Q *and branch probabilities* p *be prescribed for a network* Λ. *The following are equivalent:*

(i) *There exists* (X_t) *in* $\mathcal{D}_0$ *with hitting probabilities* Q *and branch probabilities* p *on* Λ.

(ii) *For every finite closed directed circuit* C *in* Λ, $\prod_{I\in C} \frac{Q(I)p(-I)}{Q(-I)p(I)} = 1.$

(By (4.1), *if* $p(I) = 0$ *then* $Q(I) = 0$. *In this case we replace* $Q(I)/p(I)$ *by* 1.)

PROOF. If (i) holds then (ii) follows at once from (4.1).

Conversely, let (ii) hold. Define $\lambda(x)$ for every junction point as follows: let $\lambda(x_o) = 1$ for some fixed junction point x_o, and for any junction point x, let

$$\lambda(x) = \prod_{I\in\alpha} \frac{Q(I)p(-I)}{Q(-I)p(I)}, \quad \text{where } \alpha \text{ is} \tag{4.2}$$

any directed path in Λ from x_o to x.

Let I be any directed branch of Λ, with origin x and terminus y. By (4.2)

$$\lambda(x)\,\frac{Q(I)}{p(I)} = \lambda(y)\,\frac{Q(-I)}{p(-I)}\,. \tag{4.3}$$

Let C be any Riemannian metric on the branches of Λ such that the C-length of each I is the value given in (4.3). (We note that this quantity is independent of the sense of I.) Let (Y_t) be that diffusion with generator Δ_C and branch probabilities p. (4.1) and (4.3) imply that (Y_t) has hitting probabilities Q, so the theorem is proved.

COROLLARY. *Let branch probabilities* p *be prescribed on* Λ. *Let* (X_t) *be a process in* $\mathcal{D}$ *with hitting probabilities* Q. *Then there exists a time change* τ *such that* (X_τ) *is in* $\mathcal{D}_0$ *and* (X_τ) *has branch probabilities* p *on* Λ, *if and only if condition* (ii) *of theorem* 2 *holds*.

PROOF. By making a preliminary time change, we may assume the (X_t) is in $\mathcal{D}_0$, and that the associated metric gives unit length to every branch in Λ. If condition (ii) holds, then there is a process (Y_t) in $\mathcal{D}_0$ with hitting probabilities Q and branch probabilities p. Let τ be the time change such that (X_τ) is in $\mathcal{D}_0$ and the associated metrics for (X_τ) and (Y_t) give the same length to every branch in Λ. Theorem 1 then shows that (X_τ) has branch probabilities p. The rest of the corollary follows at once from theorem 2.

The corollary to theorem 2 seems to be the natural analogue of Feller's theorem for networks. As an example, we may consider a diffusion (X_t) on the circle, regarded as a network with one branch and one junction point (chosen arbitrarily on the circle). The corollary to theorem 2 asserts that the diffusion can be time changed to Brownian motion on the circle with respect to some metric (with branch probabilities $p = \frac{1}{2}, \frac{1}{2}$ at the junction point) if and only if the probability of first winding around the circle in a counterclockwise sense equals $\frac{1}{2}$, which is also the probability of first winding around the circle in a clockwise sense. Since a circle is also a manifold, we may compare this result to theorem 1 of section 2. The present result is a bit more general, since the initial (X_t) is allowed to have unequal branch probabilities at the chosen junction point. Other than that, the two theorems agree, as they must. If (X_t) has generator $\Delta_A + v \cdot \nabla_A$ on the circle, theorem 1 of section 2 says that (X_t) can be time changed to Brownian motion on the circle with respect to some metric if and only if the line integral of v around the circle using A-metric is 0. It is easy to show that this condition is equivalent to the previous one.

We may also regard a circle as a network with n branches and n arbitrarily chosen junction points. The corollary to theorem 2 states that a diffusion on this network can be time changed to Brownian motion on the circle with respect to some metric (with branch probabilities $p = \frac{1}{2}, \frac{1}{2}$ at each junction point) if and only if $\Pi Q(I) = \Pi Q(-I)$, where the product is over all branches oriented in the clockwise sense, say. It is easy to show that this condition is again equivalent requiring that the probability of winding around the circle first counterclockwise is $\frac{1}{2}$. Indeed, one may regard the circle as an interval on the line with endpoints identified. Let $[0,1]$ be the interval. Let $I_1,\ldots,I_n$ be the branches of $[0,1]$. Let $I'_1,\ldots,I'_n$ be corresponding branches for $[-1,0]$, and define $Q(I'_i) = Q(I_i)$. We may regard Q as giving Markov transitions for a random walk on the endpoints of the intervals $I'_1,\ldots,I'_n, I_1,\ldots,I_n$. The probability of winding around the circle first in the counterclockwise direction, starting from 0, is just the probability of the random walk hitting 1 before -1, starting from 0. An easy induction shows that the probability of hitting the right end of I_k before the left end of I'_{n-k+1}, starting from 0, is $c_k \prod_{i=1}^{k} Q(I_i)$, and the probability of hitting the left end of I'_{n-k+1} before the right end of I_k, starting from 0, is $c_k \prod_{i=1}^{k} Q(I'_{n-i+1})$, so the result follows.

We now investigate the question of what can be said when condition (ii) of theorem 2 fails. Let nonzero branch probabilities p be prescribed on Λ, and let (X_t) be a process in $\mathfrak{D}$ with nonzero hitting probabilities. It follows easily from theorem 2 of section 2 (using an appropriate invariant density on each branch) that we can find a time change τ such that (X_τ) has generator $\Delta_B + w \cdot \nabla_B$ on each branch, where $\Delta_B \cdot w = 0$, and (X_τ) has branch probabilities p. Unlike the situation of section 2, however, we cannot then conclude that the length measure associated with the B-metric is an invariant measure for (X_τ). Indeed this will certainly be false unless p has equal values on all the branches at eash junction point. What can be shown is the following:

THEOREM 3. Let Λ be compact. Let nonzero branch probabilities p be prescribed on Λ. Let (X_t) be a process in $\mathfrak{D}$ with nonzero hitting probabilities Q. Then there is a time change τ such that (X_τ) has branch probabilities p, and such that if $\Delta_B + w \cdot \nabla_B$ is the generator for (X_τ), and (Y_t) is the process in $\mathfrak{D}_0$ with generator Δ_B and branch probabilities p, then (X_τ) and (Y_t) have the same invariant measure.

PROOF. Let (X_t) have generator $\Delta_A + v \cdot \nabla_A$. For each directed interval I in Λ, let $r = r(I)$ be the A-distance coordinate with $r = 0$ at the origin of I. By making a preliminary time change, we may assume that each interval I has A-length 1. Thus $r = 1$ at the terminus of I. By (3.6), (X_t) has branch probabilities Q.

Let ν be the invariant measure on Λ for (X_t). It is straightforward to show that ν has a density ρ with respect to the A-length. ρ can be taken to be linear with respect to the r-coordinate on each I. Furthermore, there exists a positive function ψ on the junction points of Λ such that for each I

$$\lim_{r\to 0} \rho(r) = \psi(x)Q(I), \quad \text{where } x \text{ is the origin of } I, \tag{4.4}$$

and for each junction point x,

$$\sum_{i=1}^{k} \left(\frac{d\rho}{dr} \text{ on } I_i\right) = 0, \quad \text{where } I_1,\ldots,I_k \text{ are the} \tag{4.5}$$

branches originating at x.

The linearity of ρ follows at once from the fact that (X_t) is Brownian motion in the r-coordinate. Conditions (4.4) and (4.5) follow from the explicit form of the transition densities constructed in section 3, or from the fact that (2.1) holds for all f in $W(X_t)$, where W is defined in section 3, definition 2.

Using (4.4) and the fact that each branch has unit A-length, equation (4.5) becomes

$$\sum_{i=1}^{k} \{\psi(y_i)Q(-I_i) - \psi(x)Q(I_i)\} = 0, \tag{4.6}$$

where y_i denotes the terminus of I_i. Thus

$$\psi(x) = \sum_{i=1}^{k} \psi(y_i)Q(-I_i). \tag{4.7}$$

Equation (4.5) says that ψ is an invariant measure on the junction points of Λ for the transition matrix defined by Q.

Let τ be determined by (2.1), for some h. Let (X_τ) have generator $\Delta_B + w \cdot \nabla_B$. Let $\ell(I)$ denote the B-length of I. Let $S = S(I)$ be the B-length coordinate on I which is 0 at the origin of I and $\ell(I)$ at the terminus of I. It is easy to verify that (X_τ) has invariant density ρ/h with respect to A-length, hence invariant density $\rho/\sqrt{h}$ with respect to B-length. Let (Y_t) be the process in $\mathcal{D}_0$ with generator Δ_B and branch probabilities p. We want to choose h, such that $\rho/\sqrt{h}$ will be an invariant density for (Y_t). As above, $\rho/\sqrt{h}$ will be an invariant density for (Y_t) if and only if $\rho/\sqrt{h}$ is linear in s and such that there exists a positive function η on the junction points of Λ with

$$\lim_{s\to 0} (\rho/\sqrt{h})(s) = \eta(x)p(I) \quad \text{for every } I, \tag{4.8}$$

where x is the origin of I, and for every junction point x,

$$\sum_{i=1}^{k} \left(\frac{d}{ds}(\rho/\sqrt{h}) \text{ on } I_i\right) = 0, \tag{4.9}$$

where $I_1,\ldots,I_k$ are the branches at 0.

By (4.4), (4.8) will hold if and only if for each I with origin x,

$$\lim_{r\to 0} h(r) = \frac{\psi^2(x)Q^2(I)}{\psi^2(x)p^2(I)}. \tag{4.10}$$

This shows that if (4.8) holds, then (X_τ) automatically has branch probabilities p, by (3.6). Thus if we find h such that $\rho/\sqrt{h}$ is the invariant density for (Y_t), the theorem is proved.

$\rho/\sqrt{h}$ linear in s means there is some constant $c(I)$ such that

$$\frac{d}{ds}(\rho/\sqrt{h}) = c(I) \quad \text{on } I. \tag{4.11}$$

We note that $c(I) = -c(-I)$.

(4.11) says

$$\sqrt{h}\,\frac{d}{dr}(\rho/\sqrt{h}) = c(I) \quad \text{on } I. \tag{4.12}$$

Let $f = \rho/\sqrt{h}$. Then

$$f^{-1}\frac{df}{dr} = c(I)/\rho. \tag{4.13}$$

We need to find η, f, c, such that (4.13) holds, and for every I with origin x

(4.14) $$\lim_{r\to 0} f(r) = \eta(x)p(I),$$

and for every junction point x

(4.15) $$\sum_{i=1}^{k} c(I_i) = 0$$

where $I_1,\ldots,I_k$ are the branches originating at x.

Let, for each I,

(4.16) $$\int_0^1 \rho^{-1}(r)\,dr = \gamma(I).$$

Then by (4.13),

(4.17) $$c(I) = [\log \eta(y)p(-I) - \log \eta(x)p(I)]\ \gamma^{-1}(I)$$

Thus (4.15) will hold if, for each junction point x,

(4.18) $$\log \eta(x) - \sum_{i=1}^{k} \log \eta(y)\ \gamma^{-1}(I_i) / \sum_{i=1}^{k} \gamma^{-1}(I_i) = \sum_{i=1}^{k} [\log p(-I_i) - \log p(I_i)]\ \gamma^{-1}(I_i).$$

If we define R(I) by

(4.19) $$R(I) = \gamma^{-1}(I) / \sum_{i=1}^{k} \gamma^{-1}(I_i),$$

then R defines a (symmetric) Markov transition matrix on the junction points of Λ. Equation (4.18) says that $\log \eta$ is a constant plus the potential with respect to R of a certain measure on Λ, which we may call λ. λ has total mass 0. It follows easily, since Λ is compact, that (4.18) has solutions η, one of which we now choose. We define c(I) by (4.17). Then (4.15) holds by (4.18). By (4.17), for a branch I with origin x and terminus y,

(4.20) $$\int_0^1 c(I)\ \rho^{-1}(r)\ dr = \log \eta(y)p(-I) - \log \eta(x)p(I).$$

Thus we can define f so that (4.13) and (4.14) hold, proving the theorem.

It is clear that the compactness assumption on Λ could be considerably relaxed.

REFERENCES

1. W. Feller, On second order differential operators, Annals of Math 61, 90-105 (1955).

2. S. Kakutani, Two dimensional Brownian motion and harmonic functions, Proc. Imp. Acad. Tokyo 20, 706-714 (1944).

3. V. A. Volkonskii, Random substitution of time in strong Markov processes, Teoriya Veroyatnostei i ee Primeneniya 3, 332-350 (1958) (translated in Theory of Probability and Its Applications 3, 310-326 (1958)).

University of Minnesota
Minneapolis, MN 55455

University of British Columbia
Vancouver, B.C. Canada V6T 1Y4

Contemporary Mathematics
Volume 26, 1984

ON SEQUENCES OF DENSITY ZERO IN ERGODIC THEORY

by

A. Bellow* and V. Losert*

SECTION 1. INTRODUCTION

We denote by $\mathbb{Z}$ the set of all integers and by $\mathbb{N}$ the set of all integers ≥ 1.

We follow the notation in [2]: $\underline{m} = (m_k)$, $\underline{n} = (n_k)$ will always denote strictly increasing sequences of positive integers.

In what follows (Ω, A, μ) is a probability space. We write

$$L^1 = L^1(\Omega, A, \mu), \quad L^\infty = L^\infty(\Omega, A, \mu) \ .$$

Let τ be the group of automorphisms of (Ω, A, μ): $T \in \tau$ if $T: \Omega \to \Omega$ is a bijection which is bimeasurable and preserves μ.

Let $T \in \tau$, $\underline{m} = (m_k)$, $p \in \mathbb{N}$, $N \in \mathbb{N}$. We set:

$$T_{p,\underline{m}} f(\omega) = \frac{1}{p} \sum_{k=1}^{p} f(T^{m_k}\omega), \quad f \in L^1$$

and for each $\lambda > 0$ we set

$$A_N(\underline{m}, T, f, \lambda) = \{\omega \mid |T_{p,\underline{m}} f(\omega)| \geq \lambda \quad \text{for some} \quad 1 \leq p \leq N\}$$

$$A_\infty(\underline{m}, T, f, \lambda) = \{\omega \mid |T_{p,\underline{m}} f(\omega)| \geq \lambda \quad \text{for some} \quad p \in \mathbb{N}\}.$$

We now recall J. P. Conze's important result on the Weak Maximal Inequality.

*Research supported by the National Science Foundation (U.S.A.)

THEOREM 1.1 [8]. With each sequence $\underline{m}$ one can associate a minimal constant $0 < C(\underline{m}) \leq +\infty$ satisfying the inequality:

$$\lambda\mu(A_\infty(\underline{m},T,f,\lambda)) \leq C(\underline{m})\|f\|_1$$

for all $T \in \tau$, $f \in L^1$, $\lambda > 0$. Furthermore $C(\underline{m}) < +\infty$ if and only if there exist $T \in \tau$ aperiodic (or ergodic) such that $\lim_p T_{p,\underline{m}} f(\omega)$ exists a.e., for each $f \in L^1$ (i.e., the Individual Ergodic Theorem holds for T along the subsequence $\underline{m}$).

We now recall the notions of "good universal sequence" and "bad universal sequence" in Ergodic Theory. The pioneering work on "bad universal sequences" was done by U. Krengel [14] who introduced the notion and first proved the existence of "bad universal sequences" (relative to the Individual Ergodic Theorem). For more complete references on the subject see [3].

DEFINITION 1.2. A sequence $\underline{m} = (m_k)$ is called a "good universal sequence" relative to the Individual Ergodic Theorem if for each $T \in \tau$

$$\lim_p T_{p,\underline{m}} f(\omega) \quad \text{exists} \quad \text{a.e.}$$

for all $f \in L^1$.

DEFINITION 1.3. A sequence $\underline{m} = (m_k)$ is called a "bad universal sequence" relative to the Individual Ergodic Theorem if $C(\underline{m}) = +\infty$; equivalently if for each $T \in \tau$ aperiodic (or ergodic) there exists $f \in L^1$ such that $(T_{p,\underline{m}} f(\omega))$ fails to converge a.e.

DEFINITION 1.4. A sequence $\underline{m} = (m_k)$ is called a "good universal sequence" relative to the Mean Ergodic Theorem if for each $T \in \tau$

$$\lim_p T_{p,\underline{m}} f \quad \text{exists in} \quad L^1$$

for all $f \in L^1$.

We also recall the notion of "set of recurrence" (in the sense of Poincaré) introduced by H. Furstenberg [10]:

DEFINITION 1.5. A set $R \subset \mathbb{N}$ is called a "set of recurrence" if whenever $(Y,\mathcal{B},\nu)$ is a probability space $T : Y \to Y$ an automorphism of Y, $A \in \mathcal{B}$ with $\nu(A) > 0$, then there exists $r \in R$ such that

$$\nu(A \cap T^{-r}(A)) > 0.$$

In this paper we study a special class of sequences, which we call "block sequences", from the point of view of both the Individual and the Mean Ergodic Theorem, and we answer some questions which as far as we know were still open.

Let $\underline{n} = (n_k)$ and $\underline{\ell} = (\ell_k)$ be strictly increasing sequences of positive integers such that $\ell_k < n_{k+1} - n_k$ for all $k \in \mathbb{N}$. Let

$$C_k = \{n_k, n_k+1, \ldots, n_k+\ell_k\} \quad \text{for each } k \in \mathbb{N} .$$

Then $\underline{m}$, the "block sequence" generated by $\underline{n} = (n_k)$ and $\underline{\ell} = (\ell_k)$ is defined to be the sequence

$$n_1, n_1+1, \ldots, n_1+\ell_1, \ldots, n_k, n_k+1, \ldots, n_k+\ell_k, \ldots$$

(that is, as a set, $\underline{m}$ is just $\bigcup_{k=1}^{\infty} C_k$). We will refer to C_k as the k-th block of the sequence $\underline{m}$.

LEMMA 1.6. *Let* $\underline{m}$ *be the block sequence generated by the sequences* $\underline{n} = (n_k)$ *and* $\underline{\ell} = (\ell_k)$. *Consider the following conditions*

(1) $\underline{m}$ *has density zero.*

(2) $\dfrac{\ell_1+\ell_2+\ldots+\ell_k}{n_k} \to 0.$

(3) $\dfrac{\ell_k}{n_k} \to 0.$

Then (1) $\equiv$ (2) $\Rightarrow$ (3). *If in addition the sequence* $\underline{m}$ *satisfies also the growth condition*

$$(*) \quad \ell_k \geq C^* n_{k-1} \quad \text{for all } k > 1$$

(for some constant $C^* > 0$), *then* (1) $\equiv$ (2) $\equiv$ (3).

PROOF: (1) $\Rightarrow$ (3). This follows immediately from the inequality

$$\frac{|C_1 \cup \ldots \cup C_k|}{n_k+\ell_k} \geq \frac{\ell_k}{n_k+\ell_k} = \frac{\frac{\ell_k}{n_k}}{1+\frac{\ell_k}{n_k}}$$

(1) $\Rightarrow$ (2)

$$\frac{|C_1 \cup \ldots \cup C_k|}{n_k+\ell_k} = \frac{\sum_{j=1}^{k} (\ell_j+1)}{n_k+\ell_k} \geq \frac{\ell_1+\ell_2+\ldots+\ell_k}{n_k+\ell_k}$$

$$= \frac{\frac{\ell_1+\ell_2+\ldots+\ell_k}{n_k}}{1+\frac{\ell_k}{n_k}}$$

since the left-hand side tends to 0 and $\frac{\ell_k}{n_k} \to 0$ (by (3)), we deduce $\frac{\ell_1+\dots+\ell_k}{n_k} \to 0$.

(2) ⇒ (1) since

$$0 \le \frac{k}{n_k} \le \frac{\ell_k}{n_k} \le \frac{\ell_1+\dots+\ell_k}{n_k} ,$$

condition (2) implies

$$\frac{\ell_k}{n_k} \to 0 \text{ and } \frac{k}{n_k} \to 0 .$$

We deduce

$$\frac{|C_1 \cup \dots \cup C_k|}{n_k+\ell_k} = \frac{(\ell_1+\dots+\ell_k)+k}{n_k+\ell_k} = \frac{\frac{\ell_1+\dots+\ell_k}{n_k} + \frac{k}{n_k}}{1+\frac{\ell_k}{n_k}}$$

and thus (2) ⇒ (1) is proved.

Let us assume now that $\underline{m}$ satisfies the growth condition (*) and let us show (3) ⇒ (1).

Since $\ell_j < n_{j+1}-n_j$ we have

$$\sum_{j=1}^{k-2} (\ell_j+1) \le n_{k-1}-n_1 < n_{k-1}$$

and hence

$$\frac{|C_1 \cup \dots \cup C_k|}{n_k+\ell_k} = \frac{\sum_{j=1}^{k}(\ell_j+1)}{n_k+\ell_k} \le \frac{n_{k-1} + (\ell_{k-1}+1) + (\ell_k+1)}{n_k+\ell_k}$$

$$= \frac{\frac{n_{k-1}}{n_k} + \frac{\ell_{k-1}}{n_k} + \frac{\ell_k}{n_k} + \frac{2}{n_k}}{1+\frac{\ell_k}{n_k}} \le \frac{\frac{1}{C^*}(\frac{k}{n_k}) + 2\frac{\ell_k}{n_k} + \frac{2}{n_k}}{1+\frac{\ell_k}{n_k}}$$

and the right-hand side clearly tends to 0. This completes the proof.

REMARKS. 1) Without some additional condition on $\underline{m}$ (such as the growth condition (*)) the implication (3) ⇒ (1) is false. The following simple example illustrates this: Take $n_k = k^2$, $\ell_k = k$. Then clearly condition (3) of Lemma 1.6 is satisfied, but (1) fails, so the corresponding "block sequence" $\underline{m}$ cannot have density zero.

2) If we let $n_k = (k+1)!$, $\ell_k = k!$, then the growth condition (*) and condition (3) of Lemma 1.6 are both satisfied; hence the resulting "block sequence" $\underline{m}$ has density zero.

SECTION 2.

"GOOD UNIVERSAL SEQUENCES" RELATIVE TO THE MEAN ERGODIC THEOREM

In this section we review some essentially known facts. We begin by recalling the following known result about block sequences [5]:

PROPOSITION 2.1. Every "block sequence" is a "good universal sequence" relative to the Mean Ergodic Theorem, and the limit of the averages along such a sequence is precisely the projection operator onto the invariant functions.

PROOF. This is a familiar proof. For the sake of completeness we sketch it below.

Let $\underline{m}$ be the "block sequence" generated by the sequences (n_k) and (ℓ_k).

Let $T \in \tau$. Denote also by T the corresponding linear operator in L^1, $T : f \to f \circ T$.

Let P be the projection operator onto the subspace of T-invariant functions. Then (see [9], p. 662),

(1) $P(Tf) = T(Pf) = Pf$ for all $f \in L^1$;

if we let

$$V = \{f \in L^1 | Tf = f\} \quad \text{and} \quad W = \{(I-T)g | g \in L^\infty\},$$

then $V+W$ is dense in L^1.

Consider the averages along the subsequence $\underline{m}$

$$T_p f = T_{p,\underline{m}} f = \frac{1}{p} \sum_{k=1}^{p} f \circ T^{m_k}, \quad f \in L^1.$$

For $f \in V$, $T_p f = f$ for all p and we trivially have convergence to $f = Pf$.

Let now $f \in W$; then $f = (I-T)g$, $g \in L^\infty$, and it is easily checked that $\|T_p f\|_\infty \to 0$. Thus $Pf = 0$ for each $f \in W$.

Since (T_p) is a sequence of contractions in L^1 and since it converges to P on a dense set, the proof is finished.

REMARK. The same argument works of course for more general sequences, the crucial property here being a Følner type condition (see Section 3, condition (T2) in Theorem 3.1).

In the terminology of Blum and Reich [6], a sequence $\underline{m} = (m_k)$ which is "good universal" relative to the Mean Ergodic Theorem and such that the limit of the averages along $\underline{m}$ is precisely the projection operator onto the invariant functions is called an "ergodic sequence". Proposition 2.1 says that any "block sequence" is ergodic.

REMARKS. 1) By the invariance property of the projection operator P (see [1]) in the proof of Proposition 2.1) it is clear that if $\underline{m} = (m_k)$ is an ergodic sequence, then for each $s \in \mathbb{N}$, the sequence $(m_k+s)_{k\in\mathbb{N}}$ is also ergodic.

2) According to Blum and Reich ([6]; see also [5]) a sequence $\underline{m} = (m_k)$ is ergodic if and only if

$$\lim_p \frac{1}{p} \sum_{k=1}^{p} e^{2\pi i m_k x} = 0 \quad \text{for all} \quad x \in \mathbb{R},\ x \notin \mathbb{Z}$$

(this can also be checked directly; for fixed $T \in \tau$, decompose Hilbert space L^2 into the space spanned by the eigenfunctions and the "continuous part"). Ergodic sequences are also encountered in the literature under various other names: Veech [15] calls them "uniformly distributed" sequences of integers, Boshernitzan [7] calls them "homogeneously distributed". Let us also note that if $\underline{m} = (m_k)$ is ergodic, then for any measure-preserving system $(Y,\mathcal{B},\nu;T)$, the Mean Ergodic Theorem along $\underline{m}$ holds, with the proper limit.

The following is an easy consequence of the definition of ergodic sequence (see also [7]):

PROPOSITION 2.2. Let $\underline{m} = (m_k)$ be an ergodic sequence. Then $\{m_1,m_2,\ldots,m_k,\ldots\}$ is a "set of recurrence". Also for each $s \in \mathbb{N}$, the translate $\{m_1+s, m_2+s,\ldots,m_k+s,\ldots\}$ is a "set of recurrence".

PROOF. Suppose that $\{m_1,m_2,\ldots,m_p,\ldots\}$ were not a set of recurrence. This means there exist a measure-preserving system $(Y,\mathcal{B},\nu;T)$ and $A \in \mathcal{B}$, with $\nu(A) > 0$ such that

$$\nu(A \cap T^{-m_j}(A)) = 0 \quad \text{for all} \quad j \in \mathbb{N},$$

that is, with $f = 1_A$,

$$< f\circ T^{m_j}, f > = 0 \quad \text{for all} \quad j \in \mathbb{N}.$$

On the other hand, since $\underline{m}$ is an ergodic sequence,

$$\frac{1}{p} \sum_{k=1}^{p} f\circ T^{m_k} \xrightarrow{L^1(\nu)} Pf = E(f|\mathcal{T})$$

($E(.|\mathcal{T})$ is of course the conditional expectation operator with respect to the σ-field $\mathcal{T} \subseteq \mathcal{B}$ of all T-invariant sets). We deduce

$$0 = < \frac{1}{p} \sum_{k=1} f\circ T^{m_k}, f > \to < E(f|\mathcal{T}), f > ,$$

whence

$$0 = \int E(f|\mathcal{T}) f \, d\nu = \int E(f|\mathcal{T})^2 d\nu \to E(f,\mathcal{T}) = 0 \quad \text{a.e.}$$

Thus, $\nu(A) = \int f d\nu = \int E(f|\mathcal{T}) d\nu = 0$, a contradiction.

REMARK. There exist sets of recurrence growing arbitrarily fast, i.e., given any sequence $\underline{n} = (n_k)$, there is (r_k) with $r_k \geq n_k$ such that $R = \{r_1, r_2, \ldots, r_k, \ldots\}$ is a set of recurrence. This can be done either by using Furstenberg's "infinite difference set" (see [10], Theorem 3.1) or by using the above "block sequences". For the construction of other interesting sets of recurrence see [7].

SECTION 3.

"GOOD UNIVERSAL SEQUENCES" (OF DENSITY ZERO) RELATIVE TO THE INDIVIDUAL ERGODIC THEOREM

We begin by recalling Tempelman's Ergodic Theorem [18]; this theorem goes back to 1967 and is not nearly so well known as it deserves to be. The case of a countable amenable group suffices for our purposes. A simple, elegant proof of Tempelman's theorem in this special case can be found in [17] (For a proof of the general case see [16], Chap. VI, Section 4, Theorem 4.4).

THEOREM 3.1. Let G be a countable amenable group acting in a measure-preserving fashion $(g \to T_g)$ on $(\Omega, \mathcal{A}, \mu)$. Let (A_k) be a sequence of subsets of G satisfying the following conditions:

(T1) $0 < |A_k| < +\infty$ *for each* $k \in N$.

(T2) (*Følner's condition*) $\lim_k \frac{|gA_k \,\Delta\, A_k|}{|A_k|} = 0$ *for all* $g \in G$.

(T3) $A_1 \subset A_2 \subset \ldots \subset A_k \subset A_{k+1} \subset \ldots$

(T4) *There is a constant* $M > 0$ *such that*

$$|A_k^{-1} A_k| \le M|A_k| \quad \text{for all} \quad k \in N.$$

Then for any $f \in L^1(\Omega, A, \mu)$,

$$\lim_k \frac{1}{|A_k|} \sum_{g \in A_k} f(T_g \omega) = Pf(\omega)$$

exists a.e. *and is invariant under* $\{T_g \mid g \in G\}$.

We shall apply the above theorem to the case $G = Z$.

LEMMA 3.2. *Assume that the "block sequence"* $\underline{m}$ *generated by the sequences* (n_k) *and* (ℓ_k) *satisfies the growth condition*

$$(*) \quad \ell_k \ge C^* n_{k-1} \quad \text{for all} \quad k > 1$$

(*for some constant* $C^* > 0$). Let $A_k = \{n_1, n_1+1, \ldots, n_1+\ell_1, \ldots, n_k, n_k+1, \ldots, n_k+\ell_k\} = \bigcup_{j=1}^{k} C_j$ *for* $k \in N$. *Then the sequence* (A_k) *satisfies conditions* (T1), (T2), (T3), (T4) *above*.

PROOF. Conditions (T1), (T2), (T3) are easily seen to hold. To verify (T4) note that

$$|C_j - C_j| = |[-\ell_j, \ell_j]| = 2\ell_j + 1$$

and on the other hand

$$C_j - \left(\bigcup_{i=1}^{j-1} C_i\right) \subset \left[n_j - (n_{j-1} + \ell_{j-1}),\ n_j + \ell_j - n_1\right]$$

$$\subset \left[n_j - (n_{j-1} + \ell_j),\ n_j + \ell_j - 1\right] ,$$

$$\left|C_j - \left(\bigcup_{i=1}^{j-1} C_i\right)\right| \le n_{j-1} + 2\ell_j \le \left(\frac{1}{C^*} + 2\right) \ell_j .$$

Thus

$$\left| C_j - \left(\bigcup_{i=1}^{j} C_i \right) \right| \leq \left| C_j - C_j \right| + \left| C_j - \left(\bigcup_{i=1}^{j-1} C_i \right) \right|$$

$$\leq \left(\frac{1}{C^*} + 4 \right) \left(\ell_j + 1 \right) = K^* \left(\ell_j + 1 \right)$$

and also

$$\left| \left(\bigcup_{i=1}^{j} C_i \right) - C_j \right| = \left| C_j - \left(\bigcup_{i=1}^{j} C_i \right) \right| \leq K^* \left(\ell_j + 1 \right) .$$

Since $|A_k| = \sum_{j=1}^{k} (\ell_j + 1)$ and

$$\left| A_k - A_k \right| \leq \sum_{j=1}^{k} \left| C_j - \left(\bigcup_{i=1}^{j} C_i \right) \right| + \left| \left(\bigcup_{i=1}^{j} C_i \right) - C_j \right|$$

condition (T4) follows.

REMARK. Lemma 3.2 remains valid if the growth condition (*) is replaced by

(**) There is $p \in N$ and a constant $C^* > 0$ such that

$\ell_k \geq C^* n_{k-p}$ <u>for all</u> $k > p$.

From Theorem 3.1 and Lemma 3.2 we immediately obtain:

COROLLARY 3.3. <u>Assume that the "block sequence"</u> $\underline{m}$ <u>generated by the sequences</u> (n_k) <u>and</u> (ℓ_k) <u>satisfies the growth condition</u> (*) <u>of</u> Lemma 3.2. <u>Then</u> $\underline{m}$ <u>is a "good universal sequence" relative to the Individual Ergodic Theorem (in fact</u> $\underline{m}$ <u>is an "ergodic sequence"</u>).

COROLLARY 3.4. <u>There are sequences of density zero which are "good universal sequences" relative to the Individual Ergodic Theorem (in fact "ergodic sequences"</u>).

PROOF. The "block sequence" $\underline{m}$ generated by $n_k = (k+1)!$ and $\ell_k = k!$ is such an example. More generally, so is any "block sequence" $\underline{m}$, generated by (n_k) and (ℓ_k) satisfying the growth condition (*) and satisfying $\frac{\ell_k}{n_k} \to 0$ (see Lemma 1.6).

REMARKS. 1) As far as we know the existence of sequences of density zero which are "good universal sequences" relative to the Individual Ergodic Theorem has until now been an open question (see [16], Chap. VIII' - Subsequences and generalized means -, Section 2, Good and bad sequences, p. 14).

2) For sequences of positive density which are "good universal sequences" relative to the Individual Ergodic Theorem there is an extensive literature. A fairly complete account may be found in [3] and [6].

SECTION 4.

"BAD UNIVERSAL SEQUENCES" RELATIVE TO THE INDIVIDUAL ERGODIC THEOREM

In this section the term "bad universal" will always refer to "bad universal sequences" relative to the Individual Ergodic Theorem. (As a matter of fact there are no bad universal sequences relative to the Mean Ergodic Theorem: vide the Blum-Hanson Theorem [4]).

As mentioned earlier, it was Krengel [14] who first introduced and proved the existence of "bad universal sequences". Subsequently a number of mathematicians obtained the existence of "bad universal sequences" by various other methods; see for instance [8], [12], [1], [2], [3]. Let us recall that if $\underline{n} = (n_k)$ is a "bad universal sequence", then its lower density must be 0, $\ell d(\underline{n}) = 0$ (see [8], Lemma 4, p. 11). The question then naturally arose whether any of the classical sequences in literature having density 0, such as (2^k), $(k!)$, (k^2), is "bad universal". It was shown in [2] that every lacunary sequence (n_k) is a bad universal sequence (we recall that (n_k) is lacunary if there exists $\lambda > 1$ such that

$$\frac{n_{k+1}}{n_k} \geq \lambda \quad \text{for all} \quad k \in \mathbb{N}).$$

In particular, the sequences (2^k), $(k!)$ are bad universal. But in fact more is true [2]; even the "block sequences" whose k-th block is given by

$$2^k,\ 2^k+1, \ldots, 2^i+k \qquad (k \geq 1)$$

$$k!,\ k!+1, \ldots, k!+k \qquad (k \geq 2)$$

are bad universal. This answered a question raised by Blum and Reich in [6].

A more careful analysis and a refinement of the method of proof used in [2] yields:

THEOREM 4.1. *Let* $\underline{n} = (n_k)$ *be a lacunary sequence, that is, such that*

$$\frac{n_{k+1}}{n_k} \geq \lambda \quad \text{for all} \quad k \in \mathbb{N}$$

and some constant $\lambda > 1$. *Let* $\underline{\ell} = (\ell_k)$ *be such that* $\ell_k < n_{k+1} - n_k$ *for all* $k \in \mathbb{N}$ *and*

$$\frac{\ell_k}{\lambda^k} \to 0.$$

Then the "block sequence" $\underline{m}$ *generated by* (n_k) *and* (ℓ_k), *namely the sequence*

$$n_1, n_1+1, \ldots, n_1+\ell_1, \ldots, n_k, n_k+1, \ldots, n_k+\ell_k, \ldots$$

is a bad universal sequence.

The details of proof will be given in [3].

CONCLUDING REMARK. It is known - and not difficult to check - that (k^2) is a "good universal sequence" relative to the Mean Ergodic Theorem, though *not* an ergodic sequence (in the sense of Section 2). It is also known that $\{1, 2^2, 3^2, \ldots, k^2, \ldots\}$ is a "set of recurrence"; this result is due to Furstenberg (see [10], Theorem 3.5, p. 220). The question whether or not (k^2) is "good universal" relative to the Individual Ergodic Theorem remains open.

BIBLIOGRAPHY

[1] A Bellow, Sur la structure des suites "mauvaises universelles" en théorie ergodique, C. R. Acad. Sci. Paris 294 (1982), 55-58.

[2] A. Bellow, On "bad universal" sequences in ergodic theory (II), Proc. Univ. of Sherbrooke Workshop on Measure Theory and Applications, to appear in Springer-Verlag Lecture Notes.

[3] A. Bellow and V. Losert, The weighted pointwise ergodic theorem and the individual ergodic theorem along subsequences, in preparation.

[4] J. R. Blum and D. L. Hanson, On the mean ergodic theorem for subsequences, Bull. Amer. Math. Soc. 66 (1960), 308-311.

[5] J. R. Blum and B. Eisenberg, Generalized summing sequences and the mean ergodic theorem, Proc. Amer. Math. Soc. 42 (1974),423-429.

[6] J. R. Blum and J. I. Reich, Strongly ergodic sequences of integers and the Individual Ergodic Theorem, Proc. Amer. Math. Soc 86 (1982), 591-595.

[7] M. Boshernitzan, Homogeneously distributed sequences and Poincaré sequences of integers of sublacunary growth, to appear in Monatshefte für Math.

[8] J. P. Conze, Convergence des moyennes ergodiques pour des sous-suites, Bull Soc. Math. France 35 (1973), 7-15.

[9] N. Dunford and J. Schwartz, Linear Operators I, John Wiley, New York 1958.

[10] H. Furstenberg, Poincaré recurrence and number theory Bull. Amer. Math. Soc. 5 (1981), 211-234.

[11] P. R. Halmos, Lectures on Ergodic Theory, Publ. Math. Soc. Japan, 3, Tokyo 1956.

[12] A. Del Junco and J. Rosenblatt, Counterexamples in ergodic theory and number theory, Math. Ann. 245 (1979), 185-197.

[13] S. Kakutani and K. Yosida, Birkhoff's ergodic theory and the maximal ergodic theorem, Proc. Imp. Acad. Tokyo 15 (1939), 165-168.

[14] U. Krengel, On the individual ergodic theorem for subsequences, Ann. Math. Stat. 42 (1971), 1091-1095.

[15] W. A. Veech, Well distributed sequences of integers, Trans. Amer. Math. Soc. 161 (1971), 63-70.

Added in proof. We are grateful to Ulrich Krengel for bringing Templeman's Theorem to our attention. This made our original, lengthy combinatorial argument in Section 3 unnecessary.

[16] U. Krengel, monograph on "Ergodic Theory", in preparation.

[17] D. Ornstein and B. Weiss, The Shannon-McMillan-Breiman theorem for a class of amenable groups, Israel J. Math. 44, No. 1 (1983), 53-60.

[18] A. A. Templeman, Ergodic theorems for general dynamical systems. Sov. Math. Dokl. 8 (1967), 1213-1216 (Engl. Transl.)

Correction. In Lemma 3.2 one should set $A_k = \{m \in \underline{m} \mid m \leq k\}$ and make the corresponding changes in the proof.

A. Bellow
Department of Mathematics
Northwestern University
Evanston, Illinois 60201
USA

V. Losert
Institute of Mathematics
University of Vienna
A-1090 Vienna
Austria

Contemporary Mathematics
Volume 26, 1984

ON THE CONTINUITY OF THE PERCOLATION PROBABILITY FUNCTION

by
J. van den Berg[+] and M. Keane

ABSTRACT. Let G be a countably infinite, connected, locally finite graph, and let s_0 be a designated vertex of G. Denote by $\theta(p)$ the percolation probability function for bond percolation on the pointed graph (G,s_0). We show that

$$\theta(p-) = \theta(p) - \Pr\left\{\begin{array}{l} s_0 \text{ belongs to an infinite p-open} \\ \text{connected component of } G\text{, which itself} \\ \text{has critical probability one.}\end{array}\right\}$$

As a corollary, we deduce that if p is strictly larger than the critical probability for bond percolation on G, and if for p there is a unique infinite p-open cluster (e.g. if G is $\mathbb{Z}^2$ with nearest neighbor bonds), then θ is continuous at p.

§1. DEFINITIONS AND NOTATIONS

Let S be a finite or a countably infinite set, and let B be a collection of two-element subsets of S. Thus $G = (S,B)$ is a (finite or infinite) undirected graph. The elements of S are called sites (= vertices) of G and those of B bonds (= edges) of G. If the bond b contains the sites s and t, then we say that b links s and t (or t and s).

A path π in G is a (finite or infinite) sequence $\pi = (b_1,b_2,\ldots)$ of elements of B such that there exists a sequence $\pi^s = (s_0,s_1,\ldots)$ of elements of S such that for each $i \geq 1$, b_i links s_{i-1} and s_i. Clearly, π^s is determined uniquely by π. If a path $\pi = (b_1,b_2,\ldots,b_n)$ is finite, with $\pi^s = (s_0,s_1,\ldots,s_n)$, then s_0 is called the initial site of π and s_n the terminal site of π. If π is infinite, then s_0 is the initial site, and it is convenient to call ∞ the terminal site (although it is not a site). If π is a path with initial site s and terminal site t, then we say that π joins s with t, and write $s \overset{\pi}{\to} t$.

Let $s_0 \in S$. The connected component of G containing s_0 is the subgraph of G whose sites are s_0 together with all $t \in S$ for which there

[+]Financially supported by the Netherlands Organization for the Advancement of Pure Research (ZWO).

0271-4132/84 $1.00 + $.25 per page

exists a path π which joins s_0 and t, and whose bonds are all bonds in B which link any of these sites with another of these sites. G is connected if for some s_0 ($\equiv$ for each s_0), the connected component of G containing s_0 is G.

§2. PERCOLATION PROBABILITY AND CRITICAL PROBABILITY.

In this paragraph, we suppose that $G = (S,B)$ is a given infinite connected graph. Let $(X_b)_{b \in B}$ be a collection of independent identically distributed random variables indexed by the bonds of G, with common distribution given by

$$\Pr\{X_b \le x\} = \begin{cases} 0 & \text{if } x \le 0 \\ x & \text{if } 0 \le x \le 1 \\ 1 & \text{if } x \le 1. \end{cases}$$

Let $0 \le p \le 1$. For a given realization of the process $(X_b)_{b \in B}$, we say that the bond $b \in B$ is p-open if $X_b < p$, and p-closed if $X_b \ge p$. A path $\pi = (b_1, b_2, \ldots)$ is p-open if for each b_i in π, b_i is p-open. We denote by $G^p = (S^p, B^p)$ the (random) subgraph of G given by $S^p = S$ and

$$B^p = \{b \in B\colon b \text{ is p-open}\} .$$

Now choose and fix a site $s_0 \in S$, so that we are dealing with a pointed graph $\dot{G} = (S,B,s_0)$ and random pointed subgraphs $\dot{G}^p = (S^p, B^p, s_0)$. Let $\dot{G}_0^p$ be the (pointed) connected component of $\dot{G}^p$ containing s_0. We call $\dot{G}_0^p$ the p-open cluster (graph) containing s_0.

DEFINITION

1) The <u>percolation probability</u> function of the pointed graph $\dot{G} = (S,B,s_0)$ is given by

$$\theta(\dot{G},p) = \theta(p) = \Pr\{\dot{G}_0^p \text{ is infinite}\} .$$

2) The <u>critical probability</u> of the graph $\dot{G}$ (see remark 4) is given by

$$p_H(\dot{G}) = p_H = \sup\{p\colon \theta(p) = 0\} = \inf\{p\colon \theta(p) > 0\} .$$

REMARKS

1. These definitions are due to Broadbent and Hammersley [2] where we have modified the notation to suit our purposes.

2. It is clear that $\theta(p)$ is a non-decreasing function of p.

3. It is not hard to deduce that if the graph G is locally finite ($\equiv$ each $s \in S$ is contained in only a finite number of bonds), then $\theta(p)$ is continuous from the right (see [7]).

4. In general, the percolation probability function depends on the choice of s_0. However (see e.g. [2]), it is quickly shown that the critical probability p_H is the same for each choice of s_0 as G is supposed connected.

5. Note that if G is locally finite,

$$\theta(p) = \Pr\{\text{there exists a p-open path } \pi \text{ with } s_0 \overset{\pi}{\to} \infty\} .$$

6. Though we restrict to bond percolation on undirected graphs, it is easily seen that analogs of the results in section 3 hold for site percolation, and for percolation on directed graphs.

§3. STATEMENT OF THE RESULTS.

Recently, some interest has been shown in determining the value of p_H ([5], [8], [9]) and the behavior of the function $\theta(p)$ and related functions for a variety of graphs. Except for regular two-dimensional graphs (see [7] and [8]) and examples where percolation is identical with "infinite life" in birth-death processes, it does not seem to be known whether the function $\theta(p)$ is continuous, although this is expected to be true for a wide class of graphs (e.g. $\mathbb{Z}^d$ with nearest neighbor bonds, $d \geq 3$). On the other hand, Harris [4] has shown for $\mathbb{Z}^2$ that for p above the critical probability p_H ($= \frac{1}{2}$, Kesten [5]), the random graph G^p possesses exactly one infinite connected component with probability one, (for a generalization to regular two-dimensional graphs see Fisher [3]), and Newman-Schulman ([6]), have investigated the possibility of existence of more than one infinite connected component of G^p in a general setting

We hope that the following result, which links the continuity of θ^p with the number and types of infinite connected components, will help to clarify the situation.

THEOREM. <u>Let</u> $\dot{G} = (S,B,s_0)$ <u>be a countably infinite, connected, locally finite graph. Then for any</u> $0 \leq p \leq 1$,

$$\theta(p-) = \theta(p) - \Pr\{\dot{G}_0^p \text{ is infinite and its critical probability is one}\} .$$

COROLLARY. <u>Let</u> $\dot{G}$ <u>be as in the theorem. Furthermore, suppose that</u> $p > p_H(\dot{G})$ <u>is such that with probability one,</u> G^p <u>contains exactly one infinite connected component. Then</u> $\theta(p)$ <u>is continuous at</u> p.

Note that regular two-dimensional graphs([3] and [4]) satisfy the conditions of the corollary (see also Russo ([7] and [8]), who proves in addition continuity at $p = p_H(\dot{G})$ and differentiability at $p \neq p_H(\dot{G})$). Kesten (private communication), has shown that for $p > \frac{1}{2}$, $\mathbb{Z}^3$ with nearest neighbor bonds satisfies the conditions of the corollary.

§4. PROOFS

To prove the theorem, we must show that

$$(*)\qquad \lim_{\substack{p' \uparrow p \\ p' \neq p}} \theta(p') = \Pr\{\dot{G}_0^p \text{ is infinite and } p_H(\dot{G}_0^p) < 1\} .$$

Let $\dot{H} = (\tilde{S},\tilde{B},s_0)$ be a pointed infinite connected subgraph of G (with point s_0), and note that under the condition $\dot{G}_0^p = \dot{H}$, the joint (conditional) distribution of the process $(X_b)_{b \in \tilde{B}}$ is i.i.d. with common distribution given by

$$\Pr\{X_b \le x | \dot{G}_0^p = \dot{H}\} = \begin{cases} 0 & \text{if } x \le 0 \\ \frac{x}{p} & \text{if } 0 \le x \le p \\ 1 & \text{if } x \ge p \end{cases} .$$

Noting that

$$\lim_{\substack{p' \uparrow p \\ p' \neq p}} \theta(p') = \Pr\{\exists\, p' < p \text{ with } \dot{G}_0^{p'} \text{ infinite}\} ,$$

and conditioning both sides of (*) by $\dot{G}_0^p = \dot{H}$, we see that it is sufficient to show that

$$\begin{aligned} &\Pr\{\exists\, p' < p \text{ with } \dot{G}_0^{p'} \text{ infinite} | \dot{G}_0^p = \dot{H}\} \\ &= \Pr\{p_H(\dot{G}_0^p) < 1 | \dot{G}_0^p = \dot{H}\} \\ &= \begin{cases} 0 & \text{if } p_H(\dot{H}) = 1 \\ 1 & \text{if } p_H(\dot{H}) < 1 \end{cases} . \end{aligned}$$

Using now the information on the joint distribution of the process $(X_b)_{b \in \tilde{B}}$ under the condition $\dot{G}_0^p = \dot{H}$, this translates to the requirement that

$$\Pr\{\exists\, p'' < 1 \text{ with } \dot{H}_0^{p''} \text{ infinite}\} = \begin{cases} 0 & \text{if } p_H(\dot{H}) = 1 \\ 1 & \text{if } p_H(\dot{H}) < 1 \end{cases} .$$

The case $p_H(\dot{H}) = 1$ is obvious, and if $p_H(\dot{H}) < 1$, then choose $\tilde{p}$ with

$p_H(\dot{H}) < \tilde{p} < 1$. By definition of $p_H(\dot{H})$, for almost every realization the random subgraph $H^{\tilde{p}}$ of H contains an infinite connected component, since this event is a tail event with positive probability. This component may not contain s_0, but since H is connected, there is a finite path π joining s_0 with an infinite component, and since we may assume with no loss of generality that $X_b < \underset{\sim}{1}$ for all b, it follows that $\dot{H}_0^{p''}$ is infinite for some $p'' < 1$ ($p'' = \max(\tilde{p}, \max_{b \in \pi} X_b)$) almost surely. This finishes the proof of the theorem.

To prove the corollary, we show that under the given hypotheses,

$$\Pr\{\dot{G}_0^p \text{ is infinite and } p_H(\dot{G}_0^p) = 1\} = 0.$$

Choose $\tilde{p}$ with $p_H(\dot{G}) < \tilde{p} < p$. Then for almost every realization, $G^{\tilde{p}}$ contains an infinite connected component (by the 0-1 law for tail events) which is contained in the (unique) infinite connected component of G^p. Thus, choosing $\dot{H}$ and conditioning on $\dot{G}_0^p = \dot{H}$ as in the proof of the theorem, we see that

$$p_H(\dot{G}_0^p) = \frac{p_H(\dot{G})}{p}$$

almost surely (see also [1]), and this (together with remark 3) proves the corollary.

REFERENCES

[1] van den Berg, J., (1982). J. Phys. A: Math. Gen. 15, 605-610.

[2] Broadbent, S.R. and Hammersley, J.M., (1957), Proc. Camb. Phil. Soc. 53, 629-641.

[3] Fisher, M.E., (1961), J. Math. Phys. 2, 620-627.

[4] Harris, T.E., (1960). Proc. Camb. Phil. Soc. 56, 13-20.

[5] Kesten, H., (1980), Commun. Math. Phys. 74, 41-59.

[6] Newman, C.M. and Schulman, L.S., (1981). J. Stat. Phys. 26, 613-628.

[7] Russo, L., (1978). Z. Wahrscheinlichkeitstheorie verw. Gebiete 43, 39-48.

[8] Russo, L., (1981). Z Wahrscheinlichkeitstheorie verw. Gebiete 56, 229-237.

[9] Wierman, J. C., (1981). Adv. Appl. Prob. 13, 298-313.

DEPARTMENT OF MATHEMATICS
DELFT UNIVERSITY OF TECHNOLOGY
JULIANALAAN 132, 2628 BL DELFT,
THE NETHERLANDS

Contemporary Mathematics
Volume 26, 1984

COINCIDENCE THEOREMS, MINIMAX THEOREMS, AND VARIATIONAL INEQUALITIES

Felix E. Browder

INTRODUCTION

Since its beginning in the early 1940's the fixed point theory of multi-valued mappings has had a vital connection with the study of minimax theory in game theory and the theory of general equilibrium in economics. In Kakutani's pioneering paper [9] where he extended the Brouwer fixed point theorem to upper-semi-continuous convex valued multivalued maps of the Euclidean n-ball into itself, the result sharpened the arguments of Von Neumann [11] in mathematical economics and has had continual application to similar results in the ensuing years (e.g. Debreu [6]). It has been clear as well since the paper of Debrunner-Flor [7] in 1964 that the fixed point theory of multi-valued mappings has an intimate relation with the structure theory of operators of monotone type from a reflexive Banach space X to its conjugate space X^*. In 1967, the writer in [3] developed a systematic treatment of the interconnections between multi-valued fixed point theorems, minimax theorems, variational inequalities, and monotone extension theorems, including a number of fixed point theorems obtained by Ky Fan. More recently, Brezis, Nirenberg, and Stampacchia [2], basing their argument upon an extension of a minimax result of Ky Fan [8], have shown that a simple extension of Fan's result to an infinite dimensional context can be made to yield a very direct proof of the existence of solutions of variational inequalities involving single-valued pseudo-monotone mappings.

In the present paper, we shall take up this theme once more in order to extend the earlier results in several interesting directions. First of all, we wish to deal with one drawback of the Brezis-Nirenberg-Stampacchia result, namely that it is not obvious how it can be applied to obtain the appropriate general form of the existence theorem for variational inequalities when the pseudo-monotone operators involved themselves multi-valued. Since the theory of operators of monotone type (including pseudo-monotone operators) is intrinsically cast in the framework of multi-valued mappings, an existence theorem for multi-valued pseudo-monotone maps (which was in fact obtained by Brezis [1]

0271-4132/84 $1.00 + $.25 per page

in his original work on pseudo-monotone operators) ought to be a natural part of the general treatment. Moreover, from the present writer's point of view, the Brouwer fixed point theorem is considerably more intuitive and more natural than the formulation of the theorem of Knaster-Kuratowski-Mazurikiewicz [10], and furnishes a more natural base for a transparent exposition of the underlying principles of argument.

In Section 1, we give an elementary proof of the basic approximation and selection theorems for convex-valued mappings. In Section 2, we establish the basic results for convex-valued upper-semi-continuous mappings on compact convex sets and some of their simplest applications. In Section 3, we consider the extension to mappings which are no longer upper-semi-continuous and obtain results on pseudo-monotone maps. In Section 4, we obtain a fixed-point result for multi-valued mappings which generalizes the results of the writer's paper [5]. Here, a multi-valued fixed point theorem follows from a result on variational inequalities, rather than the converse.

SECTION 1. If T is a mapping of the set C into subsets of the space E which contains C, so that each point x of C, there corresponds a set $T(x)$ of E, then by a fixed point x_0 of T, we mean a point x_0 such that $x_0 \in T(x_0)$. Similarly, if T is a mapping of C into subsets of C_1 and S is a mapping of C_1 into subsets of C, then by a coincidence of the two maps, we mean a point $[x,y]$ of $C \times C_1$ such that $y \in T(x)$, and $x \in S(y)$. In the simplest context which we consider, fixed point and coincidence theorems for multi-valued mappings with convex sets as values will be obtained by applying the classical Brouwer fixed point theorem to approximating continuous single-valued mappings. In the discussion of Section 1, we show how to generate such approximating mappings by elementary use of a partition of unity argument.

DEFINITION 1. *If* E *and* F *are Hausdorff topological vector spaces, then by a multi-valued mapping* T *from a subset* C *of* E *into* F, *we shall mean an assignment to each point* x *of* C *of a subset* $T(x)$ *of* F. *We shall say that* T *is convex-valued if* $T(x)$ *is non-empty and convex for each* x *in* C.

DEFINITION 2. T *is said to be upper-semi-continuous from* C *to* F *if for each neighborhood* V *of* 0 *in* F, *and each point* x_0 *in* C, *there exists a neighborhood* U *of* 0 *in* E *such that for all* x *in* $x_0 + U$, $T(x) \subset T(x_0) + V$.

PROPOSITION 1. *Let* C *be a closed subset of the Hausdorff topological vector space* E, T *a convex valued map of* C *into* F. *Suppose that for each* y *in* F, $T^{-1}(y)$ *is open in* C. *Suppose further that there exists a compact subset* L *of* C *and an element* y_0 *of* F *such that for each* x *in* $C \backslash L$ $y_0 \in T(x)$.

Then there exists a continuous map ζ *of* C *into a subset of* F *which is the convex span of a finite number of points such that for each* x *in* C, $\zeta(x) \subset T(x)$.

PROOF OF PROPOSITION 1. Since L is compact, it must be covered by a finite family of the open sets $\{T^{-1}(y_1),\ldots,T^{-1}(y_r)\}$. Hence C can be covered by this collection together with $T^{-1}(y_0)$. Since E and hence C are completely regular, we may find a partition of unity corresponding to this finite open cover of C, i.e. a family of continuous functions $\{\beta_0,\ldots,\beta_r\}$ from C to $[0,1]$ with the support of each β_j lying in $T^{-1}(y_j)$ such that $\Sigma\beta_j = 1$ on C. We form the map ζ by setting

$$\zeta(x) = \sum_{j=0}^{r} \beta_j(x)y_j .$$

The map ζ is obviously a continuous mapping of C into the convex span of the finite set $\{y_0,\ldots,y_r\}$. For each x in C, if $\beta_j(x) \neq 0$, then $x \in T^{-1}(y_j)$, i.e. $y_j \in T(x)$. Thus $\zeta(x)$ is a convex linear combination of points of $T(x)$. Since $T(x)$ is assumed to be convex for each x, it follows that $\zeta(x) \in T(x)$.

PROPOSITION 2. *Let* C *be a compact subset of the locally convex Hausdorff topological vector space* E, T *an upper-semi-continuous convex-valued mapping of* C *into* F *such that for a given convex subset* A *of* F, $T(x) \cap A \neq 0$. *Let* V_0 *be a convex neighborhood of* 0 *in* F, U_0 *a neighborhood of* 0 *in* E. *Then there exists a continuous single-valued mapping* ξ *of* C *into the convex span of a finite subset of* A *such that for each* x *in* C, *there exists* u *in* C *such that* $x \in u + U_0$ *and* $\xi(x) \in T(u) + V_0$.

PROOF OF PROPOSITION 2. Since T is upper-semi-continuous, for each x in C, there exists a convex symmetric neighborhood U_x of 0 in E such that if v lies in $x + U_x$, then $T(v) \subset T(x) + V_0$. We may assume without loss of generality that each $U_x \subset U_0$. Since C is compact, we can cover C by a finite family of open sets of the form $\{x_j + \frac{1}{2} U_{x_j} ; j = 1,\ldots,s\}$, and setting $W = \bigcap_{j=1}^{s} \frac{1}{2} U_{x_j}$, we may consider finite covering of C of the form

$$\{v_k + W;\ k = 1,\ldots,r\}$$

For each k, we choose a point y_k in $T(v_k) \cap A$, and we construct a partition of unity $\{\alpha_1, \dots, \alpha_r\}$ for the last covering. The map ξ is now defined by setting

$$\xi(x) = \sum_{k=1}^{r} \alpha_k(x) y_k .$$

Let x be any point of C. There exists an index j, with $1 \le j \le s$, such that $x \varepsilon x_j + \frac{1}{2} U_{x_j}$ (and in particular $x \varepsilon x_j + U_0$). Let k be any index, $1 \le k \le s$, such that $\alpha_k(x) \ne 0$. Then $x \varepsilon v_k + W \subset v_k + \frac{1}{2} U_{x_j}$. Hence $v_k \in x_j + U_{x_j}$. Since $y_k \in T(v_k)$, it follows by the choice of U_{x_j} that $y_k \in T(x_j) + V_0$. Hence $\xi(x)$ is a convex linear combination of points of the convex set $T(x_j) + V_0$. Hence $\xi(x) \varepsilon T(x_j) + V_0$, and the assertion of Proposition 2 follows with $u = x_j$. q.e.d.

SECTION 2. Fixed-point and coincidence theorems for convex-valued mappings:

We now apply Propositions 1 and 2 to obtain basic coincidence theorems for multi-valued mappings.

THEOREM 1. *Let C be a compact convex subset of the locally convex topological vector space E, C_1 a compact convex subset of the locally convex topological vector space F. Let T be an upper-semi-continuous convex-valued mapping of C into C_1, S an upper-semi-continuous convex-valued mapping of C_1 into C. Then T and S have a coincidence.*

COROLLARY. *Let $C = C_1$ with $E = F$. Then T has a fixed point.*

The Corollary follows by taking S to be the identity map of C.

PROOF OF THEOREM 1. By Proposition 2, for any neighborhood U of 0 in E and any neighborhood V of 0 in F, there exists a finite-dimensional mapping γ of C into C_1 and a finite dimensional mapping γ of C_1 into C, with both ξ and γ continuous, such that for each x in C, there exists a point $[u,v]$ of $G(T)$, the graph of T, such that

$$x \in u + U, \ \xi(x) \in v + V.$$

while for each y in C_1, there exists a point $[v_1, u_1]$ in $G(S)$ such that

$$y \in v_1 + V, \ \gamma(y) \in u_1 + U.$$

The composed mapping $\gamma \circ \xi$ maps C continuously into the span of a finite subset of C. By the Brouwer fixed point theorem, this mapping has a fixed point x_0. If we set $y_0 = \xi(x_0)$, we have $\gamma(y_0) = x_0$. If we assume the neighborhoods U and V convex and symmetric, then for the corresponding elements in $G(S)$ and $G(T)$, we have

$$u_1 - u \in 2U, \; v_1 - v \in 2V.$$

Thus the two compact subsets $G(T)$ and $G(S^{-1})$ of $C \times C_1$ have property that

$$G(T) \cap [G(S^{-1}) + 2(U \times V)] \neq \emptyset.$$

It follows that $G(T) \cap G(S^{-1}) \neq \emptyset$, i.e. T and S have a coincidence.

q.e.d.

THEOREM 2. Let C be a closed convex subset of the Hausdorff topological space E, T a convex-valued mapping of C into C such that for each y in C, $T^{-1}(y)$ is open in C. Suppose further that there exists a compact subset L of C and an element y_0 of C such that $y_0 \in T(x)$ for all x in $C \setminus L$

Then T has a fixed point in C.

PROOF OF THEOREM 2. By Proposition 1, there exists a continuous mapping ζ of C into the span of a finite subset of C such that $\zeta(x) \in T(x)$ for all x in C. By the Brouwer theorem, ζ has a fixed point x_0, and this is automatically a fixed point of T. q.e.d.

COROLLARY (Ky Fan's Minimax Lemma). Let C be a compact convex subset of Hausdorff topological vector space E, and let f be a real-valued function on $C \times C$ for which:

(1) $f(x,x) \leq 0$, for all x in C.

(2) For each x in C, $\{y | f(x,y) > 0\}$ is convex.

(3) For each y in C, the function $f(\cdot,y)$ is lower-semi-continuous (in the sense of real-valued functions) on C.

Then: There exists x_0 in C such that for all y in C, $f(x_0,y) \leq 0$.

PROOF OF THE COROLLARY FROM THEOREM 2. For each x in C, let $T(x) = \{y | y \in C, f(x,y) > 0\}$. By (2), $T(x)$ is convex for each x in C. By (3), $T^{-1}(y)$ is open in C for each y in C. If $T(x)$ were non-empty for each x in C, then by Theorem 2, T would have a fixed point x_0, for which $f(x_0,x_0) > 0$. Since this contradicts assumption (1), it follows that for some x_1, $T(x_1) = 0$. Hence for all y in C, $f(x_1,y) \leq 0$. q.e.d.

THEOREM 3. Let C be a compact convex subset in the locally convex topological vector space E, C_1 a compact subset of the locally convex topological vector space F, T a convex-valued mapping of C into C_1, S a convex valued mapping of C_1 into C. Suppose that S is upper-semi-continuous, and that for each y in C, $T^{-1}(y)$ is open in C.

Then T and S have a coincidence.

PROOF OF THEOREM 3. By Proposition 2, there exists a continuous mapping ζ of C into C_1 such that for all x in C, $\zeta(x) \in T(x)$. Let R be the convex-valued mapping of C into C given by $R(x) = S(\zeta(x))$. Then R is obviously upper-semi-continuous. Hence by Theorem 1, R has a fixed point x_0, i.e. $x_0 \in S(\zeta(x_0))$. Since $\zeta(x_0) \in T(x_0)$, it follows that the pair $[x_0, \zeta(x_0)]$ yields a coincidence of T and S. q.e.d.

SECTION 3. Extensions to larger classes of mappings:

Some of the most important applications in analysis of the coincidence arguments of the type developed in Section 2 concern mappings defined on weakly compact subsets of infinite dimensional Banach spaces which in the weak topology are neither upper-semi-continuous nor have the property that for each y, $T^{-1}(y)$ is weakly open. The basic technique of argument to handle these contexts is embodied in the following abstract form:

THEOREM 4. *Let* C *be a convex subset of the locally convex topological vector space* E, C_1 *a convex subset of the locally convex topological vector space* F, S *a convex valued mapping of* C *into* C_1, T *a convex-valued mapping of* C_1 *into* C, K *a compact subset of* $C \times C_1$.

Let Λ *be a partially ordered set, and for each* α *in* Λ, *let* C_α *be a convex subset of* C, $C_{1,\alpha}$ *a convex subset of* C_1, S_α *a convex-valued map of* C_α *into* $C_{1,\alpha}$, T_α *a convex-valued mapping of* $C_{1,\alpha}$ *into* C_α. *Suppose further that:*

(a) $K \cap G(S_\alpha) \cap G(T_\alpha^{-1}) \neq 0$ *for all* α *in* Λ.

(b) If $\alpha > \alpha_1$, *then*

$$G(S_\alpha) \cap G(T_\alpha^{-1}) \subset G(S_{\alpha_1}) \cap G(T_{\alpha_1}^{-1}).$$

(c) $$\bigcap_\alpha \mathrm{cl}[(G(S_\alpha) \cap G(T_\alpha^{-1})] \neq \phi \Rightarrow G(S) \cap G(T^{-1}) \neq \phi$$

where cl *denotes closure in* $C \times C_1$.

Then T *and* S *have a coincidence*.

PROOF OF THEOREM 4. For each α in Λ, $K \cap G(S_\alpha) \cap G(T_\alpha^{-1})$ is a non-empty subset of the compact set K. By (b), this family of sets has the finite intersection property. Hence

$$K \cap \bigcap_\alpha \mathrm{cl}(G(S_\alpha) \cap G(T_\alpha^{-1}) \neq \emptyset.$$

By property (c), this implies $K \cap G(S) \cap G(T^{-1}) \neq \emptyset$. Hence T and S have a coincidence. q.e.d.

As a first application, we derive a reformulation of the basic theorem of Brezis-Nirenberg-Stampacchia [2].

THEOREM 5. *Let* C *be a convex subset of the Hausdorff topological vector space* E, T *a convex-valued mapping of* C *into* C *having the following properties:*

(a) *There exists a compact subset* L *of* C *and an element* y_0 *of* L *such that* y_0 *lies in* $T(x)$ *for all* x *in* $C \setminus L$.

(b) *For each finite-dimensional subspace* F *of* E, *the mapping* T_F *of* $C \cap F$ *into* $C \cap F$ *given by* $T_F(x) = T(x) \cap F$, *has the property that*

$$T_F^{-1}(y)$$

is open in C_F *for each* y *in* C_F.

(c) *If* x_β *is a convergent filter in* C *with limit* x *and if for all* β *and a line segment* D *ending at* x, $T(x_\beta)$ *is disjoint from* D, *then* $T(x)$ *is disjoint from* D.

Then: T *has a fixed point.*

PROOF OF THEOREM 5. We assume that T has no fixed point and derive a contradiction. Let Λ be the family of finite dimensional subspaces F of E which contain y_0 and intersect C. For each F in Λ, let $C_F = C \cap F$, $L_F = L \cap F$, and T_F as above. By hypothesis (b), $T^{-1}(y)$ is open in C_F for each y in C_F. Since $y_0 \in T_F(x)$ for all x in $C_F \setminus L$ and since T_F has no fixed points, it follows from Theorem 2 that $T_F(x)$ must be empty for some point x in C_F. By (a), such a point x_F must lie in L_F.

For each F in Λ, let

$$N_F = \{x \mid x \in L,\ T(x) \cap F = \emptyset\}$$

By the preceding argument, each N_F is non-empty. Since $F \supset F_1$ implies that $N_F \subset N_{F_1}$, the family of sets N_F have the finite intersection property. Since L is compact, $\bigcap_F \mathrm{cl}(N_F) \neq \emptyset$. Let x be a point in this intersection and let y be an arbitrary point of C. Choose F in Λ such that F contains both x and y. Then for any point u in N_F, $T(u)$ is disjoint from the line segment D joining x to y. Since x lies in $\mathrm{cl}(N_F)$, there exists a filter u_β in N_F converging to x. For each β, $T(u_\beta)$ is disjoint from D. Hence $T(x)$ is disjoint from D by property (c). Since y was arbitrary, $T(x)$ must be empty, contradicting the assumption that $T(x)$ is non-empty for each x. q.e.d.

REMARK. *If each* x *of the closure of a subset* N_F *of* L *is the limit of a sequence from* N_F, *it suffices in the above argument to have the property* (c) *only for sequences. (This is always the case for bounded sets in reflective Banach spaces in their weak topology.)*

COROLLARY TO THEOREM 5. (Brezis-Nirenberg-Stampacchia minimax theorem). Let C be a convex subset of a Hausdorff topological vector space, f a real-valued function on $C \times C$. Suppose that f has the following properties:

(1) $f(x,x) \leq 0$ for all x in C.

(2) For all y in C, $\{x | x \in C, f(x,y) > 0\}$ is convex.

(3) For each fixed y in C, $f(x,y)$ is a lower-semi-continuous function of x on $C \cap F$ for each finite-dimensional subspace F of E.

(4) If x and y are distinct points of C and if $\{u_\beta\}$ is a filter on C converging to x, then $f(u_\beta, (1-t)x + ty) \leq 0$ for every t in $[0,1]$ implies that $f(x,y) \leq 0$.

(5) There exists a compact subset L of C and $y_0 \in L$ such that $f(x,y_0) > 0$ for all x in $C \setminus L$.

Then: There exists x_0 in L such that

$$f(x_0,y) \leq 0, \ (y \in C).$$

PROOF OF THE COROLLARY. For each x in C, we set $T(x) = \{y | y \in C, f(x,y) > 0\}$. By (1), T has no fixed points in C. By (2), $T(x)$ is convex for all x. By (3), for each finite dimensional subspace F of E, $T^{-1}(y) \cap F$ is open in $C \cap F$. Property (4) of the hypothesis implies condition (c) of Theorem 5, while property (5) implies condition (a). Applying the contra-positive of Theorem 5, we see that $T(x)$ must be empty for some point x in L. Hence the Corollary follows. q.e.d.

DEFINITION 3. Let C be a closed convex subset of the reflexive Banach space X, g a mapping of C into X^*, the conjugate space of X. Then g is said to be pseudo-monotone if it satisfies the following two conditions:

(a) g is continuous from any finite dimensional subset of C to the weak topology of X^*.

(b) For any sequence $\{x_j\}$ in C converging weakly to x for which $\overline{\lim}\langle g(x_j), x_j - x\rangle \leq 0$, it follows for each y in C that $\underline{\lim}\langle g(x_j), x_j - y\rangle \geq \langle g(x), x - y\rangle$.

THEOREM 6. Let C be a reflexive Banach space, C a closed convex subset of X, g a pseudo-monotone mapping of C into X^*. Let φ be a lower-semi-continuous convex real-valued function on C, and suppose that for some y_0 in C

$$\langle g(x)\ x - y\rangle + \varphi(x) - \varphi(y_0) > 0$$

for all x in C with $\|x\| > R_0$

Then: There exists x in C such that for all u in C

$$\langle g(x), x - u\rangle + \varphi(x) - \varphi(u) \leq 0.$$

PROOF OF THEOREM 6. We set

$$f(x,y) = \langle g(x), x - y\rangle + \varphi(x) - \varphi(y)$$

for all x and y in C. The hypotheses of the Corollary to Theorem 5 follow from the assumptions of Theorem 6. We note first that for all x in C, $f(x,x) = 0$. The condition that $f(x,y) > 0$ in y is equivalent to

$$\varphi(y) + \langle g(x),y\rangle < c$$

which is a convex condition since φ is convex. The condition (3) in the weak topology follows from the continuity of g on finite-dimensional subsets and the fact that each lower-semi-continuous convex function φ is also lower-semi-continuous in the weak topology. The set L is taken simply as the intersection of C with the ball B_R about 0 with $R > \|y_0\| + R_0$.

Finally, suppose u_j converges to x in the weak topology and $f(x_j,x) \leq +0$, $f(x_j,y) \leq 0$. Then we have

$$\langle g(x_j), x_j - x\rangle + \varphi(x_j) - \varphi(x) \leq 0.$$

Hence

$$\overline{\lim}\langle g(x_j), x_j - x\rangle \leq \overline{\lim}\{\varphi(x) - \varphi(x_j)\} \leq 0.$$

Moreover,

$$\langle g(x_j), x_j - y\rangle + \varphi(x_j) - \varphi(y) \leq 0.$$

Then by the pseudo-monotonicity of g,

$$\langle g(x), x - y\rangle \leq \underline{\lim}\langle g(x_j), x_j - y\rangle \leq \underline{\lim}\{\varphi(y) - \varphi(x_j)\} \leq \varphi(y) - \varphi(x),$$

so that $f(x,y) \leq 0$. Thus all the hypotheses of the Corollary to Theorem 5 are satisfied. By its conclusion, there exists x in C such that for all u in C, $F(x,u) \leq 0$, i.e.

$$\langle g(x), x - u\rangle + \varphi(x) - \varphi(u) \leq 0.$$

q.e.d.

DEFINITION 4. Let C be a closed convex subset of the reflexive Banach space X, A a convex-valued mapping of C into X^*. Then A is said to be pseudo-monotone if the following conditions hold:

(a) A maps bounded sets of C into bounded subsets of X^*. For any finite-dimensional subspace F of X, A is an upper-semi-continuous mapping of $C \cap F$ into the weak-topology of X^*.

(b) if $\{u_j\}$ is a sequence in C, and for each j, $w_j \in A(u_j)$ and if u_j converges weakly to x in X, w_j converges weakly to w in X^* and if

$$\overline{\lim}\langle w_j,u_j - x\rangle \leq 0$$

then there exists y in $A(x)$ such that for all u in C,

$$\underline{\lim}\langle w_j, x_j - u\rangle \geq \langle y, x - u\rangle .$$

THEOREM 7. *Let C be a closed convex subset of the reflexive Banach space X, A a convex-valued mapping of C into X^* which is pseudo-monotone in the sense of Definition 4. Let φ be a lower-semi-continuous convex real-valued function on C. Suppose that there exists u_0 in C and $R_0 > 0$ such that for all x in C with $\|x\| \geq R_0$, and all y in $A(x)$, we have*

$$\langle y, x - u_0\rangle + \varphi(x) - \varphi(u_0) > 0.$$

Then there exists x_0 in C and y_0 in $A(x_0)$ such that for all u in C,

$$\text{(I)} \qquad \langle y_0, x_0 - u\rangle + \varphi(x_0) - \varphi(u) \leq 0.$$

PROOF OF THEOREM 7. We shall consider the solutions $[x,u]$ of the inequality (I) as coincidence points of the mapping A and of the mapping S of X^* into C given by

$$S(y) = \{x \mid x \in C, \text{ for all } u \text{ in } C, \langle y, x - u\rangle + \varphi(x) - \varphi(u) \leq 0. .$$

Let $C_R = C \cap B_R$ with $R > R_0 + \|u_0\|$. Then C_R is compact in the weak topology on X. By hypothesis, A maps C_R into a weakly compact subset v_R of X^*.

We note that it suffices to prove Theorem 7 with C replaced by C_R. Indeed, suppose that $[u_R, y_R]$ is a solution of the variational inequality

$$\text{i)} \qquad \langle y_R, u_R - u\rangle + \varphi(u_R) - \varphi(u) \leq 0.$$

for all u in C_R with $[u_R, y_R]$ in $G(A)$. Then u_R cannot lie on the boundary S_R of B_R because of the coercivity condition of the hypothesis and since $\|u_0\| < R$. Let v be any point of C. Then there exists t with $0 < t < 1$ such that $u = (1-t)u_R + tv$ lies in C_R. Then $u - u_R = t(v - v_R)$, while

$$\varphi(u) \leq (1-t)\,\varphi(u_R) + t\varphi(v).$$

The inequality (I) for u implies the corresponding inequality for v. Thus we may assume without loss of generality that C is bounded, and that A maps C into the weakly compact convex subset K of X^*. We interpret A as a convex-valued mapping of C into K, while by the compactness of C, S is a convex-valued mapping of K into C. Indeed, by its definition, $S(y)$ is a convex subset of C. The point x of C lies in $S(y)$ if it is a solution of a single-valued variational inequality with $f(x) = y$, a constant function. Hence $S(y)$ is non-empty and S is upper-semi-continuous from the weak topology on K to the weak topology on C.

Let Λ be the partially ordered set consisting of the finite dimensional subspaces of X containing u_0, ordered by inclusion. For each such F, we may construct the convex set C_F and the mappings S_F of K into the convex subsets of C_F. For each F, $A_F = A|_{C_F}$ is a convex-valued upper-semi-continuous mapping of C_F into K. Hence by Theorem 1, A_F and S_F have a coincidence.

To complete the argument using Theorem 4, we must verify that

$$\bigcap_{\Lambda \in F} \text{cl}(G(A_F) \cap G(S_F^{-1}) \neq \phi \Rightarrow G(A) \cap G(S^{-1}) \neq \emptyset.$$

Let $[x_0,y_0]$ be a point of the intersection on the intersection on the left and consider F in Λ which contains x_0. Then there exists a sequence $[x_j,y_j]$ in $G(A_F) \cap G(S_F^{-1})$ which converges weakly to $[x_0,y_0]$, i.e. $y_j \in A(x_j)$, and

$$<y_j,x_j - x_0> + \varphi(x_j) - \varphi(x_0) \leq 0.$$

It follows immediately that there exists w_0 in $A(x_0)$ such that for all u in C_F

$$<w_0,x_0 - u> \leq \underline{\lim}<y_j,x_j - u> \leq \underline{\lim}\{\varphi(u) - \varphi(x_j)\} \leq \varphi(u) - \varphi(x_0).$$

Hence for all u in C_F, $[x_0,w_0]$ satisfies the variational inequality

$$<w_0,x_0 - u> + \varphi(x_0) - \varphi(u) \leq 0.$$

The element w_0 of $A(x_0)$ depends upon the choice of the space F. For any finite subset $M = \{u_1,\ldots,u_r\}$ of C, we may choose such a space F and obtain an element w_M of $A(x_0)$ such that the variational inequality holds for all u_j in M. If we set

$$V_M = \{w | w \in A(x_0), <w,x_0 - u> + \varphi(x_0) - \varphi(u) \leq 0, (u \in M)\},$$

we see from the weak compactness of $A(x_0)$ and the finite intersection property of the sets V_M as well as their weak-closedness that $\bigcap_{M \subset C} V_M \neq \emptyset$. However, w lies in this last intersection if and only if

$$<w,x_0 - u> + \varphi(x_0) - \varphi(u) \leq 0, (u \in C)$$

and $w \in A(x_0)$. Hence, the proof of property (d) of Theorem 4 is complete, and our conclusion follows. q.e.d.

SECTION 4. We now establish a generalization of a fixed point theorem for upper-semi-continuous convex-valued self-mappings of a compact convex set of a locally convex topological vector space by an argument based on the use of variational inequalities.

DEFINITION 5. *If* C *is a convex subset of a locally convex topological vector space* E, x *a point of* C, *then*

$$I_C(x) = \{u \mid u \in E; \text{ There exists } v \text{ in } C,\ \delta > 0 \text{ such that } u = x + \delta(v - x)\}.$$

THEOREM 8. *Let* C *be a compact convex set in the locally convex topological vector space* E, T *a convex-valued upper-semi-continuous mapping of* C *into* E. *Suppose that* T *satisfies the following condition: There exists a convex neighborhood* V *of* 0 *in* E *such that for each point* x *of* C *which is not a fixed point of* T,

$$x \notin \mathrm{cl}(T(x) + V)$$

while

$$(T(x) + V) \cap I_C(x) \neq \emptyset.$$

Then: T *has a fixed point in* C.

THEOREM 9. *Let* X *be a Banach space*, C *a compact convex subset of* X, T *a convex-valued upper-semi-continuous mapping of* C *into* X. *Suppose that there exists a continuous mapping* p *of* C *into* $[0,1]$ *such that for any* x *in* C *which is not a fixed point of* T,

$$\mathrm{dist}(x, T(x)) \geq p(x),$$

$$\mathrm{dist}(T(x), I_C(x)) < p(x).$$

Then: T *has a fixed point in* C.

COROLLARY TO THEOREM 9. *Let* C *be a compact convex subset of the Banach space* X, g *a continuous mapping of* C *into* X. *Suppose that for each* x *in* C *which is not a fixed point of* g.

$$\|x - g(x)\| > \mathrm{dist}(g(x), I_C(x)).$$

Then: g *has a fixed point in* C.

PROOF OF THEOREM 8. Assume T has no fixed points in C. By the hypothesis of the Theorem, x is disjoint from the closed set $\mathrm{cl}(T(x) + V)$ for each x in C. Hence, we may find ξ in E^* such that

$$\langle \xi, x\rangle < c < \inf_{v \in (T(x) + V)} \langle \xi, v\rangle.$$

Since T is upper-semi-continuous from C to E, there exists a neighborhood $(x + U_x)$ of x in C such that for u in $(x + U_x)$, and the given ξ,

$$\langle \xi, u\rangle < c < \inf_{v \in T(u) + V} \langle \xi, v\rangle.$$

Since C is compact, we can cover C by a finite family of such neighborhoods $\{x_j + U_j;\ j = 1,\ldots,r\}$ with a corresponding family of functionals $\{\xi_j;\ j = 1,\ldots,r\}$. Corresponding to the covering of C, we take a partition of unity $\{\beta_1,\ldots,\beta_r\}$ and form the continuous mapping ξ of C into E^* given by

$$\xi(x) = \sum_{k=1}^{r} \beta_k(x)\xi_k.$$

For each x in C, $\beta_k(x) \neq 0$ only if $x \in (x_j + U_{x_j})$, so that

$$<\xi_j,x> < <\xi_j,w>$$

for any w in $T(x) + V$. Summing these inequalities, we see that

$$<\xi(x),x> < <\xi(x),w>, \quad (w \in T(x) + V).$$

By hypothesis, there exists a point w in $I_C(x) \cap (T(x) + V)$. For this w, we have the strict inequality above.

Since ξ is continuous, there exists x_0 in C such that for all u in C,

$$<\xi(x_0),x_0> \geq <\xi(x_0),u>.$$

Let w be an element of $I_C(x_0)$ as generated in the last paragraph. Then, there exists u in C, $\delta > 0$ such that $w = x_0 + \delta(u - x_0)$. Since,

$$<\xi(x_0),u - x_0> \leq 0,$$

we have

$$<\xi(x_0),w> \leq <\xi(x_0),x_0>.$$

On the other hand, we have constructed $\xi(x_0)$ so that

$$<\xi(x_0),w> > <\xi(x_0,x_0>,$$

which is a contradiction. q.e.d.

PROOF OF THEOREM 9. For each x in C, we can find a linear functional in X^* such that

$$<\xi,x> < c_x < \inf\{<\xi,u> : u \in N_{p(x)}(T(x))\}$$

The remainder of the proof is identical with that of Theorem 8. q.e.d.

BIBLIOGRAPHY

[1] H. Brézis, Equations et inéquations nonlinéaires dans les espaces vectoriels en dualité, Ann. Inst. Fourier, 18 (1968), 115-175.

[2] H. Brézis, L. Nirenberg, and G. Stampacchia, A remark on Ky Fan's minimax principle, Boll. Unione Mat. Ital., (4) 6, (1972) 293-300.

[3] F. E. Browder, The fixed point theory of multi-valued mappings in topological vector spaces, Math. Annalen, 177 (1968), 283-301.

[4] F. E. Browder, Nonlinear operators and nonlinear equations of evolution in Banach spaces, Nonlinear Functional Analysis, Proc. Symposium in Pure Math. vol. 18, Part 2, American Mathematical Society, Providence, RI, 1976.

[5] F. E. Browder, On a sharpened form of the Schauder fixed-point theorem, Proc. Natl. Academy of Sciences USA, 74 (1977) 4749-4751.

[6] G. Debreu, Theory of Value, Yale University Press, 1959.

[7] H. Debrunner and P. Flor, Ein Erweiterungssatz für monotone Mengen, Arch. Math., 15 (1964), 445-447.

[8] K. Fan, A minimax inequality and application, Inequalities, III, Academic Press, (1972), 103-113.

[9] S. Kakutani, A generalization of Brouwer's fixed point theorem, Duke Math. Jour., 8 (1941), 457-459.

[10] B. Knaster, C. Kuratowski, and S. Mazurkiewicz, Ein Beweis des Fixpunktsatzes für n-dimensionale Simplexe, Fund. Math., 14 (1929), 132-137.

[11] J. von Neumann, Über ein ökonomisches Gleichungssystem und eine Verallgemeinerung des Brouwerschen Fixpunktsatzes, Ergeb. eines Math. Kolloquium, 8 (1937), 73-83.

DEPARTMENT OF MATHEMATICS
UNIVERSITY OF CHICAGO
CHICAGO, IL 60637

Contemporary Mathematics
Volume **26**, 1984

RECENT DEVELOPMENTS ARISING OUT OF KAKUTANI'S WORK ON COMPLETION REGULARITY OF MEASURES

J. R. CHOKSI†

Let X be a locally compact Hausdorff space. We call the σ-ring $\mathcal{B}$ generated by the class $\mathcal{C}$ of compact sets, the σ-ring of Borel sets of X; the sub σ-ring $\mathcal{B}_0$ generated by the class $\mathcal{C}_0$ of compact G_δ sets we call the σ-ring of Baire sets of X. This is the terminology in Halmos [1], other authors use the respective terms for the σ-algebras generated by the closed sets and the closed G_δ sets respectively - we call these the wide Borel and wide Baire sets respectively. The two concepts coincide in σ-compact spaces and we shall be concerned almost exclusively with this case. A compact set in $\mathcal{B}_0$ is necessarily in $\mathcal{C}_0$ (this was first proved by Kodaira [1]). A measure on $\mathcal{B}$ which is finite on $\mathcal{C}$ is called a Borel measure, a measure on $\mathcal{B}_0$ which is finite on $\mathcal{C}_0$ is called a Baire measure. A Borel measure is regular if it is inner regular with respect to $\mathcal{C}$, a Baire measure is regular if it is inner regular with respect to $\mathcal{C}_0$. Every Baire measure is regular, and has a unique extension to a regular Borel measure (this was first shown by Kakutani [1] and [3], see also Kakutani and Kodaira [1]). However, irregular Borel measures exist, the first and simplest example is due to Dieudonné [1]. (Let X denote the set of ordinals less than the first uncountable ordinal ω_1, and $Y = X \cup \{\omega_1\}$. Y is compact Hausdorff in the order topology. For every Borel set E in Y, put $\mu(E) = 1$ or 0 according as E does or does not contain an unbounded closed subset of X; μ is irregular.) Regular Borel measures on a locally compact Hausdorff space are also locally finite and so Radon in the terminology of most authors.

A full account of the topics discussed in our introductory paragraph is given in Chapter X of Halmos [1]. The reference section in this book (p. 291) also gives some indication of the history - we give a slightly fuller account. The development of measure theory in not necessarily metrisable topological

---

† Research supported in part by grants from the National Science and Engineering Research Council of Canada and the FCAC programme of the Government of Quebec.

spaces began in the mid-thirties, and was done independently by several authors. A full account of the regularity of Borel measures was given in von Neumann's unpublished 1940 Princeton lectures on Invariant measures (von Neumann [2], see also Kakutani [1]). The study of Baire and Borel sets and measures was undertaken by Kodaira [1] and developed by Kakutani and Kodaira [1] (see also Kakutani [3], Theorem 2, for the extension of Baire to regular Borel measures). The relation to linear functionals on the space of continuous functions on a compact space, and the identification with regular Borel measures (the so-called "Riesz representation theorem") were given by Kakutani [1], a more restricted version appears in von Neumann [1]. Independently, Markov [1] had proved a similar result on finitely additive measures on completely regular spaces, and a little later A. D. Alexandrov [1] generalized Markov's work and also discussed the generation of regular measures. At the same time, Bourbaki developed a theory of integration on locally compact spaces treating the linear functional as primary and the countably additive measure as secondary. The first appearance (in summary) of this theory is in Chapter II of Weil [1] (see also Bourbaki [1]). It is interesting to note that the early development of the theory by many of these authors arose as a background to their construction of Haar measure on a compact or locally compact group. Histories of the Riesz representation theorem are given in Dunford and Schwartz [1], p. 373 et seq, and Hewitt and Ross [1], p. 134 see also the Notes historiques in Bourbaki [1]. Some more recent developments of the general theory deserve mention. The theory of Radon measures was adapted to (not necessarily locally compact) Hausdorff spaces by Bourbaki [1], Chapter IX, and, in a more measure theoretic way, by Schwartz [1]. Fundamental work on Baire measures in non-locally compact spaces had been done earlier by Varadarajan [1]. Various regularity properties for Borel measures have been studied by a number of authors - an extremely thorough account of these is to be found in the survey article of Gardner [2] (see also Pfeffer [1] Chapter 18). Gardner's article contains many interesting insights into the regularity of Borel measures, - §24 contains a comparison of various strong forms of regularity including completion regularity, the main topic of our article. Most of the results we discuss are mentioned briefly in Gardner's survey, however the treatment is much more condensed since he covers a much wider area. Finally, by restricting ourselves to σ-rings of σ-compact sets, we scrupulously avoid the problems of regularity that can occur with the wide Borel and Baire sets. For a discussion of these problems we again refer to Gardner [2] and Pfeffer [1].

A regular Borel measure μ (or Baire measure μ_0) is called <u>completion regular</u> if the completion of the Baire restriction μ_0 of μ coincides with the completion of μ (or the completion of μ_0 coincides with the completion of its regular Borel extension μ). Equivalently every Borel set is measurable with respect to the completion of μ_0. The terminology is due to Halmos [1], but, as we shall see, the concept was developed earlier by Kakutani. At first sight completion regularity seems a somewhat drastic restriction on a Borel measure, perhaps even a shade like perfect normality for a topological space! Quite the contrary - in two fundamental papers written in 1943-44, the second joint with Kunihiko Kodaira, Shizuo Kakutani showed that the measures most closely tied to the topology of well-behaved topological spaces are all completion regular! (Kakutani [3], Kakutani and Kodaira [1].)

We have already noted that in the first of these papers, the extension from Baire measure to regular Borel measure is discussed for product measures. The main result of the paper is however the following (Theorem 3):

Let $\{X^\gamma, \mathcal{B}^\gamma, m^\gamma) : \gamma \in \Gamma\}$ be an uncountable family of probability measure spaces, where, for each $\gamma \in \Gamma$, the X^γ are all compact metric spaces with Borel σ-algebra $\mathcal{B}^\gamma$, and $m^\gamma(U) > 0$ for each open U in X^γ, (i.e. $\mathrm{supp}(m^\gamma) = X^\gamma$). Let m_0 be the direct or independent product measure on the product σ-algebra $\mathcal{B}^*$ of $X = \Pi X^\gamma$, whose existence was shown in Kakutani [2]. Then every Borel set in the compact Hausdorff space X is measurable with respect to the completion of m_0 on $\mathcal{B}^*$.

Baire sets are not discussed here, they only appear in Kakutani and Kodaira [1], where it is remarked that $\mathcal{B}^* = \mathcal{B}_0$, the Baire σ-algebra of X, and so the direct product measure is completion regular. We sketch the proof of the above theorem, pointing out its key features. It is sufficient to prove that every open set U in X is contained in a Baire set of the same measure (in fact this Baire set, U_1, turns out to be open). It is easily seen that U contains an open Baire set U_0 (i.e. $U_0 \in \mathcal{B}^* = \mathcal{B}_0$), of the same measure. Let $\{\gamma_n : n = 1,2,3,\ldots\}$ be the countable set of coordinates determining U_0. Let $\tilde{U}_1$ be the projection of U in $\prod_{n=1}^{\infty} X^{\gamma_n}$, which is compact metric. Since projection maps on sub-products are open, $\tilde{U}_1$ is open, so $U_1 = \tilde{U}_1 \times \prod_{\gamma \neq \gamma_n} X^\gamma$ is an open Baire (or $\mathcal{B}^*$) subset of X containing U. U_1 and U_0 are both (open) Baire sets of X, determined by the same set of co-ordinates $\{\gamma_n : n = 1,2,3,\ldots\}$, and $U_0 \subset U \subset U_1$. A product measure (or Fubini) type argument, connecting the measures $m_0 = \prod_{\gamma \in \Gamma} m^\gamma$ and $\prod_{n=1}^{\infty} m^{\gamma_n}$, and using the fact that all sections of an open set are open and so have positive product measure for the appropriate sub-product (because $\mathrm{supp}(m^\gamma) = X^\gamma$, for all $\gamma \in \Gamma$), shows that $m_0(U_1 \backslash U_0) = 0$. Thus U_1 is the required Baire cover of U.

The second paper, Kakutani and Kodaira [1], is an ingenious adaptation of the ideas of the first to prove that Haar measure in a locally compact, σ-compact group is completion regular. It starts with a preliminary discussion of Borel and Baire sets and functions (here so called for the first time) which extends earlier work of Kodaira [1]. It then tackles the problem of setting up a structure in a locally compact, σ-compact group analogous to the product structure of a product space. To this end the authors prove the following important (and independently useful) structure theorem (Theorem 6):

Let G be a locally compact, σ-compact group. Given a countable sequence of Baire functions $\{f_k\}$ on G, there exists a compact normal subgroup N of G, such that the quotient group $\bar{G} = G/N$ is separable, metrisable and locally compact, and such that the $f_k(t)$ are functions on $\bar{G}$ (i.e. are constant on the fibres of the quotient map $G \to \bar{G}$). It follows that for any locally compact, σ-compact group G, every neighborhood of the identity contains a compact normal subgroup N, whose quotient group $\bar{G} = G/N$ is separable, metrisable, locally compact. It also follows that every Baire set in G is the inverse image under the quotient map of a Borel subset of such a metrisable factor group $\bar{G}$.

We shall not sketch the proof of this theorem or analyse its subsequent history and uses in other areas. For this we refer to the commentaries on the above paper in the forthcoming Kakutani selecta, and also to the books of Hewitt and Ross [1] p. 71, and Montgomery and Zippin [1]. The latter authors (§2.6 & 2.7) use the structure theorem to describe locally compact, σ-compact groups as projective limits of metrisable groups and thus (§2.17) construct Haar measure on a locally compact group from Haar measures on locally compact metrisable groups. We shall, however, sketch how the structure theorem is used to adapt the earlier proof on completion regularity of product measures, to show that every Haar measure on a locally compact σ-compact group is completion regular (Theorem 7 of Kakutani and Kodaira [1]). Again it suffices to show that every open set U is contained in an (open) Baire set U_1 with $m(U_1 \setminus U) = 0$. (Here m denotes Haar measures on $\mathcal{B}$, m_0 its restriction to the Baire sets $\mathcal{B}_0$.) Again let U_0 be an open Baire kernel of U, i.e. $U_0 \subset U$, $m(U \setminus U_0) = 0$. Let $\bar{G} = G/N$ be a separable, metrisable factor group (determined by the structure theorem) to which U_0 'belongs', i.e. for which $U_0 = \pi^{-1}(\tilde{U}_0)$, with $\tilde{U}_0$ open in $\bar{G}$, π the projection map $G \to \bar{G}$. Then $\pi(U)$ is open in $\bar{G}$ (because π is an open map), so $U_1 = \pi^{-1}(\pi(U))$ is an open Baire set in G which contains U. Now there is a Fubini type relation connecting Haar measure on G with the pair consisting of Haar measure on $\bar{G} = G/N$ and normalized Haar measure on the compact normal subgroup N. This, with the fact that Haar measures all have full support, enables one to mimic the argument of the earlier paper to show that $m_0(U_1 \backslash U_0) = 0$, so that U_1 is the required Baire cover of U.

Essentially the entire contents of the Kakutani-Kodaira paper, in a somewhat modified form is given in the last section of the book of Halmos [1], which appeared in 1950, and became a standard reference. Both the main theorems appear in slightly more general form. In particular, in Theorem 7, σ-compactness is easily removed. For every locally compact Hausdorff group contains a σ-compact open (and closed) subgroup whose cosets cover the whole group, and every (restricted, i.e. σ-compact) Borel set lies in the union of at most countably many such cosets. Hence completion regularity of Haar measure for σ-compact locally compact groups implies the same for general locally compact groups.

Of course most measures on a locally compact Hausdorff space will not be completion regular (unless the space is perfectly normal!). Even on the cube $[0,1]^A$, A uncountable, it is easy to construct such measures; it is even possible to construct ones which are non-atomic and have full support - we shall give such an example later. But the two theorems described above show that for 'good' topological spaces such as products of compact metric spaces, or locally compact groups, of no matter how large a weight, the measures most closely tied to the topology are completion regular. For good measure we add that for an arbitrary finite or σ-finite measure algebra, the induced measure on the hyperstonean representation space is easily seen to be completion regular. The unique extension of Baire to regular Borel measure and the completion regularity of product measures (Kakutani [3], Theorems 2 and 3) have considerable relevance to the construction of separable and measurable versions of stochastic processes; this was discussed by Doob [1] in 1947, a more recent and extensive discussion can be found in the articles of D. G. Kendall [1], [2] and [3].

There was not a great deal of activity in general measure theory in the early fifties. However in 1958, Dorothy Maharam [2], proved the lifting theorem - in the course of the proof she obtained an entirely different proof of Kakutani's theorem on the completion regularity of product measures. While her aim was to construct a lifting for an arbitrary σ-finite measure space, she first obtained a lifting for the power Lebesgue measure on the Cantor space $\{0,1\}^A$, A arbitrary. The lifting she obtained was (completion) Baire, and also strong (the lift of each open set contains the open set). Thus every open set has a Baire cover and completion regularity of the power Lebesgue measure on $\{0,1\}^A$ follows. Standard theorems on compact metric spaces then enable one to deduce the general Kakutani theorem, when the $\{0,1\}$ are replaced by arbitrary compact metric spaces X^{γ}. This principle that the existence of a strong completion Baire lifting implies completion regularity, has subsequently been used by other authors. It suggests a connection between strong lifting and completion regularity. A connection clearly exists, but it is somewhat elusive. As we shall see later, the mere existence of a strong (not necessarily Baire) lifting does not imply completion regularity; nor does completion regularity imply the existence of a strong lifting. But all the good measures which are completion regular (products of measures with full support on compact metric spaces, and Haar measures on locally compact groups) also have the strong lifting property (A. & C. Ionescu Tulcea [1], pp. 118-119, and [2]); so, trivially, does the hyperstonean representation space of a σ-finite measure space.

In 1965, Berberian [1], p, 231, introduced the notion of a monogenic Baire measure, one which admits a unique (necessarily regular) Borel extension. At about the same time Bourbaki introduced the notion of a Radon space, one in which every finite Borel measure is inner regular with respect to the compact sets. On locally compact, σ-compact Hausdorff spaces, it is clear that every Baire measure is monogenic if and only if the space is Radon. A completion regular measure is trivially monogenic. However a monogenic measure need not be completion regular even on a compact Radon space. It was noticed by several authors that a classical theorem of Ulam implies that the one point compactification of a discrete space of cardinal $\aleph_1$ is a Radon space. (See Schwartz [1], p, 120, Martin [1]; Gardner [1], Example 6.2 has a larger class of compact Radon spaces including this one.) Martin [1] remarked that since the Baire sets of this space consist only of countable sets which exclude the point at infinity, and complements of such sets, the point measure at the point at infinity is not completion regular, although of course it is monogenic. Berberian [1] (p. 176 Ex. 1 and p. 233 Ex. 5) remarked that a locally compact Hausdorff space in which every point is a G_δ is first countable, and so if every Borel measure is completion regular, then the space must be first countable.

In the mid-seventies Babikar [1] and [2] introduced a concept called uniform regularity for measures. It is essentially different from completion regularity - uniformly regular measures are not too far from being separable - but there are some relations, see Babikar [3] and Gardner [2], §§22, 23 and 24.

Since the mid-seventies there has been a revival of interest in problems on completion regular measures more directly related to Kakutani's original work. In 1976, Fremlin [1] showed that the product of two completion regular measures, each on a compact space, need not be completion regular: the product of the hyperstonean space of Lebesgue measure on [0,1] with itself yields a counterexample. Fremlin's proof makes ingenious use of an earlier construction of Erdős and Oxtoby [1]. It is not known what happens for the product of two compact spaces, each a product of compact metric spaces, each with a completion regular measure different from a product measure. However Choksi and Fremlin [1] have proved the following extension of the theorem of Kakutani [3]: Let X be a compact Hausdorff space, μ a completion regular measure on X. If Γ is countable or uncountable, and for each $\gamma \in \Gamma$, X_γ is a compact metric space, μ_γ a Borel probability measure of full support on X_γ, then the product measure $\mu \times \prod_{\gamma\in\Gamma} \mu_\gamma$, on $X \times \prod_{\gamma\in\Gamma} X_\gamma$ is completion regular. If $\Gamma = \{\gamma_0\}$, the proof depends on the fact that every open set in $X \times X_{\gamma_0}$ is a countable union of product open sets. The general case is proved by a modified version of Kakutani's argument.

The 1979 paper of Choksi and Fremlin [1] is a study of how many distinct, i.e., non-isomorphic completion regular measures can exist on a product of compact metric spaces. The authors start by asking when a probability measure μ on such a product space $\prod X_\gamma$, $\gamma\in\Gamma$, is 'isomorphic' in some sense to a (necessarily completion regular) product measure with full support. If 'isomorphic' means homeomorphic, the measure μ is necessarily completion regular, but homeomorphism of measures is not their main concern. (We shall discuss this problem near the end of the present article). They ask instead when there is a completion Baire measurable isomorphism of $\prod X_\gamma$, $\gamma\in\Gamma$, onto itself, taking μ to a product measure. Such a measure μ need not be completion regular even if it has full support. Indeed they show, using earlier work of Choksi [1] and the fundamental work of Maharam [1], that if the measure algebra of μ is homogeneous of Maharam type Γ, then μ is completion Baire isomorphic to such a product measure; clearly the condition is also necessary, for such isomorphism. [We recall that a measure algebra is <u>homogeneous</u> if every non-zero principal ideal has a minimal σ-basis of the same cardinal, called the <u>Maharam type</u> of the homogeneous measure algebra. A famous theorem of Maharam [1] states that a homogeneous probability measure algebra of Maharam type Γ is measure preserving isomorphic to the measure algebra $\mathcal{E}_\Gamma$ of the power Lebesgue measure on $[0,1]^\Gamma$ or $\{0,1\}^\Gamma$. The result of Choksi [1] suitably modified shows that if two Baire probability measures on $\prod X_\gamma$ have isomorphic measure algebras, then they are completion Baire isomorphic.] Choksi and Fremlin are thus led to consider which measures in $\prod_{\gamma\in\Gamma} X_\gamma$, have measure algebras which are homogeneous of Maharam type Γ. Positivity on open sets plus non-atomicity is not sufficient. Wiener measure (defined on $\mathbb{R}^{[0,1]} = \mathbb{R}^c$, but easily injected to $[0,1]^c$) gives a counter-example. Apart from being non-atomic and positive on open sets, it is of Maharam type $\aleph_0 < c$. It has the strong lifting property. Note however that it is carried by a Borel set of cardinal c (the image under the injection of the continuous functions) and so cannot be completion regular, since the Baire sets of $[0,1]^c$ have cardinal 2^c. Choksi and Fremlin are thus led to two conjectures:

(i) If μ is a completion regular probability measure on $[0,1]^\Gamma$ or $\{0,1\}^\Gamma$, then μ is homogeneous of Maharam type Γ and so is completion Baire isomorphic to the power Lebesgue measure.

(ii) Any two completion regular probability measures on ΠX_γ, $\gamma \in \Gamma$ (each X_γ compact metric with at least two points), are completion Baire isomorphic.

They show that the two conjectures are equivalent, and that any additional assumption of positivity on open sets (full support) is unnecessary because the support of a completion regular measure on ΠX_γ is always Baire. On product spaces with uncountably many factors, since one point sets are not G_δ sets, completion regularity clearly implies non-atomicity. Incidentally they show that the existence of a completion regular measure of Maharam type A on ΠX_γ, $\gamma \in \Gamma$, depends only on Γ and not on the particular X_γ; this shows in particular, that the two conjectures are equivalent. They show next that for a large, in fact, co-final, class of cardinals Γ, the conjectures are true: this class K is defined by

$$K = \{\Gamma: \alpha < \Gamma \quad \text{implies} \quad \alpha^{\aleph_0} < \Gamma\} .$$

The proof is essentially a counting argument. Note that $\aleph_1$ and c are not in K, but the successor of c is in K.

The remaining results in the paper are somewhat surprising. If the continuum hypothesis holds, there exists a completion regular measure of Maharam type $\aleph_0$ on $\{0,1\}^{\aleph_1} = \{0,1\}^c$, so the conjectures are then false for these cardinals. This result follows from the following (unconditional) result.

If λ, κ are cardinals with $\lambda \le \kappa$, and (E_λ, μ) is the homogeneous probability measure algebra of Maharam type λ, then the following are equivalent:

(i) There is a completion regular measure on $\{0,1\}^\kappa$ whose measure algebra is (E_λ, μ).

(ii) There is a family $\{b_i: i \in \kappa\}$ in E_λ, weakly independent, with the property that for every $a \in E_\lambda$, there is a countable set $I_a \subset \kappa$ such that the sub-algebras of E_λ generated by $\{a\} \cup \{b_\xi: \xi \in I_a\}$ and $\{b_\xi: \xi \in \kappa \setminus I_a\}$ are weakly independent.

For each compact set K with corresponding measure algebra element a, it is the countable set of co-ordinates I_a which determines its Baire kernel.

Next it is shown that if Martin's Axiom is true and the continuum hypothesis is false, then the conjectures are true for $\aleph_1$. In fact a measure of type $\aleph_0$ on $\{0,1\}^{\aleph_1}$ cannot then be completion regular - here use is made of the fact that under these set-theoretic hypotheses such a measure is '$\aleph_1$-continuous from below'. Finally it is shown that under the same hypotheses the conjectures are true for $\aleph_n (n \leq \omega)$, so with the additional (consistent) hypothesis that $c = \aleph_2$, the conjectures are true for c. Thus for $\aleph_1$ and c, the conjectures are, somewhat surprisingly, undecidable!

In view of the Kakutani-Kodaira theorem on the completion regularity of Haar measure, it is interesting to ask whether analogues of the results of Choksi and Fremlin without additional set-theoretic hypotheses are valid for compact (or locally compact) groups, and possibly even for homogeneous spaces. Analogues of Choksi's result on the Baire isomorphism for measures on products of compact metric spaces are known (Choksi [2], Choksi and Simha [1]), and the Kakutani-Kodaira structure theorem gives a good description of the Baire sets of locally compact groups and even of homogeneous spaces; Choksi and Simha [1] do this for homogeneous spaces and also use the structure theorem to describe homogeneous spaces as projective limits of metrisable homogeneous spaces indexed by compact normal subgroups. So far as we know, no one has attempted seriously to prove such analogues.

We conclude this article with a further discussion of the relation between completion regularity and strong lifting and of the homeomorphism problem for measures on product spaces. We have already noted that the measure induced on $[0,1]^c$ by Wiener measure is non-atomic, has full support, and has the strong lifting property, but it is not completion regular. In 1978, Losert [1] finally settled the strong lifting conjecture by constructing a Radon measure on $\{0,1\}^{\aleph_2}$ of full support which has no strong lifting. However Losert's example was not completion regular and it still seemed possible that all completion regular measures of full support on products of compact metric spaces possessed strong liftings. Late in 1979, however, Fremlin [2] showed that this is false, he constructed a completion regular measure of full support, but with no strong lifting, on $\{0,1\}^{\aleph_2}$. Since this example is interesting, and Fremlin's paper has only appeared in preprint form, we outline the construction (but not the proof). Let $S = \{0,1\}^{\mathbb{N}}$ and ν the power Lebesgue measure on S. Let M be a closed, nowhere dense subset of S with $\nu(M) > 0$. Let I be a set of cardinal $\geq \aleph_2$, μ the power measure of ν on S^I. For $\xi \in I$, put $M_\xi = \{t: t \in S^I, t(\xi) \in M\}$. Let

$$A = \{(\xi,\eta) : \xi \neq \eta\} \subset I \times I.$$

For $t \in S^I$, $(\xi,\eta) \in A$, let $\nu^t_{\xi,\eta}$ be the Radon probability measure on S given by

$$\nu^t_{\xi,\eta} = \nu \quad \text{if} \quad t \notin M_\xi \cap M_\eta$$

$$= \delta_{t(\xi)}, \text{ the point mass at } t(\xi), \text{ if } t \in M_\xi \cap M_\eta.$$

For each $t \in S^I$ let $\lambda_t = \prod_{(\xi,\eta)\in A} \nu^t_{\xi,\eta}$ on S^A. Let $X = S^I \times S^A$.

It can be shown that if D is a Baire set in X and for $t \in S^I$, $D_t = \{u: u \in S^A, (t,u) \in D\}$, then

$$\lambda(D) = \int \lambda_t(D_t)\mu(dt)$$

is defined and gives a unique Radon measure λ on X. The proof is straightforward, though not trivial. λ is easily seen to have full support. By ingenious arguments, Fremlin shows that λ is completion regular, but has no strong lifting. To give the proofs would take us too far afield; we content ourselves with the remarks that (i) because M is nowhere dense, the product measures λ_t have too few atomic factors to prevent the completion regularity of λ; (ii) because M is topologically small but $\nu(M) > 0$, we can find a closed Baire set $F \subset X$, of positive measure, which is not the whole space X and cannot contain its own lift, thus there is no strong lifting. We also remark that each λ_t, and so also λ is homogeneous of Maharam type $\aleph_2$.

As Fremlin himself remarks, his construction can be carried out with any compact metric space in place of S, in particular it can be carried out with $[0,1]$ and Lebesgue measure. We thus obtain a Radon probability measure on $[0,1]^{\aleph_2}$ which has full support, is homogeneous of Maharam type $\aleph_2$, is completion regular, but has no strong lifting. Thus although completion Baire isomorphic to the power Lebesgue measure; it is not homeomorphic to that measure. For the finite, n-dimensional cube $[0,1]^n$ and the Hilbert cube $[0,1]^{\aleph_0}$, the homeomorphic measures problem has been completely solved. It is a classical result of von Neumann, Oxtoby and Ulam (see Oxtoby and Ulam [1]) that a Borel probability measure ν is homeomorphic to Lebesgue measure on $[0,1]^n$, if and only if it is positive on open sets, non-atomic and vanishes on the boundary of $[0,1]^n$. In 1977 Oxtoby and Prasad [1] generalized this to the Hilbert cube $[0,1]^{\aleph_0}$, which has no boundary, here ν is homeomorphic to the power Lebesgue measure if and only if it is positive on open sets and non-atomic. Similar results hold for tori, but the situation is much more complex for the Cantor set $\{0,1\}^{\mathbb{N}}$, many good non-homeomorphic measures exist; the situation is certainly no better for $\{0,1\}^{\Gamma}$, Γ arbitrary. Thus the class of uncountable product spaces for which the homeomorphic measures problem (as opposed to the Baire isomorphic measures problem) is interesting, is quite restricted, including only such spaces as $[0,1]^{\Gamma}$ and the torus T^{Γ}, Γ arbitrary. Perhaps this has something to do with the universality of these as topological spaces (Urysohn's embedding theorem)! Fremlin's example given above shows that a probability measure on $[0,1]^{\Gamma}$ can be of full support, Maharam type Γ and completion regular, but fail to be homeomorphic to the power Lebesgue measure on $[0,1]^{\Gamma}$ because it fails to have a strong lifting. Theorem 7 of Choksi and Fremlin [1] shows that, assuming the continuum hypothesis, there exists a probability measure on $[0,1]^{c}$ which is completion regular, of full support and has a strong lifting but is not homeomorphic to the power Lebesgue measure on $[0,1]^{c}$ because it is of Maharam type $\aleph_0$. (It is known that the continuum hypothesis implies the existence of a strong lifting for all measures on spaces of weight $\leq \aleph_1$.) Similar remarks hold for tori. We conclude by asking what happens if one assumes all the known necessary conditions. Specifically if ν is a Radon probability measure on $[0,1]^{\Gamma}$ (or T^{Γ}) which is (i) of full support, (ii) homogeneous of Maharam type Γ, (iii) completion regular, (iv) has a strong lifting, then is it homeomorphic to the power Lebesgue measure? A clear answer to this question would, in a sense, close the subject. We suspect that, if there is a clear answer it is no; however it is much more likely that set theory intervenes, the answer may again be different for different Γ, and may well be dependent on additional set-theoretic hypotheses!

P.S. I have recent information from David Fremlin and Gryllakis Constantinos, that, at least when Martin's Axiom is true and the continuum hypothesis is false, the product of two completion regular measures, each on a product of compact metric spaces, is itself completion regular.

REFERENCES

A.D. Alexandrov [1]. Additive set functions in abstract spaces I, Mat. Sb. 8 (50), (1940), 307-348, II, ibid 9 (51), (1941), 563-628, III, ibid 13 (55), (1943), 169-238.

A.G.A.G. Babikar [1]. Uniform regularity of measures on compact spaces, J. Reine Angew. Math. 289 (1977), 188-198.

__________ [2]. On uniformly regular topological measure spaces, Duke Math. J. 43 (1976), 775-789.

__________ [3]. Lebesgue measures on topological spaces, Mathematika 24 (1977), 52-59.

N. Bourbaki [1]. Intégration, Chap. I-IX, Eléments de Mathématiques Liv. VI, Hermann, Paris, 1952-1969.

S. Berberian [1]. Measure and integration, Macmillan, New York, 1965.

J.R. Choksi [1]. Automorphisms of Baire measures on generalized cubes I, Z. Wahrscheinlichkeitstheorie 22 (1972), 195-204, II, ibid 23 (1972), 97-102.

__________ [2]. Measurable transformations on compact groups, Trans. Amer. Math. Soc. 184 (1973), 101-124.

J.R. Choksi and D.H. Fremlin [1]. Completion regular measures on product spaces, Math. Ann. 241 (1979), 113-128.

J.R. Choksi and R.R. Simha [1]. Measurable transformations on homogeneous spaces, Studies in probability and ergodic theory, Adv. in Math. Suppl. Stud. 2 (1978), 269-286.

J. Dieudonné [1]. Un exemple d'espace normal non susceptible d'une structure uniforme d'espace complet, C.R. Acad. Sci. Paris 209 (1939), 145-147.

J.L. Doob [1]. Probability in function space, Bull. Amer. Math. Soc. 53 (1947), 15-30.

N. Dunford and J.T. Schwartz [1]. Linear operators, I, Wiley-Interscience, New York, 1958.

P. Erdős and J.C. Oxtoby [1]. Partitions of the plane into sets having positive measure in every non-null measurable product set, Trans. Amer. Math. Soc. 79 (1955), 91-102.

D.H. Fremlin [1]. Products of Radon measures: a counterexample, Canadian Math. Bull. 19 (1976), 285-289.

__________ [2]. Losert's example: Note of 18/9/79, Preprint.

R.J. Gardner [1]. The regularity of Borel measures and Borel measure compactness, Proc. London Math. Soc. [3] 30 (1975), 95-113.

__________ [2]. The regularity of Borel measures, Measure theory Oberwolfach 1981, Proceedings, Lecture Notes in Math. 945, Springer-Verlag, Berlin-Heidelberg-New York, 1982, 42-100.

P.R. Halmos [1]. Measure theory, Van Nostrand, New York, 1950.

E. Hewitt and K.A. Ross [1]. Abstract harmonic analysis, I, Springer, Berlin-Heidelberg-New York, 1963.

A. and C. Ionescu Tulcea [1]. Topics in the theory of lifting, Springer, Berlin-Heidelberg-New York, 1969.

__________ [2]. On the existence of a lifting commuting with the left translations of an arbitrary locally compact group, Proc. Fifth Berkeley Symposium, II - pt. I, Berkeley, 1967, 63-97.

S. Kakutani [1]. Concrete representation of abstract [M] spaces, Ann. Math. [2] 42 (1941), 994-1024.

__________ [2]. Notes on infinite product measure spaces, I, Proc. Imperial Acad. Tokyo 19 (1943), 148-151.

S. Kakutani [3]. Notes on infinite product measure spaces, II, Proc. Imperial Acad. Tokyo 19 (1943), 184-188.

S. Kakutani and K. Kodaira [1]. Über das Haarsche Mass in der lokal bikompacten Gruppe, Proc. Imperial Acad. Tokyo 20 (1944), 444-450.

D.G. Kendall [1]. An introduction to stochastic analysis, Stochastic Analysis, a tribute to the memory of Rollo Davidson, Wiley, London, 1973, 3-43.

____________ [2]. Separability and measurability of stochastic processes: a survey, ibid, 415-443.

____________ [3]. Foundations of a theory of random sets. Stochastic Geometry, a tribute to the memory of Rollo Davidson, Wiley, London, 1974, 322-376.

K. Kodaira [1]. Über die Beziehung zwischen den Masse und Topologien in einer Gruppe, Proc. Phys.-Math. Soc. Japan 23 (1941), 67-119.

V. Losert [1]. A measure space without the strong lifting property. Math. Ann. 239 (1979), 119-128.

D. Maharam [1]. On homogeneous measure algebras, Proc. Nat. Acad. Sci. Washington 28 (1942), 108-111.

__________ [2]. On a theorem of von Neumann, Proc. Amer. Math. Soc. 9 (1958), 987-994.

A. Markov [1]. On mean values and exterior densities, Mat. Sb. 4 [46] (1938), 165-191.

A.F. Martin [1]. A note on monogenic Baire measures, Amer. Math. Monthly 84 (1977), 554-555.

D. Montgomery and L. Zippin [1]. Topological transformation groups, Interscience New York, 1955.

J. von Neumann [1]. Zum Haarschen Mass in topologischen Gruppen, Compositio Math. 1 (1934), 106-114.

_____________ [2]. Lectures on invariant measures (unpublished), Princeton, 1940.

J.C. Oxtoby and V.S. Prasad [1]. Homeomorphic measures in the Hilbert cube, Pacific J. Math. 77 (1978), 483-497.

J.C. Oxtoby and S.M. Ulam [1]. Measure preserving homeomorphisms and metrical transitivity, Ann. Math. [2] 42 (1941), 874-920.

W.F. Pfeffer [1]. Integrals and measures, Dekker, New York, 1977.

L. Schwartz [1]. Radon measures on arbitrary topological spaces and cylindrical measures, Oxford, London & Bombay, 1973.

V.S. Varadarajan [1]. Measures on topological spaces (Russian), Mat. Sb. 55 [97] (1961), 35-100; English transln. in Amer. Math. Soc. Transln Ser. 2, 48, 161-228.

A. Weil [1]. _L'intégration dans les groupes topologiques et ses applications_, Hermann, Paris 1939, 2nd. ed. 1953.

Department of Mathematics
McGill University
Burnside Hall
805 Sherbrooke St. West
Montreal, P.Q. Canada H3A 2K6

Contemporary Mathematics
Volume 26, 1984

AN AUTOMORPHISM OF A HOMOGENEOUS MEASURE ALGEBRA WHICH DOES NOT FACTORIZE INTO A DIRECT PRODUCT.

J.R. Choksi* & S.J. Eigen

A homogeneous measure algebra (see Maharam [Ma 1] for the definition) is the natural setting for the generalization of those aspects of ergodic theory usually only studied for Lebesgue spaces: such a measure algebra has many ergodic and mixing automorphisms, further all the usual approximation theorems and many of the category theorems generalize to it (Choksi & Prasad [Ch.Pr.].) Further a homogeneous measure algebra $(\mathcal{E},\mu)$, (of total measure 1) is by Maharam's famous structure theorem ([Ma 1], Theorem 1) a direct product of Lebesgue algebras (i.e. measure algebras of Lebesgue spaces). One might suspect therefore that all automorphisms of such a measure algebra are also direct products of automorphisms of Lebesgue algebras. Specifically one can ask if given an automorphism τ of $(\mathcal{E},\mu)$ do there exist independent Lebesgue algebras $\{L_i : i \in A\}$ of $(\mathcal{E},\mu)$ such that $(\mathcal{E},\mu) = \bigotimes_{i\in A} (L_i,\mu)$ and $\tau L_i = L_i$ for all $i \in A$. (The cardinal of the index set A will necessarily be the Maharam type of $(\mathcal{E},\mu)$.) If this property holds we say τ is <u>factorizable</u>. Factorizability is clearly a conjugacy invariant: if τ is factorizable, so is $\sigma^{-1}\tau\sigma$ for every automorphism σ. Since each periodic automorphism is conjugate to one which is factorizable, it follows that every periodic automorphism is factorizable; thus factorizable automorphisms are dense in both coarse and uniform topologies. (See [Ch.Pr]§§3 & 4).

The <u>factorization conjecture</u> asserts that every automorphism of $(\mathcal{E},\mu)$ is factorizable. One could make an even stronger conjecture, that there is one representation of $(\mathcal{E},\mu)$ as a product of independent Lebesgue algebras, such that some countable sub-products will factorize <u>every</u> automorphism of $(\mathcal{E},\mu)$. If this stronger conjecture were true then in the notation of [Ma 2] or [Ch], given any measure algebra automorphism τ of the cube $[0,1]^A$, every countable subset C of A would be contained in a countable subset $\tilde{C}$ of A for which both $\tilde{C}$ and $A\setminus\tilde{C}$ were invariant under τ. The proofs in both these papers could then be greatly simplified. It was pointed out by S. Kakutani in 1975

*Both authors were partially supported by Grants from the National Science and Engineering Research Council of Canada, and the FCAC Programme of the Government of Quebec.

that although it was tacitly assumed in both papers that this could not be done, no counter example was given. A counter-example was given soon after, and this led to the formulation of the weaker factorization conjecture. This seemed so plausible that for several years most people leaned towards believing it to be true. The purpose of this paper is however to give a counter-example to the conjecture; like the earlier counter-example it is measure preserving and uses skew products, however it uses new ideas and several deep results from Lebesgue space ergodic theory. The example is strongly mixing and so, by the conjugacy property, non-factorizable automorphisms are dense in the coarse topology. There is of course an obvious and interesting question concerning the Baire categories of factorizable and non-factorizable automorphisms. (Much of the above discussion is repeated from [Ch.Pr], §7.)

We follow the notation of [Ma 2] and [Ch], and of [Ch.Pr] except that we use greek letters τ, σ for automorphisms and roman letters T, S for measurable point transformations. α, β, γ are usually ordinals, Ω is the first uncountable ordinal. I (with or without a suffix) denotes a copy of the closed unit interval $[0,1]$; for any ordinal $\alpha \geq 1$, $S(\alpha) = \prod_{\beta<\alpha} I_\beta$, μ_α the power Lebesgue measure on $S(\alpha)$, $(E(\alpha),\mu_\alpha)$ the measure algebra of $(S(\alpha),\mu_\alpha)$; note that $S(1) = I_0$.

THEOREM: <u>There exists a measure preserving automorphism</u> τ <u>of the homogeneous measure algebra</u> (E,μ) <u>of type</u> $\aleph_1$ <u>(and total measure</u> 1) <u>which does not factorize</u>.

PROOF: <u>Step 1</u>. It is sufficient to prove that there exists a τ-invariant Lebesgue sub-algebra L such that no other τ-invariant Lebesgue sub-algebra is independent of L. For suppose $\bigotimes_{i \in A} L_i$ factorizes τ (i.e. the L_i are independent Lebesgue algebras such that $E = \bigotimes_{i\in A} L_i$ and $\tau L_i = L_i$ for each $i \in A$). Let $\{a_n : n \in \mathbb{N}\}$ be a σ-basis of L. For each n, there exist $i_{n,j}$ in A such that $a_n \in \bigotimes_{j=1}^{\infty} L_{i_{n,j}}$, so

$$L \subset \bigotimes_{n=1}^{\infty} \bigotimes_{j=1}^{\infty} L_{i_{n,j}} .$$

Since A must be uncountable (in fact of cardinal $\aleph_1$), there exists $i \in A$, $i \neq i_{n,j}$ $\forall n,j$. L_i is a τ-invariant Lebesgue algebra independent of L, contradiction.

<u>Step 2</u>. LEMMA. <u>Let</u> T_0 <u>be a Bernoulli</u> 2-<u>shift on</u> $[0,1]$. <u>There exists an extension</u> T <u>of</u> T_0 <u>to</u> $[0,1] \times [0,1]$, <u>which is also a Bernoulli</u> 2-<u>shift, such that no factor of</u> T <u>is independent of</u> T_0. (For basic facts about entropy and about Bernoulli shifts see e.g. [Sh] or [Pa].)

PROOF OF LEMMA: Let $\{P_0,P_1\}$ be a generating partition for T_0, and let $R_\theta = x + \theta \pmod 1$, $x, \theta \in [0,1]$, θ irrational. For $(x,y) \in [0,1] \times [0,1]$ put

$$T(x,y) = \begin{cases} (T_0x, R_\theta y), & \text{if} \quad x \in P_0\,, \\ (T_0x, y), & \text{if} \quad x \in P_1\,. \end{cases}$$

Then T is a Bernoulli shift by [Ad.Sh] (Theorem 2 on p. 219). Further $h(T) = h(T_0)$, since we always have

$$h(T) = h(T_0) + h_{T_0}(T)$$

and $h_{T_0}(T) = 0$, since the fibre transformations have discrete spectrum ([Ab.Ro], p. 261 line 12 and p. 264 line 27). Hence T is also a Bernoulli 2-shift.

Now let S be a factor of T, since T is Bernoulli so is S, and so $h(S) > 0$. But if S is independent of T_0, then

$$h(T) \geq h(S) + h(T_0) > h(T_0),$$

contradiction.

Step 3. Let Ω be the first uncountable ordinal, i.e. the least ordinal with cardinal $\aleph_1$. For each (countable) ordinal α, $1 \leq \alpha < \Omega$ we construct an invertible measure preserving transformation T_α on the Lebesgue space $(S(\alpha),\mu_\alpha)$ (with induced automorphism τ_α on $(E(\alpha),\mu_\alpha)$) such that

(i) if $1 \leq \beta < \alpha$, then T_α extends T_β (so τ_α extends τ_β), and no factor of T_α is independent of T_β (so no factor of τ_α is independent of τ_β);

(ii) each T_α is a Bernoulli 2-shift.

Clearly, T_1 can be chosen to be an arbitrary Bernoulli 2-shift on $S(1) = I_0$. Suppose that for all $1 \leq \beta < \alpha$, T_β (respectively τ_β) have been constructed to satisfy properties (i) and (ii).

Case (1): α has an immediate predecessor, say $\alpha = \gamma + 1$. Since T_γ is a Bernoulli 2-shift on the Lebesgue space $(S(\gamma),\mu_\gamma)$ and $(S(\alpha),\mu_\alpha) = (S(\gamma),\mu_\gamma) \otimes ([0,1], \mu_0)$ (where μ_0 is of course Lebesgue measure), by the Lemma in Step 2, there exists T_α on $(S(\alpha),\mu_\alpha)$ extending T_γ, such that T_α is also a Bernoulli 2-shift and such that no factor of T_α is independent of T_γ. If $1 \leq \beta < \gamma$, then since $h(T_\alpha) = h(T_\beta)$, and all factors of T_α are Bernoulli and so have positive entropy, the same argument as in the proof of the Lemma in Step 2 shows that no factor of T_α can be independent of T_β.

Case (2): α has no immediate predecessor.

For $1 \le \beta < \alpha$, the transformations T_β define a bijective map T_α of $S(\alpha)$, which is clearly an invertible μ_α measure preserving point transformation of the Lebesgue space $(S(\alpha),\mu_\alpha)$, and which extends each T_β. Since $\{\beta:\beta < \alpha\}$ has a cofinal increasing sequence say $\{\beta_n\}$, by [Or] (p. 340 Theorem 5), T_α, being the limit of an increasing sequence of Bernoulli transformations, is itself Bernoulli; further since the measure algebras $E(\beta_n)$ generate $E(\alpha)$, by [Ro] (p. 28 Theorem 9.6), $h(T_{\beta_n}) \uparrow h(T_\alpha)$, so $h(T_\alpha) = h(T_{\beta_n}) = \log 2$, i.e. T_α is a Bernoulli 2-shift. Now, again by the same argument that concludes the proof of the Lemma of Step 2, since all factors of T_α are Bernoulli and so have positive entropy, no factor of T_α can be independent of T_β for any $1 \le \beta < \alpha$. Thus T_α has the required properties.

By transfinite induction we have thus constructed T_α (with induced automorphisms τ_α) for all $1 \le \alpha < \Omega$, with the required properties.

Step 4: We now construct the required transformation T on $(S(\Omega),\mu_\Omega)$, such that the induced automorphism τ on $(E(\Omega),\mu_\Omega)$ is not factorizable. Exactly as in Case 2 of the construction of Step 3 above, for $1 \le \alpha < \Omega$ the transformations T_α define a bijective map T of $S(\Omega)$ which is clearly an invertible μ_Ω measure preserving point transformation of $(S(\Omega),\mu_\Omega)$. Clearly also T extends T_α for each $1 \le \alpha < \Omega$, and so its induced automorphism τ of $(E(\Omega),\mu_\Omega)$ extends each τ_α.

It remains to show that τ has the property of Step 1, i.e. that there exists a τ-invariant Lebesgue algebra L, such that no other τ-invariant Lebesgue algebra can be independent of L: it will then follow by Step 1 that τ is not factorizable. We show that for any $1 \le \alpha < \Omega$, $E(\alpha)$ is such a Lebesgue algebra. For $E(\alpha)$ is clearly τ invariant and Lebesgue. Suppose for some α, L_0 is a Lebesgue algebra which is τ invariant and independent of $E(\alpha)$. Let $\{b_n : n \in \mathbb{N}\}$ be a σ-basis of L_0. There exist $\beta_n < \Omega$, such that $b_n \in E(\beta_n)$, and (since $\aleph_1$ has cofinality $\aleph_1$), there exists $\lambda < \Omega$, such that $\alpha < \lambda$, $\beta_n < \lambda$ for all $n \in \mathbb{N}$. Then $E(\alpha)$, L_0 are both τ_λ invariant Lebesgue sub-algebras of the Lebesgue algebra $E(\lambda)$, and since they are independent by assumption, τ_λ has a factor (on L_0) which is independent of τ_α. This contradicts property (i) for τ_α and τ_λ (see Step 3), and this completes the proof of the theorem.

Note that τ, being the projective limit of Bernoulli shifts is certainly strongly mixing ([Br] p. 28, Proposition 1.11).

References

[Ab.Ro] L. Abramov and V.A. Rohlin, "Entropy of a skew product of transformations with invariant measure" A.M.S. Translations Series (2) 48 (1965), 255-265.

[Ad.Sh] R.L. Adler and P. Shields, "Skew products of Bernoulli transformations with rotations", Israel J. Math. 12 (1972), 215-220.

[Br] J.R. Brown, Ergodic Theory and Topological Dynamics, Academic Press, New York, 1976.

[Ch] J.R. Choksi, Automorphisms of Baire measures on generalized cubes I and II, Z. Wahrscheinlichkeitstheorie 22 (1972), 195-204 and 23 (1972), 97-102.

[Ch.Pr] J.R. Choksi and V.S. Prasad, Ergodic theory on homogeneous measure algebras, Measure Theory Oberwolfach 1981, Springer Lecture Notes in Math. No. 945, Springer, Berlin-Heidelberg-New York 1982, 366-408.

[Ma 1] D. Maharam, On homogeneous measure algebras, Proc. Nat. Acad. Sci. (Washington) 28 (1942), 108-111.

[Ma 2] D. Maharam, Automorphisms of products of measure spaces, Proc. Amer. Math. Soc. 9 (1958), 702-707.

[Or] D.S. Ornstein, Two Bernoulli shifts with infinite entropy are isomorphic, Adv. in Math. 5 (1970), 339-348.

[Pa] W. Parry, Topics in ergodic theory, Cambridge Univ. Press, Cambridge, 1981.

[Ro] V.A. Rohlin, Lectures on entropy theory of measure preserving transformations, Usp. Mat. Nauk. 22 (1967), 3-56, translated in Russian Math. Surveys 22 (1967), 1-52.

[Sh] P. Shields, The theory of Bernoulli shifts, Univ. of Chicago Press, Chicago 1973.

J.R. Choksi
Department of Mathematics
McGill University
Montreal, Quebec
CANADA H3A 2K6

S.J. Eigen
Department of Mathematics
Northeastern University
Boston, Mass. 02115

Contemporary Mathematics
Volume 26, 1984

ANOTHER GENERALIZATION OF THE BROUWER FIXED POINT THEOREM

Daniel I. A. Cohen

In 1941 Shizuo Kakutani (4) introduced a generalization of Brouwer's Fixed Point Theorem which is now universally referred to as the Kakutani Fixed Point Theorem. At that time Kakutani was working closely with von Neumann at Princeton. Von Neumann was developing the celebrated Minimax Theorem, and in its early forms of the crucial lemma in the proof was the Kakutani Fixed Point Theorem. Kakutani's Theorem is currently employed in numerous other applications e.g. the calculation of equilibria in finite economies, optimization and the nonlinear complementarity problem. It has become a cornerstone of contemporary mathematical economics (cf. 2, 3, 7, 9, 11).

Proofs of the Kakutani Fixed Point Theorem generally depend on finding piecewise linear approximations and determining the fixed points of each. Since the actual calculation of Kakutani fixed points is an important and often performed task, the difficulties which arise in knowing only approximately the locations of the fixed points of the approximation functions must be handled carefully. The variation of this theorem which we present here avoids the difficulty of second order errors, although it does not necessarily apply to as wide a variety of cases.

THEOREM:

<u>Let</u> Δ <u>be an n-simplex and let</u> ϕ <u>be a point-to-set map associating to each</u> $x \in \Delta$ <u>the non-empty compact set</u> $\phi(x) \subset \Delta$ <u>such that</u>

i) ϕ <u>is upper semi-continuous</u>

ii) <u>for all</u> $x \in \Delta$ <u>either</u> $x \in \phi(x)$ <u>or else there is a vertex</u> v <u>of the carrier of</u> x <u>such that</u> v <u>is closer to</u> x <u>than it is to</u> $\phi(x)$,

<u>then</u> ϕ <u>has a fixed point i.e. there is some</u> x^* <u>such that</u> $x^* \in \phi(x^*)$.

REMARK:

The difference between this theorem and the Kakutani Fixed Point Theorem is that the latter assumes that $\phi(x)$ must always be convex and then it drops

0271-4132/84 $1.00 + $.25 per page

condition ii. In both cases if we restrict ϕ to be a point-to-point map, the u.s.c. condition reduces to continuity and we have the Brouwer Fixed Point Theorem.

Proof:

Let us assume that ϕ does not have such a fixed point.

Let us take a triangulation T of Δ. We can label every vertex of T with a vertex of its carrier in Δ which is closer to it than to its image under ϕ. This is guaranteed by condition ii. This is a proper Sperner labeling of the vertices of the triangulation and by Sperner's Lemma (1, 10) one triangle in the triangulation must have a complete set of labels i.e. representing each vertex of Δ. As we take finer and finer subtriangulations of T we arrive at a point x^* such that x^* is not farther from any vertex of Δ than $\phi(x^*)$ is. This x^* must be a fixed point. □

For the purposes of actual calculation we can now apply the excellent constructive algorithms for finding complete Sperner triangles (1, 5, 6, 7, 8, 9, 11).

In practice the best way of discovering whether ϕ satisfies condition ii may be to try to run the procedure which may succeed even when the theorem does not technically apply.

References

1. Cohen, D. I. A. "On the Sperner lemma", J. Comb. Th. 2 (1967) 585-7.

2. Eaves, B. C. "Nonlinear Programming via Kakutani Fixed Points" Working paper No. 294, Center for Res. in Management Sci. U. of Calif., Berkeley, (1970).

3. _____ "Computing Kakutani Fixed Points", SIAM J. of Appl. Math., 21 2 (1971) 236-44.

4. Kakutani, S., "A generalization of Brouwer's fixed point theorem", Duke Math J., 8 (1941) 457-9.

5. Kuhn, H. W. "Simplicial Approximations of fixed points", Proc. Nat. Acad. Sci. USA, 61 (1968) 1238-42.

6. _____ "Approximate search for fixed points", Computing Methods in Optimization Problems - 2, Academic Press, NY, (1969).

7. _____ "How to Compute Economic Equilibria by Pivotal Methods", Dep of Econ and Math, Princeton U., (1975).

8. Scarf, H. E. "The approximation of fixed points of a continuous mapping", SIAM J. Appl. Math, 15 5 (1967), 1328-1343.

9. _____ and T. Hansen, Computation of Economic Equilibria, Yale U. Press, New Haven, (1973).

10. Sperner, E., "Neuer Bewis fur die Invarianz der Dimensionzahl und des Gebiets", Abh. Math Sem Univ. Hamburg 6, (1928).

11. Todd, M. J., The Computation of Fixed Points and Applications, Lecture Notes in Econ. and Math. Systems, No. 124, Berlin, (1976).

Daniel I. A. Cohen
Hunter College, CUNY
The Rockefeller University

Contemporary Mathematics
Volume **26**, 1984

AN ERGODIC THEOREM

Yael Naim Dowker

INTRODUCTION

An ergodic theorem is proved under conditions which involve only minimal restrictions of a measure theoretical nature and no involvement of integrals or norms. The method of proof is such that it seems to apply to a variety of situations and the actual statement of the theorem is only one of several possibilities. The one stated here was chosen more for simplicity of presentation and a focus on the method of proof than for generality and refinement of the conditions. The author may pursue some of these possibilities at a later date.

Let X be a space and T a bijection of X to X. Let $p(x)$ be a positive real valued function on X and let $Uf(x)$ be defined for any real valued function $f(x)$ on X by

$$Uf(x) = f(Tx)p(x).$$

Let $f_0 = f,\ f_1 = f + Uf, \ldots, f_N = f + Uf + \cdots + U^N f$.

LEMMA. Let $f(x)$ be a real valued function on X. Put

$$\bar{F}_N = \sup(0, f, f_1, \ldots, f_N)$$

$$\phi_0 = \sup(f,0),\ \phi_1 = \sup(f,-Uf), \ldots,$$

$$\phi_N = \sup(f, -\sum_{i=1}^{N} U^i f) \quad \text{and let}$$

$$g_N = \inf(\phi_1, \phi_2, \ldots \phi_N).$$

Then $$g_N(x) + U\bar{F}_N(x) \geq \bar{F}_N(x) \quad \forall x \in X.$$

The proof is a very slight modification of the usual proof of this kind of lemma.

NOTE. $g_N(x) \geq f(x) \forall x \in X$ and $g_N(x) = f(x)$ if x is such that $\bar{F}_N(x) > 0$. Moreover $g_N(x) \downarrow \forall x \in X$.

COROLLARY 1. With the same notation and conditions as above we have $\sum_{i=0}^{n} U^i g_N(x) \geq \bar{F}_N(x) - U^{n+1}\bar{F}_N(x)$ for $\forall x \in X$ and $n = 0,1,2,\ldots$.

PROOF. $\sum_{i=0}^{n} U^i g_N(x) \geq \bar{F}_N(x) - U\bar{F}_N(x) + U\bar{F}_N(x) - U^2\bar{F}_N(x) + U^2\bar{F}_N(x)$

$$+\dots+ U^n\bar{F}_N(x) - U^{n+1}\bar{F}_N(x).$$

Let us now assume that (X,β,m) is a measure space, and let T be a non-singular, measurable bijection of (X,β,m) to (X,β,m).

Let V, f, p, $\bar{F}_N$, g_N be defined as before and let us assume that f and p are measurable functions. We say that f satisfies condition 1, if

$$\frac{U^{n+1}f(x)}{\sum_{i=0}^{n} U^i p(x)} \to 0 \quad \text{a.e. in } X.$$

Let us assume that $\sum_{i=0}^{\infty} U^i p(x) = \infty$ a.e. in X and that f and p satisfy condition 1.[†]

COROLLARY 2.
$$\underline{\lim} \frac{\sum_{i=0}^{n} U^i g_N(x)}{\sum_{i=0}^{n} U^i p(x)} \geq 0 \quad \underline{\text{a.e. in}} \ X.$$

The proof is immediate.

Let $\bar{F} = \sup(0,\frac{f}{p},\frac{f_1}{p_1},\dots,\frac{f_n}{p_n},\dots)$

($\bar{F}(x)$ may be $= \infty$ for some $x \in X$).

LEMMA. Let $\bar{F}(x) > 0$ a.e. in X. Then there exists a sequence of functions $g_N(x)$ such that

$$g_N(x) \downarrow f(x) \text{ and}$$

$$\frac{\sum_{i=0}^{n} U^i g_N(x)}{\sum_{i=0}^{n} U^i p(x)} \geq 0 \quad \underline{\text{a.e. in}} \ X.$$

PROOF. The proof follows directly from Corollary 1, and the note preceding it, and Corollary 2.

Let $h(x)$ be a measurable function such that $Uhp = hUp$.

With the same notations and conditions as before we have

COROLLARY 3. Let $\bar{F}(x) > h(x)$ a.e. in X. Then there exist a sequence of functions $\bar{h}_N(x)$ s.t. $\bar{h}_N(x) \downarrow f(x)$, and

[†]It can be easily proved that if $f(x)$ satisfies condition 1 then so do $\bar{F}_N(x)$ for $N = 1,2,\dots$.

$$\underline{\lim}\frac{\Sigma U^i \bar{h}_N(x)}{\Sigma U^i p(x)} \geq h(x) \quad \underline{\text{a.e. in}} \quad X.$$

PROOF. Put $\bar{f} = f - hp$. Then $\bar{f}$ satisfies condition 1. By Corollary 2, there exist $\bar{g}_N \downarrow \bar{f}$ and

$$\underline{\lim}\frac{\sum_{i=0}^{n} U^i \bar{g}_N}{\sum_{i=0}^{n} U^i p} \geq 0 \quad \text{a.e.} \quad \text{in} \quad X.$$

Let $\bar{h}_N = \bar{g}_N + hp$. Then

$$\underline{\lim}\frac{\Sigma U^i \bar{g}_N}{\Sigma U^i p} = \underline{\lim}\frac{\Sigma U^i \bar{h}_N - h\Sigma U^i p}{\Sigma U^i p}$$

$$= \underline{\lim}\left(\frac{\Sigma U^i \bar{h}_N}{\Sigma U^i p} - h\right)$$

$$\underline{\lim}\frac{\Sigma U^i \bar{h}_N(x)}{\Sigma U^i p(x)} \geq h(x) \quad \text{a.e.} \quad \text{in} \quad X.$$

Let $\underline{F} = \inf(0, \frac{f}{p}, \frac{f_1}{p_1}, \ldots, \frac{f_n}{p_n}, \ldots)$

($\underline{F}(x)$ may be $-\infty$ for some $x \in X$).

COROLLARY 4. Let $\underline{F}(x) < r(x)$ a.e. in X. Then there exist a sequence of functions $\underline{h}_N(x)$ s.t. $\underline{h}_N \uparrow f$ and

$$\overline{\lim}\frac{\Sigma U^i \underline{h}_N}{\Sigma U^i p} \leq r \qquad \underline{\text{a.e. in}} \quad X.$$

PROOF. The proof is obtained by the usual obvious modifications to that of Corollary 3.

Let f, p satisfy the conditions stated before and let

$$f^*(x) = \overline{\lim}\frac{\Sigma U^i f(x)}{\Sigma U^i p(x)} \quad \text{and} \quad f_*(x) = \underline{\lim}\frac{\Sigma U^i f(x)}{\Sigma U^i p(x)} \quad .$$

Notice that $f^*(x)$ and $f_*(x)$ may be ∞ or $-\infty$ for some $x \in X$.

LEMMA. $U(f^*p) = f^*Up$ and $u(f_*p) = f_*Up$.

PROOF. It is easy to see that $f^*(Tx) = f^*(x)$ and $f_*(Tx) = f_*(x)$. Hence

$$U(f^*(x)p(x)) = f^*(Tx))p(Tx)p(x) = f^*(x)Up(x)$$

and similarly for $f_*(x)$.

We say that $g(x)$ satisfies condition 2 if $g^*(x)$ and $g_*(x)$ are finite a.e.

LEMMA. *Let* $f(x)$ *satisfy condition* 2 *and let* $\varepsilon > 0$. *Then there exist sequences* $\bar{h}_N(x)$ *and* $\underline{h}_N(x)$ *(depending on* ε*) such that* $\bar{h}_N(x) \downarrow f(x)$, $\underline{h}_N(x) \uparrow f(x)$, $(\bar{h}_N(x))_* \geq f^*(x) - \varepsilon$ *and* $\underline{h}_N^*(x) \leq f_*(x) + \varepsilon$ *a.e. in* X, $N = 1,2,\ldots$.

PROOF. Let $h(x) = f^*(x) - \varepsilon$ and $r(x) = f_*(x) + \varepsilon$, then

$$\bar{F}(x) \geq f^*(x) > f^*(x) - \varepsilon \quad \text{and}$$

$$\underline{F}(x) \leq f_*(x) < f_*(x) + \varepsilon \quad \text{a.e. in } X.$$

Apply Corollary 3 and Corollary 4 and the lemma is proved.

REMARK. It is easy to see that if $f(x)$ satisfies condition 2, then so do $\bar{h}_N(x)$ and $\underline{h}_N(x)$ for $N = 1,2,\ldots$.

Let Ω be a linear space of measurable functions which closed under the operation of taking finite supremums and infimums. We say that Ω satisfies condition * if $\psi_n(x) \in \Omega$ $n = 1,2,\ldots$ and $\psi_n(x) \downarrow 0$ a.e. implies that $(\psi_n(x))_* \downarrow 0$ a.e. in X. Let $p(x) \in \Omega$ $(\Sigma U^i p = \infty$ a.e.).

THEOREM. *Let the functions of* Ω *satisfy conditions* 1, 2 *and* *. *Let* $f(x) \in \Omega$ *then*

$$\frac{\sum_{i=0}^{n} U^i f(x)}{\sum_{i=0}^{n} U^i p(x)} \quad \text{converges a.e. in } X.$$

PROOF. Let $\varepsilon > 0$. Then by the previous lemma there exist sequences $\bar{h}_N(x)$ and $\underline{h}_N(x)$ such that $\bar{h}_N(x) \downarrow f(x)$, $\underline{h}_N(x) \uparrow f(x)$ and such that $(\bar{h}_N(x))_* \geq f^*(x) - \varepsilon$ and $\underline{h}_N^*(x) \leq f_*(x) + \varepsilon$ a.e. in X for $N = 1,2,\ldots$.

$$\bar{h}_N(x), \underline{h}_N(x) \in \Omega \quad \text{and} \quad \bar{h}_N(x) - \underline{h}_N(x) \downarrow 0 \quad \text{a.e. in } X.$$

By condition * we have that $(\bar{h}_N(x) - \underline{h}_N(x))_* \downarrow 0$ a.e. in X. Now

$$(\bar{h}_N(x) - \underline{h}_N(x))_* \geq (\bar{h}_N(x))_* - \underline{h}_N^*(x) \geq f^*(x) - \varepsilon - f_*(x) - \varepsilon$$

$$= f^*(x) - f_*(x) - 2\varepsilon, \; N = 1,2,\ldots$$

$$\therefore \; 0 \geq f^*(x) - f_*(x) - 2\varepsilon \quad \text{a.e. in } X.$$

Since ε is arbitrary we obtain

$$f^*(x) - f_*(x) = 0 \quad \text{a.e. in } X. \qquad \text{q.e.d.}$$

REMARKS. If T is measure preserving and $\Omega = L_1(m)$, $p(x) = 1$, then conditions 1, 2, and * hold as well as $\sum_{i=0}^{n} p(x) = n+1$ a.e. in X. Their proofs are quite elementary. Our theorem then becomes the classical Birkhoff's ergodic theorem.

If T is non-singular and $p(x) = w(x)$ where $w(x)$ is given by $m(TA) = \int_A w(x)dm$, then our theorem is a Hopf-Hurewicz-Oxtoby type theorem. In this case condition 1 can be proved in an elementary way (cf. A. Garsia, Topics in Almost Everywhere Convergence, p. 32). Our method of proof can also be modified to proved theorems of the Chacon-Ornstein type.

DEPARTMENT OF MATHEMATICS
IMPERIAL COLLEGE OF SCIENCE AND TECHNOLOGY
UNIVERSITY OF LONDON
LONDON, SW7 2BZ, ENGLAND

Contemporary Mathematics
Volume **26**, 1984

HIGHER ORDER PARTIAL MIXING

Nathaniel A. Friedman*

ABSTRACT. Our purpose is to study higher order partially mixing transformations. For each positive integer k, a transformation is constructed that is partially k-mixing but not partially (k+1)-mixing. These transformations have finite rank and therefore the equivalence theory implies each one is induced by every zero entropy loose Bernoulli transformation.

1. INTRODUCTION. Partially mixing transformations were introduced in [5] and α-mixing transformations were constructed in [6] for $0 < \alpha \le 1$, where $\alpha = 1$ corresponds to mixing. Our purpose is to study partially k-mixing transformations, $k \ge 1$, and the corresponding α-k-mixing transformations, $0 < \alpha < 1$ (see §2 for definitions). In §3 we will construct a transformation that is 1/2-1-mixing and 0-2-mixing. In §4 a transformation is constructed that is 2/3-1-mixing, 1/3-2-mixing, and 0-3-mixing. This construction is extended in §5 to any positive k. A transformation is constructed that is $(k+1-j)/(k+1)$-j-mixing $1 \le j \le k+1$. Induced transformations are considered in §6. The case of transformations that are partially mixing of all orders will be discussed in a subsequent paper.

I would like to thank Jean-Paul Thouvenot for several helpful conversations and the hospitality of the University of Paris.

2. PRELIMINARIES. Let $(X,\mathcal{B},m)$ be a measure space isomorphic to the unit interval with Lebesgue measure. Let T be an invertible measure preserving transformation mapping X into X. Let k be a positive integer. T is <u>partial</u> k-<u>mixing</u> if there exists $\beta > 0$ such that for all $B_i \in \mathcal{B}$, $0 \le i \le k$,

$$\liminf m\left(\bigcap_{i=0}^{k} T^{n_i}B_i\right) \ge \beta \prod_{i=0}^{k} m(B_i), \tag{2.1}$$

as $n_i - n_{i-1} \to \infty$, $1 \le i \le k$, $n_0 = 0$. Let $0 < \alpha \le 1$. A partial k-mixing transformation is α-k-<u>mixing</u> if (2.1) holds with $\beta = \alpha$ but (2.1) does not hold for $\beta = \alpha + \varepsilon$, $\varepsilon > 0$. The case $\alpha = 1$ corresponds to k-mixing, in which case the limit exists in (2.1). Partial mixing was introduced in [5] and α-1-mixing

* Partially supported by N.S.F. Grant No. MCS-8102101.

(α-mixing) transformations were constructed in [6] for each α, $0 < \alpha < 1$.

We will use independent cutting and stacking to construct examples. This construction is described in [2,3,12] and is reviewed briefly as follows. A <u>column</u> is an ordered set of disjoint left-closed right-open intervals of the same length. A <u>tower</u> G is an ordered set of disjoint columns. The transformation T_C maps an interval in a column to the next interval in the column. We view a column with h intervals as h rungs on a ladder. An interval is referred to as a level. Thus T_C maps a level to the level above; hence T_C is not defined on the top level. The transformation T_G consists of T_C for C in G.

Given a tower G, we define the tower SG by independent cutting and stacking as follows. Let G have k columns and let G_0 be the half-size copy of G consisting of the left halves of the columns in G. Cut the right halves of each column into k subcolumns where the subcolumns have the same relative width as the columns in G. Form k copies of G by grouping the jth subcolumns, $1 \le j \le k$. Place the jth copy of G above the jth column in G_0, $1 \le j \le k$. This yields a tower SG consisting of a copy G_0 of G and a copy of G above each column in G_0. In particular, SG has k^2 columns.

The transformation T_{SG} extends T_G to map the upper left halves of each column in G proportionally into the lower right halves of each column in G. If $w(G)$ is the sum of the measures of the top intervals in G, then $w(SG) = w(G)/2$. In particular, T_{SG} extends T_G to half of where it was not defined. Let $S^nG = S(S^{n-1}G)$, $n > 1$.

We also let G denote the union of the levels in the columns in G. A set A that is a union of some levels in columns in G is simply called a set A in G. In general, $m(G) \le 1$. A transformation $T(G)$ is defined on G by

$$T(G)(x) = \lim_{n\to\infty} T_{S^nG}(x), \quad x \in G. \tag{2.2}$$

If G contains two columns with k and ℓ intervals with k and ℓ relatively prime, then we say G is an M-tower. In particular, G is an M-tower if two column heights differ by one. If G is an M-tower, then $T(G)$ is a mixing Markov shift relative to the measure $m_G(A) = m(A)/m(G)$ [3,12]. In particular, $T(G)$ is mixing of all orders.

To simplify notation, we will say that T^t $(1-\varepsilon)$-k-<u>mixes levels in</u> G, $t \ge t^*$, if for any $k+1$ levels I_i in G, $0 \le i \le k$, we have

$$\left| 1 - m_G\left(\bigcap_{i=0}^{k} T^{n_i} I_i\right) \middle/ \prod_{i=0}^{k} m_G(I_i) \right| < \varepsilon \tag{2.3}$$

for $n_i - n_{i-1} \ge t^*$, $1 \le i \le k$, $n_0 = 0$. Since $T(G)$ is mixing of all orders, we have the following lemma.

LEMMA 2.0. Let G be an M-tower, k a positive integer, and $\varepsilon > 0$. There exists t^* such that if $T = T(G)$, then

(a) T^t $(1-\varepsilon)$-k-mixes levels in G, $t \geq t^*$.

Let $t^{**} > t^*$. There exists r such that if T extends T_{S^rG}, then

(b) T^t $(1-\varepsilon)$-k-mixes levels in G, $t^* \leq t \leq t^{**}$.

PROOF. Part (b) follows from (2.2).

A tower can be converted to a column and then periodicidity can be introduced, as in the following lemma.

LEMMA 2.1. Let G be a tower and $\varepsilon > 0$. There exists a column C with $C = G$ and a positive integer p such that if T extends T_C, then $m(T^pI \cap I) > (1-\varepsilon)m(I)$ for each level I in G.

PROOF. We will consider only towers with columns of rational widths. Hence each column in G can be divided into subcolumns, all of the same width w. We now stack these subcolumns to form one column C_1 with width w. Choose n so that $1/n < \varepsilon$. Cut C_1 into n subcolumns and stack them to form one column C. Let p be the height of C_1. Note that each level I in G is a union of levels in C_1. The conclusion holds for each level J in C_1; hence the conclusion holds for each level I in G.

The next lemma converts a column into an M-tower and then Lemma 2.0 is applied. It is stated separately for convenience.

LEMMA 2.2. Let C be a column, k a positive integer, and $\varepsilon > 0$. There exists an M-tower $G \supset C$ with $m(G - C) < \varepsilon$ and t^* such that if $T = T(G)$, then

(a) T^t $(1-\varepsilon)$-k-mixes levels in G, $t \geq t^*$.

Let $t^{**} > t^*$. There exists r such that if T extends T_{S^rG}, then

(b) T^t $(1-\varepsilon)$-k- mixes levels in G, $t^* \leq t \leq t^{**}$.

PROOF. Choose n so large that $1/n < \varepsilon$. Cut C into n subcolumns of equal width w. Add one interval of width w to the last column. This yields an M-tower G with n columns where two heights differ by one. The remainder follows by Lemma 2.0.

In the constructions below we will frequently apply Lemma 2.1 to obtain a column C_1. Lemma 2.2(a) will be applied to C_1 to obtain G_1 and t_1^*. Lemma 2.1 will be applied again to another tower, disjoint from G_1, to obtain a column C_2. Lemma 2.2(a) will be applied to C_2 to obtain G_2 and $t_2^* > t_1^*$. Then Lemma 2.2(b) will be applied to G_1 with $t^* = t_1^*$ and $t^{**} = t_2^*$.

Let G be a tower and let A be in G. That is, A is a union of levels in G. If $A \subset B$ and $m(A) = am(B)$, then we say aB is in G. For example, we may have $(1/3)B = B/3$ in G.

Suppose aA and aB are in G and $m(G) = a$. Assuming G is an M-tower, we have

$$\lim_{n\to\infty} m(T(G)^n A \cap B) = m(aA)m(aB)/m(G) = am(A)m(B).$$

Thus mixing of aA and aB in a set of measure a yields a-mixing. In general, k-mixing of aA_i, $0 \le i \le k$, in a set of measure a yields a-k-mixing.

In the constructions in §3, §4, §5 we will have separate mixing occurring for certain time intervals in disjoint towers. The total mixing for a time interval is the sum of the separate mixing in each tower. For example, if we have 1/5-2-mixing occurring for a certain time interval in three disjoint towers, then we have 3/5-2-mixing occurring for a certain time interval.

For notational convenience, we denote

$$R(T,t,s,A,B,C) = m(T^t(T^s A \cap B) \cap C)/m(A)m(B)m(C).$$

In general, we have

$$R(T,t_i,A_i,0 \le i \le k) = m(T^{t_k}(\cdots(T^{t_1}A_0 \cap A_1) \cap\cdots) \cap A_k)/\prod_{i=0}^{k} m(A_i)$$

We also denote $T_{ij} = T_{G_{ij}}$ or $T_{ij} = T_{C_{ij}}$ in the constructions.

3. PARTIAL MIXING AND ZERO TWO-MIXING. A transformation will be constructed that is .5-mixing and 0-2-mixing. At each stage in the construction we will alternately mix half of the space and introduce periodicity in the other half. The following diagram illustrates the induction step.

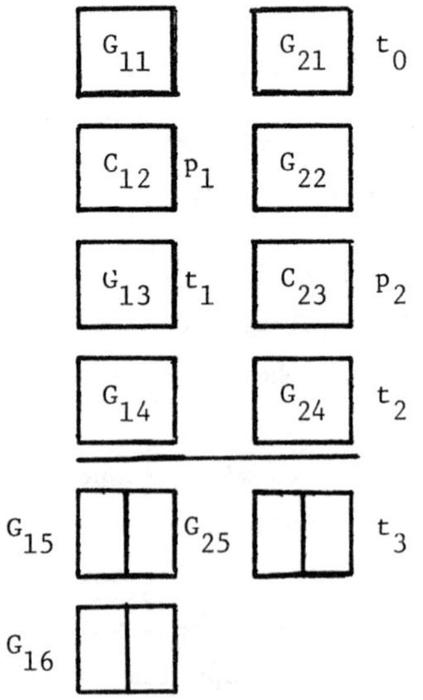

Figure 1

The induction step will be described first. There are two disjoint M-towers G_{11} and G_{21} of equal measure. We have $\varepsilon > 0$ and t_0 such that if $T = T(G_{21})$, then

(0) $$T^t \text{ (1-}\varepsilon\text{)-mixes levels in } G_{21},\ t \geq t_0.$$

Periodicity is first introduced in G_{11}. Apply Lemma 2.1 with $G = G_{11}$ to obtain $p_1 = p$ and $C_{12} = C$. If T extends T_{12} and I is a level in G_{11}, then

(1) $$m(T^t I \cap I) > (1-\varepsilon)m(I),\ t = p_1.$$

Lemma 2.2(a) is now applied for $k = 1$ with $C = C_{12}$. We obtain $G_{13} = G$ and $t_1 \geq t^*$, where $t_1 = t_0 + 1$ if $t^* \leq t_0$.

Now t_1 tells us how long to mix G_{21} until mixing can start in G_{13}. Apply Lemma 2.0 for $k = 1$ with $G = G_{21}$, $t^* = t_0$, and $t^{**} = t_1$. We obtain r so large that if $G_{22} = S^r G_{21}$ and T extends T_{22}, then

(2) $$T^t \text{ (1-}\varepsilon\text{)-mixes levels in } G_{21},\ t_0 \leq t \leq t_1.$$

From (2) we conclude that if $A/2$ and $B/2$ are in G_{21}, then

(3) $$m(T^t A \cap B) \geq (1-\varepsilon)m(A)m(B)/4m(G_{21}),\ t_0 \leq t \leq t_1.$$

Lemma 2.1 is now applied with $G = G_{22}$ to obtain $p_2 = p$ and $C_{23} = C$. If T extends T_{23} and I is a level in G_{22}, then

(4) $$m(T^t I \cap I) \geq (1-\varepsilon)m(I),\ t = p_2.$$

Lemma 2.2(a) is now applied for $k = 1$ with $C = C_{23}$. We obtain $G_{24} = G$ and $t_2 \geq t^*$, $t_2 > t_1$.

Now t_2 tells us how long to mix G_{13}. Apply Lemma 2.2(b) for $k = 1$ with $G = G_{13}$, $t^* = t_1$, and $t^{**} = t_2$. We obtain r so large that if $G_{14} = S^r G_{13}$ and T extends T_{14}, then

(5) $$T^t \text{ (1-}\varepsilon\text{)-mixes levels in } G_{13},\ t_1 \leq t \leq t_2.$$

If $A/2$ and $B/2$ are in G_{11}, then $A/2$ and $B/2$ are in G_{13}. From (5) we conclude

(6) $$m(T^t A \cap B) \geq (1-\varepsilon)m(A)m(B)/4m(G_{13}),\ t_1 \leq t \leq t_2.$$

For each $j = 1,2$, we now form two equal copies of G_{j4} denoted by $g_{ij} = G_{j4}/2$, $i = 1,2$. Let $G_{i5} = (g_{i1}, g_{i2})$, $i = 1,2$. Apply Lemma 2.0(a) for $k = 1$ with $\varepsilon = \eta$ and $G = G_{25}$. We obtain $t_3 > t_2$ such that if $T = T(G)$, then

(7) $$T^t \text{ (1-}\eta\text{)-mixes levels in } G_{25},\ t > t_3.$$

Now t_3 tells us how long to mix g_{11} and g_{12} until mixing can start in G_{25}. Since g_{1j} is a copy of G_{j4}, $j = 1,2$, the mixing for levels I in

G_{j4} can be continued for $I/2$ in g_{1j}, $j = 1,2$. Choose r so large that if $C_{16} = (S^r g_{11}, S^r g_{12})$ and T extends T_{16}, then

(8) T^t $(1-\varepsilon)$-mixes $I/2$ in g_{1j} for I in G_{j4}, $j = 1,2$, $t_2 \le t \le t_3$.

If $A/2$ and $B/2$ are in G_{i1}, $i = 1,2$, then $A/2$ and $B/2$ are in G_{i4}, $i = 1,2$. Hence $A/4$ and $B/4$ are in g_{1j}, $j = 1,2$. Therefore (8) implies

$$m(T^t A \cap B) \ge (1-\varepsilon)\frac{m(A)m(B)}{16}\left[\frac{1}{m(g_{11})} + \frac{1}{m(g_{12})}\right],\ t_2 \le t \le t_3. \tag{9}$$

Now consider a set A consisting of levels in G_{i1}, $i = 1,2$. Let $A_i = A \cap G_{i1}$ and $A_i^c = A^c \cap G_{i1}$, $i = 1,2$. From (1) we obtain

$$m(T^t A_1 \cap A_1) \ge (1-\varepsilon)m(A_1),\ t = p_1. \tag{10}$$

Hence (10) implies

$$m(T^t A_1 \cap A_1^c) \le \varepsilon,\ t = p_1. \tag{11}$$

From (4) we obtain

$$m(T^t A_2^c \cap A_2^c) \ge (1-\varepsilon)m(A_2^c),\ t = p_2. \tag{12}$$

Hence (12) implies

$$m(T^t A_2^c \cap A_2) \le \varepsilon,\ t = p_2. \tag{13}$$

Now Figure 1 indicates there is essentially no mixing between levels in G_{11} and G_{21} until G_{i4}, $i = 1,2$, are formed and divided. Thus the construction can be made to guarantee

$$m(T^t G_{11} \cap G_{21}) < \varepsilon,\ m(T^t G_{21} \cap G_{11}) < \varepsilon, \tag{14}$$

for $t = p_1$ and $t = p_2$. Thus (14) and (11) imply

$$\begin{aligned} m(T^{p_1} A \cap A^c) &= m(T^{p_1}(A_1 \cup A_2) \cap (A_1^c \cup A_2^c)) \\ &\le m(T^{p_1} A_1 \cap A_1^c) + m(T^{p_1} A_2 \cap A_2^c) + 2\varepsilon \\ &\le m(T^{p_1} A_2 \cap A_2^c) + 3\varepsilon. \end{aligned} \tag{15}$$

Now (15), (14), and (13) imply

$$\begin{aligned} m(T^{p_2}(T^{p_1} A \cap A^c) \cap A) &\le m(T^{p_2}(T^{p_1} A_2 \cap A_2^c) \cap A) + 3\varepsilon \\ &\le m(T^{p_2} A_2^c \cap A_2) + 4\varepsilon \le 5\varepsilon. \end{aligned} \tag{16}$$

Thus for any set A consisting of levels in G_{i1}, $i = 1,2$, we have

$$m(T^{p_2}(T^{p_1} A \cap A^c) \cap A) \le 5\varepsilon. \tag{17}$$

We now have G_{16}, G_{25}, and t_3 which play the roles of G_{11}, G_{21}, and t_0 respectively, in the induction step of the next stage. The construction of the transformation will now be described using the induction step.

We begin with an M-tower G. Divide G into two disjoint copies $G^1_{i1} = .5G$, $i = 1,2$. Let $\varepsilon_1 > 0$. Apply Lemma 2.0 for $k = 1$ with $\varepsilon = \varepsilon_1$ and $G = G^1_{21}$ to obtain t_{10} such that if $T = T(G^1_{21})$, then

(1,0) $$T^t \ (1-\varepsilon_1)\text{-mixes levels in } G^1_{21},\ t \geq t_{10}.$$

We now apply the induction step where $G_{i1} = G^1_{i1}$, $i = 1,2$. Let $\eta = \varepsilon_2 < \varepsilon_1$. We obtain G_{16} and G_{25}. Let $G^2_{11} = G_{16}$ and $G^2_{21} = G_{25}$. We now repeat the induction step with $G_{i1} = G^2_{i1}$, $i = 1,2$, $\varepsilon = \varepsilon_2$, and $\eta = \varepsilon_3 < \varepsilon_2$.

In general, let $G^n_{11} = G_{16}$ and $G^n_{21} = G_{25}$ be the corresponding towers at the end of the $(n-1)^{st}$ stage of the construction. We now repeat the induction step with $G_{i1} = G^n_{i1}$, $i = 1,2$, $\varepsilon = \varepsilon_n$, and $\eta = \varepsilon_{n+1} < \varepsilon_n$, where $\varepsilon_n \to 0$.

If A is in either G^k_{i1}, $i = 1$ or 2, then the construction implies $A/2$ is in G^n_{i1}, $i = 1,2$, for $n > k$. We also have m normalized so that $\lim_{n\to\infty} m(G^n_{i1}) = 1/2$, $i = 1,2$. We denote $t_{nj} = t_j$, $j = 0,1,2,3$, and $p_{nj} = p_j$, $j = 1,2$ for the nth stage of the construction. Hence if $A/2$ and $B/2$ are in G^n_{21}, then (3) of the induction step implies

(n,3) $$m(T^tA \cap B) \geq (1-2\varepsilon_n)m(A)m(B)/2,\ t_{n0} \leq t \leq t_{n1}.$$

If $A/2$ and $B/2$ are in G^n_{11}, then (6) of the induction step implies

(n,6) $$m(T^tA \cap B) \geq (1-2\varepsilon_n)m(A)m(B)/2,\ t_{n1} \leq t \leq t_{n2}.$$

If $A/2$ and $B/2$ are in G^n_{i1}, $i = 1,2$, then (9) of the induction step implies

(n,9) $$m(T^tA \cap B) \geq (1-2\varepsilon_n)m(A)m(B)/2,\ t_{n2} \leq t \leq t_{n3}.$$

Now $G^{n+1}_{2,1} = G^n_{2,5}$. If $A/2$ and $B/2$ are in G^n_{i1}, $i = 1,2$, then $A/2$ and $B/2$ are in $G^n_{2,5}$. By (7) of the induction step, we have $t_{(n+1)0} = t_{n3}$. Thus mixing of A and B will continue in the $(n+1)^{st}$ stage. Since $\varepsilon_n \to 0$, (n,3), (n,6), and (n,9) imply

(18) $$\liminf_{n\to\infty} m(T^nA \cap B) \geq .5m(A)m(B).$$

Each set in $\mathcal{B}$ can be approximated arbitrarily well by sets consisting of levels in G^n_{i1}, $i = 1,2$, for n sufficiently large. Thus (18) extends to all sets A and B in $\mathcal{B}$. Hence T is α-mixing for $\alpha \geq .5$. It follows from (11) with $\varepsilon_n \to 0$ and $p_{1n} \to \infty$ that $\alpha = .5$.

From (17) with $p_{in} \to \infty$, $i = 1,2$, and $\varepsilon_n \to 0$, we conclude that for any set A,

(19) $$\liminf_{n,k\to\infty} m(T^n(T^kA \cap A^c) \cap A) = 0.$$

In particular, T is 0-2-mixing.

An open problem is to construct a transformation that is α-1-mixing but 0-2-mixing for some $\alpha > 1/2$.

4. PARTIAL TWO-MIXING AND ZERO THREE-MIXING. A transformation will be constructed that is 2/3-mixing, 1/3-2-mixing, and 0-3-mixing. At each stage in the construction we will alternately mix two-thirds of the space and introduce periodicity in the other third. The following diagram illustrates the induction step.

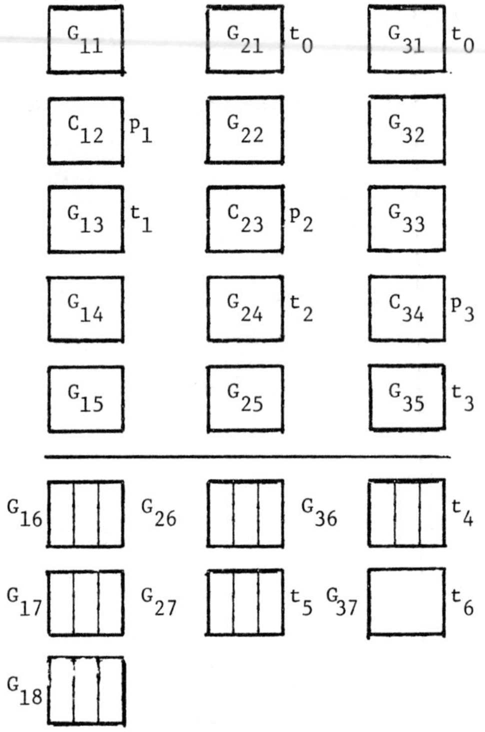

Figure 2

The induction step begins with three M-towers G_{i1}, $i = 1,2,3$, of equal measure. We have $\varepsilon > 0$ and t_0 such that if $T = T(G_{i1})$, $i = 2,3$, then

(0) $$T^t \text{ } (1-\varepsilon)\text{-2-mixes levels in } G_{i1},\ t \geq t_0.$$

Periodicity is first introduced in G_{11}. Apply Lemma 2.1 with $G = G_{11}$ to obtain $p_1 = p$ and $C_{12} = C$. If T extends T_{12} and I is a level in G_{11}, then

(1) $$m(T^t I \cap I) > (1-\varepsilon)m(I),\ t = p_1.$$

Lemma 2.2(a) is now applied for $k = 2$ with $C = C_{12}$. We obtain G_{13} and $t = t^* > t_0$.

Now t_1 tells us how long to mix G_{i1}, $i = 2,3$, until mixing can start in G_{13}. Apply Lemma 2.0 for $k = 2$ with $G = G_{i1}$, $t^* = t_0$, and $t^{**} = t_1$, $i = 2,3$. We obtain r so large that if $G_{i2} = S^n G_{i1}$ and T extends T_{i2}, $i = 2,3$, then

(2) T^t $(1-\varepsilon)$-2-mixes levels in G_{i1}, $t_0 \le t \le t_1$.

Lemma 2.1 is now applied with $G = G_{22}$ to obtain $p_2 = p$ and $C_{23} = C$. If T extends T_{23} and I is a level in G_{22}, then

(3) $$m(T^t I \cap I) > (1-\varepsilon)m(I),\ t = p_2.$$

Lemma 2.2(a) is now applied for $k = 2$ with $C = C_{23}$. We obtain $G_{24} = G$ and $t_2 = t^* > t_1$.

Now t_2 tells us how long to mix G_{13}. Apply Lemma 2.2(b) for $k = 2$ with $G = G_{13}$, $t^* = t_1$ and $t^{**} = t_2$. We obtain r so large that if $G_{14} = S^r G_{13}$ and T extends T_{14}, then

(4) T^t $(1-\varepsilon)$-2-mixes levels in G_{13}, $t_1 \le t \le t_2$.

We also continue mixing in (2) for levels in G_{31} by choosing r so large that if $G_{33} = S^r G_{32}$ and T extends T_{33}, then

(5) T^t $(1-\varepsilon)$-2-mixes levels in G_{31}, $t_1 \le t \le t_2$.

Lemma 2.1 is now applied with $G = G_{33}$ to obtain $p_3 = p$ and $C_{34} = C$. If T extends T_{34} and I is a level in G_{33}, then

(6) $$m(T^t I \cap I) > (1-\varepsilon)m(I),\ t = p_3.$$

Lemma 2.2(a) is now applied for $k = 2$ with $C = C_{34}$. We obtain G_{35} and $t_3 = t^* > t_2$.

Now t_3 tells us how long to mix G_{24}. Apply Lemma 2.2(b) for $k = 2$ with $G = G_{24}$, $t^* = t_2$ and $t^{**} = t_3$. We obtain r so large that if $G_{25} = S^r G_{24}$, and T extends T_{25}, then

(7) T^t $(1-\varepsilon)$-2-mixes levels in G_{24}, $t_2 \le t \le t_3$.

We also continue the mixing in (4). Choose r so large that if $G_{15} = S^r G_{14}$ and T extends T_{15}, then

(8) T^t $(1-\varepsilon)$-2-mixes levels in G_{13}, $t_2 \le t \le t_3$.

We now have G_{j5}, $j = 1,2,3$. Form three equal copies $G_{j5}/3$ for each j denoted by $g_{ij} = G_{j5}/3$, $i = 1,2,3$. Let $G_{i6} = (g_{i1}, g_{i2}, g_{i3})$, $i = 1,2,3$.

Let $0 < \eta < \varepsilon$. Since G_{36} is an M-tower, there exists $t_4 > t_3$ such that if $T = T(G_{36})$, then

(9) T^t $(1-\eta)$-2-mixes levels in G_{36}, $t \ge 4$.

We now continue mixing separately in g_{ij}, $i = 1,2$, $j = 1,2,3$. Let r be a positive integer and let $G_{i7} = (S^r g_{i1}, S^r g_{i2}, S^r g_{i3})$, $i = 1,2$.

By (7) r can be chosen so large that if T extends T_{i7}, $i = 1,2$, then

(10) T^t $(1-\varepsilon)$-2-mixes $I/3$ in g_{i2} for I in G_{24}, $t_3 \le t \le t_4$, $i = 1,2$.

By (8) r can be chosen so large that if T extends T_{i7}, $i = 1,2$, then

(11) T^t $(1-\varepsilon)$-2-mixes $I/3$ in g_{i1} for I in G_{13}, $t_3 \le t \le t_4$, $i = 1,2$.

Since g_{i3} is a copy of G_{35}, r can be chosen so large that if T extends T_{i7}, $i = 1,2$, then

(12) T^t $(1-\varepsilon)$-2-mixes $I/3$ in g_{i3} for I in G_{35}, $t_3 \le t \le t_4$, $i = 1,2$.

Let r be so large that (10)-(12) are satisfied.

Since G_{27} is an M-tower, there exists $t_5 > t_4$ such that if $T = T(G_{27})$, then

(13) $$T^t \ (1-\eta)\text{-2-mixes levels in } G_{27},\ t \ge t_5.$$

Now t_5 tells us how long to mix G_{36}. By (9) r can be chosen so large that if $G_{37} = S^r G_{36}$ and T extends T_{37}, then

(14) $$T^t \ (1-\eta)\text{-2-mixes levels in } G_{36},\ t_4 \le t \le t_5.$$

Since G_{37} is an M-tower, there exists $t_6 > t_5$ such that if $T = T(G_{37})$, then

(15) $$T^t \ (1-\eta)\text{-2-mixes levels in } G_{37},\ t \ge t_6.$$

We also continue the mixing in G_{17} for $i = 1$ in (10)-(12). Choose s so large that if $G_{18} = (S^{r+s} g_{11}, S^{r+s} g_{12}, S^{r+s} g_{13})$ and T extends T_{18}, then

(16) T^t $(1-\varepsilon)$-2-mixes $I/3$ in g_{12} for I in G_{24},
$I/3$ in g_{11} for I in G_{13},
$I/3$ in g_{13} for I in G_{35}, $t_4 \le t \le t_6$.

This completes the induction step. The M-towers G_{18}, G_{27}, and G_{37} play the role of G_{11}, G_{21}, and G_{31}, respectively, in the next stage, where ε is replaced by η and t_0 is replaced by t_6.

We begin with an M-tower G. Divide G into three disjoint copies $G^1_{i1} = G/3$, $i = 1,2,3$. Let $\varepsilon_1 > 0$. Apply Lemma 2.0 for $k = 2$ with $\varepsilon = \varepsilon_1$ and G replaced by $G/3$ to obtain t_{10} such that if $T = T(G^1_{i1})$, $i = 2,3$, then

(1,0) $$T^t \ (1-\varepsilon_1)\text{-mixes levels in } G^1_{i1},\ t \ge t_{10}.$$

We now apply the induction step where $G_{i1} = G^1_{i1}$, $i = 1,2,3$. Let $\eta = \varepsilon_2 < \varepsilon_1$. We obtain G_{18}, G_{27}, and G_{37} and t_6. Let $G^2_{11} = G_{18}$ and $G^2_{i1} = G^2_{i7}$, $i = 2,3$. We now repeat the induction step with $G_{i1} = G^2_{i1}$, $i = 1,2,3$, $\varepsilon = \varepsilon_2$, $\eta = \varepsilon_3 < \varepsilon_2$, and $t_0 = t_{20} = t_6$.

In general, let $G^n_{11} = G_{18}$ and $G^n_{i1} = G_{i7}$, $i = 2,3$, be the corresponding towers at the end of the $(n-1)^{st}$ stage of the construction. We now repeat the induction step with $G_{i1} = G^n_{i1}$, $i = 1,2,3$, $\varepsilon = \varepsilon_n$, $\eta = \varepsilon_{n+1} < \varepsilon_n$, $t_0 = t_{n0} = t_6$, where $\varepsilon_n \to 0$.

If A is in some G_{i1}^{k}, $i = 1,2,3$, then the construction implies $A/3$ is in G_{i1}^{n}, $i = 1,2,3$, $n > k$. We also have m normalized so that $\lim_{n\to\infty} m(G_{i1}^{n}) = 1/3$, $i = 1,2,3$. Denote $t_{nj} = t_j$, $0 \le j \le 6$, and $p_{nj} = p_j$, $j = 1,2,3$ for the nth stage of the construction.

We now assume $A/3$, $B/3$, and $C/3$ are in G_{i1}^{n}, $i = 1,2,3$. From (2) we obtain

$$(n,0) \qquad R(T,t,s,A,B,C) \ge \frac{1-\varepsilon_n}{27} \sum_{i=2}^{3} m(G_{i1}^{n})^{-2}, \quad t_{n0} \le t,\ s \le t_{n1}.$$

From (4) and (5) we obtain

$$(n,1) \qquad R(T,t,s,A,B,C) \ge \frac{1-\varepsilon_n}{27} \sum_{i=1,3} m(G_{i3}^{n})^{-2}, \quad t_{n1} \le t,\ s \le t_{n2}.$$

From (7) and (8) we obtain

$$(n,2) \qquad R(T,t,s,A,B,C) \ge \frac{1-\varepsilon_n}{27} \sum_{i=1}^{2} m(G_{i5}^{n})^{-2}, \quad t_{n2} \le t,\ s \le t_{n3}.$$

From (10)-(12) we obtain

$$(n,3) \qquad R(T,t,s,A,B,C) \ge \frac{1-\varepsilon_n}{729} \sum_{i=1}^{2} \sum_{j=1}^{3} m(g_{ij}^{n})^{-2}, \quad t_{n3} \le t,\ s \le t_{n4}.$$

From (14)-(16) we obtain

$$(n,4) \qquad R(T,t,s,A,B,C) \ge \frac{1-\varepsilon_n}{27}\Big[\sum_{j=1}^{3} (3m(g_{1j}^{n}))^{-2} + m(G_{36}^{n})^{-2}\Big], \quad t_{n4} \le t,\ s \le t_{n5}.$$

From the continuation of mixing in $G_{i1}^{n+1} = G_{i7}^{n}$, $i = 2,3$, we obtain

$$(n,5) \qquad R(T,t,s,A,B,C) \ge \frac{1-\varepsilon_n}{27} \sum_{i=2}^{3} m(G_{i7}^{n})^{-2}, \quad t_{n5} \le t,\ s \le t_{n6}.$$

If $A/3$, $B/3$, $C/3$ are in G_{i1}^{n}, $i = 1,2,3$, then $A/3$, $B/3$, $C/3$ are in G_{i1}^{n+1}, $i = 1,2,3$. By (16) of the induction step, we have $t_{(n+1)0} = t_{n6}$. In particular, mixing of A and B will continue in the $(n+1)^{st}$ stage. Since $\varepsilon_n \to 0$, (n,0)-(n,5) imply

$$(17) \qquad \liminf_{t\to\infty} m(T^t A \cap B) \ge \frac{2}{3} m(A)m(B).$$

By approximation, it follows that (17) holds for A,B,C in $\mathcal{B}$. Thus T is α-mixing for $\alpha \ge 2/3$. From (1) with $p_{1n} \to \infty$, it follows that $\alpha = 2/3$.

Let A be in G_{i1}^{k} for some $i = 1,2$, or 3. As in the previous Section 3, from (1), (3), and (6) with $p_{in} \to \infty$, $i = 1,2,3$, we obtain

$$(18) \qquad \lim_{n\to\infty} m(T^{p_{3n}}(T^{p_{2n}}(T^{p_{1n}}A \cap A^c) \cap A) \cap A^c) = 0.$$

By approximation, it follows from (18) that for any set A,

$$(19) \qquad \liminf_{n,k,\ell\to\infty} m(T^n(T^k(T^\ell A \cap A^c) \cap A) \cap A^c) = 0.$$

In particular, T is 0-3-mixing.

It remains to verify that T is 1/3-2-mixing. Let $I^n = [t_{n0}, t_{n6})$ and $I_i^n = [t_{ni-1}, t_{ni})$, $1 \le i \le 6$.

Let $\delta > 0$. Assume n is so large that $(1-\varepsilon_n)m(G^n_{ij}) > (1-\delta)/3$. Consider A,B,C such that $A/3$, $B/3$, and $C/3$ are in G^n_{i1}, $i = 1,2,3$. For $s,t \geq t_{n0}$ it will be shown that

$$R(T,t,s,A,B,C) \geq p(1-\delta),\ p \geq 1/3. \tag{20}$$

Now n is fixed and we denote $I_i = I^n_i$, $t_i = t_{ni}$, and $G_{ij} = G^n_{ij}$ for convenience. The towers $G_{i1}, G_{i2},\ldots$ in Figure 2 will be called the ith column in Figure 2. Mixing occurs in G_{ij} in $(n,0)$-$(n,5)$ for certain t,s in I^n. We refer to this as mixing in the ith column. We proceed by cases.

If $s,t \in I_i$, then $(n,i-1)$ implies (20) with $p = 2/3$, $1 \leq i \leq 6$. This is the simplest case.

If $s \in I_1$ and $t \in I_3$, then we apply mixing in the second column as follows. By 1-mixing in G_{22}, $T^sA/3 \cap B/3$ is essentially a union of levels in G_{24} and is close in measure to $m(A)m(B)/3$. The 1-mixing of levels in G_{24} by T^t implies (20) with $p = 1/3$. If $t \in I_1$ and $s \in I_3$, then we consider $T^sA/3 \cap (B/3 \cap T^{-t}C/3)$ in the same manner.

If $s,t \in I_1 \cup I_2$, then 2-mixing in the third column in G_{33} implies (20) with $p = 1/3$. If $s,t \in I_2 \cup I_3$, then 2-mixing in the first column in G_{15} implies (20) with $p = 1/3$.

If $s,t \in I_i$, $i = 4$, 5, or 6, then separate 2-mixing in the first column in G_{18} implies (20) with $p = 1/3$.

If $s \leq t_3$ and $t \in I_4$, then 2/9-2-mixing occurs in both G_{16} and G_{26}. Hence (20) holds with $p = 4/9$. If $s \leq t_3$ and $t \in I_5 \cup I_6$, then 2/9-2-mixing occurs in both G_{17} and G_{37}. If $t \leq t_3$ and $s \in I_i$, $i = 4$, 5, or 6, then consider $T^sA \cap (B \cap T^{-t}C)$ in a similar manner to obtain (20) with $p = 4/9$.

If $s \in I^n$ and $t \in I^j$, $j > n$, then 2/3-1-mixing in the nth and j^{th} stages implies

$$\begin{aligned} m(T^t(T^sA \cap B) \cap C) &\geq \tfrac{2}{3}(1-\varepsilon_j)m(T^sA \cap B)m(C) \\ &\geq \tfrac{2}{3}(1-\varepsilon_j)\tfrac{2}{3}(1-\varepsilon_n)m(A)m(B)m(C). \end{aligned}$$

Thus we obtain (20) with $p = 4/9$. If $t \in I^n$ and $s \in I^j$, then consider $T^sA \cap (B \cap T^{-t}C)$ in the same manner.

If $s,t \in I^{n+1}$, then we have the same situation as above since $A/3$, $B/3$, and $C/3$ are in G^{n+1}_{1i}, $1 \leq i \leq 3$. Thus induction yields (20). Since $\varepsilon_n \to 0$, we obtain 1/3-2-mixing for A,B,C that are in some G^k_{1i}, $i = 1,2,3$. By approximation, we obtain 1/3-2-mixing for A,B,C in $\mathcal{B}$.

5. PARTIAL k-MIXING AND ZERO (k+1)-MIXING. Let k be a positive integer. A transformation will be constructed that is $(k+1-j)/(k+1)$-j-mixing,

$1 \le j \le k+1$. At each stage in the construction we will alternately mix $k/(k+1)$ of the space and introduce periodicity in the other $1/(k+1)$ of the space. The following diagram illustrates the induction step.

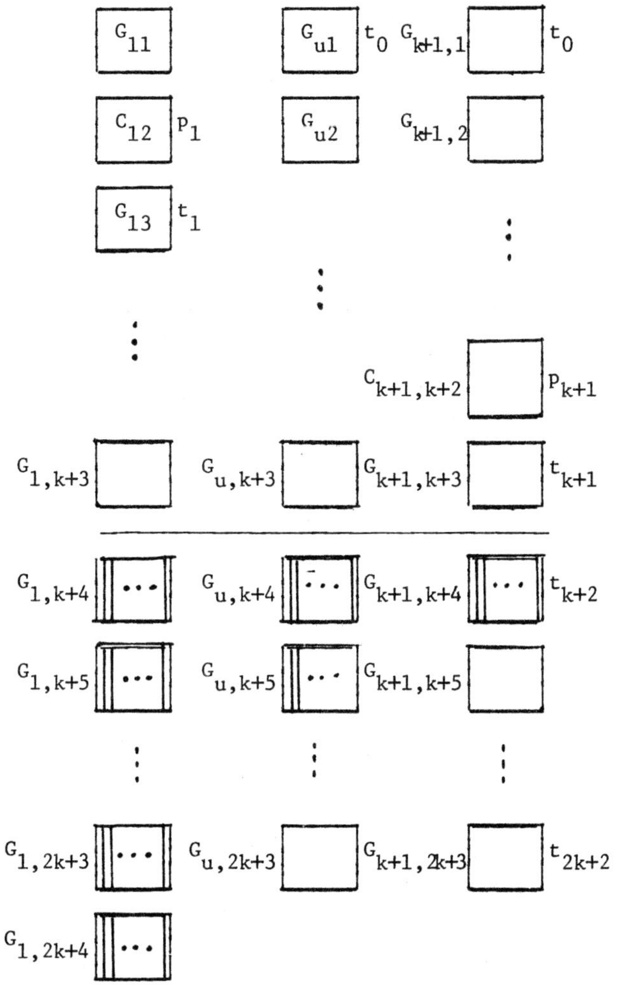

Figure 3

The induction step begins with $k+1$ disjoint M-towers G_{i1}, $1 \le i \le k+1$, of equal measure. We have $\varepsilon > 0$ and t_0 such that if $T = T(G_{i1})$, $2 \le i \le k$, then

(0) $$T^t \ (1-\varepsilon)\text{-}k\text{-mixes levels in } G_{i1}, \ t \ge t_0.$$

Apply Lemma 2.1 with $G = G_{11}$ to obtain $p_1 = p$ and $C_{12} = C$. If T extends T_{12} and I is a level in G_{11}, then

(1) $$m(T^t I \cap I) > (1-\varepsilon)m(I), \ t = p_1.$$

Lemma 2.2(a) is now applied for $k = 2$ with $C = C_{12}$. We obtain G_{13} and $t_1 \geq t^*$, where $t_1 = t_0 + 1$ if $t^* \leq t_0$.

We now proceed inductively. Assume we have increasing t_j, $0 \leq j \leq u < k+1$. We also have $G_{i,u+1}$, $1 \leq i \leq u-1$, $G_{u,u+2}$, and $G_{i,u}$, $u+1 \leq i \leq k+1$. If $T = T(G_{u,u+2})$, then

(2) $$T^t \text{ (1-}\varepsilon\text{)-k-mixes levels in } G_{u,u+2},\ t \geq t_u.$$

We choose r so large that if $G_{i,u+2} = S^r G_{i,u+1}$, $1 \leq i \leq u-1$, and T extends $T_{i,u+2}$, then for $1 \leq i \leq u-1$,

(3) $$T^t \text{ (1-}\varepsilon\text{)-k-mixes levels in } G_{i,u+1},\ t_{u-1} \leq t \leq t_u.$$

We also choose r so large that if $G_{i,u+1} = S^r G_{iu}$, $u+1 \leq i \leq k+1$, and T extends $T_{i,u+1}$, then for $u+1 \leq i \leq k+1$,

(4) $$T^t \text{ (1-}\varepsilon\text{)-k-mixes levels in } G_{i,u},\ t_{u-1} \leq t \leq t_u.$$

We now apply Lemma 2.1 to $G = G_{u+1,u+1}$ to obtain $p_{u+1} = p$ and $C_{u+1,u+2}$. If T extends $T_{u+1,u+2}$ and I is a level in $G_{u+1,u+1}$, then

(5) $$m(T^t I \cap I) > (1-\varepsilon)m(I),\ t = p_{u+1}.$$

Lemma 2.2(a) is now applied with $C = C_{u+1,u+2}$. We obtain $G_{u+1,u+3}$ and $t_{u+1} \geq t^*$, where $t_{u+1} = t_u + 1$ if $t^* \leq t_u$.

Proceeding by induction, we obtain t_{k+1} and $G_{j,k+3}$, $1 \leq j \leq k+1$. Form $k+1$ equal copies $G_{j,k+3}/(k+1)$ for each j denoted by $g_{ij} = G_{j,k+3}/(k+1)$, $1 \leq i \leq k+1$. Let $G_{i,k+4} = (g_{i1},\ldots,g_{ik+1})$, $1 \leq i \leq k+1$.

Let $0 < \eta < \varepsilon$. Since $G_{k+1,k+4}$ is an M-tower, there exists $t_{k+2} > t_{k+1}$ such that if $T = T(G_{k+1,k+4})$, then

(6) $$T^t \text{ (1-}\varepsilon\text{)-k-mixes levels in } G_{k+1,k+4},\ t \geq t_{k+2}.$$

We also continue mixing separately in g_{ij}, $1 \leq i \leq k$, $1 \leq j \leq k+1$. Let r be a positive integer and let $G_{i,k+5} = (S^r g_{i1},\ldots,S^r g_{ik+1})$, $1 \leq i \leq k$.

We can now choose r so large that if T extends $T_{i,k+5}$, then

(7) $$T^t \text{ (1-}\eta\text{)-k-mixes } I/(k+1) \text{ in } g_{ij} \text{ for}$$
$$I \text{ in } G_{j,j+2},\ t_{k+1} \leq t \leq t_{k+2},\ 1 \leq i \leq k,\ 1 \leq j \leq k+1.$$

We now proceed inductively. Assume we have increasing t_j, $0 \leq j \leq u < 2k+2$. Let $v = 2k + 3 - u$. If $T = T(G_{v,u+2})$, then

(8) $$T^t \text{ (1-}\eta\text{)-k-mixes levels in } G_{v,u+2},\ t \geq t_u.$$

Now t_u tells us how long to continue to mix $G_{i,u+1}$ for $v < i \leq k+1$. Choose r so large that if $G_{i,u+2} = S^r G_{i,u+1}$ and if T extends $T_{i,u+2}$, $v < i \leq k+1$, then

(9) $$T^t \text{ (1-}\eta\text{)-k-mixes levels in } G_{i,5+2k-1},\ t_{u-1} \leq t \leq t_u.$$

Also t_u tells us how long to continue to mix $G_{i,u+2}$, $1 \le i \le v$. We have s such that $G_{i,u+2} = (S^s g_{i1},\dots,S^s g_{ik+1})$. Choose r so large that if $G_{i,u+3} = (S^{r+s}g_{i1},\dots,S^{r+s}g_{ik+1})$ and T extends $T_{i,u+3}$, then for $1 \le i < v$,

(10) $\qquad T^t$ $(1-\varepsilon)$-k-mixes $I/(k+1)$ in g_{ij} for I in $G_{j,j+2}$, $1 \le j \le k+1$, $t_{u-1} \le t \le t_u$.

Now let $w = v - 1$. Since $G_{w,u+3}$ is an M-tower there exists $t_{u+1} > t_u$ such that if $T = T(G_{w,u+3})$, then

(11) $\qquad T^t$ $(1-\eta)$-k-mixes levels in $G_{w,u+3}$, $t \ge t_{u+1}$.

Proceeding by induction, we obtain $G_{1,2k+4}$ and $G_{i,2k+3}$, $2 \le i \le k+1$. Choose $t_{2k+2} > t_{2k+1}$ so that if $T = T(G_{i,2k+3})$, $2 \le i \le k+1$, then

(12) $\qquad T^t$ $(1-\eta)$-k-mixes levels in $G_{i,2k+3}$, $t > t_{2k+2}$.

This completes the induction step. The M-towers $G_{1,2k+4}$ and $G_{i,2k+3}$, $2 \le i \le k+1$, play the role of $G_{i,1}$, $1 \le i \le k+1$, respectively, in the next state, where ε is replaced by η and t_0 is replaced by t_{2k+2}. Note that separate mixing is continued in the first column until t_{2k+2}; hence (10) holds for $i = 1$ and $u = 2k+2$.

We begin with an M-tower G. Divide G into $k+1$ disjoint copies $G^1_{i1} = G/(k+1)$, $1 \le i \le k+1$. Let $\varepsilon_1 > 0$. Apply Lemma 2.0 with $\varepsilon = \varepsilon_1$ and $G = G^1_{i1}$ to obtain t_{10} such that if $T = T(G^1_{i1})$, $2 \le i \le k+1$, then

(1,0) $\qquad T^t$ $(1-\varepsilon_1)$-k-mixes levels in G^1_{i1}, $t \ge t_{10}$.

Now apply the induction step where $G_{i1} = G^1_{i1}$, $1 \le i \le k+1$. Let $\eta = \varepsilon_2 < \varepsilon_1$. We obtain $G^2_{11} = G_{1,2k+4}$ and $G^2_{i1} = G_{i,2k+3}$, $2 \le i \le k+1$.

In general, let $G^n_{11} = G_{1,2k+4}$ and $G^n_{i1} = G_{i,2k+3}$, $2 \le i \le k+1$, be the corresponding towers at the end of the $(n-1)$st stage of the construction. Now repeat the induction step with $G_{i1} = G^n_{i1}$, $1 \le i \le k+1$, $\varepsilon = \varepsilon_n$, and $\eta = \varepsilon_{n+1} < \varepsilon_n$, where $\varepsilon_n \to 0$.

If A is in some G^j_{i1}, $1 \le i \le k+1$, then the construction implies $A/(k+1)$ is in G^n_{i1}, $1 \le i \le k+1$, for $n > j$. Also m is normalized so that $\lim_{n\to\infty} m(G^n_{i1}) = 1/(k+1)$. Denote $t_{nj} = t_j$, $0 \le j \le 2k+2$, and $p_{nj} = p_j$, $1 \le j \le k+1$, for the nth stage of the construction. Let $I^n = [t_0, t_{2k+2})$, $I_j = [t_{j-1}, t_j)$, $1 \le j \le 2k+2$, and $G_{ij} = G^n_{ij}$. Let $\delta > 0$ and assume n is so large that $(1-\varepsilon_n)m(G_{ij}) > (1-\delta)/(k+1)$.

To verify $k/(k+1)$-mixing, we consider A and B such that $A/(k+1)$ and $B/(k+1)$ are in G_{i1}, $1 \le i \le k+1$. Assume $t_0 \le t \le t_{k+1}$. Hence $t \in I_a$ for one a, $1 \le a \le k+1$. Thus $t \notin I_i$ for k i's. We refer to mixing in $G_{i1}, G_{i2}, \dots$ as <u>mixing in the ith column</u>. From (3) and (4) with $u + 1 = a$ we conclude T^t is mixing in k columns; hence

(13) $$m(T^t A \cap B) \geq (1-\varepsilon_n)km(A)m(B)/(k+1)^2 m(G_{i1})$$

$$> (1-\delta)km(A)m(B)/(k+1).$$

The same analysis holds for $t_{k+2} \leq t \leq t_{2k+2}$, where (9) and (10) are applied to obtain mixing in the ith column for k columns. Now $A/(k+1)$ in G_{i1}^n implies $A/(k+1)$ is in G_{i1}^{n+1}, $1 \leq i \leq k+1$, and similarly for B. Thus by induction (13) holds for $t \geq t_{n0}$; hence

(14) $$\liminf_{t\to\infty} m(T^t A \cap B) \geq km(A)m(B)/(k+1).$$

Approximation of sets in $\mathcal{B}$ by sets in G_{i1}^n, $1 \leq i \leq k+1$, $n \geq 1$, yields (14) for $A,B \in \mathcal{B}$. Thus T is α-mixing for $\alpha \geq k/(k+1)$. From (1) with $p_{n1} \to \infty$, it follows that $\alpha = k/(k+1)$.

We can extend the method in Section 3 to verify

(15) $$\lim_{n\to\infty} m(T^{p_{k+1,n}}(\ldots(T^{p_{1n}}A \cap A^c) \cap A)\ldots) \cap B) = 0,$$

where $B = A^c$ if k is even and $B = A$ if k is odd. In particular, T is 0-(k+1)-mixing.

We now consider ℓ, $1 < \ell < k+1$, and will verify that T is $(k+1-\ell)/(k+1)$-ℓ-mixing. Fix sets A_j, $0 \leq j \leq \ell$, such that $A_j/(k+1) \in G_{1i}^n$, $1 \leq i \leq k+1$. Let $\delta > 0$ and assume n is so large that

(16) $$(1-\varepsilon_n)/(k+1)^\ell m(G_{11}^n)^\ell > (1-\delta).$$

Fix ℓ integers $s_j \geq t_0$ and first assume $s_1 \leq s_2 \leq \cdots \leq s_\ell \leq t_{k+1}$. There are at least $k+1 - \ell$ i's, $1 \leq i \leq k+1$, such that $s_j \notin I_i$ for all ℓ j's. By (3) and (4) we have $(1-\varepsilon_n)$-ℓ-mixing in the ith columns for these $k+1-\ell$ i's for the sets $A_j/(k+1)$, $0 \leq j \leq \ell$. Hence (3), (4), and (16) imply

(17) $$R(T,s_j,A_j,0 \leq j \leq \ell) > (1-\delta)(k+1-\ell)/(k+1).$$

If $t_{k+1} < s_1 \leq s_2 \leq \cdots \leq s_\ell < t_{2k+2}$, then there are at least $k+1-\ell$ i's, $k+2 \leq i \leq 2k+2$, such that $s_j \notin I_i$ for all ℓ j's. By (9) and (10) we obtain (17) as in the preceding case.

Now assume $s_a \leq t_{k+1} < s_{a+1}$ and $s_\ell \leq t_{2k+2}$. Denote

(18) $$B_a = T^{s_a}(\ldots(T^{s_1}A_1 \cap A_0)\ldots) \cap A_a,$$

hence (17) with $\ell = a$ implies

(19) $$m(B_a) > (1-\delta)(k+1-a)\prod_{i=0}^{a} m(A_i)/(k+1)$$

Now B_a appears as $B_a/(k+1)$ in the $k+1$ towers G_{ik+4}, $1 \leq i \leq k+1$. There are at least $k+1-(\ell-a)$ i's, $k+2 \leq i \leq 2k+2$, such that $s_j \notin I_i$ for $a+1 \leq j \leq \ell$. From (19), (9), (10), and (18) with a replaced by ℓ, we obtain

(20) $$m(B_\ell) > (1-\delta)^2(k+1-a)(k+1-\ell+a)\prod_{i=0}^{\ell} m(A_i)/(k+1)^2 .$$

Thus we need to check that

(21) $$\frac{(k+1-a)(k+1-\ell+a)}{(k+1)^2} > \frac{(k+1-\ell)}{k+1} .$$

Now (21) reduces to $a\ell - a^2 > 0$, which holds since $a < \ell$. Thus (20) implies

(22) $$R(T,s_j,A_j,0 \le j \le \ell) > (1-\delta)^2(k+1-\ell)/(k+1).$$

This is the estimate in this case.

Next, suppose the s_j's are increasing but some $s_j > t_{2k+2}$. Thus we need to consider I^p for $p > n$. Assume there are R p_r's, $1 \le r \le R$, with a_r s_j's in I^{p_r}, $1 \le r \le R$; hence $a_1 + a_2 + \cdots + a_R = \ell$.

In (18) denote $B^r = B_{a_r}$, $1 \le r \le R$. The above case (22), where all s_j are in the same I^r, implies

(23) $$R(T,s_j,A_j,0 \le j \le a_1) > (1-\delta)^2(k+1-a_1)/(k+1).$$

Now B^1 appears as a union of levels in the towers corresponding to I^{p_2}. From (22) and (23) we obtain

(24) $$R(T,s_j,A_j,0 \le j \le a_2) > (1-\delta)^4 \prod_{i=1}^{2} (k+1-a_i)/(k+1).$$

Proceeding inductively, we obtain

(25) $$R(T,s_j,A_j,0 \le j \le \ell) > (1-\delta)^{2R} \prod_{i=1}^{R} (k+1-a_i)/(k+1).$$

From (25) with $z = k + 1$, it suffices to check that

(26) $$\prod_{i=1}^{R} (z-a_i)/z > (z-\ell)/z.$$

Now (26) can be written as

(27) $$\prod_{i=1}^{R} (z-a_i) > z^{R-1}(z-\ell),$$

where $\ell = a_1 + \cdots + a_R$. Induction on r implies

$$\begin{aligned}\prod_{i=1}^{r-1} (z-a_i)(z-a_r) &> z^{r-2}(z-(\ell-a_r))(z-a_r) \\ &= z^{r-1}(z-\ell)+z^{r-2}a_r(\ell-a_r) \\ &> z^{r-1}(z-\ell).\end{aligned}$$

Hence (25) and (26) imply

(28) $$R(T,s_j,A_j,0 \le j \le \ell) > (1-\delta)^{2R}(k+1-\ell)/(k+1).$$

This is the estimate for the general case of increasing s_j's. Since $\delta > 0$ is arbitrary, it follows that T is $(k+1-\ell)/(k+1)$-ℓ-mixing.

The case when the s_j's are not increasing can be reduced to the increasing case by extending the application of T^{-1} to certain powers as in writing

$$m(T^s(T^tA \cap B) \cap C) = m(T^tA \cap D),$$

where $D = B \cap T^{-s}C$ and $s < t$.

For another example, we write

$$T^{t_0}(T^{t_{k+2}}(T^{t_2}A \cap B) \cap C) \cap D = T^{t_{k+2}}E \cap F,$$

where $E = T^{t_2}A \cap B$ and $F = C \cap T^{-t_0}D$. Now $m(E) \geq km(A)m(B)/(k+1)$ and $m(F) \geq km(C)m(D)/(k+1)$. Hence it follows that

$$m(T^{t_{k+2}}E \cap F) \geq km(E)m(F)/(k+1)$$

$$\geq \left(\frac{k}{k+1}\right)^3 m(A)m(B)m(C)m(D).$$

Lastly $(k/k+1)^3 > (k+1-3)/(k+1)$. In general, we apply T^{-1} successively to reduce to the case of increasing s_j's. The details will be omitted.

6. INDUCING PARTIALLY MIXING AND PARTIALLY RIGID. A transformation T is rigid if there exists an increasing sequence of positive integers (k_n) such that

(6.1) $$\lim_{n\to\infty} m(T^{k_n}B \cap B) = m(B), \quad B \in \mathcal{B}.$$

A transformation is mild mixing [8] if it does not admit any rigid factors. Partial mixing implies mild mixing.

A transformation T will be called partially rigid if there exists an increasing sequence (k_n) and γ, $0 < \gamma \leq 1$, such that

(6.2) $$\liminf_{n\to\infty} m(T^{k_n}B \cap B) \geq \gamma m(B), \quad B \in \mathcal{B}.$$

A transformation T is η-rigid, $0 < \eta \leq 1$, if there exists (k_n) such that (6.2) holds for $\gamma = n$ but there does not exist (k_n) such that (6.2) holds for $\gamma = \eta + \varepsilon$, $\varepsilon > 0$. The α-mixing transformations constructed in [6] are $(1-\alpha)$-rigid. Thus they are mild mixing and partially rigid.

The following lemma will be useful for verifying that a transformation is α-mixing and $(1-\alpha)$-rigid.

(6.3) LEMMA. *If T is β-mixing and η-rigid, where $\beta \geq \alpha$ and $\eta \geq 1 - \alpha$, then $\beta = \alpha$ and $\eta = 1 - \alpha$.*

PROOF. Since T is β-mixing, $B \in \mathcal{B}$ implies

(1) $$\liminf_{n\to\infty} m(T^nB \cap B^c) \geq \beta m(B)(1-m(B)).$$

Suppose $\eta = (1-\beta) + \delta$, $\delta > 0$. Hence for each set B there exist arbitrarily large n such that

(2) $$m(T^nB \cap B) \geq ((1-\beta) + \delta)m(B).$$

From (1) and (2) we conclude that for $B \in \mathcal{B}$ there exists n such that

$$(3) \qquad m(B) = m(T^n B) \geq m(B)(1+\delta-\beta m(B)).$$

If $m(B) < \delta/\beta$, then (3) is a contradiction. Thus $\delta \leq 0$; hence $\eta \leq 1 - \beta$. Since $\beta \geq \alpha$, we have $\eta \leq 1 - \alpha$ so $\eta = 1 - \alpha$. Therefore $1 - \alpha \leq 1 - \beta$ implies $\alpha \geq \beta$ so $\alpha = \beta$.

A transformation that is α-mixing and $(1-\alpha)$-rigid will simply be called an *α-transformation*. The transformations T_k that are $(k+1-j)/(k+1)$-j-mixing, $0 \leq j \leq k+1$, are α-transformations for $\alpha = k/(k+1)$. One can use (p_{1n}) in (6.2) with $\gamma = 1/(k+1)$ since $\varepsilon_n \to 0$. These transformations have rank $k+1$. This follows from the observation that the levels in the columns $C^n_{i,i+1}$, $1 \leq i \leq k+1$, $n \geq 1$, generate $\mathcal{B}$. The equivalence theory [10] therefore implies each T_k induces all T_j, $j \geq 1$. In general, [10] implies the following result.

(6.4) THEOREM. *If T is an ergodic loose Bernoulli transformation with zero entropy, then T induces T_k, $k \geq 1$.*

(6.5) COROLLARY. *Every ergodic loose Bernoulli transformation with zero entropy induces transformations that are partially k-mixing but not partially k+1-mixing,* $k \geq 1$.

The construction in Section 5 can be modified to obtain finite rank α-transformations for $0 < \alpha < 1$. It has also been pointed out by A. del Junco that one can obtain rank one α-transformations by modifying Ornstein's rank one mixing construction [9]. The modification consists in adding levels to an α fraction of the columns at each stage in the construction, rather than adding levels to all columns. The heights of the levels added at stage n are also chosen to be uniformly distributed with respect to the length of n-2-names. The existence of finite rank α-transformations and [10] imply the following result.

(6.6) THEOREM. *Every ergodic loose Bernoulli transformation with zero entropy induces α-transformations,* $0 < \alpha < 1$.

The rank one α-transformations obtained by modifying Ornstein's construction are α_k-k-mixing, where $\alpha_k > 0$ and $\lim_{k\to\infty} \alpha_k = 0$. Examples of transformations that are partially mixing of all orders will be discussed in a subsequent paper.

The transformation T_k has rank $k+1$ and therefore Theorem 3.2 [1] implies the spectral multiplicity of T_k is at most $k+1$. We do not know if the spectral multiplicity is $k+1$. Transformations with spectral multiplicity k were recently constructed in [11] for each $k \geq 1$.

REFERENCES

[1] Chacon, R.V., Approximation and spectral multiplicity, Contributions to Ergodic Theory and Probability, Springer Lecture Notes 160(1970), 18-27.

[2] Friedman, N.A., Introduction to Ergodic Theory, Van Nostrand Reinhold, New York, 1970.

[3] Friedman, N.A., Bernoulli shifts induce Bernoulli shifts, Advances in Math. 10(1973), 39-48.

[4] Friedman, N.A., Mixing on sequences, to appear in Canadian J. Math.

[5] Friedman, N.A., and D.S. Ornstein, On partial mixing transformations, Indiana University Math. J. 20, (1971), 767-775.

[6] Friedman, N.A., and D.S. Ornstein, On mixing and partial mixing, Illinois J. Math 16(1972), 61-68.

[7] Friedman, N.A., and D.S. Ornstein, Ergodic transformations induce mixing transformations, Advances in Math. 10(1973), 147-163.

[8] Furstenberg, H., and B. Weiss, The finite multipliers of infinite ergodic transformations, Springer Lecture Notes 668(1977), 127-132.

[9] Ornstein, D.S., On the root problem in ergodic theory, Proc. Sixth Berkeley Symp. Math. Stat. Prob. (1970), 347-356.

[10] Ornstein, D.S., Rudolph, D.J., and B. Weiss, Equivalence of Measure Preserving Transformations, A.M.S. Memoirs 262 (1982).

[11] Robinson, E.A., Ergodic measure preserving transformations with arbitrary finite spectral multiplicities, to appear.

[12] Shields, P., Cutting and independent stacking of intervals, Math. Systems Theory 7 (1973), 1-4.

DEPARTMENT OF MATHEMATICS AND STATISTICS
STATE UNIVERSITY OF NEW YORK AT ALBANY
1400 WASHINGTON AVENUE
ALBANY, NEW YORK 12222

Contemporary Mathematics
Volume **26**, 1984

IP-SYSTEMS IN ERGODIC THEORY

by

Hillel Furstenberg

INTRODUCTION

Many aspects of the theory of dynamical systems and ergodic theory can be placed in the framework of general group actions on topological spaces and measure spaces. Some of the theory also goes over to semigroups of transformations on these spaces. Here we shall discuss a further generalization, where the family of transformations acting on a space forms what we shall call an "IP-system". The purpose of this is not generalization for its own sake; rather it appears that for certain questions, IP-systems form a natural framework.

IP-systems of transformations possess two important features. One can define the notion of a limit "in the IP-sense" for such systems, and when it exists for an IP-system of transformations, the limit is an idempotent operator. Hence the name "IP". In the measure theoretic framework this idempotent operator plays a role similar to that of the ergodic average. In the topological dynamics setup, convergent IP-systems lead to points of recurrence. The second crucial feature is that in many situations a given IP-system will possess some convergent subsystem, giving rise to an idempotent limit operator. This is related to the following striking property of IP-systems: no matter how an IP-system is partitioned into finitely many subsets, one of these again contains an IP-system.

For an IP-system of measure preserving transformations one has an analogue of pure point spectrum and of weak mixing. The corresponding properties will be called "rigidity" and "mixing". These notions have implications for ordinary ergodic systems generated by a single transformation T by restricting to the various IP-sub-systems of the semigroup $\{T^n\}$. Typically the IP-theory leads to assertions that can be made regarding T^n for "every IP-subset of integers n."

1. SCHUR'S LEMMA AND HINDMAN'S THEOREM.

In a paper dealing with the Fermat Theorem [7], Schur proved a combina-

0271-4132/84 $1.00 + $.25 per page

torial lemma which can be formulated as follows:

SCHUR'S LEMMA: <u>If the natural numbers</u> $\mathbb{N}$ <u>are partitioned into finitely many sets,</u> $\mathbb{N} = C_1 \cup C_2 \cup \cdots \cup C_r$, <u>then one of these sets contains numbers</u> x,y,z <u>satisfying</u> $z = x+y$.

In fact Schur proved a "finite" version of this in which a sufficiently large interval $\{1,2,3,\ldots,T\}$ is partitioned into a fixed number of sets. From this he proved that for given m, if p is a sufficiently large prime, then one can always solve in a non-trivial way the congruence $X^m + Y^m \equiv Z^m \pmod p$.

Graham and Rothschild conjectured a far reaching generalization of this and their conjecture was established by Hindman [6]. In this conjecture appears the notion of an IP-set.

DEFINITION 1: Let S be a commutative semigroup. An IP-<u>subset</u> of S consists of a sequence of (not necessarily distinct) elements $\sigma_i \in S$, together with all products $\sigma_{i_1}\sigma_{i_2}\cdots\sigma_{i_k}$ that can be formed using distinct indices $i_1 < i_2 < \cdots < i_k$.

For example, if S is the additive group of integers, the set of those numbers having only 0's and 1's in their decimal expansion forms an IP-set, the generators of the set being $\{1,10,10^2,\ldots,10^i,\ldots\}$.

If α denotes a finite subset of $\mathbb{N} = \{1,2,3,\ldots\}$, $\alpha = \{i_1,i_2,\ldots,i_k\}$, we denote by σ_α the product $\sigma_{i_1}\sigma_{i_2}\cdots\sigma_{i_k}$. An IP-subset generated by $\{\sigma_i\}$ contains all σ_α as α ranges over finite subsets of $\mathbb{N}$. If $\alpha \cap \beta = \phi$, then together with σ_α, σ_β, also the product $\sigma_\alpha\sigma_\beta = \sigma_{\alpha\cup\beta}$ lies in the IP-set. In general $\sigma_\alpha\sigma_\beta$ need not belong to the set and so an IP-set need not be a semigroup. On the other hand every countable semigroup is an IP-set. If each generator σ_i of an IP-set occurs infinitely often in the sequence $\{\sigma_j\}$, then the IP-set generated is a semigroup.

We can now state Hindman's theorem.

THEOREM 1.1: <u>If the natural numbers</u> $\mathbb{N}$ <u>are partitioned into finitely many sets,</u> $\mathbb{N} = C_1 \cup C_2 \cup \cdots \cup C_n$, <u>then one of these sets contains an IP-set</u>.

Clearly this contains Schur's result. Both results are reminiscent of van der Waerden's theorem regarding arithmetic progressions, (if $\mathbb{N} = C_1 \cup C_2 \cup \cdots \cup C_n$, some C_i contains arbitrarily long arithmetic progressions). Note however that unlike van der Waerden's theorem, Hindman's result assures the existence of an infinite configuration of a certain type in some cell of an arbitrary partition. For a proof of Hindman's theorem see [5], [6], or [3]. The last of these gives a proof based on topological dynamics considerations.

Every IP-set contains many other IP-sets. Let $\{\sigma_\alpha\}$ be an IP-set, where α ranges over the finite subsets of $\mathbb{N}$. If $\alpha_1,\alpha_2,\alpha_3,\ldots$ is a sequence of disjoint finite subsets in $\mathbb{N}$, and we set $\sigma'_i = \sigma_{\alpha_i}$, then the σ'_i generate an IP-subset of $\{\sigma_\alpha\}$, since $\sigma'_{i_1}\sigma'_{i_2}\cdots\sigma'_{i_k} = \sigma_{\alpha_{i_1}\cup\alpha_{i_2}\cup\cdots\cup\alpha_{i_k}}$ for $i_1 < i_2 <\cdots< i_k$. This procedure for forming IP-subsets will be used frequently.

LEMMA 1.2: Let $m \neq 0$ be an integer. Any IP-subset of $\mathbb{Z}$ contains an IP-subset consisting of integers divisible by m.

PROOF: Let the IP-subset be generated by $\{s_i\}$. We can find infinitely many s_i in the same congruence class, say a, modulo m. Let $\{\alpha_n\}$ be disjoint subsets with m elements in each, each of which consists only of indices j for which $s_j \equiv a$. Then $s_{\alpha_1}, s_{\alpha_2}, \ldots, s_{\alpha_n}, \ldots$ generate an IP-subset of $m\mathbb{Z}$. □

Using this lemma we can reformulate Hindman's theorem as a combinatorial theorem relating to the family of all finite subsets $\alpha \subset \mathbb{N}$. Let us denote by $\mathcal{F}$ the set of all finite subsets of $\mathbb{N}$.

THEOREM 1.3: If $\mathcal{F}$ is partitioned into finitely many sets, $\mathcal{F} = C_1 \cup C_2 \cup\cdots\cup C_r$, then there exists a sequence of disjoint sets, $\alpha_1, \alpha_2, \ldots, \alpha_n, \ldots \in \mathcal{F}$ such that for some j, all finite unions $\alpha_{i_1} \cup \alpha_{i_2} \cup\cdots\cup \alpha_{i_k}$ belong to the same C_j.

PROOF: To each set $\alpha = (\ell_1, \ell_2, \ldots, \ell_k) \in \mathcal{F}$ we associate the integer $\varphi(\alpha) = 2^{\ell_1-1} + 2^{\ell_2-1} +\cdots+ 2^{\ell_k-1}$. φ establishes a 1-1 correspondence between $\mathcal{F}$ and $\mathbb{N}$, and we obtain a partition $\mathbb{N} = \varphi(C_1) \cup \varphi(C_2) \cup\cdots\cup \varphi(C_r)$. By Hindman's theorem some $\varphi(C_j)$ contains an IP-set Q. We shall show that C_j contains the desired array of sets $\{\alpha_{i_1} \cup \alpha_{i_2} \cup\cdots\cup \alpha_{i_k}\}$. First let us remark that for any IP-subset Q of $\mathbb{N}$, if $x_1, x_2, \ldots, x_t \in Q$, the set $Q(x_1, x_2, \ldots, x_t) = \{y : x_i + y \in Q,\ i = 1,2,\ldots,t\}$ contains an IP-set, namely the IP-set generated by those generators of Q not occurring in any of the x_i. Now let α_1 be any element in C_j and let $m_1 = 2^{n_1}$ be some power of 2 larger than $\varphi(\alpha_1)$. If α_2 is chosen so that $\varphi(\alpha_2) \in Q(\varphi(\alpha_1)) \cap m_1\mathbb{Z}$, then α_2 will be disjoint from α_1, $\alpha_2 \in \varphi^{-1}(Q) \subset C_j$, and $\alpha_1 \cup \alpha_2 \in C_j$ as well, since $\varphi(\alpha_1 \cup \alpha_2) = \varphi(\alpha_1) + \varphi(\alpha_2)$, α_1 and α_2 being disjoint, and $\varphi(\alpha_1) + \varphi(\alpha_2) \in Q$. We continue in this way inductively, choosing $\varphi(\alpha_n)$ to be divisible by a sufficiently high power of 2 so that α_n will be disjoint from $\alpha_1, \alpha_2, \ldots, \alpha_{n-1}$, furthermore choosing $\varphi(\alpha_n)$ inside $Q(\{\varphi(\alpha_{i_1} \cup \alpha_{i_2} \cup\cdots\cup \alpha_{i_k}) : i_1 < i_2 <\cdots< i_k \leq n-1\})$, thereby ensuring that $\varphi(\alpha_{i_1} \cup \alpha_{i_2} \cup\cdots\cup \alpha_n) = \varphi(\alpha_{i_1} \cup \alpha_{i_2} \cup\cdots\cup \alpha_{i_k}) + \varphi(\alpha_n) \in Q$, for $i_1 < i_2 <\cdots< i_k < n$. This procedure enables us to construct the sequence $\{\alpha_n\}$ as asserted. □

This set-theoretic version of Hindman's theorem implies Theorem 1.1. In fact we have more.

THEOREM 1.4: Let Q be an IP-subset of a semigroup S. If Q is partitioned

<u>into finitely many sets</u>, $Q = C_1 \cup C_2 \cup \cdots \cup C_r$, <u>then some</u> C_j <u>contains an IP-set</u>.

PROOF: Say $Q = \{\sigma_\alpha\}_{\alpha \in \mathcal{F}}$. The given partition of Q determines a covering of $\mathcal{F}$ by not necessarily disjoint sets, $\mathcal{F} = C_1^* \cup C_2^* \cup \cdots \cup C_r^*$, where $C_i^* = \{\alpha : \sigma_\alpha \in C_i\}$. Since the C_i^* could be cut down to form a partition of $\mathcal{F}$, Theorem 1.3 applies and we can find $\alpha_1, \alpha_2, \ldots, \alpha_n, \ldots \in \mathcal{F}$ which are mutually disjoint, and for which all $\alpha_{i_1} \cup \alpha_{i_2} \cup \ldots \cup \alpha_{i_k}$, $i_1 < i_2 < \cdots < i_k$, are in the same C_j^*. If we set $\sigma_i' = \sigma_{\alpha_i}$ we find that $\{\sigma_i'\}$ generate an IP-subset of Q contained in C_j. □

Theorem 1.4 shows that if a set does <u>not</u> contain an IP-set then it is in some sense a "null set", for the union of any finite number of these will never contain an IP-set. Other such notions in various contexts are: sets of measure 0, sets of first category, sets of integers of density 0, finite subsets of an infinite set. It is often useful to introduce the dual notion of a set whose complement is a null set.

DEFINITION 2. A subset of a semigroup S is an <u>IP*-set</u> if it has a non-empty intersection with every IP-subset of S.

Lemma 1.2 shows that the ideal $m\mathbb{Z}$, $m \neq 0$, is an IP*-subset of $\mathbb{Z}$. Some more examples of IP*-sets will occur in the course of our discussion.

The following properties of IP*-sets are readily deduced from the definition and from Theorem 1.4.

LEMMA 1.5. <u>The intersection of an IP*-set and an IP-set contains an IP-set. The intersection of finitely many IP*-sets is an IP-set</u>.

On the other hand, the intersection of two IP-sets may be empty. If Q_a, $a = 1,2,\ldots,q$, is the IP-set of integers generated by $\{a, a.10, \ldots, a.10^n, \ldots\}$ then clearly $Q_a \cap Q_b = \phi$ for $a \neq b$. Note that 0 is not a member of any of these IP-sets.

2. $\mathcal{F}$-SEQUENCES AND IP-LIMITS

In this section we will formulate still another version of Hindman's theorem. Recall that $\mathcal{F}$ denotes the set of all finite subsets of $\mathbb{N} = \{1,2,3,\ldots\}$. A sequence indexed by the (multi-indices) $\alpha \in \mathcal{F}$ will be called an $\mathcal{F}$-sequence.

DEFINITION 3. Let $\{x_\alpha\}_{\alpha \in \mathcal{F}}$ be an $\mathcal{F}$-sequence of points in a topological space X and let $x_0 \in X$. We say that x_0 is the <u>IP-limit</u> of $\{x_\alpha\}$, i.e., $\text{IP-}\lim_\alpha x_\alpha = x_0$, if for every neighborhood V of x there exists an index $\beta = \beta(V)$ so that whenever $\alpha \cap \beta = \phi$ we have $x_\alpha \in V$.

In other words, for α sufficiently "far out" x_α is close to x_0. Note that we may have $\text{IP-}\lim x_\alpha = x_0$ and nonetheless infinitely many x_α may be outside of the neighborhood V. We will often abbreviate $\text{IP-}\lim_\alpha x_\alpha = x_0$ to $x_\alpha \to x$ where by using indices $\alpha \in \mathcal{F}$ it will be implicit that the convergence is in the IP-sense.

Next we define an $\mathcal{F}$-<u>subsequence</u> of an $\mathcal{F}$-sequence $\{x_\alpha\}$ in the following way. We suppose that $\alpha_1, \alpha_2, \ldots, \alpha_n, \ldots$ is a sequence of disjoint sets of $\mathcal{F}$, and we let $y_{\{i_1, i_2, \ldots, i_k\}} = x_{\alpha_{i_1} \cup \alpha_{i_2} \cup \cdots \cup \alpha_{i_k}}$. The sequence $\{y_\alpha\}$ is then an $\mathcal{F}$-subsequence of $\{x_\alpha\}$. Note that an $\mathcal{F}$-subsequence of an IP-set $\{\sigma_\alpha\}$ is again an IP-set, since $\sigma_{\alpha_{i_1} \cup \alpha_{i_2} \cup \cdots \cup \alpha_{i_k}} = \sigma_{\alpha_{i_1}} \sigma_{\alpha_{i_2}} \cdots \sigma_{\alpha_{i_k}}$.

THEOREM 2.1. *If $\{x_\alpha\}$ is an $\mathcal{F}$-sequence in a compact metric space X, there exists an $\mathcal{F}$-subsequence $\{y_\alpha\}$ which possesses an IP-limit in X.*

PROOF: Note that if X is a finite space then Theorem 2.1 is just a reformulation of Theorem 1.3. For the general case we let $\varepsilon_n \searrow 0$, and we consider finite coverings of X, $X = \bigcup_{i=1}^{r_n} A_i^{(n)}$, by sets $A_i^{(n)}$ of diameter $< \varepsilon_n$. By Theorem 1.3 we see that for each n we can find an $\mathcal{F}$-subsequence of $\{x_\alpha\}$ lying wholly inside some $A_j^{(n)}$. More specifically, let $\alpha_1^{(1)}, \alpha_2^{(1)}, \ldots, \alpha_n^{(1)}, \ldots$ be a sequence of disjoint sets of $\mathcal{F}$ and let j_1 be an index so that $x_\alpha \in A_{j_1}^{(1)}$ whenever α is a union of $\alpha_m^{(1)}$. Using the $\alpha_m^{(1)}$ as atoms we repeat the procedure for the covering $X = \bigcup_{i=1}^{r_2} A_i^{(2)}$. Namely, let $\alpha_1^{(2)}, \alpha_2^{(2)}, \ldots, \alpha_n^{(2)}, \ldots$ be disjoint unions of the $\alpha_m^{(1)}$ and let j_2 be an index such that $x_\alpha \in A_{j_2}^{(2)}$ whenever α is a union of $\alpha_m^{(2)}$. Proceed in this way obtaining a double array $\alpha_m^{(m)}$, each row of which consists of disjoint unions from the preceding row, and such that for any finite union α taken from the n th row $x_\alpha \in A_{j_n}^{(n)}$ for a suitable j_n. We now choose $\{m_n\}$ so that the $\alpha_{m_n}^{(n)}$ are disjoint and we let these generate an $\mathcal{F}$-subsequence $\{y_\alpha\} \subset \{x_\alpha\}$. We will now have that if $\alpha \cap \{1,2,\ldots,n\} = \phi$, then $y_\alpha \in A_{j_n}^{(n)}$. So if y_0 is the unique point in $\bigcap \overline{A_{j_n}^{(n)}}$ we see that $\text{IP-}\lim_\alpha y_\alpha = y_0$. This proves the theorem. □

We next turn to $\mathcal{F}$-sequences of transformations of a space.

DEFINITION 4. An IP-set $\{T_\alpha\}$ of commuting transformations of a space will be called an <u>IP-system</u>.

$\{T_\alpha\}$ is an IP-set in a commuting semigroup of transformations of the space, and so for $\alpha \cap \beta = \phi$ we will have $T_{\alpha \cup \beta} = T_\alpha T_\beta$. If $\{T'_\alpha\}$ is an $\mathcal{F}$-subsequence of $\{T_\alpha\}$ then $\{T'_\alpha\}$ is again an IP-set and we refer to $\{T'_\alpha\}$ as an IP-<u>subsystem</u> of $\{T_\alpha\}$.

The following two results play an important role in the sequel.

THEOREM 2.2. *Let* $\{T_\alpha\}$ *be an IP-system of continuous transformations of a metric space* X. *Assume that for some* $x,y \in X$,

$$\text{IP-}\lim_\alpha T_\alpha x = y.$$

Then

$$\text{IP-}\lim_\alpha T_\alpha y = y.$$

PROOF: Let V be a neighborhood of y and let V' be a smaller neighborhood so that $\bar{V}' \subset V$. Suppose that for $\alpha \cap \beta = \phi$, $T_\alpha x \in V'$. We shall show that under the same condition $T_\alpha y \in V$. For fixed T_α, if z is sufficiently close to y and $T_\alpha z \in V'$, we will have $T_\alpha y \in V$. Take $z = T_\gamma x$ with $\gamma \cap (\alpha \cup \beta) = \phi$; then $T_\alpha z = T_{\alpha\cup\gamma} x \in V'$. Since z can be chosen arbitrarily close to y, we will have $T_\alpha y \in V$. □

Theorem 2.2 implies that if an IP-system converges pointwise, $T_\alpha x \to Tx$, then the limit transformation T satisfies $T^2 = T$. This idempotent property of limits of IP-systems is at the basis of the name IP.

THEOREM 2.3. *Let* $\{T_\alpha\}$ *be an IP-system of transformations of a compact metric space* X. *For any point* $x_0 \in X$ *there exists an IP-subsystem* $\{T'_\alpha\} \subset \{T_\alpha\}$ *such that the IP-limit of* $\{T'_\alpha x_0\}$ *exists*.

This follows immediately from Theorem 2.1.

COROLLARY. *Let* $\{T_\alpha\}$ *be an IP-system of continuous maps of a compact metric space* X. *Then there exists a point* $x_0 \in X$ *and an IP-subsystem* $\{T'_\alpha\} \subset \{T_\alpha\}$ *such that*

$$\text{IP-}\lim_\alpha T'_\alpha x_0 = x_0.$$

This corollary ensures the existence of "recurrent" points for IP-systems of transformations of a compact metric space. For the proof let $x_0 = \text{IP-}\lim_\alpha T' y_0$ for any arbitrary $y_0 \in X$, and for an appropriate subsystem $\{T'\}$. The result follows from Theorem 2.2.

3. RECURRENCE AND IP-SETS

In this section we shall see how IP-sets occur in ordinary dynamical systems, that is, dynamical systems for which the acting group or semigroup consists of powers of a single transformation. Let X be a metric space and let T be a continuous map of X to itself, $T : X \to X$. T generates a semigroup of transformations of X, and we speak of a *dynamical system* (X,T). We say $x_0 \in X$ is a *recurrent point* of (X,T) if, for some sequence $n_k \to \infty$ we have $T^{n_k} x_0 \to x_0$. Equivalently, x_0 is a recurrent point if for every neighborhood V of x, $\{n : T^n x \in V\}$ is non-empty. When this is the case, the set of "return times", $\{n : T^n x \in V\}$, will not be an arbitrary set. Namely, we have

THEOREM 3.1. *If* x_0 *is a recurrent point of* (X,T), *then for any neighborhood* V *of* x_0, *the set* $\{n : T^n x_0 \in V\}$ *contains an (additive) IP-subset of* $\mathbb{N}$. *Conversely if* $Q \subset \mathbb{N}$ *is an IP-set, there exists a dynamical system* (X,T) *and a point* $x_0 \in X$ *recurrent for* (X,T) *and a neighborhood* V *of* x_0 *such that* $Q \supset \{n : T^n x_0 \in V\}$.

PROOF: For the first part of the theorem let p_1 be such that $T^{p_1} x_0 \in V$. Then $V_1 = T^{-p_1} V \cap V$ is again a neighborhood of x_0, and we can find p_2 with $T^{p_2} x_0 \in V_1$. We form $V_2 = T^{-p_2} V_1 \cap V_1$, obtaining again a neighborhood of x_0. We continue obtaining successively smaller neighborhoods V_n of x_0, with $T^{p_n} x_0 \in V_{n-1}$ and $V_n = T^{-p_n} V_{n-1} \cap V_{n-1}$. We have $V_n \subset V_{n-1}$ and $T^{p_n} V_n \subset V_{n-1}$ and so $T^{\varepsilon_n p_n} T^{\varepsilon_{n-1} p_{n-1}} \cdots T^{\varepsilon_1 p_1} V_n \subset V$ for every choice of the ε's each ε_i being chosen as either 0 or 1. It follows that $T^{p_{i_1} + p_{i_2} + \cdots + p_{i_k}} x_0 \in V$ whenever $i_1 < i_2 < \cdots < i_k$.

For the other direction, let Q be an IP-subset of $\mathbb{N}$. Consider first the case in which the series $\{p_1, p_2, \ldots, p_n, \ldots\}$ of generators of Q contains only finitely many distinct terms. Then Q contains a set of the form $p\mathbb{N}$ and this contains the return times of a periodic dynamical system on a space with p points, proving our assertion. Otherwise we can assume $p_n \to \infty$, and by refining this set, we may suppose $p_{n+1} > p_1 + p_2 + \cdots + p_n$. Now let $X = \{0,1\}^{\mathbb{N}}$ be the space of $\{0,1\}$-sequences, and let $T : X \to X$ be the shift map $(Tx)(n) = x(n+1)$. Let $x_0(1) = 1$ and for $n > 1$ let $x_0(n) = 1$ if $n-1 \in Q$ and let $x_0(n) = 0$ if $n-1 \notin Q$. We claim x_0 is a recurrent point. A basis of neighborhoods of x_0 is given by $V_m = \{x : x(i) = x_0(i),\ i = 1,2,\ldots,m\}$. Consider $T^{p_n} x_0$ for p_n large. $T^{p_n} x_0(1) = x_0(p_n + 1) = x_0(1)$ since $p_n \in Q$. Also, for $i > 1$, $T^{p_n} x_0(i) = 1$ if and only if $i + p_n - 1 \in Q$. Now if $i + p_n - 1 = p_{i_1} + p_{i_2} + \cdots + p_{i_k}$ for $i_1 < i_2 < \cdots < i_k$ and for some $i < p_1 + \cdots + p_{n-1}$, then on account of the rapid growth of the p_j, we must have $p_n = p_{i_k}$. This gives $i-1 = p_{i_1} + p_{i_2} + \cdots + p_{i_{k-1}} \in Q$. Thus we find $T^{p_n} x_0(i) = x_0(i)$ for $i < p_1 + p_2 + \cdots + p_{n-1}$. Hence for given m we will have $T^{p_n} x_0 \in V_m$ if

$p_1 + p_2 + \cdots + p_{n-1} > m$, and so x_0 is a recurrent point of (X,T). Finally, it is clear that $T^{p_{i_1}+p_{i_2}+\cdots+p_{i_k}} x_0 \in V_1$ for $i_1 < i_2 < \cdots < i_k$ since $x_0(p_{i_1} + p_{i_2} + \cdots + p_{i_k} + 1) = 1$. This proves the theorem. □

It was first shown by Birkhoff that any dynamical system (X,T) with X a compact metric space possesses some recurrent point. Using the results of the preceding section we can deduce this by restricting to IP-subsystems of $\{T^n\}$. More precisely, let $\{n_\alpha\}$ be an IP-set in $\mathbb{N}$. Then setting $T_\alpha = T^{n_\alpha}$ we obtain an IP-system of commuting transformations of X. Apply the corollary to Theorem 2.3. There exists accordingly a point x_0 and a subsystem $\{T'_\alpha\}$ with $\text{IP-}\lim_\alpha T'_\alpha x_0 = x_0$. Clearly x_0 is a recurrent point of (X,T). Moreover we see that the point x_0 can be chosen so that its return times meet a pre-assigned IP-set. From this we obtain

THEOREM 3.2. <u>If</u> X <u>is a compact metric space</u>, T <u>a continuous map of</u> X <u>to itself, then for any</u> $\varepsilon > 0$, <u>the set of</u> $n \in \mathbb{N}$ <u>for which there exists a (recurrent) point</u> $x \in X$ <u>with</u> $d(T^n x, x) < \varepsilon$ <u>is an IP*-set</u>.

Given two dynamical systems, (X,T) and (X',T'), we consider the product $(X \times X', T \times T')$ where $T \times T'(x,x') = (Tx, T'x')$. If x_0 is a recurrent point of (X,T) and x'_0 is a recurrent point of (X',T'), it does not follow that (x_0,x'_0) is a recurrent point of $(X \times X', T \times T')$. To see this choose two IP-subsets $Q,Q' \subset \mathbb{N}$ with $Q \cap Q' = \phi$. (For example, let Q be generated by $\{10^i\}$ and let $Q' = 2Q$.) By Theorem 3.1 we can find systems (X,T), (X',T') and points $x_0 \in X$, $x'_0 \in X'$ and respective neighborhoods V of x_0 and V' of x'_0 such that $Q \supset \{n : T^n x_0 \in V\}$ and $Q' \supset \{n : T^n x'_0 \in V'\}$. Then the orbit of (x_0,x'_0) never returns to $V \times V'$ and (x_0,x'_0) is not recurrent.

DEFINITION 5. We shall say that x_0 is a <u>strongly recurrent point</u> of a system (X,T) if whenever x'_0 is a recurrent point of (X',T'), then (x_0,x'_0) is a recurrent point of $(X \times X', T \times T')$.

From Theorem 3.1 we obtain the following criterion for strong recurrence.

THEOREM 3.3. <u>A point</u> x_0 <u>is strongly recurrent for a system</u> (X,T) <u>iff for every neighborhood</u> V <u>of</u> x_0, <u>the set of return times</u> $\{n : T^n x_0 \in V\}$ <u>forms an IP*-set in</u> $\mathbb{N}$.

In [1, Chapter 9] strong recurrence is referred to as IP*-recurrence on account of this criterion. It is also shown there how this notion relates to the property of <u>distality</u>. Here we shall content ourselves with exhibiting a family of systems for which each point is strongly recurrent.

Let $\mathbb{T}$ denote the group $\mathbb{R}/\mathbb{Z}$ of reals modulo 1, and let (X,T) be some dynamical system. If $\varphi : X \to \mathbb{T}$ is a continuous function we define a system in $X \times \mathbb{T}$ by setting

$$\tilde{T}(x,t) = (Tx, t + \psi(x)).$$

We call this system a <u>skew product</u> of (X,T) with T.

THEOREM 3.4. *If $(X \times \mathbb{T}, \tilde{T})$ is a skew product of (X,T) with $\mathbb{T}$, and if $x_0 \in X$ is a recurrent point for (X,T), then for any t, (x_0,t) is recurrent for $(X \times \mathbb{T}, \tilde{T})$.*

PROOF: We shall use Kronecker's theorem to the effect that for any $\beta \in \mathbb{T}$, there is a sequence $m_k \in \mathbb{N}$ with $m_k\beta \to 0$ in $\mathbb{T}$. Now suppose $T^{n_k}x_0 \to x_0$. Passing to a subsequence if necessary we suppose that $\tilde{T}^{n_k}(x_0,t)$ also converges, and we write $\tilde{T}^{n_k}(x_0,t) \to (x_0,t+\beta)$. From the form of $\tilde{T}$ it follows that $\tilde{T}^{n_k}(x_0,t+\beta) \to (x_0,t+2\beta),\ldots\tilde{T}^{n_k}(x_0,t+(m-1)\beta) \to (x_0,t+m\beta)$. Now let $Y \subset X \times \mathbb{T}$ denote the closure of the $\tilde{T}$-orbit of (x_0,t). Y contains $(x_0,t+\beta)$ and so it also contains the orbit of this point. It follows that Y contains $(x_0,t+2\beta)$, and proceeding in this way we find $(x_0,t+m\beta) \in Y$ for each m. By Kronecker's theorem Y contains (x_0,t) and so (x_0,t) is recurrent. □

THEOREM 3.5. *If $(X \times \mathbb{T}, \tilde{T})$ is a skew product of (X,T) with $\mathbb{T}$, and if $x_0 \in X$ is a strongly recurrent point of (X,T), then for any $t \in \mathbb{T}$, (x_0,t) is strongly recurrent for $(X \times \mathbb{T}, \tilde{T})$.*

PROOF: Suppose $'_0$ is a recurrent point for the system (X',T'). Note that $(X \times \mathbb{T} \times X', \tilde{T} \times T')$ is a skew product of $(X \times X', T \times T')$ and $\mathbb{T}$. Since x_0 is strongly recurrent, (x_0,x_0') is recurrent. Now apply Theorem 3.4. It follows that (x_0,t,x_0') is recurrent for the skew product $(X \times \mathbb{T} \times X', \tilde{T} \times T)$. It follows that (x_0,t) is strongly recurrent. □

Theorem 3.5 provides us with a procedure for constructing systems with the property that every point is strongly recurrent. We begin with a one-point system. The skew product with $\mathbb{T}$ gives us a system on the circle (rotation by a fixed angle) and taking further skew products leads to dynamical systems on the torus. More specifically we conclude that if $\mathbb{T}^r$ is the r-dimensional torus, and we set

$$T(\theta_1,\theta_2,\ldots,\theta_d) = (\theta_1+\alpha_1, \theta_2+\psi_1(\theta_1),\ldots,\theta_d+\psi_{d-1}(\theta_1,\theta_2,\ldots,\theta_{d-1})),$$

then $(\mathbb{T}^r,T)$ is a system with each point of $\mathbb{T}^r$ strongly recurrent. This will be true in particular if we let the ψ_i be linear functions. In that case we find that the orbit of a point has the form $T^n(\theta_1,\theta_2,\ldots,\theta_d) = (p_1(n),p_2(n),\ldots,p_d(n))$ where $p_i(t)$ is a polynomial of degree t. Using this together with Theorem 3.3 we obtain:

THEOREM 3.6. <u>For any polynomial</u> $p(t)$ <u>with real coefficients, and any</u> $\rho > 0$, <u>the inequality</u> $|e^{2\pi i p(n)} - e^{2\pi i p(0)}| \leq \rho$ <u>defines an</u> IP^*<u>-set of integers</u> $n \in \mathbb{N}$.

So, for example, the set of n for which $n^2\alpha$ has fractional part either on $[0,\rho)$ or in $(1-\rho,1]$, is an IP^*-set.

4. IP-SYSTEMS OF UNITARY OPERATORS AND MEASURE PRESERVING TRANSFORMATIONS

Suppose $\mathcal{H}$ is a separable Hilbert space and that $\{U_\alpha\}_{\alpha\in F}$ is an IP-system of unitary operators on $\mathcal{H}$. Each closed ball $B_R = \{x : \|x\| \leq R\}$ is a compact metrizable space in the weak topology and $\{U_\alpha\}$ acts on B as an IP-system of homeomorphisms. Suppose that the weak limit of the operators $\{U_\alpha\}$ exists in the IP-sense, i.e., assume that $\text{IP-}\lim_\alpha \langle U_\alpha x,y\rangle = \langle Px,y\rangle$ for an operator P and for every $x,y \in \mathcal{H}$. P is a linear operator, and by Theorem 2.2, P is an idempotent. We claim that in this case, P is an orthogonal projection onto a subspace of $\mathcal{H}$.

LEMMA 4.1. <u>If</u> $\{U_\alpha\}$ <u>is an IP-system of unitary operators on</u> $\mathcal{H}$ <u>and</u> $U_\alpha \to P$ <u>weakly, then</u> P <u>is the orthogonal projection onto its range</u> $P\mathcal{H}$.

PROOF: Clearly if $U_\alpha x \to Px$ weakly for all x, then $U_\alpha^* y \to P^* y$ weakly for all y. Now $U_\alpha x \to x$ weakly iff $\langle U_\alpha x,x\rangle \to \|x\|^2$ which is also equivalent to $U_\alpha^* x \to x$ weakly. It follows that $P\mathcal{H} = P^*\mathcal{H}$. If now $x \perp P\mathcal{H}$, then $x \perp P^*\mathcal{H}$ and so $Px = 0$. But this is precisely the condition for P to be an orthogonal projection. □

Given an IP-system $\{U_\alpha\}$, it need not be the case that $\text{IP-}\lim_\alpha U_\alpha x$ exists for each $x \in \mathcal{H}$. However for each $x \in \mathcal{H}$ we can find an IP-subsystem $\{U'_\alpha\} \subset \{U_\alpha\}$ such that $\text{IP-}\lim_\alpha U'_\alpha x$ does exist, according to Theorem 2.1. Using the same "diagonal procedure" used in the proof of Theorem 2.1, we may go further and show the existence of an IP-subsystem $\{U'_\alpha\}$ for which $\text{IP-}\lim_\alpha U'_\alpha x_n$ exists for each x_n of some countable set $\{x_n\} \subset \mathcal{H}$. Choosing $\{x_n\}$ dense we can assume that $\text{IP-}\lim_\alpha U'_\alpha x$ exists for <u>every</u> $x \in \mathcal{H}$.

LEMMA 4.2. <u>If</u> $\{U_\alpha\}$ <u>is an IP-system of unitary operators on a Hilbert space</u> $\mathcal{H}$, <u>we may find an IP-subsystem such that</u>

$$\text{IP-}\lim_\alpha U'_\alpha x = Px$$

<u>exists for each</u> $x \in \mathcal{H}$ <u>and defines an orthogonal projection</u> P <u>onto a subspace of</u> $\mathcal{H}$.

We remark that the range $P\mathcal{H}$ of the limit projection of an IP-system $\{U_\alpha\}$ includes not only fixed vectors for all the U_α, but also simultaneous eigenvectors of $\{U_\alpha\}$. For if we have $U_\alpha v = \lambda_\alpha v$, $\alpha \in F$, then since $|\lambda_\alpha| = 1$, $\|Pv\| = \|v\|$. Since P is an orthogonal projection this implies $Pv = v$. We can, if we like, regard $P\mathcal{H}$ as a generalization of the discrete spectrum component of a unitary operator.

We next consider an IP-system of measure preserving transformations $\{T_\alpha\}$ of a measure space $(X,\mathcal{B},\mu)$. These determine an IP-system of unitary operators on the Hilbert space $L^2(X,\mathcal{B},\mu)$. We shall denote the unitary operator induced by a transformation of the space X by the same symbol; thus the IP-system of unitary operators is given by $T_\alpha f(x) = f(T_\alpha x)$. Now suppose that we have $T_\alpha \to P$ weakly as before. P is the orthogonal projection onto the subspace of functions satisfying $T_\alpha f \to f$. The subspace $P\mathcal{H} \subset L^2(X,\mathcal{B},\mu)$ has the following property: if $f,g \in P\mathcal{H} \cap L^\infty(X,\mathcal{B},\mu)$, then the product $fg \in P\mathcal{H}$. For, if $T_\alpha f \to f$ weakly, then since $\|T_\alpha f - f\|^2 = 2\|f\|^2 - 2\mathcal{R}\langle T_\alpha f,f\rangle \to 0$, we also have $T_\alpha f \to f$ strongly. Hence for $f,g \in P\mathcal{H} \cap L^\infty(X,\mathcal{B},\mu)$,

$$\|T_\alpha(fg) - fg\|_2 \le \|(T_\alpha f-f)g\|_2 + \|T_\alpha f(g-T_\alpha g)\|_2$$

$$\le \|T_\alpha f - f\|_2\|g\|_\infty + \|f\|_\infty\|g-T_\alpha g\|_2 \to 0.$$

We also see that if $f \in P\mathcal{H}$ and φ is a continuous bounded function on the range of f, then $\varphi\circ f \in P\mathcal{H}$. These two properties of the subspace $P\mathcal{H}$ imply that the latter has the form $P\mathcal{H} = L^2(X,\mathcal{B}_p,\mu)$, where $\mathcal{B}_p$ consists of all sets A for which the indicator function $1_A \in P\mathcal{H}$.

DEFINITION 6. An IP-system $\{T_\alpha\}$ of measure preserving transformations of a space $(X,\mathcal{B},\mu)$ is (strongly) mixing if for each $f \in L^2(X,\mathcal{B},\mu)$ $\text{IP-}\lim_\alpha T_\alpha f = \int f d\mu$, in the weak topology of $L^2(X,\mathcal{B},\mu)$.

Equivalently, $\{T_\alpha\}$ is mixing if for $f,g \in L^2(X,\mathcal{B},\mu)$ we have $\int f T_\alpha g d\mu \to \int f d\mu \int g d\mu$. This condition, in turn, is equivalent to $\mu(A \cap T_\alpha^{-1}B) \to \mu(A)\mu(B)$ for $A,B \in \mathcal{B}$.

DEFINITION 7. An IP-system $\{T_\alpha\}$ of measure preserving transformations of a space $(X,\mathcal{B},\mu)$ is rigid if $\text{IP-}\lim_\alpha T_\alpha f = f$ for all $f \in L^2(X,\mathcal{B},\mu)$.

Rigidity implies that functions on X, and hence subsets of X, "return to themselves" under the application of T_α. If $A \Delta B$ denotes the symmetric difference of sets, then rigidity is equivalent to

$$\text{IP-}\lim_\alpha \mu(A \Delta T_\alpha^{-1}A) = 0$$

for all $A \in \mathcal{B}$.

Given an IP-system $\{T_\alpha\}$ of measure preserving transformations, we can pass to a subsystem $\{T'_\alpha\}$ for which $\text{IP-}\lim_\alpha T'_\alpha = P$ exists (Lemma 4.2). The range of P will have the form $L^2(X,\mathcal{B}_p,\mu)$. If $\mathcal{B}_p$ is the trivial algebra of sets of measure 0 and 1, then $Pf = \int f d\mu$. In this case the system $\{T'_\alpha\}$ is mixing. On the other hand, if $\mathcal{B}_p$ is non-trivial, it makes sense to consider the restriction of the system $\{T'_\alpha\}$ to $(X,\mathcal{B}_p,\mu)$. For every $f \in L^2(X,\mathcal{B}_p,\mu)$ $T'_\alpha f \to Pf = f$, and so the actions of $\{T'_\alpha\}$ on $(X,\mathcal{B}_p,\mu)$ is rigid. We thus have

THEOREM 4.3. <u>If $\{T_\alpha\}$ is an IP-system of measure preserving transformations of a space</u> $(X,\mathcal{B},\mu)$, <u>then there is an IP-subsystem</u> $\{T'_\alpha\} \subset \{T_\alpha\}$ <u>for which one of the following alternatives holds</u>: (i) $\{T'_\alpha\}$ <u>is mixing on</u> $(X,\mathcal{B},\mu)$, <u>or</u> (ii) <u>for a non-trivial algebra</u> $\mathcal{B}' \subset \mathcal{B}$, $\{T'_\alpha\}$ <u>acts rigidly on</u> $(X,\mathcal{B}',\mu)$.

Often we fix a commuting group Γ of measure preserving transformations of $(X,\mathcal{B},\mu)$ and we suppose the IP-systems are chosen from within Γ. We say that $(X,\mathcal{B}',\mu)$ is a <u>factor</u> of $(X,\mathcal{B},\mu)$ if $\mathcal{B}'$ is a Γ-invariant subalgebra. Now suppose $S \in \Gamma$ and that $T'_\alpha f \to f$ for some $f \in L^2(X,\mathcal{B},\mu)$. Then $T'_\alpha Sf = ST'_\alpha f \to Sf$. It follows that $\mathcal{B}_p$ is a Γ-invariant algebra. In this case we can reformulate Theorem 4.3 more precisely as saying that for an appropriate IP-subsystem of $\{T_\alpha\}$ the action is either mixing, or one can find a non-trivial factor in which the action is rigid.

We conclude this section with the remark that unlike the conventional notion of mixing, if an IP-system $\{T_\alpha\}$ is mixing, it may happen that a power $\{T^k_\alpha\}$ of the system is not mixing. Indeed it is possible to construct a mixing IP-system $\{T_\alpha\}$ for which T^2_α = identity for each α. We sketch this construction. To begin with we find an IP-system of unitary operators $\{U_\alpha\}$ on a Hilbert space $\mathcal{H}$ with $U^2_\alpha = I$, and with $\text{IP-lim}_\alpha U_\alpha = 0$ weakly. Having done this we attach to $\mathcal{H}$ a system of Gaussian variables, $u \to X_u$, $u \in \mathcal{H}$ with $E(X_u X_v) = \langle u,v\rangle$. The unitary operators $\{U_\alpha\}$ will induce measure preserving transformations T_α satisfying T^2_α = identity and $E(T_\alpha X_u \cdot X_v) = \langle U_\alpha u,v\rangle \to 0$, and since for Gaussian variables orthogonality implies independence, we will have $T_\alpha X \to E(X)$ for every variable X, and so $\{T_\alpha\}$ is mixing. To construct $\{U_\alpha\}$, let $\mathcal{H} = L^2[0,1]$ and let φ_n be the nth Rademacher function, $\varphi_n(t) = 2\phi_n(t) - 1$ where $\varphi_n(t) = 0,1$ and $t = \Sigma\varphi_n(t)2^{-n}$. The φ_n are independent and their products span $\mathcal{H}$. For $\alpha \in \mathcal{F}$, $\alpha = \{i_1,i_2,\dots,i_k\}$ set $\varphi_\alpha = \varphi_{i_1}\varphi_{i_2}\cdots\varphi_{i_k}$. We can check that $\text{IP-lim}_\alpha \int\varphi_\alpha g\,dt = 0$ for each $g \in L^2[0,1]$. Hence $U_\alpha f = \varphi_\alpha f$ defines an IP-system of unitary operators with the desired properties.

5. MIXING OF HIGHER ORDER

An ordinary ergodic system generated by a single measure preserving transformation $(X,\mathcal{B},\mu,T)$ is mixing if $\int fT^n g\,d\mu \to \int f d\mu \int g d\mu$ for all $f,g \in L^2(X,\mathcal{B},\mu)$. It is an open question whether this implies the joint asymptotic independence of f, $T^{n_1}g$, $T^{n_2}h,\dots$ provided n_1, n_2, n_2-n_1, etc., are large. It is not even known if mixing implies that $\int fT^n gT^{2n}h\,d\mu \to \int f d\mu \int g d\mu \int h d\mu$, as $n \to \infty$, for $f,g,h \in L^\infty(X,\mathcal{B},\mu)$. In this section we shall consider the analogous question for IP-systems. As remarked at the end of the preceding section, if we merely assume that $\{T_\alpha\}$ is mixing, we may have T^2_α = identity, and we cannot expect to have $\int fT_\alpha gT_\alpha^2 hd\mu \to \int fd\mu \int gd\mu \int hd\mu$. What we shall be able to prove is the following

THEOREM 5.1. *Let* $\{T_\alpha\}$ *be an IP-system of measure preserving transformations of a space* $(X,\mathcal{B},\mu)$ *and assume that the systems* $\{T_\alpha\},\{T_\alpha^2\},\cdots\{T_\alpha^k\}$ *are mixing. Then we can pass to an IP-subsystem* $\{T'_\alpha\}$ *for which*

$$\text{(5.1)} \qquad \text{IP-}\lim_\alpha \int f_0 T'_\alpha f_1 T'^2_\alpha f_2 \cdots T'^k_\alpha f_k d\mu = \int f_0 d\mu \int f_1 d\mu \int f_2 d\mu \cdots \int f_k d\mu,$$

whenever $f_0, f_1, f_2, \ldots, f_k \in L^\infty(X,\mathcal{B},\mu)$.

We begin with a lemma about IP-convergence in a Hilbert space.

LEMMA 5.2. *Suppose* $\{x_\alpha\}_{\alpha\in\mathcal{F}}$ *is an* $\mathcal{F}$*-sequence of bounded vectors in a Hilbert space* $\mathcal{H}$. *If*

$$\text{IP-}\lim_\beta \ \text{IP-}\lim_\alpha <x_\alpha, x_{\alpha\cup\beta}> = 0,$$

then for some $\mathcal{F}$*-subsequence* $\{x'_\alpha\}$, *we will have*

$$\text{IP-}\lim_\alpha x'_\alpha = 0$$

in the weak topology of $\mathcal{H}$.

PROOF: Passing to a subsequence we can assume to begin with that $x_2 \to u$ (Theorem 2.1). Assume $u \neq 0$; we can find $\delta > 0$ and α arbitrarily far out, say $\alpha \cap \gamma_0 = \phi$, for which $<x_\alpha, u> > \delta$. Choose $\varepsilon > 0$. The condition of the lemma can be restated as follows. There exists $\gamma_1 \in \mathcal{F}$ such that if $\beta \cap \gamma_1 = \phi$ there exists $\gamma_2(\beta) \in \mathcal{F}$ such that $\alpha \cap \gamma_2(\beta) = \phi$ implies that $<x_\alpha, x_{\alpha\cup\beta}> < \varepsilon$. Choose $\alpha_1, \alpha_2, \ldots, \alpha_k$ inductively with $\alpha_i \cap (\gamma_1 \cup \gamma_0) = \phi$ and with $\alpha_j \cap \gamma_2(\alpha_1 \cup \alpha_2 \cup \cdots \cup \alpha_i) = \phi$ for $j > i$, then $<x_{\alpha_1\cup\alpha_2\cup\ldots\cup\alpha_i}, x_{\alpha_1\cup\alpha_2\cup\ldots\cup\alpha_j}> < \varepsilon$ for $i < j$. Write $y_i = u_{\alpha_1\cup\ldots\cup\alpha_i}$. Then $<y_i, u> > \delta$ and at the same time $<y_i, y_j> < \varepsilon$ for $i \neq j$. But

$$< \frac{y_1 + \cdots + y_k}{k}, u > > \delta$$

implies

$$\left\|\frac{y_1 + \cdots + y_k}{k}\right\| > \delta \|u\|^{-1},$$

and, on the other hand, $<y_i, y_j> < \varepsilon$ implies

$$\left\|\frac{y_1 + \cdots y_k}{k}\right\|^2 < \frac{\max\{\|y_i\|^2\}}{k} + \varepsilon.$$

Choosing $\varepsilon > 0$ small and k large one is led to a contradiction which proves the lemma. □

PROOF OF THE THEOREM. The theorem is proved by induction on k. At each stage it suffices to show that (5.1) holds for some subsystems when $f_0,f_1,\ldots,f_k$ are given. By a diagonal procedure as was used in Theorem 2.1 we can then obtain a single subsystem for which (5.1) holds for all $f_0,f_1,\ldots,f_k$. Suppose we wish to establish (5.1) for given $f_0,f_1,\ldots,f_k$. We can assume $\int f_k d\mu = 0$ (since the case f_k = constant corresponds to (5.1) for a smaller value of k). The assertion then is that some $\mathcal{F}$-subsequence of $T_\alpha f_1 T_\alpha^2 f_2 \cdots T_\alpha^k f_k$ converges weakly to 0. We shall use the criterion of the lemma. For this we should show that

$$\underset{\alpha}{\text{IP-lim}}\ \underset{\alpha}{\text{IP-lim}} \langle T_\alpha f_1 T_\alpha^2 f_2 \cdots T_\alpha^k f_k, T_{\alpha\cup\beta} f_1 T_{\alpha\cup\beta}^2 f_2 \cdots T_{\alpha\cup\beta}^k f_k \rangle = 0.$$

In this expression we can assume $\alpha \cap \beta = \phi$ since for fixed β we assume that α is far out. The foregoing inner product becomes

$$\begin{aligned}\int T_\alpha\{(f_1 T_\beta f_1) T_\alpha (f_2 T_\beta^2 f_2) \cdots T_\alpha^{k-1}(f_k T_\beta^k f_k)\} d\mu &= \\ \int (f_1 T_\beta f_1) T_\alpha (f_2 T_\beta^2 f_2) \cdots T_\alpha^{k-1}(f_k T_\alpha^k f_k) d\mu.\end{aligned} \tag{5.2}$$

Since we assume the theorem is valid for $k - 1$, we can suppose that the IP-system is such that (5.1) is valid for $k - 1$ and for the system itself. Accordingly the integral in (5.2) converges as an IP-lim with respect to α to

$$\int f_1 T_\beta f_1 d\mu \int f_2 T_\beta^2 f_2 d\mu \cdots \int f_k T_\beta^k f_k d\mu.$$

We now take the IP-lim with respect to β. Since $\{T_\beta^k\}$ is mixing and $\int f_k d\mu = 0$ we find that the limit in question is 0. This proves the theorem. □

In the next section we shall see applications of mixing IP-systems to ordinary dynamical systems.

6. MILD MIXING

In this section we consider a classical measure preserving system, i.e., a measure space $(X,\mathcal{B},\mu)$, with $\mu(X) = 1$, and a measure preserving transformation $T : X \to X$. Each IP-subset $\{n_\alpha\} \subset N = \{1,2,3,\ldots\}$ determines an IP-system $T_\alpha = T^{n_s}$ on $(X,\mathcal{B},\mu)$.

DEFINITION 8. The measure preserving system $(X,\mathcal{B},\mu,T)$ is said to be <u>mildly mixing</u> if every IP-set in $\mathbb{Z}$ contains an IP-subset $\{n_\alpha\}$ such that the IP-system $\{T^{n_\alpha}\}$ is mixing on $(X,\mathcal{B},\mu)$.

We shall find a number of analogies between mildly mixing and weakly mixing systems. A system $(X,\mathcal{B},\mu,T)$ is weakly mixing if it satisfies one of the following three equivalent properties.

(a) For any $A,B \in \mathcal{B}$ and $\varepsilon > 0$, the set of $n \in N$ for which $|\mu(A \cap T^{-n}B) - \mu(A)\mu(B)| < \varepsilon$ has density 1.

(b) If $(Y,\mathcal{D},\nu,S)$ is any ergodic measure preserving system, then $(X \times Y,\mathcal{B} \times \mathcal{D},\mu \times \nu,T \times S)$ is ergodic.

(c) The only functions $f \in L^2(X,\mathcal{B},\mu)$ satisfying $f(Tx) = \lambda f(x)$ with λ a constant are the constant functions.

Property (a) is the reason for calling these systems weakly mixing, since strong mixing can be defined by the requirement that $\{n:|\mu(A \cap T^{-n}B) - \mu(A)\mu(B)| < \varepsilon\}$ is the complement of a finite set. Recall that a subset $R \subset \mathbb{N}$ has density 1 if the function defined by $\pi_R(n) = \text{card}(R \cap \{1,2,\ldots,n\})$ satisfies $\pi_R(n) \sim n$.

Property (c), or the absence of non-trivial eigenfunctions for the system can be rephrased in terms of factors. A system $(X',\mathcal{B}',\mu',T')$ is a <u>factor</u> of $(X,\mathcal{B},\mu,T)$ if there is a measurable, measure preserving map $\varphi : X \to X'$ with $T'\circ\varphi = \varphi\circ T$. A system is <u>Kronecker</u> if the underlying space is a compact abelian group and T is defined by a translation within the group $Tx = ax$. Thus a rotation of the unit circle defines a Kronecker system. We can then reformulate (c) as:

(c') $(X,\mathcal{B},\mu,T)$ has no non-trivial Kronecker factor.

Note that whereas (a) and (b) are positive statements regarding a weakly mixing system, (c) implies a positive property for systems that fail to be weak mixing - namely, the existence of an eigenfunction. In a sense we have a certain dichotomy! either a system behaves in a manner in which past and future tend to become independent, or the system exhibits some phenomenon of periodicity. We find a similar dichotomy for the notion of mild mixing.

THEOREM 6.1. <u>A system</u> $(X,\mathcal{B},\mu,T)$ <u>is mild mixing if any of the following equivalent conditions are fulfilled</u>.

(A) <u>For any</u> $A,B \in \mathcal{B}$ <u>and</u> $\varepsilon > 0$, <u>the set of</u> n <u>for which</u> $|\mu(A \cap T^{-n}B) - \mu(A)\mu(B)| < \varepsilon$ <u>is an</u> IP^*<u>-set</u>.

(B) <u>If</u> $(Y,\mathcal{D},S,\nu)$ <u>is a non-atomic non-singular ergodic system, i.e.,</u> S <u>is a measurable map of</u> $Y \to Y$ <u>such that</u> $\nu(A) = 0 \iff \nu(S^{-1}A) = 0$, <u>and, moreover</u> $A = S^{-1}A$ <u>only if</u> $\nu(A) = 0$ <u>or</u> 1, <u>then</u> $(X \times Y,\mathcal{B} \times \mathcal{D},\mu \times \nu,T \times S)$ <u>is again ergodic</u>.

(C) <u>The only functions satisfying</u> $T^{n_\alpha}f \to f$ <u>for</u> $\{n_\alpha\}$ <u>any IP-set in</u> $\mathbb{N}$ <u>are</u> $f =$ <u>constant</u>. (<u>The mode of convergence here is immaterial</u>).

PROOF: We begin with (C). If $(X,\mathcal{B},\mu,T)$ is mildly mixing then $\{T^{n_\alpha}\}$ has a mixing subsystem $\{T^{n'_\alpha}\}$. We then have $T^{n'_\alpha}f \to f$ and at the same time $T^{n'_\alpha}f \to \int f d\mu$. This shows that mild mixing implies (C). Assume now that (C) holds and consider some IP-system $\{T^{n_\alpha}\}$. According to Theorem 4.3 applied to $\{T^{n_\alpha}\}$ we see that either some subsystem is mixing or for a subsystem, some non-trivial factor is rigid. But the latter is excluded by (C); hence a subsystem is mixing and so $(X,\mathcal{B},\mu,T)$ is mildly mixing.

Using the fact that intersections of IP^*-sets are IP^*-sets, it can be shown that (A) is equivalent to

(A') For $f,g \in L^2(X, ,\mu)$, $\varepsilon > 0$, the set of n for which

$$\left|\int fT^n g d\mu - \int f d\mu \int g d\mu\right| < \varepsilon \text{ is an } IP^*\text{-set.}$$

Clearly (A') $\Rightarrow$ (C), for if $T^{n_\alpha} f \to f$ then we will have

$$\left|\int \bar{f} T^n f d\mu - \int \bar{f} f d\mu\right| < \varepsilon$$

for any $\varepsilon > 0$ and some n in any IP^*-set. But then $\int \bar{f} f d\mu = \left|\int f d\mu\right|^2$ which implies f = constant. Conversely suppose that (A') does not hold, and for some IP-set $\{n_\alpha\}$,

$$\left|\int gT^{n_\alpha} f d\mu - \int f d\mu \int g d\mu\right| \geq \varepsilon. \tag{6.1}$$

Pass to a subsystem for which $T^{n'_\alpha} f \to Pf$. By (6.1) Pf is not constant. On the other hand $T^{n'} Pf$ $P^2 f = Pf$ contradicting (C).

As regards (B) we will only show that mild mixing implies (B) referring the reader to [4] for the converse direction. Assume $(X,\mathcal{B},\mu,T)$ is mild mixing and that $(Y,\mathcal{D},\nu,S)$ is non-singular and ergodic. Suppose that the product system $(X \times Y,\mathcal{B} \times \mathcal{D},\mu \times \nu,T \times S)$ is not ergodic and let $f(x,j)$ be a non-constant invariant function in $L^2(X \times Y)$. For $y \in Y$ let $f_y \in L^2(X)$ be defined by $f_y(x) = f(x,y)$. Since $(Y,\mathcal{D},\nu,S)$ is ergodic we can suppose that for a.e. y, f_y is non-constant. We can find a non-constant function $\varphi \in L^2(X)$ such that for any $\varepsilon > 0$, the set of $y \in Y$ with $\|f_y - \varphi\|_{L^2(X)} < \varepsilon$ has positive measure. Choose ε so that $\int |\varphi|^2 d\mu - \left|\int \varphi d\mu\right|^2 > 2\varepsilon\|\varphi\|$, and let $B = \{y : \|f_y - \varphi\| < \varepsilon\}$. Since $(Y,\mathcal{D},\nu,S)$ is ergodic and non-atomic we can find $p_1 > 0$ with $\nu(B \cap S^{-p_1}B) > 0$. Set $B_1 = B \cap S^{-p_1}B$; find $p_2 > 0$ with $\nu(B_1 \cap S^{-p_2}B_1) > 0$, and continue this procedure. In this way we obtain a sequence $\{p_i\}$ that generates an IP-subset of N, $\{p_\alpha\}$, such that for each α, $\nu(B \cap S^{-p_\alpha}B) > 0$. Let $y \in B \cap S^{-p_\alpha}B$; then $\|f_y - \varphi\| < \varepsilon$ and $\|f_{S^{p_\alpha}y} - \varphi\| < \varepsilon$. The invariance of $f(x,y)$ implies that $T^{p_\alpha} f_{S^{p_\alpha}y} = f_y$. It follows that $\|f_y - T^{p_\alpha}\varphi\| < \varepsilon$ so that $\|T^{p_\alpha}\varphi - \varphi\| < 2\varepsilon$. Now by (A') we can find p_α such that

$$\left|\int \bar{\varphi} T^{p_\alpha} \varphi d\mu - \left|\int \varphi d\mu\right|^2\right| < \int |\varphi|^2 d\mu - \left|\int \varphi d\mu\right|^2 - 2\varepsilon\|\varphi\|.$$

But then

$$\begin{aligned}\int \bar{\varphi}\varphi d\mu - \left|\int \varphi d\mu\right|^2 &< \left|\int \bar{\varphi} T^{p_\alpha} \varphi d\mu - \left|\int \varphi d\mu\right|^2\right| + \|\varphi\|\|T^{p_\alpha}\varphi - \varphi\| \\ &< \int |\varphi|^2 d\mu - \left|\int \varphi d\mu\right|^2,\end{aligned}$$

a contradiction. This proves (B). □

Let us mention two more properties of weak mixing that carry over to analogous properties for mild mixing.

THEOREM. (i) $(X,\mathcal{B},\mu,T)$ <u>is weakly mixing iff for every</u> $F \in L^2(X \times X,\mathcal{B} \times \mathcal{B}, \mu \times \mu)$,

$$\frac{1}{N}\sum_{1}^{N} T^n F \to \iint F(x,y)\,d\mu(x)\,d\mu(y)$$

<u>weakly in</u> $L^2(X \times X,\mathcal{B} \times \mathcal{B},\mu \times \mu)$.

(ii) <u>If</u> $(X,\mathcal{B},\mu,T)$ <u>is weakly mixing, then for any</u> $f_0,f_1,\dots,f_k \in L^\infty(X,\mathcal{B},\mu)$ <u>and for any</u> $\varepsilon > 0$

$$\{n : |\int f_0 T^n f_1 T^{2n} f_2 \cdots T^{kn} f_k\,d\mu - \int f_0\,d\mu \int f_1\,d\mu \cdots \int f_k\,d\mu| < \varepsilon\}$$

<u>has density</u> 1.

The necessity of the condition in (i) follows from the ergodicity of $(X \times X,\mathcal{B} \times \mathcal{B},\mu \times \mu,T \times T)$ which is guaranteed by property (b) of weakly mixing systems. Both (i) and (ii) can be found in [1]. We turn to the analogous theorem for mildly mixing systems.

THEOREM 6.2.: (I) $(X,\mathcal{B},\mu,T)$ <u>is mildly mixing iff for every</u> $F \in L^2(X \times X, \mathcal{B} \times \mathcal{B},\mu \times \mu)$, <u>and for every IP-set</u> $\{n_\alpha\}$, <u>there exists an IP-subset</u> $\{n'_\alpha\}$ <u>with</u>

$$\text{IP-}\lim_{\alpha} T^{n'_\alpha} F = \iint F(x,y)\,d\mu(x)\,d\mu(y)$$

<u>weakly in</u> $L^2(X \times X,\mathcal{B} \times \mathcal{B},\mu \times \mu)$.

(II) <u>If</u> $(X,\mathcal{B},\mu,T)$ <u>is mildly mixing, then for any</u> $f_0,f_1,\dots,f_k \in L^\infty(X,\mathcal{B},\mu)$ <u>and for</u> $\varepsilon > 0$

$$\{n : |\int f_0 T^n f_1 T^{2n} f_2 \cdots T^{kn} f_k\,d\mu - \int f_0\,d\mu \int f_1\,d\mu \cdots \int f_k\,d\mu| < \varepsilon\}$$

<u>is an</u> IP^*<u>-set</u>.

Thus the ergodic average in the weak mixing case is replaced by an IP-limit in the mild mixing case. In addition sets of density one carry over to IP^*-sets. Both families represent "substantial" sets having the property that arbitrary finite intersections in the family lead to sets in the family.

PROOF OF THE THEOREM: (I) Suppose that $(X,\mathcal{B},\mu,T)$ is mildly mixing and let $\{n_\alpha\}$ be an IP-set. By definition there is an IP-subset with $T^{n'_\alpha} f \to \int f\,d\mu$ for every $f \in L^2(X,\mathcal{B},\mu)$. But then $T^{n'_\alpha}(f(x)g(y)) \to \iint f(x)g(y)\,d\mu(x)\,d\mu(y)$ and this implies the property in (I). On the other hand restricting the property in (I) to functions of a single variable is precisely the definition of mild mixing.

(II) Follows directly from Theorem 5.1. □

We remark that (I) becomes significant when we extend the notion of mild mixing to that of a mild mixing extension $(X,\mathcal{B},\mu,T)$ of a factor $(X',\mathcal{B}',\mu',T')$.

It then turns out that the analogue of (I) for the fibre product of X with itself, relative to X', gives the correct definition of a mild mixing extension.

REFERENCES

[1] H. Furstenberg, Recurrence in Ergodic Theory and Combinatorial Number Theory, Princeton University Press 1981, Princeton, NJ.

[2] H. Furstenberg, Y. Katznelson, and D. Ornstein, The Ergodic Theoretic Proof of Szemeredi's Theorem, to appear in Bull. Amer. Math. Soc.

[3] H. Furstenberg and B. Weiss, Topological Dynamics and Combinatorial Number Theory, Jour. D'Analyse Math. 34 (1978), 61-85.

[4] ______, The Finite Multipliers of Infinite Ergodic Transformations, in Structure of Attractors in Dynamical Systems. Springer Lecture Notes #668, Springer (1978), 127-133.

[5] S. Glazer, Ultrafilters and Semigroup Combinatorics, J. Combinatorial Theory (A), to appear.

[6] N. Hindman, Finite Sums From Sequences Within Cells of a Partition of IN, J. Combinatorial Theory (A) 17 (1974), 1-11.

[7] I. Schur, Uber die Kongruenz $x^m + y^m \equiv z^m \pmod p$, Jahresbericht der deutschen Math. Ver. 25 (1916), 114-117.

The Hebrew University
Jerusalem, ISRAEL

Contemporary Mathematics
Volume 26, 1984

INDUCED TRANSFORMATIONS ON A SECTION

Arshag Hajian and Yuji Ito

We denote by $G(X)$ the group of all 1-1, onto, measurable, and non-singular transformations T defined on a measure space $(X, \mathcal{B}, m)$. In the sequel we shall consider only measurable subsets of X, and often we shall make statements disregarding sets of measure zero; we also assume that all measures are σ-finite. An important subset of the group $G(X)$ consists of the ergodic transformations; namely, $E = E(X) = \{T \in G(X): TA = A$ implies $m(A) = 0$ or $m(X - A) = 0\}$. We say that a transformation $T \in G(X)$ is measure preserving if $m(TA) = m(A)$ for all $A \in \mathcal{B}$, and two measures μ and m are equivalent, $\mu \sim m$, if they have the same null sets. In a natural way the set E decomposes into three subsets E_{II_1}, E_{II_∞}, E_{III}; namely, the set of all transformations $T \in E$ which preserve a finite, an infinite, or no invariant measure $\mu \sim m$, respectively. In the study of the factors in the theory of W^*-algebras these subsets provide important and canonical examples and are crucial in the classification of these factors.

In [6] and [3] we constructed examples of transformations belonging to E_{II_∞}. An important property of these examples was exhibited by the behavior of their centralizers. As a consequence of this fact we were able to construct in a systematic way examples of transformations $T \in E_{III}$. Subsequently, it was noticed in [4] that the significant feature for our construction was the orbit structure of the transformation $T \in G(X)$. In particular, for a transformation $T \in G(X)$ by the orbit of a point $x \in X$ we mean $\mathrm{Orb}_T(x) = \{T^n x: n \in Z\}$, by the full group of T we mean $[T] = \{T' \in G(X): \mathrm{Orb}_{T'}(x) \subset \mathrm{Orb}_T(x)$ for all $x \in X\}$, and by the normalizer of T we mean $N(T) = \{S \in G(X) : S[T] = [T]S\}$. The normalizer of T is a generalization of the centralizer of T; namely, $C(T) = \{S \in G(X): ST = TS\}$. If S and T are two transformations that commute then it is easy to see that T maps the set of orbits of S onto itself while preserving the order structure of these orbits. It was noticed in [4] that if we drop this latter property of preserving the order structure of the orbits then many of the key arguments of the results in [3] still carry through. As a consequence of this fact we were led to consider and study in some detail the normalizer $N(S)$ of

0271-4132/84 $1.00 + $.25 per page

a dissipative transformation $S \in G(X)$. We say that $S \in G(X)$ is a dissipative transformation if there exists a set A such that $X = \coprod_{n \in Z} S^n A$ (disj). The main feature of dissipative transformations is the fact that they possess sections; namely, a set A is an S-section if every orbit under S intersects the set A in exactly one point. In a sense the sections of a transformation S represent the space of orbits of S, and when S is a dissipative transformation then the space of orbits happens to have a nice and manageable structure. As a consequence, in [4] we generalized the definition of induced transformations as introduced in [7] and formulated a general method of inducing a transformation R on a section A of a dissipative transformation S by a transformation $T \in N(S)$. Using this general method of inducing we were able to induce a transformation $R \in E_{III}$ by a pair T and S where $T \in E_{II_\infty}$ and S is dissipative with $T \in N(S)$. Soon it became clear that a general study of the relationship between the induced transformations R and the pairs T and S was desirable. Some preliminary results in this direction were obtained in [4], and in the present article we continue the study of this relationship further. For a dissipative transformation S we show in Theorem 1 that the correspondence between the transformations $T \in N(S)$ and $R \in G(X)$ induced by the pair T and S on an S-section X defines an isomorphism between the groups $N(S)/[S]$ and $G(X)$. In Theorem 2 we show that this relationship carries further to the full groups $[T]$ and $[R]$. Two transformations $T_1, T_2 \in G(X)$ are said to be weakly equivalent if their full groups are conjugate subgroups of $G(X)$. Theorem 3 then, under certain restrictions, shows that if $T_1 T_2 \in N(S)$ and R_1, R_2 are the corresponding transformations they induce on a section of S, respectively, then R_1 and R_2 are weakly equivalent if and only if T_1 and T_2 are weakly equivalent in a restricted way. As a corollary to Theorem 3, we obtain a short proof of the following fact: If R_1 and R_2 are ergodic and if W_1 and W_2 are the W^*-algebras constructed from R_1 and R_2, respectively, using von Neumann's group measure space construction, then W_1 and W_2 are strongly $*$-isomorphic (for definitions see below) if and only if R_1 and R_2 are weakly equivalent.

For the purposes of Theorem 1 that follows we recall the general definition of inducing a transformation by a pair of transformations as described in [4]. For a dissipative transformation $S \in G(Y)$ and a transformation $T \in G(Y)$ with $T \in N(S)$ we let X be an S-section. Then for $x \in X$ we define $Rx = S'Tx$ where $S' \in [S]$ maps the S-section TX onto X.

THEOREM 1: Let S be a dissipative transformation defined on the measure space (Y, B, m), and let X be an S-section. For every $T \in N(S)$ we let $\sigma(T) = R$ where $R \in G(X)$ is the transformation induced by the pair T and S

as described above. Then σ is a homomorphism from $N(S)$ onto $G(X)$, with $[S]$ as its kernel.

PROOF: Let X be an S-section and $\sigma: N(S) \to G(X)$ be defined as above. Then if one looks at the orbits of S it becomes clear that $\sigma(T)$ = the identity transformation on X if and only if $T \in [S]$. To show that σ is onto we let $R \in G(X)$. We represent $Y = X \times Z$ and let $\tilde{R}(x,n) = (Rx,n)$ for $(x,n) \in Y$. It follows that $\tilde{R}S = S\tilde{R}$, which implies that $\tilde{R} \in N(S)$. From the definition of σ follows that $\sigma(\tilde{R}) = R$, and this completes the proof of Theorem 1.

Next we introduce the following map $\pi: Y \to X$ which is useful in what follows:

We represent $Y = X \times Z$ and define $\pi(x,n) = x$ for $(x,n) \in Y$. It follows that if $S(x,n) = (x, n+1)$, $T \in N(S)$, and $R = \sigma(T)$ where σ is the map defined in Theorem 1 then $\pi \cdot T = R \cdot \pi$. Furthermore, $\pi(\mathrm{Orb}_T(y)) = \mathrm{Orb}_R(\pi y)$ for $y \in Y$.

We say that a transformation R is antiperiodic if $R^n x \neq x$ for $n \neq 0$ and $x \in X$.

THEOREM 2: Let the map $\sigma: N(S) \to G(X)$ be defined as above, and let $T \in N(S)$ with $\sigma(T) = R$. If R is an antiperiodic transformation then σ restricted to $N(S) \cap [T]$ defines an isomorphism from $N(S) \cap [T]$ onto $[R]$.

PROOF: Let $T' \in N(S) \cap [T]$, and $R' = \sigma(T')$. Then for $x \in X$ let $y \in Y$ be such that $\pi y = x$. Since $T' \in [T]$ we have $\mathrm{Orb}_{R'}(x) = \pi(\mathrm{Orb}_{T'}(y)) \subset \pi(\mathrm{Orb}_T(y)) = \mathrm{Orb}_R(x)$, and therefore, $R' \in [R]$. To show that the image of $N(S) \cap [T]$ by σ is $[R]$ we let $R' \in [R]$, and let $A_n = \{x \in X: R'x = R^n x\}$. Then $X = \cup_{n \in Z} A_n(\text{disj}) = \cup_{n \in Z} R^n A_n(\text{disj})$. If $y \in Y$ then $\pi y = x \in A_n$ for some $n \in Z$, and we define $T'y = T^n y$. To show that T' is onto we let $y \in Y$. Then $\pi y \in R^n(A_n)$ for some n, and we let $y' = T^{-n}y$. It follows that $\pi y' = \pi(T^{-n}y) = R^{-n}(\pi y) \in A_n$, and $T'y' = T^n(T^{-n}y) = y$. Next we suppose that $T'y_1 = T'y_2$ for $y_1 \in \pi^{-1}(A_p)$ and $y_2 \in \pi^{-1}(A_q)$. Then $T'y_1 = T^p y_1$, and $T'y_2 = T^q y_2$. Therefore,

$$(1) \qquad R^p(\pi y_1) = \pi(T^p y_1) = \pi(T^q y_2) = R^q(\pi y_2).$$

We also have from the definition of R'

$$(2) \qquad R'(\pi y_1) = R^p(\pi y_1) \quad \text{and} \quad R'(\pi y_2) = R^q(\pi y_2).$$

Since R' is 1-1 from (1) and (2) we conclude that $p = q$. Also, since $T^p y_1 = T^p y_2$ we conclude that T' is 1-1. Thus $T' \in G(Y)$, and clearly $T' \in [T]$ with $\sigma(T') = R'$. Finally, from Theorem 1, since $[S]$ is the kernel of σ it follows that the restriction to σ to $N(S) \cap [T]$ will have $[S] \cap [T]$ for its kernel. Since R is antiperiodic we conclude that $[S] \cap [T]$ = the identity element, and this completes the proof of the Theorem.

We remark that the construction of T' above shows that if $T \in C(S)$ then T' is also in $C(S)$; hence if $T \in C(S)$ then $N(S) \cap [T] = C(S) \cap [T]$.

COROLLARY 1: Let $\sigma: N(S) \to G(X)$ be defined as above, and let $T \in N(S)$ with $\sigma(T) = R$. Then $[T] \cap \sigma^{-1}(R') \neq \phi$ if and only if $R' \in [R]$. Moreover, $[T] \cap \sigma^{-1}(R')$ consists of a single element in case $R' \in [R]$ and R' is antiperiodic.

PROOF: If $T' \in [T] \cap \sigma^{-1}(R')$ then $R' = \sigma(T') \in [R]$ by Theorem 2. Conversely, if $R' \in [R]$, then again by Theorem 2 there exists $T' \in [T] \cap N(S)$ such that $\sigma(T') = R'$; in fact, in case R' is antiperiodic such a T' is unique.

COROLLARY 2: Let $\sigma: N(S) \to G(X)$ be defined as above, and let $R \in E_{II}(X)$. Then there exists a transformation $T \in \sigma^{-1}(R)$ such that $T \in E_{II_\infty}(Y) \cap C(S)$.

PROOF: We let $R \in E_{II_1}(X)$. Then by Dye's Theorem there exists a von Neumann transformation $R_1 \in [R]$ such that $[R_1] = [R]$; see [5]. A similar construction as in [3] shows the existence of $T_1 \in E_{II_\infty}(Y)$ such that $\sigma(T_1) = R_1$; in fact T_1 commutes with S. Since R_1 is ergodic, by Theorem 2, $[R_1]$ is isomorphic with $N(S) \cap [T_1]$. This says that there exists $T \in [T_1]$ such that $\sigma(T) = R$. Since T_1 commutes with S, the construction of $T \in [T_1]$ as described in Theorem 2 shows that T also commutes with S.

Next we prove that T is ergodic by showing $T_1 \in [T]$. This follows from the fact that R_1 is ergodic and $R_1 \in [R]$; by corollary 1 there exists an element $T_2 \in [T] \cap \sigma^{-1}(R_1) \subset [T_1] \cap \sigma^{-1}(R_1)$. But since $[T_1] \cap \sigma^{-1}(R_1)$ consists of the unique element T_1 we conclude that $T_1 = T_2 \in [T]$.

Finally we assume $R \in E_{II_\infty}(X)$, and let $m_1 \sim m$ be the invariant measure for R. We choose $A \subset X$ with $m_1(A) = 1$, and let R_A be the induced transformation by R on A defined by the first return time to A. It follows that $R_A \in E_{II_1}(A)$, and hence there exists a transformation $T_1 \in E_{II_\infty}(A \times Z)$ such that $\sigma(T_1) = R_A$. We define $T \in G(Y)$ by

$$Ty = T(x,n) = \begin{cases} (Rx,n) & \text{if } x \notin R^{-1}A \\ T_1(R^{-k}x,n) & \text{if } x \in R^kA_{k+1} \quad \text{for } k=0,1,2,\ldots \end{cases}$$

where $A_k = \{x \in A:$ k is the smallest positive integer such that $R^kx \in A\}$. Then $A = \bigcup_{1 \le k < \infty} A_k$ (disj), $X = \bigcup_{1 \le k < \infty} \bigcup_{0 \le j < k} R^jA_k$ (disj), and $R^{-1}(A) = \bigcup_{1 \le k < \infty} R^{k-1}(A_k)$ (disj). This shows that T is well defined. It is clear that $\sigma(T) = R$, and T induces the transformation T_1 on $A \times Z$. Since T_1 is ergodic we get $T \in E_{II_\infty}(Y)$, and $T_1 \in C(S)$ implies that T also belongs to $C(S)$. This completes the proof of the Corollary.

We mention here that a similar result as the above Corollary was obtained in [1] using different methods. Combining results in [4] with the above we conclude the following Corollary and state it without proof:

COROLLARY 3: Let $\sigma: N(S) \to G(X)$ be defined as above and let $R \in G(X)$ be ergodic. Then $\sigma^{-1}(R) \cap E_{II_\infty}(Y) \neq \phi$ if and only if $R \notin E_{III_0}$.

In reference to Corollary 3 we mention that E_{III_0} consists of all ergodic transformations $R \in G(X)$ such that for any $\mu \sim m$ the set $V = \{V \in [R]: V \text{ preserves } \mu\}$ is not ergodic. In the terminology of [8] those are the transformations in E_{III} which do not contain a measure.

THEOREM 3: Let $\sigma: N(S) \to G(X)$ be defined as in Theorem 1, and let $T_1, T_2 \in N(S)$ with $\sigma(T_1) = R_1$ and $(T_2) = R_2$. Suppose that T_1 and T_2 are both dissipative transformations with A_1 a T_1-section, A_2 a T_2-section and such that both A_1 and A_2 are S-sections also. Then

(I) There exists $S_1 \in [S]$ with $S_1[T_1]S_1^{-1} = [T_2]$ if and only if $[R_1] = [R_2]$,

and

(II) There exists $P \in N(S)$ with $P[T_1]P^{-1} = [T_2]$ if and only if there exists $Q \in G(X)$ with $Q[R_1]Q^{-1} = [R_2]$.

PROOF: Suppose for a $P \in N(S)$ we have $P[T_1] = [T_2]P$. Then by Theorem 2 we get $\sigma(P[T_1] \cap N(S)) = \sigma(P)\,\sigma([T_1] \cap N(S)) = \sigma(P)[R_1]$ and $\sigma([T_2]P \cap N(S)) = \sigma([T_2] \cap N(S))\,\sigma(P) = [R_2]\,\sigma(P)$. We let $Q = \sigma(P)$; then it follows that $Q[R_1] = [R_2]Q$. In the above argument we let $P = S' \in [S]$. Since $\sigma(S') = $ Identity, then if $S'[T_1] = [T_2]S'$ we get $[R_1] = [R_2]$. This proves the if parts of (I) and (II) of the Theorem.

Conversely, we assume $[R_1] = [R_2]$. For every $y \in Y$, since A_1 is a T_1-section, there exists a unique $i = i(y)$ such that $T_1^i y \in A_1$. Since $[R_1] = [R_2]$ there exists a unique $k = k(y)$ such that $\pi(T_1^i y) = R_1^i(\pi y) = R_2^k(\pi y) = \pi(T_2^k y)$. Similarly, interchanging the roles of T_1 and T_2, for every $y \in Y$ we get a unique j and a unique n such that $T_2^j y \in A_2$ and $\pi(T_2^i y) = R_2^j(\pi y) = R_1^n(\pi y) = \pi(T_1^n y)$. We let $S_o \in [S]$ be such that $S_o A_1 = A_2$, and for $y \in Y$ we define $S_1 y = T_2^{-k} S_o T_1^i y$ and $S_1' y = T_1^{-n} S_o^{-1} T_2^j y$. From the uniqueness of all these integers follows that $S_1 S_1' y = S_1' S_1 y = y$. This shows that S_1 is 1-1 and onto transformation. Furthermore, $\pi(S_1 y) = \pi(T_2^{-k} S_o T_1^i y) = R_2^{-k} R_1^i(\pi y) = \pi y$ shows that $S_1 \in [S]$.

Finally, from the definition of S_1, for $y \in Y$ we get $S_1(\mathrm{Orb}_{T_1}(y)) = \mathrm{Orb}_{T_2}(S_0 T_1^i y) = \mathrm{Orb}_{T_2}(T_2^{-k} S_0 T_1^i y) = \mathrm{Orb}_{T_2}(S_1 y)$, which shows that $S_1[T_1] \subset [T_2]S_1$. Similarly, using S_1' instead of S_1, we obtain the reverse inclusion. Therefore, we conclude $S_1[T_1] = [T_2]S_1$, and this proves (I).

Next we suppose $Q[R_1]Q^{-1} = [R_2]$ for some $Q \in G(X)$. We let $QR_1Q^{-1} = R_1' \in G(X)$; then $[R_1'] = [R_2]$. We define $\hat{Q} \in G(Y)$ by $\hat{Q}y = \hat{Q}(x,n) = (Qx,n)$ for $y = (x,n) \in Y$. Then $\hat{Q}$ commutes with S and $\sigma(\hat{Q}) = Q$. Since A_1 is a T_1 and an S-section it follows that $\hat{Q}(A_1)$ is a T_1' and an S section also, where $T_1' = \hat{Q}T_1\hat{Q}^{-1}$. Thus by part (I) there exists $S' \in [S]$ such that $S'[T_1']S'^{-1} = [T_2]$. If we let $P = S'\hat{Q}$, then $P \in N(S)$ and $P[T_1]P^{-1} = S'\hat{Q}[T_1]\hat{Q}^{-1}S'^{-1} = S'[\hat{Q}T_1\hat{Q}^{-1}]S'^{-1} = S'[T_1']S'^{-1} = [T_2]$. This completes the proof of the Theorem .

In the remainder of this article we present an application of Theorem 3. We give a different proof to a known result in the theory of W^*-algebras, see [2].

We let $(X, \mathcal{B}, m)$ be a σ-finite measure space, $Y = X \times Z$, $\mathcal{A} = \mathcal{B} \times 2^Z$, and $\nu = m \times$ counting measure on Z. We denote by $\mathcal{H}$ the Hilbert space $L^2(Y, \mathcal{A}, \nu)$ and by $\mathcal{B}(\mathcal{H})$ the algebra of all bounded linear operators on $\mathcal{H}$. An element $g \in L^\infty(X) = L^\infty(X, \mathcal{B}, m)$ can be identified with an element $L_g \in \mathcal{B}(\mathcal{H})$ defined by $(L_g\xi)(y) = (L_g\xi)(x,n) = g(x)\,\xi(x,n)$ for $\xi \in \mathcal{H}$. With this identification $L^\infty(X)$ can be regarded as an abelian $*$-subalgebra of $\mathcal{B}(\mathcal{H})$.

For $R \in G(X)$ we define $\tilde{R} \in G(Y)$ by $\tilde{R}y = (Rx, n+1)$ where $y = (x,n) \in Y$. It is clear that $\tilde{R} \in C(S) \subset N(S)$, $\sigma(\tilde{R}) = R$, and X is an $\tilde{R}$-section as well as an S-section. We define $U_{\tilde{R}} \in \mathcal{B}(\mathcal{H})$ by

$$(U_{\tilde{R}}\xi)(x,n) = \xi(\tilde{R}^{-1}(x,n))\left(\frac{d\nu\tilde{R}^{-1}}{d\nu}(x,n)\right)^{1/2} = \xi(R^{-1}x,n-1)\left(\frac{dmR^{-1}}{dm}(x)\right)^{1/2}$$

for $\xi \in \mathcal{H}$. Then, $U_{\tilde{R}}$ is a unitary operator in $\mathcal{B}(\mathcal{H})$. We denote by $\mathcal{W}_R$ the W^*-subalgebra of $\mathcal{B}(\mathcal{H})$ generated by $L^\infty(X)$ and $U_{\tilde{R}}$. By the well known theorem of Murray - von Neumann, $\mathcal{W}_R$ is precisely the double commutant $\{L^\infty(X), U_{\tilde{R}}\}''$. We also note that $\mathcal{W}_R$ is the W^*-algebra constructed from the cyclic group $\{R^n\colon n \in Z\}$ by means of the so called group measure space construction.

We let $T \in G(Y)$ and assume $R \in G(X)$ to be antiperiodic, then the following facts are known and are not difficult to prove; see [10]:

(A) If $U_T \in \mathcal{B}(\mathcal{H})$ is defined by $U_T\xi(y) = \xi(T^{-1}y)\left(\frac{d\nu T^{-1}}{d\nu}(y)\right)^{1/2}$ for $\xi \in \mathcal{H}$, then $U_T \in \mathcal{W}_R$ if and only if $T \in C(S) \cap [\tilde{R}]$.

(B) If V is a unitary operator in $\mathcal{W}_R$ and if $VL_gV^{-1} = L_{g\cdot R'}$ for every $g \in L^\infty(X)$ and for some $R' \in G(X)$, then $R' \in [R]$.

Suppose $R_1, R_2 \in G(X)$ and $\mathcal{W}_1$ and $\mathcal{W}_2$ are the corresponding W^*-algebras as defined above. We say that $\mathcal{W}_1$ and $\mathcal{W}_2$ are strongly *-isomorphic if there exists a *-isomorphism Φ of $\mathcal{W}_1$ and $\mathcal{W}_2$ such that $\Phi(L^\infty(X)) = L^\infty(X)$. As a corollary to Theorem 3, we then have

COROLLARY 4: <u>Let</u> $R_1, R_2 \in G(X)$ <u>be antiperiodic and let</u> $\mathcal{W}_1$ <u>and</u> $\mathcal{W}_2$ <u>be the corresponding</u> W^*<u>-algebras. Then,</u> $\mathcal{W}_1$ <u>and</u> $\mathcal{W}_2$ <u>are strongly *-isomorphic if and only if there exists a</u> $Q \in G(X)$ <u>such that</u> $Q[R_1] = [R_2]Q$.

PROOF: Suppose $Q[R_1] = [R_2]Q$. Since $\sigma(\tilde{R}_1) = R_1$, $\sigma(\tilde{R}_2) = R_2$, and since X is a section for $\tilde{R}_1$, $\tilde{R}_2$ and S, using Theorem 3 we obtain $P \in N(S)$ such that $P[\tilde{R}_1] = [\tilde{R}_2]P$, and $\sigma(P) = Q$. We let $V = U_P \in \mathcal{B}(H)$ be defined by

$$V\xi(y) = \xi(P^{-1}y)\left(\frac{d\nu P^{-1}}{d\nu}(y)\right)^{1/2}$$

and let $\Phi\colon \mathcal{B}(H) \to \mathcal{B}(H)$ be defined by

$$\Phi(T) = VTV^{-1} \quad \text{for} \quad T \in \mathcal{B}(H).$$

Then, Φ is a *-isomorphism of $\mathcal{B}(H)$, and if $g \in L^\infty(X)$ then $\Phi(L_g) = VL_gV^{-1} = L_{g\cdot Q^{-1}}$ where $g\cdot Q^{-1}(x) = g(Q^{-1}(x))$. Since $g\cdot Q^{-1} \in L^\infty(X)$ if and only if $g \in L^\infty(X)$, we get $\Phi(L^\infty(X)) = L^\infty(X)$. Next; $\Phi(U_{\tilde{R}_1}) = VU_{\tilde{R}_1}V^{-1} = U_{P\tilde{R}_1P^{-1}}$. Since $P\tilde{R}_1P^{-1} \in [\tilde{R}_2]$ and since $P, \tilde{R}_1 \in N(S)$ we have $P\tilde{R}_1P^{-1} \in [\tilde{R}_2] \cap N(S)$. Since $R_2 \in C(S)$ we have $[\tilde{R}_2] \cap N(S) = [\tilde{R}_2] \cap C(S)$ as was pointed out in the Remark after Theorem 2. Therefore, $P\tilde{R}_1P^{-1} \in [\tilde{R}_2] \cap C(S)$, and by (A) we get that $\Phi(U_{\tilde{R}_1}) \in \mathcal{W}_2$. Therefore, $\Phi(\mathcal{W}_1) \subset \mathcal{W}_2$. Replacing Φ by Φ^{-1} we obtain $\Phi(\mathcal{W}_1) = \mathcal{W}_2$. This shows that Φ is a strong *-isomorphism of $\mathcal{W}_1$ and $\mathcal{W}_2$.

Conversely, suppose that there exists a strong *-isomorphism Φ of $\mathcal{W}_1$ onto $\mathcal{W}_2$. If $B \in \mathcal{B}$ then $\chi_B \in L^\infty(X)$, and $\Phi(L_{\chi_B})^2 = \Phi(L_{\chi_B}^2) = \Phi(L_{\chi_B})$, so that $\Phi(L_{\chi_B})$ corresponds to an element $g \in L^\infty(X)$ which takes values in $\{0,1\}$. If we let $B' = \{x\colon g(x) = 1\}$, then $\Phi(L_{\chi_B}) = L_{\chi_{B'}}$. Therefore, we define a map Q of the Boolean algebra $(\mathcal{B}, m)$ onto itself by setting $Q(B) = B'$. Since Φ is an algebra isomorphism, Q is a Boolean automorphism of $(\mathcal{B}, m)$, and hence by a theorem of von Neumann [11], since $(X, \mathcal{B}, m)$ is a Lebesgue space, we may assume that Q is induced by an element of $G(X)$, which we denote again by Q.

Next, we let $V = \Phi(U_{\tilde{R}_1})$. Then V is a unitary operator belonging to $\mathcal{W}_2$ and for every $B \in \mathcal{B}$ we have

$$VL_{\chi_B}V^{-1} = \Phi(U_{\tilde{R}_1})\Phi(L_{\chi_{Q^{-1}B}})\,\Phi(U_{\tilde{R}_1})^{-1} = \Phi(U_{\tilde{R}_1}L_{\chi_{Q^{-1}B}}U_{\tilde{R}_1^{-1}})$$

$$= \Phi(L_{\chi_{R_1Q^{-1}B}}) = L_{\chi_{QR_1Q^{-1}B}} .$$

Therefore, if we denote by $R_2' = QR_1Q^{-1}$, then

$$VL_{\chi_B}V^{-1} = L_{\chi_{R_2'B}} = L_{\chi_B \circ R_2'^{-1}} \qquad \text{for every } B \in \mathcal{B} .$$

From this it follows that for every $g \in L^\infty(X)$, $VL_gV^{-1} = L_{g\circ R_2'^{-1}}$ holds, and therefore, by (B) $R_2'^{-1} \in [R_2]$. Then $R_2' = QR_1Q^{-1} \in [R_2]$ and consequently, $Q[R_1]Q^{-1} \subset [R_2]$. Using Φ^{-1} and reversing the roles of R_1 and R_2, we obtain the opposite inclusion, and therefore, $Q[R_1] = [R_2]Q$. This completes the proof of the Corollary.

REMARK: A result similar to Corollary 4 has been shown by H. Choda in [2] using techniques of cross products. We should point out also that the main result of W. Krieger in [9] proves that if $R_1, R_2 \in E(X)$ then there exists a $Q \in G(X)$ such that $Q[R_1] = [R_2]Q$ if and only if $\mathcal{W}_1$ and $\mathcal{W}_2$ are *-isomorphic. From this it follows that if there exists a *-isomorphism between $\mathcal{W}_1$ and $\mathcal{W}_2$ then there exists a strong *-isomorphism between them. However, a direct proof of this fact seems to be difficult.

REFERENCES

[1] A. LoBello; Ergodic Transformations of Lebesgue Spaces, Studies in Probability Theory and Ergodic Theory, Advances in Math. Suppl. Studies 2 (1978), 287-293.

[2] H. Choda; On the Crossed Product of Abelian von Neumann Algebras, I, II Proc. Japan Acad. 43 (1967), 111-116, 198-201.

[3] A. Hajian, Y. Ito, and S. Kakutani; Invariant Measures and Orbits of Dissipative Transformations., Adv. in Math. 9 (1972), 52-65.

[4] A. Hajian, Y. Ito and S. Kakutani; Orbits, Sections, and Induced Transformations., Israel Jour. of Math. 18 (1974), 97-115.

[5] A. Hajian, Y. Ito and S. Kakutani; Full Groups and a Theorem of Dye., Advances in Math. 17 (1975), 48-59.

[6] A. Hajian and S. Kakutani; Example of an Ergodic Measure Preserving Transformation on an Infinite Measure Space, Contributions to Ergodic Theory and Probability, Lecture Notes in Mathematics 160, Springer-Verlag, Berlin, 1970; pp. 45-52.

[7] S. Kakutani; Induced Measure Preserving Transformations., Proc. Japan Acad., 19 (1943), 635-641.

[8] W. Krieger; On Non-Singular Transformations of a Measure Space, I, II, Z. Wahr. Verw. G., 11 (1969), 83-97, 98-119.

[9] W. Krieger ; On Ergodic Flows and Isomorphism of Factors, Math. Annalen. 223 (1976), 19-70.

[10] F. J. Murray and J. von Neumann; On Rings of Operators II, Trans. Amer. Math. Soc. 41 (1937), 208-248.

[11] J. von Neumann; Einige Sätze über messbare Abbildungen, Annals of Math. 33 (1932), 574-586.

DEPARTMENT OF MATHEMATICS
NORTHEASTERN UNIVERSITY
BOSTON, MA 02115

DEPARTMENT OF MATHEMATICS
RIKKYO UNIVERSITY
TOKYO, JAPAN

Contemporary Mathematics
Volume 26, 1984

CONJUGATE FOURIER SERIES ON THE CHARACTER GROUP OF THE ADDITIVE RATIONALS

Edwin Hewitt

We consider first an arbitrary noncyclic subgroup of the additive group $\mathbb{Q}$ of rational numbers. Let $\underline{\underline{a}} = (a_o,a_1,a_2,\dots)$ be any infinite sequence of integers all exceeding 1. Let $A_o = 1$ and $A_k = a_oa_1a_2\dots a_{k-1}$ for $k = 1,2,\dots$. Let $\mathbb{Q}_{\underline{\underline{a}}}$ be the set of all numbers $\frac{\ell}{A_n}$ where ℓ runs through $\mathbb{Z}$ and n through $0,1,2,\dots$. Plainly $\mathbb{Q}_{\underline{\underline{a}}}$ is a noncyclic subgroup of $\mathbb{Q}$. It is a classical fact that every such group is isomorphic to a group $\mathbb{Q}_{\underline{\underline{a}}}$ (Beaumont & Zuckerman 1951).

By Pontrjagin's duality theorem, the character group of the discrete group $\mathbb{Q}_{\underline{\underline{a}}}$ is a compact Abelian group $\Sigma_{\underline{\underline{a}}}$, and the group of continuous characters of $\Sigma_{\underline{\underline{a}}}$ is isomorphic with $\mathbb{Q}_{\underline{\underline{a}}}$. We realize the group $\Sigma_{\underline{\underline{a}}}$ as follows. First let $\Delta_{\underline{\underline{a}}}$ be the set of all sequences $\underline{\underline{x}} = (x_o,x_1,x_2,\dots)$ with x_j $\{0,1,2,\dots,a_j-1\}$. Define the sum $\underline{\underline{x}} + \underline{\underline{y}} = \underline{\underline{z}}$ by induction. First write

$$x_o + y_o = z_o + q_oa_o, \quad \text{where } q_o = 0 \text{ or } 1.$$

Define z_n by

$$x_n + y_n + q_{n-1} = z_n + q_na_n,$$

where again q_n is 0 or 1. It is easy to see that $\Delta_{\underline{\underline{a}}}$ is an Abelian group. With the product topology, taking every set $\{0,1,2,\dots,a_j-1\}$ as a discrete space, $\Delta_{\underline{\underline{a}}}$ is a zero-dimensional compact Abelian group. Haar measure λ_o on $\Delta_{\underline{\underline{a}}}$ is the product of the uniform distribution measures on the individual factors. Let $\underline{\underline{u}}$ be the element $(1,0,0,\dots)$ of $\Delta_{\underline{\underline{a}}}$.

We realize $\Sigma_{\underline{\underline{a}}}$ as the set

$$[-\tfrac{1}{2},\tfrac{1}{2}[\times \Delta_{\underline{\underline{a}}},$$

with addition defined by

$$(s,\underline{\underline{x}}) + (t,\underline{\underline{y}}) = (s + t - [s + t + \tfrac{1}{2}],\ \underline{\underline{x}} + \underline{\underline{y}} + [s + t + \tfrac{1}{2}]\underline{\underline{u}}).$$

(For a real number r, $[r]$ denotes the greatest integer not exceeding r.) This definition makes $\Sigma_{\underline{\underline{a}}}$ into an Abelian group. For positive integers n, we define the neighborhood U_n of the neutral element $(0,\underline{\underline{0}})$ of $\Sigma_{\underline{\underline{a}}}$ by

0271-4132/84 $1.00 + $.25 per page

$$]\frac{-1}{2n},\frac{1}{2n}[\times \Lambda_n,$$

where Λ_n is the subgroup of $\Delta_{\underline{\underline{a}}}$ consisting of all $\underline{\underline{x}}$ such that $x_0 = x_1 = x_2 = \cdots = x_{n-1} = 0$. With this definition, $\Sigma_{\underline{\underline{a}}}$ is a compact connected Abelian group. It is torsion-free if and only if $\mathbb{Q}_{\underline{\underline{a}}}$ is the entire group $\mathbb{Q}$, and this occurs if and only if every positive integral power of every prime occurs among the numbers A_k.

For $\alpha = \frac{\ell}{A_k}$ in $\mathbb{Q}_{\underline{\underline{a}}}$ and $(t,\underline{\underline{x}})$ in $\Sigma_{\underline{\underline{a}}}$, we define

(1) $$\chi_\alpha(t,\underline{\underline{x}}) = \exp[2\pi i\frac{\ell}{A_k}(t + \sum_{\nu=0}^{\infty} x_\nu A_\nu)],$$

with the natural convention that we ignore all terms in the infinite series with index $\geq k$. First regard (1) as defining a function on $\Sigma_{\underline{\underline{a}}}$ for a fixed α in $\mathbb{Q}_{\underline{\underline{a}}}$. The mapping $\alpha \mapsto \chi_\alpha$ is an isomorphism of $\mathbb{Q}_{\underline{\underline{a}}}$ onto the (multiplicative!) group of continuous characters of $\Sigma_{\underline{\underline{a}}}$. Next regard (1) as defining a function on $\mathbb{Q}_{\underline{\underline{a}}}$ for a fixed $(t,\underline{\underline{x}})$ in $\Sigma_{\underline{\underline{a}}}$. Every character of the discrete group $\mathbb{Q}_{\underline{\underline{a}}}$ has the form $\alpha \mapsto \chi_\alpha(t,\underline{\underline{x}})$ for some fixed $(t,\underline{\underline{x}})$ in $\Sigma_{\underline{\underline{a}}}$.

Normalized Haar measure μ on $\Sigma_{\underline{\underline{a}}}$ is the product of Lebesgue measure λ on $[-\frac{1}{2},\frac{1}{2}[$ and normalized Haar measure λ_0 on $\Delta_{\underline{\underline{a}}}$ (note that $\Delta_{\underline{\underline{a}}}$ can also be thoguht of as Λ_0).

The writer and Dr. Gunter Ritter (Universitat Erlangen-Nurnberg, Federal Republic of Germany) have in two papers ([1], [2]) studied Fourier series on $\Sigma_{\underline{\underline{a}}}$. Given a function f in $\mathfrak{L}_1(\Sigma_{\underline{\underline{a}}})$, let $\hat{f}$ be the function on $\mathbb{Q}_{\underline{\underline{a}}}$ defined by

$$\hat{f}(\alpha) = \int_{\Sigma_{\underline{\underline{a}}}} f(t,\underline{\underline{x}})\overline{\chi_\alpha(t,\underline{\underline{x}})}d\mu(t,\underline{\underline{x}}) \quad \text{for } \alpha \text{ in } \mathbb{Q}_{\underline{\underline{a}}}.$$

The paper [2] is the theme of this announcement. The group $\mathbb{Q}_{\underline{\underline{a}}}$ has a natural order: we take the unique order in $\mathbb{Q}_{\underline{\underline{a}}}$ that makes 1 positive. Define the function sgn on $\mathbb{Q}_{\underline{\underline{a}}}$ as usual by the rule that $\operatorname{sgn} \alpha$ is 1, 0, or −1 according as α is positive, zero, or negative. Given f in $\mathfrak{L}_1(\Sigma_{\underline{\underline{a}}})$, we define the <u>conjugate Fourier transform</u> g <u>of</u> f as the function on $\mathbb{Q}_{\underline{\underline{a}}}$ such that

$$g(\alpha) = -i\operatorname{sgn} \alpha\, \hat{f}(\alpha) \qquad \text{for } \alpha \text{ in } \mathbb{Q}_{\underline{\underline{a}}}.$$

An abstract form of a classical theorem of Marcel Riesz, due to Helson and Rudin ([3], pp. 216-220) shows that if $p > 1$ and f is in $\mathfrak{L}_p(\Sigma_{\underline{\underline{a}}})$, then g is the Fourier transform of a certain function $\tilde{f}$ in $_p(\Sigma_{\underline{\underline{a}}})$, <u>i.e.</u>,

(2) $$\hat{\tilde{f}}(\alpha) = -i\operatorname{sgn} \alpha\, \hat{f}(\alpha) \qquad \text{for } \alpha \text{ in } \mathbb{Q}_{\underline{\underline{a}}},$$

and furthermore

$$\|\tilde{f}\|_p \leq A_p\|f\|_p,$$

where A_p depends only upon p.

In the classical case $\mathbb{Q}_{\underline{\underline{a}}}$ is replaced by the group $\mathbb{Z}$ of all integers and $\Sigma_{\underline{\underline{a}}}$ by the circle group T. Riesz showed in this case that the conjugate function $\tilde{f}$ is almost everywhere on T equal to a limit:

$$\text{(3)} \qquad \tilde{f}(\theta) = \lim_{r\uparrow 1} \int_{-\frac{1}{2}}^{\frac{1}{2}} f(\theta - t)\frac{2r\sin(2\pi t)}{1 - 2r\cos(2\pi t) + r^2}dt,$$

where we realize T as the interval $[-\frac{1}{2},\frac{1}{2}[$ with addition modulo 1. The abstract version of Marcel Riesz's theorem yields no such pointwise realization of $\tilde{f}$: it is an existence theorem pure and simple.

In [2], a pointwise realization of $\tilde{f}$ in the case $\Sigma_{\underline{\underline{a}}}$ is obtained. We proceed as follows. For f in $\mathfrak{L}_1(\Sigma_{\underline{\underline{a}}})$, write

$$\text{(4)} \quad P_u * f(t,\underline{\underline{x}}) = \frac{1}{\pi}\int_{-\infty}^{\infty} f(v - [\tfrac{1}{2} + v],\ \underline{\underline{x}} + [\tfrac{1}{2} + v]\underline{\underline{u}})\frac{u}{u^2 + (t - v)^2}\,dv$$

One proves by standard arguments that the integral in (4) exists as a Lebesgue integral for all positive real numbers u, all real numbers t, and all $\underline{\underline{x}}$ in a subset E_f of $\Delta_{\underline{\underline{a}}}$ which has the property that $\lambda_o(\Delta_{\underline{\underline{a}}}\setminus E_f) = 0$. The function (4) is called <u>the Poisson integral of</u> f. It is very like the classical Poisson integral except for being singular with respect to Haar measure on $\Sigma_{\underline{\underline{a}}}$. The analogue of the conjugate Poisson integral (3) is

$$\text{(5)} \qquad K_u f(t,\underline{\underline{x}}) =$$

$$\frac{1}{\pi}\int_{-\infty}^{\infty} f(v - [\tfrac{1}{2} + v],\ \underline{\underline{x}} + [\tfrac{1}{2} + v]\underline{\underline{u}})\left[\frac{t - v}{u^2 + (t - v)^2} + \frac{v}{1 + v^2}\right]dv.$$

The summand $\dfrac{v}{1 + v^2}$ is required in (5) to make the integral exist. The integral in (5) exists as a Lebesgue integral for all $\underline{\underline{x}}$ in E_f, all positive real u, and all real t. The function $K_u f(t,\underline{\underline{x}})$ is <u>not</u> convolution of f with a measure, nor indeed can it be. One proves by standard arguments that

$$\text{(6)} \qquad \lim_{u\downarrow 0} K_u f(t,\underline{\underline{x}}) = Kf(t,\underline{\underline{x}})$$

exists for all $\underline{\underline{x}}$ in E_f and almost all real numbers t. Rather delicate but classical estimates show that the mapping $f \mapsto Kf$ is a bounded linear mapping of $\mathfrak{L}_p(\Sigma_{\underline{\underline{a}}})$ into itself for all $p > 1$. The function Kf is <u>not</u> however the conjugate function $\tilde{f}$. The summand $\dfrac{v}{1 + v^2}$ in the integral in (5) prevents this. To obtain $\tilde{f}$, we must introduce a correction term.

To this end, consider the subgroup $\{0\} \times \Lambda_n$ $(n = 1,2,3,\ldots)$ of $\Delta_{\underline{\underline{a}}}$ and normalized Haar measure λ_n on this subgroup. Given f in $\mathfrak{L}_1(\Sigma_{\underline{\underline{a}}})$, the convolution $f*\lambda_n$ has the property that

$$\lim_{k\to\infty} \frac{1}{\pi} \int_{-kA_k}^{kA_k} f*\lambda_n(v - [\tfrac{1}{2} + v], \underline{\underline{x}} + [\tfrac{1}{2} + v]\underline{\underline{u}}) \frac{v}{1 + v^2} dv = C_{f*\lambda_n}(\underline{\underline{x}})$$

exists for all $\underline{\underline{x}}$ in $\Delta_{\underline{\underline{a}}}$. This is proved by Poisson's summation formula. We then prove that the sequence of functions $(C_{f*\lambda_n})_{n=1}^{\infty}$ is a martingale on $\Delta_{\underline{\underline{a}}}$. The nth σ-algebra of sets is the finite family of subsets of $\Delta_{\underline{\underline{a}}}$ that depend only on the first n coordinates in $\Delta_{\underline{\underline{a}}}$. For f in $\mathfrak{L}_p(\Sigma_{\underline{\underline{a}}})$ $(p > 1)$, a classical theorem of Doob shows that

$$\lim_{n\to\infty} C_{f*\lambda_n}(\underline{\underline{x}}) = C_f(\underline{\underline{x}})$$

exists for λ_o-almost all $\underline{\underline{x}}$ in $\Delta_{\underline{\underline{a}}}$ and is in $\mathfrak{L}_p(\Delta_{\underline{\underline{a}}})$. Finally we show that

$$(Kf + C_f)^{\wedge}(\alpha) = -i\,\mathrm{sgn}\,\alpha\, \hat{f}(\alpha) \quad \text{for} \quad \alpha \quad \text{in} \quad \mathbb{Q}_{\underline{\underline{a}}}.$$

That is, we construct $\tilde{f}$ as the function $Kf + C_f$.

Therefore the group $\Sigma_{\underline{\underline{a}}}$ admits a complete pointwise analogue of Marcel Riesz's celebrated theorem on conjugate functions.

LITERATURE

[1] Hewitt, Edwin and Ritter, Gunter. Fourier series on certain solenoids. Math. Annalen 257 (1981), 61-83.

[2] Hewitt, Edwin and Ritter, Gunter. Conjugate Fourier series on certain solenoids. Trans. Amer. Math. Soc. 276(1983), 817-840.

[3] Rudin, Walter. Fourier analysis on groups. New York: John Wiley & Sons, 1962.

DEPARTMENT OF MATHEMATICS
UNIVERSITY OF WASHINGTON
SEATTLE, WA 98195

Contemporary Mathematics
Volume 26, 1984

A STOCHASTIC DIFFERENTIAL EQUATION IN INFINITE DIMENSIONS

Kiyosi Itô

§1. NOTATION AND INTRODUCTION.

Throughout this paper we use the following notation.

(Ω, P)	:	the basic probability space,
ω	:	a generic point of Ω ,
m	:	the Lebesgue measure on the real numbers R,
C	:	the continuous functions: $[0,\infty) \to R$,
γ	=	$(\gamma_t,\ t \in [0,\infty))$: a generic element of C,
P_x	:	the Wiener measure on C starting at x,
E_x	:	the expectation based on P_x,
$L_2(\Omega)$	=	$L_2(\Omega,P)$, $L_2(R) = L_2(R,m)$,
S	:	the rapidly decreasing functions,
S'	:	the tempered distributions,
$g_t(x)$	:	the Gauss density with mean 0 and variance t,
1_E	:	the indicator of a set E.

Suppose that we are given an infinite particle system satisfying the following conditions:

(C.1) The counting measure of the starting points is a Poisson random measure with intensity $\lambda > 0$,

(C.2) Under the condition that the starting points are given, the particles make independent Brownian motions.

The counting measure of the positions occupied by the particles at time t, $N_t^\lambda(A,\omega)$, is regarded as a measure-valued stochastic process which corresponds to an S'-valued process:

$$N_t^\lambda(\varphi,\omega) = \int_R \varphi(x) N_t^\lambda(dx,\omega), \quad \varphi \in S \tag{1.1}$$

Normalizing this we define

$$X_t^\lambda(\varphi,\omega) = \lambda^{-1/2}\,(N_t^\lambda(\varphi,\omega) - E(N_t^\lambda(\varphi,\omega)))$$
$$= \lambda^{-1/2}\,(N_t^\lambda(\varphi,\omega) - \lambda \int_R \varphi(x)dx). \tag{1.2}$$

As λ tends to ∞, the system of random variables $\{X_t(\varphi)\}_{t,\varphi}$ converges in distribution to a centered Gaussian system $\{X_t(\varphi)\}_{t,\varphi}$ with covariance functional:

$$E(X_s(\varphi)\,X_t(\psi)) = \int_R dxE_x(\varphi(\gamma_s)\psi(\gamma_t)) \tag{1.3}$$

The purpose of this paper is to prove the following:

(i) $X_t(\varphi)$ is regarded as a sample continuous S'-valued process by taking a regular version.
(ii) This version satisfies a stochastic differential equation in infinite dimensions.

A similar problem was discussed in our previous paper [1]. Also E. B. Dynkin discussed stochastic differential equations governing infinite dimensional Gaussian processes [2].

§2. THE REGULAR VERSION OF $\{X_t(\varphi)\}_{t,\varphi}$

Fix t for the moment. By virtue of (1,3) the map $\varphi \to X_t(\varphi)$ is linear and continuous as a map: $S \to L_2(\Omega)$. But the sample functional $\varphi \to X_t(\varphi,\omega)$ is neither linear nor continuous in general. Hence $X_t(\cdot,\omega)$ is not an S'-valued random variable. However, modifying $X_t(\varphi,\omega)$ on a P-null ω-set which may depend on φ, we can make $X_t(\cdot,\omega)$ be an S'-valued random variable, as we can prove below. This modified one is called an S' regularization of $\{X_t(\varphi,\omega)\}_\varphi$.

THEOREM 2.1. <u>There exists an S'-valued random variable</u> $\tilde{X}_t = (\tilde{X}_t(\varphi),\ \varphi \in S)$ <u>such that</u> $\tilde{X}_t(\varphi) = X_t(\varphi)$ <u>a.s. for every</u> $\varphi \in S$.

PROOF. The map $\varphi \to X_t(\varphi)$ is a linear map: $S \to L_2(\Omega)$. This map is norm-preserving, because (1,3) implies that

$$E(X_t(\varphi)^2) = \int_R dx\, E_x(\varphi(\gamma_t)^2) = \int_R dx \int_R dy\ g_t(y-x)\varphi(y)^2$$
$$= \int_R dy\ \varphi(y)^2$$

i.e.

$$\|X_t(\varphi)\| = \|\varphi\| \tag{2.1}$$

Since S is dense in $L_2(R)$, the map $\varphi \to X_t(\varphi)$ can be extended to a linear norm-preserving map: $L_2(R) \to L_2(\Omega)$, which we will denote by the same notation X_t. Define a stochastic process $Y_t(x)$, $-\infty < x < \infty$, x being regarded as a time parameter, as follows:

$$Y_t(x) = X_t(1_{(0,x]}) \ (x \geqq 0), \ -X_t(1_{(x,0)}) \ (x < 0).$$

Then $Y_t(x)$ is a centered Gaussian process and (2,1) implies that

$$E((Y_t(y) - Y_t(x))^2) = E(X_t(1_{(x,y]})^2) = y - x.$$

Since $Y_t(y) - Y_t(x)$ is centered Gauss distributed, we have

$$E((Y_t(y) - Y_t(x))^4) = 3(y - x)^2.$$

Hence the Kolmogorov continuous version theorem ensures that $Y_t(x)$, $-\infty < x < \infty$, has a continuous version, which we will denote by the same notation. Since

$$E(\int_{-\infty}^{\infty} \frac{Y(x)^2}{1 + |x|^3} dx) = \int_{-\infty}^{\infty} \frac{x}{1+|x|^3} dx < \infty,$$

the sample function $(Y(x,\omega), -\infty < x < \infty)$ defines the following distribution a.s. :

$$Y_t(\varphi,\omega) = \int_R Y_t(x)\varphi(x)dx, \quad \varphi \in S .$$

Hence the distribution derivative of $Y_t(\cdot,\omega)$ is defined by

$$\partial Y_t(\varphi,\omega) = - \int_R Y_t(x) \varphi'(x)dx.$$

Since every $\varphi \in S$ is approximated uniformly by linear combinations of $1_{(x,y]}$, $-\infty < x < y < \infty$, we can easily prove

$$X_t(\varphi,\omega) = - \int_R Y_t(x) \varphi'(x)dx \quad \text{a.s.},$$

where the integral is to be understood in the sense of Bochner integrals in $L_2(\Omega)$. Thus we obtain

$$X_t(\varphi) = \partial Y_t(\varphi) \quad \text{a.s.} \quad \text{for every } \varphi \in S .$$

Since $\partial Y_t = (\partial Y_t(\varphi,\omega), \ \varphi \in S)$ is clearly a tempered distribution for every ω, $\tilde{X}_t = \partial Y_t$ is what we wanted to construct. ∎

As we can see in the discussion above, $Y_t(x)$, $-\infty < x < \infty$, is a Wiener process with time parameter $x \in (-\infty,\infty)$. Hence $\tilde{X}_t = \partial Y_t$ is a white noise.

From now on we assume that $(X_t(\varphi), \varphi \in S)$ is already regularized, so that $\{X_t, t \in [0,\infty)\}$ is an $\mathcal{S}'$-valued stochastic process. We will modify X_t on a P-null ω-set for every t to obtain a sample continuous version of X_t. To do this we will mention some known facts on the topology in $\mathcal{S}'$. Let $H_n(x)$ be the Hermite polynomial of degree n and $h_n(x)$ the corresponding Hermite functions, i.e.

$$h_n(x) = c_n H_n(x)\, e^{-x^2/2}, \quad n = 0, 1, 2, \ldots,$$

where c_n is a constant for which $\|h_n\| = 1$. The Hermite functions form an orthonormal base in $L_2(R)$ and belong to $\mathcal{S}$. The $(-p)$-norm of $f \in \mathcal{S}'$ are defined by

$$\|f\|_{-p}^2 = \Sigma_{n=0}^{\infty} f(h_n)^2 (2n+1)^{-p} \in [0,\infty] \tag{2.2}$$

Let $\mathcal{S}'_p$ be the subspace of $\mathcal{S}'$ consisting of all elements in $\mathcal{S}'$ with finite $(-p)$-norm. $\mathcal{S}'_p$ is a Hilbert space with norm $\| \;\|_{-p}$, $\mathcal{S}'_1 \subset \mathcal{S}'_2 \subset \ldots \to \mathcal{S}'$ and the $(-p)$-norm topology in $\mathcal{S}'_p$ coincides with the topology induced from that in $\mathcal{S}'$.

Using (2.1) and (2.2), we obtain

$$\begin{aligned} E(\|X_t\|_{-2}^2) &= \Sigma_n E(X_t(h_n)^2)(2n+1)^{-2} = \Sigma_n \|h_n\|^2 (2n+1)^{-2} \\ &= \Sigma_n (2n+1)^{-2} < \infty . \end{aligned}$$

Hence

$$X_t \in \mathcal{S}'_2 \quad \text{a.s.} \quad \text{for every} \quad t, \tag{2.3}$$

and so X_t is regarded as an $\mathcal{S}'_p$-valued process for every $p \geqq 2$.

Now we will evaluate $E(\|X_t - X_s\|_{-4}^2)$ when $0 \leqq t - s \leqq 1$. By the definition of $(-p)$-norm we have

$$\begin{aligned} &E(\|X_t - X_s\|_{-4}^2) \\ &= \Sigma_n E((X_t(h_n) - X_s(h_n))^2)(2n+1)^{-4}. \end{aligned}$$

But

$$\begin{aligned} &E((X_t(h_n) - X_s(h_n))^2) \\ &= \int_R dx\, E_x[(h_n(\gamma_t) - h_n(\gamma_s))^2] \\ &= \int_R dx\, E_x\left(\left(\int_s^t h_n'(\gamma_u)\, d\gamma_u + \int_s^t \tfrac{1}{2} h_n''(\gamma_u) du\right)^2\right), \end{aligned}$$

because γ_u, $\gamma \in (C, P_x)$, is a Wiener process starting at x on the probability space (C, P_x). The last integral is no more than

$$2 \int_R dx \, [E_x((\int_s^t h_n'(\gamma_u)d\gamma_u)^2) + E_x((\int_s^t \frac{1}{2} h_n''(\gamma_u)du)^2)]$$

$$\leqq 2 \int_R dx \, [\int_s^t E_x(h_n'(\gamma_u)^2)du + \int_s^t \frac{1}{4} E_x(h_n''(\gamma_u)^2)du]$$

$$= 2 \int_s^t \|h_n'\|^2 \, du + \frac{1}{2} \int_s^t \|h_n''\|^2 \, du$$

$$= (t - s)\,(2\,\|h_n'\|^2 + \frac{1}{2}\,\|h_n''\|^2),$$

because

$$\int_R dx \, E_x(\varphi(\gamma_u)^2) = \int_R dx \int_R g_u(y-x)\,\varphi(y)^2 \, dy = \|\varphi\|^2.$$

Since $h_n' = a_n h_{n-1} + b_n h_{n+1}$ with $a_n, b_n = O(n^{1/2})$, we have

$$\|h_n'\| = O(n^{1/2}) \quad \text{and} \quad \|h_n''\| = O(n).$$

Therefore

$$E(\|X_t - X_s\|_{-4}^2) \leqq c_1(t-s) \;\; \Sigma_m \, n^2(4n+1)^{-4} = c_2(t-s) \tag{2.4}$$

where c_1 and c_2 are positive constants and $0 \leqq t - s \leqq 1$.

Next we will evaluate $E(\|X_t - X_s\|_{-4}^4)$ when $0 \leqq t-s \leqq 1$. By the definition of $(-p)$-norm we have

$$E(\|X_t - X_s\|_{-4}^4) = E((\Sigma_n(X_t(h_n) - X_s(h_n))^2(2n+1)^{-4})^2)$$

$$\leqq (\Sigma_n \, E((X_t(h_n) - X_s(h_n))^4)^{1/2} \, (2n+1)^{-4})^2$$

$$= 3(\Sigma_n \, E((X_t(h_n) - X_s(h_n))^2)(2n+1)^{-4})^2$$

$$= 3(E(\|X_t - X_s\|_{-4}^2))^2,$$

where we used that fact that if $X(\omega)$ is centered Gauss distributed, then $E(X^4) = 3(E(X^2))^2$. Thus we obtain

$$E(\|X_t - X_s\|_{-4}^4) \leqq 3C_2^2 \, (t - s)^2 \tag{2.5}$$

whenever $0 \leqq t - s \leqq 1$.

Since S'_4 is a complete metric space, the Kolmogorov continuous version theorem holds for S'_4-valued processes. The inequality (2.5) ensures that the assumption of this theorem is satisfied for the process $\{X_t\}$. Hence we have the following.

THEOREM 2.2. The S'_4-valued process $\{X_t\}$ has a sample continuous version.

From now on we will denote this version by the same notation X_t. It is obvious that X_t is regarded as a sample continuous S'-valued process.

§3. A STOCHASTIC DIFFERENTIAL EQUATION IN INFINITE DIMENSIONS.

In this section we will prove that the sample continuous S'-valued process X_t satisfies the following stochastic differential equation:

$$dX_t = \partial d\, b_t + \frac{1}{2} \partial^2 X_t\, dt \tag{3.1}$$

where b_t is a standard Wiener S'-valued process whose definition will be given below.

A sample continuous S'-valued centered Gaussian process $b_t = b_t(\varphi, \omega)$ is called a standard Wiener S'-valued process if its covariance functional is given by

$$E(b_s(\varphi)\, b_t(\psi)) = (s \wedge t)(\varphi, \psi) \tag{3.2}$$

The existence and uniqueness (in distributions) of such process b_t is known [1]. Since

$$E(b_t(\varphi)^2) = t\, \|\varphi\|^2 \tag{3.3}$$

b_t is a white noise with intensity t for each t. The S'-valued process b_t, $t \in [0,\infty)$ has independent stationary increments. Hence b_t plays the same role in infinite dimensions as the Wiener process does in finite dimensions.

We define a measure μ on C by

$$\mu(B) = \int_R dx\, P_x(B), \quad B \in \mathcal{B}(C) \tag{3.4}$$

and denote $L_2(C,\mu)$ by $L_2(C)$. $\varphi(\gamma_t)$, as a function of $\gamma \in C$, belongs to $L_2(C)$ for every $t \in [0,\infty)$ and every $\varphi \in S$. The closed linear span of $\{\varphi(\gamma_t)\}_{t,\varphi}$ in $L_2(C)$ is denoted by M_C. Similarly the closed linear span of $\{X_t(\varphi)\}_{t,\varphi}$ in $L_2(\Omega)$ is denoted by M_Ω. Since (1.3) can be written as

$$E(X_s(\varphi)\, X_t(\psi)) = \int_C \varphi(\gamma_s)\, \psi(\gamma_t)\, \mu(d\gamma) \quad , \tag{3.5}$$

the map $\theta : M_C \to M_\Omega$, $\varphi(\gamma_s) \to X_s(\varphi)$, defines an isomorphism.

Since γ_t is a Wiener process starting at x on the probability space (C, P_x) for every x, we have

$$\varphi(\gamma_t) = \varphi(\gamma_0) + \int_0^t \varphi'(\gamma_s)\, d\gamma_s + \int_0^t \frac{1}{2} \varphi''(\gamma_s) ds \tag{3.6}$$

a.s. (P_x) for every x. Hence this holds a.e. (μ) on C. As $\varphi(\gamma_t)$, $\varphi(\gamma_0)$ and the last integral belong to M_C, so does the stochastic integral. Let $I_t(\varphi)$ denote the image of the stochastic integral under θ. Since θ carries $\varphi(\gamma_t)$, $\varphi(\gamma_0)$ and $\varphi''(\gamma_s)$ to $X_t(\varphi)$, $X_s(\varphi)$ and $X_s(\varphi'')$ respectively, the equation (3.6) goes over into

$$X_t(\varphi) = X_0(\varphi) + I_t(\varphi) + \int_0^t \frac{1}{2} X_s(\varphi'')\, ds. \tag{3.7}$$

It is also obvious that

$$\begin{aligned} E(I_s(\varphi)\, I_t(\psi)) &= \int_R dx\, E_x\Big(\int_0^t \varphi'(\gamma_u)\, d\gamma_u \int_0^s \psi'(\gamma_u)\, d\gamma_u\Big) \\ &= \int_R dx \int_0^{s\wedge t} E_x(\varphi'(\gamma_u)\, \psi'(\gamma_u))\, du \\ &= \int_0^{s\wedge t} \int_R dx \int_R g_u(y-x)\, \varphi'(y)\, \psi'(y)\, dy \\ &= (s \wedge t)(\varphi', \psi'). \end{aligned} \tag{3.8}$$

Comparing (3.2) and (3.8), we obtain

$$E(\partial b_s(\varphi) \cdot \partial b_t(\psi)) = E(b_s(\varphi')\, b_t(\psi')) = E(I_s(\varphi)\, I_t(\psi)). \tag{3.9}$$

Since I_t is a sample continuous centered Gaussian $\mathcal{S}'$-valued process by virtue of (3.8), the sample process of I_t has the same probability law as that of ∂b_t. Hence we can identify I_t with ∂b_t by extending the probability space (Ω, P) so that b_t can be defined on the extended probability space. Roughly speaking, the extended probability space is obtained as follows:

$$P(\omega \in A,\ b. \in B) = \int_A P(d\omega) P(b. \in B \mid \partial b. = \beta)\Big|_{\beta = I.(\omega)}. \tag{3.10}$$

Thus we obtain

$$X_t(\varphi) = X_0(\varphi) + \partial b_t(\varphi) + \int_0^t \frac{1}{2}\, \partial^2 X_s(\varphi)\, ds\ , \tag{3.11}$$

which can be written in the form (3.1).

REFERENCES

[1] K. Itô, Distribution-valued processes arising from independent Brownian motions, to appear in the forthcoming Math. Zeitschrift, 1982.

[2] E. B. Dynkin, Gaussian random fields and Gaussian evolutions, preprint.

DEPARTMENT OF MATHEMATICS
GAKUSHUIN UNIVERSITY
MEJIRO, TOKYO 171, JAPAN

Contemporary Mathematics
Volume 26, 1984

ERGODIC THEORY AND COMBINATORICS

by

Konrad Jacobs

INTRODUCTION

The purpose of this paper is a thorough investigation of the connections between two mathematical disciplines which at first sight seem to have little to do with each other. Combinatorics is discrete mathematics, and ergodic theory, with all its topological and measure theoretical techniques, is typically analytic in structure. There is, however, one link between both domains. It is one of the richest and most beautiful mathematical objects ever invented: shift space. By definition, the shift space over a finite alphabet A is the infinite product space

$$X = A^{\mathbb{Z}} = \{x = (\ldots,x_{-1},x_0,x_1,\ldots)\ x_0,x_{-1},x_1,\ldots \in A\} .$$

We shall think it always endowed

a) with the product topology, which makes it a compact metric space homeomorphic to a Cantor set.

b) with the shift transformation
$\sigma: (\ldots,x_{-1},x_0,x_1,\ldots) \to (\ldots,x_0,x_1,x_2,\ldots)$
which is a homeomorphism of X.

These structures form a good fundament for measure theory, and product or Markov measure constructions as well as averaging procedures provide X with a wealth of shift invariant measures, living in many cases not simply on X but on proper shift invariant subsets of X. Such subsets can e.g. be obtained as orbit closures of sequences like (for $A = \{0,1\}$)

...0110100110010110... = the Thue-Morse sequence
= der Geist, der stets verneint
...0010011100010011101101 10001... = the infinite Mephisto waltz

The combinatorical intricacies of these famous symbol sequences (= points in shift space) add considerably to the digital flavor of ergodic theory provided by the shift space model.

0271-4132/84 $1.00 + $.25 per page

We divide this paper into two major chapters:

I. What combinatorics does for ergodic theory.

II. What ergodic theory does for combinatorics.

Ch. I is devoted to an account of

1) occasional helpful combinatorial tricks in proofs of theorems in ergodic theory

2) combinatorial constructions of symbol sequences with ergodic properties prescribed in advance.

Ch. II reports on

1) the Goncharov-Vershik-Schmidt asymptotic theory of the behavior of cycle lengths in the symmetric group S_n, for $n \to \infty$.

2) the van der Waerden-Szemeredi-Furstenberg-Katznelson-Weiss saga of arithmetic progressions.

Ch. I, section 2 is an attempt to bring home to Shizuo Kakutani the harvest earned so far from his Berkeley papers. Kakutani [1967], [1972].

CH. 1. WHAT COMBINATORICS DOES FOR ERGODIC THEORY.

§1. Combinatorial trickery in ergodic proofs.

1. A short glance at the beginnings of ergodic theory.
Ergodic theory begins with remarks of Ludwig Boltzmann [1868], [1871] on the nature of gas particle statistics, and his coining of the word "ergodisch".

Ehrenfest-Ehrenfest [1914] explain the etymology of "ergodisch" as a composite of the greek word ἔργον ("work") and ὁδός ("path"), but a recent remark of my colleague R. Nagel in Tubingen makes it plausible that Boltzmann rather thought of the greek adjective ἐργώδης ("work-like").

The first mathematical results which can be incorporated into ergodic theory are Kronecker's density theorem (Kronecker [1884] and Poincaré's recurrence theorem (Poincaré [1889]). One could also cite Borel's theorem on the uniformity of digits in decimal expansions (Borel [1909]). See e.g. Jacobs [1972 a], [1972 b]. There is a light touch of combinatorics in the proof of Kronecker's theorem: it makes use of the pigeon hole principle. Poincaré's Recurrence Theorem has a combinatorical special case: the decomposition of a permutation into cycles. Borel's theorem has a combinatorial flavor only insofar as it deals with the digitalization of a continuum.

2. The marriage theorem and the existence of Haar measure.

The first "ergodic theorem" was Weyl's equidistribution theorem

(Weyl [1916]) which can also be viewed as a precursor to Haar measure theory (Haar [1933]). And here comes the first intervention of substantial combinatorics into ergodic theory: Maak [1936] used the marriage theorem (of which he himself proved an equivalent; the general theorem is due to Ph.Hall [1935] but was later discovered to be a special case of an older graph theoretical theorem by D. Konig [1916]) in order to prove the mean value theorem for almost periodic functions on groups, i.e. the existence of Haar measure for compact groups. Maak's beautiful paper stimulated S. Kakutani to give a proof of the existence of Haar measure also in the general locally compact case via the marriage theorem. This proof was never published but appears in the handwritten notes of P. Halmos codifying a course in measure theory given by von Neumann at Princeton in 1940/41 (Neumann [1941] ; I am indebted to Roy Adler who provided me with a photostat of it). The gist of these proofs is to obtain nearly invariant averages of continuous functions by "marrying" translated portions of them suitably.

I should certainly not fail to draw the attention of the reader to the beautiful slim book of Hlawka [1979] on equidistribution mod 1.

The marriage theorem, generalized into a Harem Theorem, proved useful also in the paper of Ornstein [1970] on isomorphy of Bernoulli schemes with equal entropy.

3. The leader lemma and the maximal ergodic theorem.

Every ergodist knows how to derive the individual ergodic theorem of G.-D. Birkhoff [1931] from the so-called maximal ergodic theorem, which appears under this name first in Yosida-Kakutani [1939]. The proof of the maximal ergodic theorem is nowadays an easy exercise, thanks to an observation of A. Garsia [1965]. In P. Halmos' delightful and seminal book Halmos [1956] it is proved by means of a combinatorial "leader lemma" which goes back to Kolmogorov [1937].

§2. Combinatorial constructions of points in shift space.

A nonempty closed shift invariant subset X_o of shift space is called strictly ergodic if it is minimal and carries exactly one invariant measure of total mass one. Given minimality, Oxtoby [1952] characterized the strict ergodicity of a minimal X_o by uniform convergence on X_o, of the means

$$\frac{1}{n}(f + Tf + \ldots + T^{n-1}f)$$

of continuous functions f on X_o (clearly, $(Tf)(x) = f(Tx)$). A minimal set X_o equals the orbit closure $X(x)$ of each of its points $x \in X_o$. Strict ergodicity of $X(x)$ can be read from x: it happens if and only if every block $B \in A^*$ which occurs in x at all, not only occurs with bounded gaps (this almost periodicity property characterizes the minimality of $X(x)$) but even with uniform frequency: if a section of x is sufficiently long, then the frequency of occurrences of B in it is practically independent of its site in x. Shift space points x, i.e. symbol sequences, are called strictly ergodic if they have this property, i.e. iff $X(x)$ is a strictly ergodic set, and the unique shift invariant measure of mass 1 carried by $X(x)$ is denoted by μ_x. It is always ergodic.

Every strictly ergodic symbol sequence x is so to say a handle by which we get hold of measure theoretical dynamical system. A famous imbedding theorem by Jewett-Krieger (Jewett [1970], Krieger [1972], Bellow-Furstenberg [1972]) states that one gets hold of every dynamical system with finite entropy that way, up to isomorphy, and given some mild regularity conditions. The question arises how many of these systems one gets if one imposes on x the condition of machinal constructibility. There is a general answer by Blanchard [1981] which claims a fairly wide range. The purpose of this section is to review a bunch of explicit combinatorial constructions of strictly ergodic points $x \in X = A^{\mathbb{Z}}$ along with the ergodic properties of the associated invariant measures.

1. Generalized Morse sequences.

The reader will surely guess what the construction principle of the Thue-Morse sequence

(1) $$0110100110010110\ldots$$

is. If one replaces 01 by 0, 10 by 1 in it, it reproduces itself. It could be generated by an obvious iteration of the substitution

$$0 \to 01$$
$$1 \to 10$$

or else as an infinite block product

$$01 \times 01 \times \ldots$$

where $(a_1 \ldots a_r)(b_1 \ldots b_s)$ means: repeat $A = a_1 \ldots a_r$ or the "contrary" $0 \leftrightarrow 1$ of A according to the pattern given by $B = b_1 \ldots b_r$. The sequence (1) appears first in Thue [1906] (see Hedlund [1967]) and was rediscovered by Marston Morse [1921]. Kakutani [1967] considered sequences that could be written

$$0a_1 \times 0a_2 \times 0a_3 \times \ldots$$

with an arbitrary 0-1-sequence $a_1 a_2 \ldots$, and Michael Keane [1968] gave the theory its final shape, at least as far as 0-1-sequences are concerned, by inventing 0-1-block multiplication (as sketched above) and investigating sequences of the type

(2) $$0B_1 \times 0B_2 \times 0B_3 \times \ldots$$

with an arbitrary sequence of o-1-blocks, $B_1, B_2, B_3 \ldots$. Sequences (2) have always minimal orbit closure. Keane [1968] gave a criterion for strict ergodicity of (2) (and constructed a new bunch of examples of "minimality but not strict ergodicity" by skillfully not fulfilling it). Making use of finite abelian groups, Gösta Poch [1976][1976a] further generalized part of Keane's results to other finite alphabets.

I should not fail to mention that there is another important strand of combinatorial constructions of symbol sequences by iterated substitutions as alluded to above: substitutions of constant or varying length. This strand comes perhaps closest to the results arising from block multiplication in the papers Coven-Keane [1971] and Dekking-Keane [1978]. But one should also compare Coven [1971], Dekking [1978], Kamae [1970], [1972], Martin [1971], [1973], Michel [1974, [1978].

As far as the ergodic properties of the attached invariant normalized measures are concerned, the papers Kakutani [1967], Keane [1968], Dekking-Keane [1978] converge upon facts of the following type: entropy 0, rational point spectrum, presence and partial analysis of a continuous part of the spectrum, various mixing properties.

2. Non-repetition.

The investigations of Thue [1906] had the goal of producing sequences with certain non-repetition properties (see Hedlund [1967]). Hedlund-Morse [1944] proved that the Thue-Morse sequence does, for any 0-1-block $B = b_o b_1, \ldots b_{r-1}$, not contain a block BBb_o, and Gottschalk-Hedlund [1964] characterized the two-sided "daughters" of the Thue-Morse sequence by this property. Poch [1976] [1976a] gave a similar property in his case of group alphabets. Dekking [1976] showed that there is an infinite 0-1-sequence which has no BB with length ≥ 4 of B, and also no BBB with any B. But he also showed that a sequence with no BBB, and no BB with a B of length ≥ 3, needs be finite. In Dekking [1979] there is an example of an infinite 0-1-sequence with no $B_1 B_2 B_3 B_4$ where B_1, B_2, B_3, B_4 are permutations of each other. See also Guy [1981], section E 21.

3. Toeplitz sequences.

In 1968/9 Michael Keane and I worked out a paper (Jacobs-Keane [1969]) in which we applied the following construction: Let

$$x^{(1)}, x^{(2)}, \ldots$$

be bilateral periodic sequences of symbols 0, 1, ∞ . The symbol ∞ stands for "hole". We assume that "holes" always really occur. Now produce a 0-1-sequence by the following method:

1) fill $x^{(2)}$ into the holes of $x^{(1)}$ and obtain $y^{(1)}$: this sequence still has periodic holes.

2) fill $x^{(3)}$ into the holes of $y^{(1)}$ and obtain $y^{(2)}$: this sequence still has periodic holes

etc. Under mild assumptions, in the end all holes are filled and an $x \in \{0,1\}^Z$ with minimal orbit closure is obtained. Since our construction goes back to an idea of Toeplitz [1928], x is called a Toeplitz sequence. The easiest special case ("the classical Toeplitz sequence") runs as follows

$$\begin{array}{l} \ldots 0 \; \infty \; 0 \; \infty \; 0 \; \infty \; 0 \; \infty \; 0 \; \infty \; 0 \; \infty \; 0 \quad \infty \quad \ldots \\ \ldots 0 \; \infty \; 0 \;\; 1 \; 0 \; \infty \; 0 \; 1 \; 0 \; \infty \; 0 \; 1 \; 0 \quad \infty \quad \ldots \\ \ldots \; \cdot \; \cdot \; \cdot \; \cdot \; \cdot \; \cdot \; \cdot \; \cdot \; \cdot \; \cdot \; \cdot \; \cdot \; \cdot \; \cdot \quad \cdot \quad \ldots \end{array}$$

(fill every other hole with 0 and 1 in alternating steps). If the filling is too slow and the $x^{(n)}$ are oddly chosen, x is not strictly ergodic. Actually, the old famous example of Oxtoby [1952] shines up here (Williams has, in her Ph.D. thesis, investigated properties of shift space points obtained as slowly filled Toeplitz sequences: prescribed number of invariant ergodic measures and the like).

If the filling is, however, fast enough to make the frequency of holes go down to 0, then the Toeplitz sequence x is ergodic, and μ_x appears to have pure rational point spectrum (and hence entropy 0). There must therefore be, by an old Theorem of von Neumann [1932], isomorphy to a translation of a compact group. Neveu [1969] immediately established such an isomorphism explicitly in a special case, and Ernst Eberlein [1971] settled the general case. Kakutani [1972] subsequently observed how to destroy, in a special case, the whole point spectrum by doubling all symbols 1 in x. The resulting invariant measure turned out to be weakly but not strongly mixing.

Further Toeplitz, if he still were alive (at 101 years by now), would certainly ask the question whether new almost periodic functions would turn out if his old device

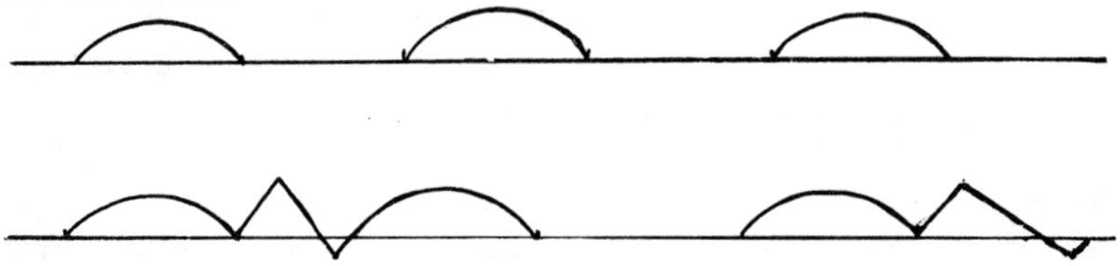

(fill every second free unit interval with the same function and let the functions f_n uniformly converse back to f_1) would be replaced by the new ones. This question was answered in the affirmative by Hans Haller [1978].

4. Positive entropy, K-systems, Bernoulli systems.

After so many symbol sequences whose attached invariant normalized measure had entropy 0, the question arose whether it would be ever possible to obtain entropy >0. This was first achieved by a very sophisticated construction in Hahn-Katznelson [1967]. Would it be possible to do it in a visibly machinal fashion? After all, machines seem unable to produce real randomness. In 1972, Christian Grillenberger [1972/1973] turned up with a construction which was as clever as it was simple. Taking suggestions from the theory of cyclic approximation of dynamical systems (Katok-Stepin [1967]), he invented the following "block permutation device": Let $A_1,\ldots,A_n$ be 0-1-blocks. For any permutation τ of $1,\ldots,n$, weld $A_{\tau(1)},\ldots, A_{\tau(n)}$ together in that order to obtain B_τ. This gives $n!$ blocks. Doing the same with the B_τ, we obtain $(n!)!$ blocks. Going on in this way with suitably chosen $A_1,\ldots,A_n$, a symbol sequence x arises such that μ_x has entropy > 0. By skillfully steering modifications of this procedure, Grillenberger [1972/1973a] could bring the entropy to any prescribed value between 0 and log 2, and his idea applied as well to other alphabet lengths and even infinite alphabets and infinite entropy. Combining his device with Parry's theory of subshifts of any order (Parry [1964]), Grillenberger [1972/73b] could even enforce the K-property for μ_x with machinality constructed x, and when Paul Shields joined bringing along Ornstein's Bernoullicity techniques, μ_x became even Bernoulli (Grillenberger-Shields [1975]). These examples do not yet demonstrate that combinatorial techniques bring you everywhere in the domain of dynamical systems, but they bring you considerably far. Wouldn't it be imaginable that one day a machinal symbol sequence x turns up such that μ_x is mixing of degree 2 but not 3, or has simple Lebesgue spectrum?

CH. II. WHAT ERGODIC THEORY DOES FOR COMBINATORICS.

§1. The Goncharov-Vershik-Schmidt theory: cycle lengths in S_n - behavior for $n \to \infty$.

After pioneering work of Goncharov [1944] and interesting contributions of Feller [1945], Erdos-Turán [1965/68] and others (see Kolchin-Sevastyanov-Chistyakov [1976], Eberlein [1962], Shepp-Lloyd [1966]), Vershik-Schmidt [1977/78] presented a comprehensive study of the asymptotics of S_n, setting a measure theoretical frame which places their work on the frontier line between probability theory and ergodic theory.

The basic layout is given by the following sequence spaces

$$I^\infty = \{x = (x_1,x_2,\dots) \mid 0 \le x_t \le 1 \quad (t = 1,2,\dots)\}$$

$$U = \{y = (y_1,y_2,\dots) \mid 0 \le y_t \le 1 \quad (t = 1,2,\dots), \ \sum_{t=1}^{\infty} y_t \le 1\}$$

$$V = \{z = (z_1,z_2,\dots) \mid z \in U, \ \sum_{t=1}^{\infty} z_t = 1\}$$

$$W = \{w = (w_1,w_2,\dots) \mid w \in V, \ w_1 \ge w_2 \ge \dots\} .$$

Thus $I^\infty \supseteq U \supseteq V \supseteq W$. The following mappings

$$\alpha: I^\infty \to U$$

$$\beta: V \to W$$

$$\gamma: V \to I^\infty$$

are considered

$$\alpha(x_1,x_2,\dots) = (y_1,y_2,\dots)$$

where

$$y_1 = x_1, \ y_2 = x_2(1 - x_1), \ y_3 = x_3(1 - x_1)(1 - x_2), \dots,$$

hence

$$(*) \qquad (1 - x_1)(1 - x_2)\dots(1 - x_n) = 1 - (y_1 + \dots + y_n).$$

If we endow the infinite-dimensional cube I^∞ with product Lebesgue measure λ^∞, the left hand of (*) diverges to 0 a.s., hence $y = \alpha(x)$ is a.s. a probability sequence from V, and α sends λ^∞ into a probability measure μ on $V: \mu = \alpha(\lambda^\infty)$. Moreover $\alpha^{-1}: V \to I^\infty$ exists μ-a.e.

γ is defined in an obvious way and β can μ-a.s. be defined by a random permutation of $\{1,2,\dots\}$ since μ-a.s. the components $z_1, z_2, \dots$ are pairwise distinct. We set

$$\nu = \beta(\mu).$$

Now the symmetric groups come in. For a given n and a given $\tau \in S_n$, let $r = r_\tau$ be the number of cycles in τ and order the cycles according to the natural order of their minimal elements in $\{1,\dots,n\}$. Let $\ell_1, \dots, \ell_r$ be the lengths of the cycles in this order. Then

$$z_1(\tau) = \frac{\ell_1}{n}, \dots, z_r(\tau) = \frac{\ell_r}{n}, \quad z_{r+1}(\tau) = z_{r+2}(\tau) = \dots = 0$$

defines a point $z(\tau) \in V$.

Equidistribution (= Haar measure) on S_n is sent into a probability measure μ_n in V, by $\tau \to z(\tau)$. Put

$$\nu_n = \beta(\mu_n)$$

- under this probability the distribution of the first component w_1 of $w \in W$ is the distribution of the relative length of the longest cycle in a permutation $\tau \in S_n$.

Vershik-Schmidt now easily establish the weak convergences

$$\begin{aligned} \mu_n &\to \tilde{\mu} && \text{in } V \\ \nu_n &\to \tilde{\nu} && \text{in } W \\ \lambda_n &\to \lambda^\infty && \end{aligned}$$

where $\lambda_n = \alpha^{-1}\mu_n$.

The distribution of w_1 under $\tilde{\nu}$ can be considered as the limit distribution of the length of the maximal cycle. As the result of an intricate analysis, Vershik-Schmidt obtain a density $\rho: [0,1] \to R_+$ for this distribution. It is the unique solution of

$$(**) \qquad t\rho(t) = \int_0^{\frac{t}{1-t}} \rho(s)\, ds \qquad (0 \le t < 1)$$

when extended to R_+ by

$$\rho(t) = 0 \qquad (t \ge 1).$$

(**) enables us to calculate ρ on intervals tending to 0 from the right, beginning with

$$\rho(t) = \frac{1}{t} \qquad (\tfrac{1}{2} \le t < 1)$$

$$\rho(t) = \frac{1}{t}[1 - \log \frac{1-t}{t}] \qquad (\tfrac{1}{3} \le t < \tfrac{1}{2})$$

Vershik-Schmidt even give formulas for the joint distributions of $w_1,\ldots,w_r$ under ν, for all $r = 1,2,\ldots$.

Going back to I^∞, they find

$$\bar{\lambda} = \text{weak} - \lim \alpha^{-1} \nu_n$$

a Markov-measure with transition kernel density

$$P(s, t) = \begin{cases} 0 & (t \ge \min[\frac{s}{1-s}, 1]) \\ \frac{\rho(t)}{s\rho(s)} & (t \le \frac{s}{1-s}) \end{cases}$$

and initial probability density ρ. This Markov measure becomes stationary when ρ is replaced by const $\frac{\rho(t)}{t}$. The resulting shift invariant Markov measure in T^∞ fulfills the mixing condition of M. Rosenblatt [1956]. Thus the central limit theorem of Ibragimov-Linnik [1971] applies, and one gets asymptotic normalcy of the length distribution of the maximal cycle plus other similar results, vastly improving the results of the predecessors of Vershik-Schmidt.

§2. Arithmetic progressions: the van-der-Waerden-Szemerédi-Furstenberg-Weiss saga.

1. Van der Waerden's theorem of 1926.

A k-element set of equidistant integers is called an arithmetic progression of length k. A set of integers is called beautiful if it contains arithmetic progressions of any length. In 1926, young B.L. van der Waerden proved the

THEOREM (van der Waerden [1927]).
<u>In any finite collection of sets covering</u> N, <u>at least one member is beautiful</u>.

Now, this is a theorem "for angels" since it doesn't tell which one of the members is beautiful. The way how the - by the way finitistic - proof of the theorem was found, was delightfully described by van der Waerden [1971] himself. He gives credit to Emil Artin and Otto Schreier for a lot of encouragement ("Wir waren zu dritt...", he writes).

2. Szemerédi's Theorem (1973).

By 1936, Erdös-Turan [1936] gave the question "which one is beautiful?" the following formulation. For any natural numbers k, n let

$r_k(n)$ = the maximal power of a subset of n consecutive integers that doesn't contain an arithmetic progression of length k.

It is easily seen that

$$r_k(m+n) \le r_k(m) + r_k(n)$$

and hence

$$r_k = \lim_{n \to \infty} \frac{r_k(n)}{n}$$

exists. What are the values of $r_3, r_4, \ldots$?

After a lot of only partly successful attacks by famous mathematicians (see e.g. Roth [1952], [1953/54], [1967]). E. Szemerédi proved, thereby winning 1000 $ from Paul Erdös, the following

THEOREM (Szemeredi [1975]).

$$r_3 = r_4 = \dots = 0$$

COROLLARY. _If_ M _is a set of natural numbers such that its upper density_

$$\limsup_{n\to\infty} \frac{|M \cap \{1,\dots,n\}|}{n}$$

_is_ > 0, _then_ M _is beautiful_.

The corollary follows simply because M has initial sections $M \cap \{1,\dots,n\}$ that are "too fat" for $r_k = 0$ if n is suitably chosen.

Do we now know which one is beautiful? Well, if $M_1,\dots,M_n$ cover the natural numbers at least one of them has positive upper density and hence, by Szemeredi's corollary, is beautiful. Sure, the superior knowledge of an angel is needed again in order to know whether a set has positive upper density or not, hence Szemeredi's result still doesn't bring us down to the earth. But it brings us closer down than we were before and it is a characteristic feature of mathematical thinking to rejoice if one knows that even a lesser angel could find out certain things for us. The prime number theorem in its pure asymptoticity, without additional explicit estimates (for such see e.g. Rosser-Schoenfeldt [1962]) is of a similarly "angelic" nature, yet we consider it as a great achievement. Reportedly, Paul Erdös has set a price of 3000 $ for the proof of the following

CONJECTURE. _Every set_ M _of natural numbers such that_

$$\sum_{n \in M} \frac{1}{n} = \infty$$

is beautiful.

The most prominent candidate is the set of all primes.

3. The achievement of Hillel Furstenberg.

By the end of 1975 Szemeredi's result became known to H. Furstenberg and the ergodist group at Jerusalem. This group was disposed to investigate sets of positive upper density. In spring 1976, Hillel Furstenberg finished the proof of the following

THEOREM (Furstenberg [1977]). _Let_ $(\Omega, \mathcal{L}, \mu, T)$ _be any dynamical system with_ $\mu(\Omega) = 1$, _and let_ $A \in \mathcal{L}$ _such that_ $\mu(A) > 0$. _Then for any natural number_ k _there is a natural number_ n _such that_

$$\mu(A \cap T^{-n} A \cap \dots \cap T^{-(k-1)n} A) > 0. \tag{1}$$

This theorem is a strong generalization of Poincaré's recurrence theorem (Poincaré [1899], case k=2). But the most astonishing fact is that Szemeredi's Corollary is also a corollary of Furstenberg's theorem: consider the indicator function of a set $M \subseteq N$ of positive upper density as a point x in shift space and obtain a shift invariant measure μ such that $\mu([1]) > 0$ for the cylinder $[1] = A$ by the usual averaging and weak convergence method. By Furstenberg's theorem, there is, for a given k, some n such that (1) holds. But this requires that some translate of x visits $[1]$ k times in equal distances n, and this means that M contains an arithmetic progression of length k. - Furstenberg's theorem was consequently also named "the ergodic Szemeredi theorem". It was extended to commuting transformations by Furstenberg-Katznelson [1978], thus bearing combinatorial consequences which seemed out of the scope of Szemeredi's combinatorial methods. See also Furstenberg-Katznelson-Ornstein [1982].

The question turned up, whether van der Waerden's theorem could not be proved by ergodic theory methods simpler than Furstenberg's theorem. The positive answer to this was given by Furstenberg-Weiss [1978]. This paper contains a not too difficult proof of the following

THEOREM (Furstenberg-Weiss [1978]). *Let* $T_1,\ldots,T_k$ *be commuting homeomorphisms of a compact metric space* X *(e.g. iterates of one given homeomorphism). Then* X *contains a point* x *such that*

$$\left.\begin{array}{l} T_1^{n_j} x \to x \\ \cdots\cdots\cdots\cdots \\ T_k^{n_j} x \to x \end{array}\right\} \quad (j \to \infty)$$

for a suitable sequence $n_1 < n_2 < \cdots$ *of natural numbers.*

Finding x in the orbit closure of an indicator function of a set $M \subseteq N$, van der Waerden's primordial result is obtained as a corollary.

One more question: the proof in Furstenberg-Weiss [1978] invokes Zorn's lemma (via minimal sets) and is thus purely existential; but in shift space one can replace Zorn's lemma by constructions; in doing this, do we get an improvement of the estimates contained in van der Waerden's proof? The answer, by Girard [1981] is: if we modify the proof in Furstenberg-Weiss a bit and then make it "constructive", we get exactly what van der Waerden got.

A comprehensive presentation of the van-der-Waerden-Szemeredi-Furstenberg-Weiss saga can be found in Furstenberg [1981]. For the original van-der-Waerden theorem and an extension of it - the theorem of Hales-Jewett [1963], which has meanwhile also been "caught" by ergodists - see Graham-Rothschild-Spencer [1980].

I consider the achievement of Hillel Furstenberg and his comrades as nothing short of the services done to classical number theory by the theory of functions of one complex variable.

REFERENCES

[1979] Bellow, A., and H. Furstenberg, An application of number theory to ergodic theory and the construction of uniquely ergodic models, Isr. J. Math. 33 (1979), 231-240.

[1931] Birkhoff, G.D., Proof of the ergodic theorem, Proc. Nat. Acad. Sci. USA 17 (1931), 656-660.

[1981] Blanchard, F., Systèmes dynamiques topologiques, associés à des automates récurrents, ZfW 58 (1981), 549-564.

[1868] Boltzmann, L., Studien über das Gleichgewicht der lebendigen Kraft zwischen bewegten materiellen Punkten, Wein Ber., 58 (1868), 517.

[1871] Boltzmann, L., Einige allgemeine Sätze über Wärmegleichgewicht, Wien Ber., 63 (1971), 679

[1909] Borel, E., Les probabilités dénombrables et leurs applications arithmétiques, R.C. Circolo Mat. Palermo 27 (1909), 247-270.

[1971] Coven, E., Endomorphisms of substitution minimal sets, ZfW 20 (1971), 129-133.

[1971] Coven, E., and M. Keane, The structure of substitution minimal sets, TAMS 162 (1971), 89-102.

[1976] Dekking, F. M., On repetitions of blocks in binary sequences, J. Comb. Th. A 20 (1976), 292-299.

[1978] Dekking, F.M., The spectrum of dynamical systems arising from substitutions of constant length, ZfW 41 (1978), 221-239.

[1979] Dekking, F.M., Strongly non-repetitive sequences and progression-free sets, J. Comb. Th. A 27 (1979), 181-185.

[1980] Dekking, F.M., Combinatorial and statistical properties of sequences generated by substitutions, Proefschrift Nijmegen 1980

[1978] Dekking, F.M., and M. Keane, Mixing properties of substitutions, ZfW 42 (1978), 23-33.

[1971] Eberlein, E., Toeplitz-Folgen und Gruppentranslationen, Arch. Math. 22 (1971), 291-301.

[1914] Ehrenfest, P., und T. Ehrenfest, Begriffliche Grundlagen der statistischen Auffassung in der Mechanik, Enz. Math. Wiss. IV, 4 (1914).

[1936] Erdös, P., and P. Turán, On some sequences of integers, J. London Math. Soc. 11 (1936), 261-264.

[1965/68] Erdös, P., and P. Turán, One some problem of statistical group theory, I. Zfw 4 (1965), 175-186 IV, Acta Math. Acad. Sci. Hung. 19 (1968), 413-435.

[1945] Feller, W., The fundamental limit theorems in probability, BAMS 51 (1945), 800-832.

[1977] Furstenberg, H., Ergodic behavior of diagonal measures and a theorem of Szemeredi on arithmetic prograssions, J. Anal. Math. 31 (1977), 204-256.

[1981] Furstenberg, H., Recurrence in ergodic theory and combinatorial number theory, Princeton (Univ. Press) 1981.

[1978] Furstenberg, H., and Y. Katznelson, An ergodic Szemeredi theorem for commuting transformations, J. Anal. Math. 34 (1978), 275-291.

[1978] Furstenberg, H., and B. Weiss, Topological dynamics and combinatorial number theory, J. Anal. Math. 34 (1978), 61-85.

[1982] Furstenberg, H., and Y. Katznelson and D. Ornstein, The ergodic theoretical proof of Szemeredi's Theorem, BAMS, 7 (1982), 527-552.

[1965] Garsia, A., A simple proof of Hopf's maximal ergodic theorem, J. Math. Mech. 14 (1965), 381-382.

[1981] Girard, J.-Y., L'analyse du théorème de van der Waerden et de sa démonstration topologique a l'aide de la théorie de la démonstration, preprint Paris 1981.

[1944] Goncharov, V. L., From the realm of combinatorics, Izv. AN SSSR ser. mat. 8 (1944), 3-48 (russian).

[1964] Gottschalk, W. H., and G. Hedlund, A characterization of the Morse minimal set, PSMA 15 (1964), 70-74.

[1980] Graham, R.L, B. L. Rothschild and J.H. Spencer, Ramsey Theory, New York (Wiley-Interscience) 1980.

[1972/3a] Grillenberger, Chr., Constructions of strictly ergodic systems, I. Given entropy ZfW 25 (1972/3), 323-334.

[1972/3b] Grillenberger, Chr., Constructions of strictly ergodic systems, II. K-systems, Zfw 25 (1972/73), 335-342.

[1975] Grillenberger, Chr., and P. Shields, Constructions of strictly ergodic systems, III., Bernoulli systems, Zfw 33 (1975), 215-217.

[1981] Guy, R.K., Unsolved problems in number theory, Berlin-Heidelberg-New York (Springer-Verlag) 1981.

[1933] Haar, A., Der Maßbegriff in der Theorie der kontinuierlichen Gruppen, Ann. of Math. (2) 34 (1933), 147-169.

[1967] Hahn, F., and Y. Katznelson, On the entropy of uniquely ergodic transformations, TAMS 126 (1967), 335-360.

[1963] Hales, A.W., and P.I. Jewett, Regularity and positional games, TAMS 106 (1963), 222-229.

[1935] Hall, Ph., On representatives of subsets, J. London Math. Soc. 10 (1935), 26-30.

[1978] Haller, H., Kombinatorische Konstruktion fastperiodischer Funktionen, 66 pp. Diss. Erlangen 1978.

[1956] Halmos, P. R., Lectures on ergodic theory, Tokyo 1956.

[1967] Hedlund, G. A., Remarks on the work of Axel Thue on sequences, Nord. Mar. Tidsskr. 15 (1967), 148-150.

[1944] Hedlund, G. A., and M. Morse, Unending chess, symbolic dynamics and a problem in semigroups, Duke Math. J. 11 (1944), 1-7.

[1979] Hlawka, E., Theorie der Gleichverteilung, Mannheim (Bibl. Inst.) 1979.

[1971] Ibragimov, I.A., and Ya.V. Linnik, Independent and stationary sequences of random variables, Groningen (Wolters-Noordhoft) 1971.

[1972a] Jacobs, K., Poincaré's Wiederkehrsatz, Selecta Mathematica IV, Berlin-Heidelberg-New York (springerverlag) 1972.

[1972b] Jacobs, K., Gleichverteilung mod 1, Selecta Mathematica IV, Berlin-Heidelberg-New York (springerverlag) 1972.

[1969] Jacobs, K., and M. Keane, 0-1-sequences of Toeplitz type, ZfW 13 (1969), 123-131.

[1970] Jewett, R., The prevalence of uniquely ergodic systems, J. Math. Mech. 19 (1970), 712-729.

[1967] Kakutani, S., Ergodic theory of shifts transformations, Proc. V. Berkeley, Symp., vol. II, part 2, 405-414 (1967).

[1972] Kakutani, S., Strictly ergodic symbolic dynamical systems, Proc. VI. Berkeley Symp., Vol. II, 319-326, (1972).

[1970] Kamae, T., Spectrum of a substitution minimal set, J. Math. Soc. Japan 22 (1970), 567-578.

[1972] Kamae, T., A topological invariant of substitution minimal sets, J. Math. Soc. Japan 24 (1972), 285-306.

[1967] Katok, A. B., and A. M. Stepin, Approximations in ergodic theory, Russ. Math. Surveys (Uspekhi) 22 (1967), 77-102.

[1968] Keane, M., Generalized Morse sequences, ZfW 10 (1968), 335-353.

[1937] Kolmogorov, A.N., Ein vereinfachter Beweis des Birkhoff-Khintchine'schen Ergodensatzes, Rec. Math. Moscow 2 (1937), 367-368.

[1976] Kolchin, V.F., B. A. Sevastyanov and V.P. Chistyakov, Random Permutations, Moscow ("Nauka") 1976 (Russian).

[1916] König, D., Über Graphen und ihre Anwendungen, Math. Ann. 77 (1916), 453-465.

[1972] Krieger, W., On unique ergodicity, Proc. VI. Berkeley Symp. I (1972), 327-346.

[1884a] Kronecker, L., Die Periodensysteme von Funktionen reeller Variablen, Werke vol. III, 31-46.

[1884b] Kronecker, L., Näherungsweise ganzzahlige Auflösung linearer Gleichungen, Werke vol. III. 47-109.

[1936] Maak, W., Eine neue Definition der fastperiodischen Funktionen, Abh. Math. Sem. Hamburg 11 (1936), 240-244.

[1971] Martin, J. C., Substitution minimal flows, Amer. J. Math. 93 (1971), 503-526.

[1973] Martin, J. C., Minimal flows arising from substitutions of non-constant length, Math. Syst. Th. 7 (1973), 73-82.

[1974] Michel, P., Stricte ergodicité d'ensembles minimaux de substitutions, C. R. Acad. Sci. Paris Ser. A-B 278 (1974), 811-813.

[1978] Michel, P., Coincidence calues and spectra of substitutions, ZfW 42 (1978), 205-227.

[1921] Morse, M., Recurrent geodesics on a surface of negative curvature, TAMS 22 (1921), 84-100.

[1932] Neumann, John von, Zur Operatorenmethode in der klassischen Mechanik, Ann. of Math. 33 (1932), 587-642.

[1941] Neumann, John von, Invariant Measures, handwritten notes by P. R. Halmos, Princeton 1940/41.

[1969] Neveu, J., Sur les suites de Toeplitz, ZfW 13 (1969), 132-134.

[1970] Ornstein, D. S., Bernoulli shifts with the same entropy are isomorphic, Adv. in Math. 4 (1970), 337-352.

[1952] Oxtoby, J., Ergodic Sets, BAMS 58 (1952), 116-136.

[1964] Parry, W., Intrinsic Markov Chains, TAMS 112 (1964), 55-66.

[1976] Poch, G., Rekurrente Folgen über abelschen Gruppen, 48 pp., Diss. Erlangen 1976.

[1976a] Poch, G., Suites récurrentes sur les groupes abéliens, C.R. Acad. Sci. Paris Ser. A-B 283 (1976), no. 16, A ii, A 1111 - A 1113.

[1899] Poincaré, H., Les nouvelles methodes de la mécanique céleste, tome III, Paris 1899.

[1956] Rosenblatt, M., A central limit theorem and strong mixing condition, Proc. Mat. Acad. Sci. USA 42 (1956), 43-47.

[1962] Rosser, J.B., and L. Schoenfeldt, Approximate formulas for some functions of prime numbers, Ill. J. Math. 6 (1962), 64-94.

[1952] Rosser, K.F., Sur quelques ensembles d'entiers, C.R. Acad. Sci. Paris 234 (1952), 388-399.

[1953/4] Roth, K. F., On certain sets of integers, London Math. Soc. I: 28 (1953), 104-109, II: 22 (1954), 20-26.

[1967] Roth, K.F., Irregularities of sequences relative to arithmetic progressions II. Math. Ann. 174 (1967), 41-52.

[1966] Shepp, L. A. and S. P. Lloyd, Ordered cycle lengths in a random permutation, TAMS 121 (1966), 340-357.

[1969] Szemerédi, E., On sets of integers containing no four elements in arithmetic progression, Acta. Math. Acad. Sci. Hung. 20 (1969), 89-104.

[1975] Szemerédi, E., On sets of integers containing no k in arithmetic progression, Acta Arith. 27 (1975), 199-245.

[1906] Thue, A., Über unendliche Zeichenreihen, Selected Mathematical Papers, Oslo (Universitetsforlaget) 1977.

[1928] Toeplitz, O., Ein Beispiel zur Theorie der fastperiodischen Funktionen, Math. Ann. 98 (1928), 281-295.

[1978] Vershik, A.M., and A. A. Schmidt, Limit measures arising in the asymptotic theory of symmetric groups, I., Th. Prob. Appl. (teor.ver. Prim) 22 (1977), 70-85, II, same journal 23 (1978), 36-

[1927] Waerden, B. L. van der, Beweis einer Baudet'schen Vermutung, Nieuw Art. Wisk. 15 (1927), 212-216.

[1971] Waerden, B.L. van der, How the proof of Baudet's conjecture was found, Stud. Pure Math. (ed. Mirsky), New York (Academic Press) 1971, 251-260.

[1973] Weiss, B., Subshifts of finite type and sofic systems, Monatsh. Math. 77 (1973), 462-474.

[1916] Weyl, H., Über die Gleichverteilung von Zahlen mod Eins, Math Ann. 77 (1916), 313-352.

[1939] Yosida, K., and S. Kakutani, Birkhoff's ergodic theorem and the maximal ergodic theorem, Proc. Imp. Acad. Tokyo 15 (1939), 165-168.

MATHEMATISCHES INSTITUT
UNIVERSITY OF ERLANGEN-NURNBERG
FEDERAL REPUBLIC OF GERMANY

Contemporary Mathematics
Volume 26, 1984

EXTENSIONS OF LIPSCHITZ MAPPINGS INTO A HILBERT SPACE

William B. Johnson[1] and Joram Lindenstrauss[2]

INTRODUCTION

In this note we consider the following extension problem for Lipschitz functions: Given a metric space X and $n = 2, 3, 4, \ldots$, estimate the smallest constant $L = L(X, n)$ so that every mapping f from every n-element subset of X into ℓ_2 extends to a mapping $\tilde{f}$ from X into ℓ_2 with

$$\|\tilde{f}\|_{\ell ip} \le L \, \|f\|_{\ell ip} .$$

(Here $\|g\|_{\ell ip}$ is the Lipschitz constant of the function g.) A classical result of Kirszbraun's [14, p. 48] states that $L(\ell_2, n) = 1$ for all n, but it is easy to see that $L(X, n) \to \infty$ as $n \to \infty$ for many metric spaces X.

Marcus and Pisier [10] initiated the study of $L(X, n)$ for $X = L_p$. (For brevity, we will use hereafter the notation $L(p, n)$ for $L(L_p(0,1), n)$.) They prove that for each $1 < p < 2$ there is a constant $C(p)$ so that for $n = 2, 3, 4, , , ,$

$$L(p, n) \le C(p) \, (\text{Log } n)^{1/p - 1/2} .$$

The main result of this note is a verification of their conjecture that for some constant C and all $n = 2, 3, 4, , , ,$

$$L(X, n) \le C(\text{Log } n)^{1/2}$$

for all metric spaces X. While our proof is completely different from that of Marcus and Pisier, there is a common theme: Probabilistic techniques developed for linear theory are combined with Kirszbraun's theorem to yield extension theorems.

The main tool for proving Theorem 1 is a simply stated elementary geometric lemma, which we now describe: Given n points in Euclidean space, what

---

[1] Supported in part by NSF MCS-7903042.

[2] Supported in part by NSF MCS-8102714.

is the smallest $k = k(n)$ so that these points can be moved into k-dimensional Euclidean space via a transformation which expands or contracts all pairwise distances by a factor of at most $1 + \varepsilon$? The answer, that $k \le C(\varepsilon)$ Log n, is a simple consequence of the isoperimetric inequality for the n-sphere in the form studied in [2].

It seems likely that the Marcus-Pisier result and Theorem 1 give the right order of growth for $L(p, n)$. While we cannot verify this, in Theorem 3 we get the estimate

$$L(p, n) \ge \delta \left(\frac{\text{Log } n}{\text{Log Log } n}\right)^{1/p - 1/2} \qquad (1 \le p < 2)$$

for some absolute constant $\delta > 0$. (Throughout this paper we use the convention that Log x denotes the maximum of 1 and the natural logarithm of x.) This of course gives a lower estimate of

$$\delta \left(\frac{\text{Log } n}{\text{Log Log } n}\right)^{1/2}$$

for $L(\infty, n)$. That our approach cannot give a lower bound of $\delta(\text{Log } n)^{1/p - 1/2}$ for $L(p, n)$ is shown by Theorem 2, which is an extension theorem for mappings into ℓ_2 whose domains are ε-separated.

The minimal notation we use is introduced as needed. Here we note only that $B_Y(y, \varepsilon)$ (respectively, $b_Y(y, \varepsilon)$) is the closed (respectively, open) ball in Y about y of radius ε. If $y = 0$, we use $B_Y(\varepsilon)$ and $b_Y(\varepsilon)$, and we drop the subscript Y when there is no ambiguity. $S(Y)$ is the unit sphere of the normed space Y. For isomorphic normed spaces X and Y, we let

$$d(X,Y) = \inf \|T\| \, \|T^{-1}\|,$$

where the inf is over all invertible linear operators from X onto Y. Given a bounded Banach space valued function f on a set K, we set

$$\|f\|_\infty = \sup_{x \in K} \|f(x)\|.$$

1. THE EXTENSION THEOREMS

We begin with the geometrical lemma mentioned in the introduction.

LEMMA 1. For each $1 > \tau > 0$ there is a constant $K = K(\tau) > 0$ so that if $A \subset \ell_2^n$, $\overline{\overline{A}} = n$ for some $n = 2, 3, \ldots$, then there is a mapping f from A onto a subset of ℓ_2^k $(k \equiv [K \log n])$ which satisfies

$$\|\tilde{f}\|_{\ell ip}\,\|\tilde{f}^{-1}\|_{\ell ip} \le \frac{1+\tau}{1-\tau}\,.$$

PROOF. The proof will show that if one chooses at random a rank k orthogonal projection on ℓ_2^n, then, with positive probability (which can be made arbitrarily close to one by adjusting k), the projection restricted to A will satisfy the condition on $\tilde{f}$. To make this precise, we let Q be the projection onto the first k coordinates of ℓ_2^n and let σ be normalized Haar measure on $O(n)$, the orthogonal group on ℓ_2^n. Then the random variable

$$f : (O(n),\sigma) \to L(\ell_2^n)$$

defined by

$$f(u) = U^* QU$$

determines the notion of "random rank k projection." The applications of Levy's inequality in the first few self-contained pages of [2] make it easy to check that $f(u)$ has the desired property. For the convenience of the reader, we follow the notation of [2].

Let $|||\cdot|||$ denote the usual Euclidean norm on $\mathbb{R}^n$ and for $1 \le k \le n$ and $x \in \mathbb{R}^n$ set

$$r(x) = r_k(x) = \sqrt{n}\left(\sum_{i=1}^{k} x(i)^2\right)^{1/2},$$

which is equal to

$$\sqrt{n}\;|||Qx|||$$

for our eventual choice of $k = [K \log n]$. Thus $r(\cdot)$ is a semi-norm on ℓ_2^n which satisfies

$$r(x) \le \sqrt{n}\;|||x||| \qquad (x \in \ell_2^n).$$

(In [2], $r(\cdot)$ is assumed to be a norm, but inasmuch as the left estimate $a|||x||| \le r(x)$ in formula (2.5) of [2] is not needed in the present situation, it is okay that $r(\cdot)$ is only a semi-norm.)

Setting

$$B = \left\{\frac{x-y}{|||x-y|||} : x, y \in A;\ x \neq y\right\} \subset S^{n-1},$$

we want to select $U \in O(n)$ so that for some constant M,

$$M(1-\tau) \le r(Ux) \le M(1+\tau) \quad (x \in B) .$$

Let M_r be the median of $r(\cdot)$ on S^{n-1}, so that

$$\mu_{n-1}[x \in S^{n-1} : r(x) \ge M_r] \ge 1/2$$

and

$$\mu_{n-1}[x \in S^{n-1} : r(x) \le M_r] \ge 1/2$$

where μ_{n-1} is normalized rotationally invariant measure on S^{n-1}.

We have from page 58 of [2] that for each $y \in S^{n-1}$ and $\varepsilon > 0$,

$$\sigma[U \in O(n) : M_r - \sqrt{n}\,\varepsilon \le r(Uy) \le M_r + \sqrt{n}\,\varepsilon] \ge 1 - 4 \exp\left(\frac{-n\varepsilon^2}{2}\right).$$

Hence

$$(1.1) \qquad \sigma[U \in O(n) : M_r - \sqrt{n}\,\varepsilon \le r(Uy)) \le M_r + \sqrt{n}\,\varepsilon \quad \text{for all } y \in B] \ge$$

$$\ge 1 - 2n(n+1) \exp\left(\frac{-n\varepsilon^2}{2}\right).$$

By Lemma 1.7 of [2], there is a constant

$$C \le 4 \sum_{m=1}^{\infty} (m+1)\, e^{-m^2/2}$$

so that

$$(1.2) \qquad \left| \int_{S_{n-1}} r(x)\, d\mu_{n-1}(x) - M_r \right| < C .$$

We now repeat a known argument for estimating $\int_{S_{n-1}} r(x)\, d\mu_{n-1}(x)$ which uses only Khintchine's inequality.

For $1 \le k \le n$ we have:

$$\operatorname{Av}_{\pm} \int_{S^{n-1}} \left| \sum_{i=1}^{k} \pm x(i) \right| d\mu_{n-1}(x) =$$

$$= \operatorname{Av}_{\pm} \int_{S^{n-1}} \left|\left\langle x, \sum_{i=1}^{k} \pm \delta_i \right\rangle\right| d\mu_{n-1}(x)$$

$$= \sqrt{k} \int_{S^{n-1}} |\langle x, \delta_1 \rangle|\, d\mu_{n-1}(x) \qquad \left[\begin{array}{l}\text{by the rotational}\\ \text{invariance of } \mu_{n-1}\end{array}\right].$$

Setting

$$\alpha_n = \int_{S^{n-1}} |\langle x, \delta_1 \rangle|\, d\mu_{n-1}(x) ,$$

we have from Khintchine's inequality that for each $1 \le k \le n$,

$$\sqrt{nk}\,\alpha_n \le \int_{S^{n-1}} r_k(x)\, d\mu_{n-1}(x) \le \sqrt{2nk}\,\alpha_n .$$

(We plugged in the exact constant of $\sqrt{2}$ in Khintchine's inequality calculated in [5] and [13], but of course any constant would serve as well.) Since obviously $r_n(x) = \sqrt{n}$, we conclude that for $1 \le k \le n$

$$\sqrt{k/2} \le \int_{S^{n-1}} r_k(x)\, d\mu_{n-1}(x) \le \sqrt{k} . \tag{1.3}$$

Specializing now to the case $k = [K \log n]$, we have from (1.2) and (1.3) that

$$\sqrt{k/3} \le M_r$$

at least for $K \log n$ sufficiently large. Thus if we define

$$\varepsilon = \tau \sqrt{k/3n}$$

we get from (1.1) that

$$\sigma\,[U \in O(n) : (1 - \tau)M_r \le r(Uy) \le (1 + \tau)M_r \quad \text{for all} \quad y \in B]$$

$$\ge 1 - 2n(n + 1) \exp\left(-\frac{\tau^2 k}{18}\right)$$

$$\ge 1 - 2n(n + 1) \exp\left(-\frac{\tau^2 K \log n}{18}\right)$$

which is positive if, say,

$$K \ge (10/\tau)^2 . \qquad \square$$

It is easily seen that the estimate $K \log n$ in Lemma 1 cannot be improved. Indeed, in a ball of radius 2 in ℓ_2^k there are at most 4^k vectors $\{x_i\}$ so that $\|x_i - x_j\| \ge 1$ for every $i \ne j$ (see the proof of Lemma 3 below). Hence for τ sufficiently small there is no map F which maps an orthonormal set with more than 4^k vectors into a k-dimensional subspace of ℓ_2 with

$$\|F\|_{\ell ip}\ \|F^{-1}\|_{\ell ip} \le \frac{1 + \tau}{1 - \tau} .$$

We can now verify the conjecture of Marcus and Pisier [10].

THEOREM 1. $\sup_{n=2,3,\ldots} (\log n)^{-1/2} L(\infty, n) < \infty$. In other words: there is a constant K so that for all metric spaces X and all finite subsets M of X (card $M = n$, say) every function f from M into ℓ_2 has a Lipschitz extension $\tilde{f} : X \to \ell_2$ which satisfies

$$\|\tilde{f}\|_{\ell ip} \le K \sqrt{\log n}\, \|f\|_{\ell ip} .$$

PROOF. Given X, $M \subset X$ with card $M = n$, and $f : M \to \ell_2$, set $A = f[M]$. We apply Lemma 1 with $\tau = 1/2$ to get a one-to-one function g^{-1} from A onto a subset $g^{-1}[A]$ of ℓ_2^k (where $k \le K \log n$) which satisfies

$$\|g^{-1}\|_{\ell ip} \le 1; \quad \|g\|_{\ell ip} \le 3 .$$

By Kirszbraun's theorem, we can extend g to a function $\tilde{g} : \ell_2^k \to \ell_2$ in such a way that

$$\|\tilde{g}\|_{\ell ip} \le 3 .$$

Let $I : \ell_2^k \to \ell_\infty^k$ denote the formal identity map, so that

$$\|I\| = 1, \quad \|I^{-1}\| = \sqrt{k} .$$

Then

$$h \equiv I g^{-1} f, \quad h : M \to \ell_\infty^k$$

has Lipschitz norm at most $\|f\|_{\ell ip}$, so by the non-linear Hahn-Banach theorem (see, e.g., p. 48 of [14]), h can be extended to a mapping

$$\tilde{h} : X \to \ell_\infty^k$$

which satisfies

$$\|\tilde{h}\|_{\ell ip} \le \|f\|_{\ell ip} .$$

Then

$$\tilde{f} \equiv \tilde{g}\, I^{-1} \tilde{h}; \quad \tilde{f} : X \to \ell_2$$

is an extension of f and satisfies

$$\|\tilde{f}\|_{\ell ip} \le 3 \sqrt{k}\, \|f\|_{\ell ip} \le 3K \sqrt{\log n}\, \|f\|_{\ell ip} . \qquad \square$$

Next we outline our approach to the problem of obtaining a lower bound for $L(\infty,n)$. Take for f the inclusion mapping from an ε-net for S^{N-1} into ℓ_2^N, and consider ℓ_2^N isometrically embedded into L_∞. A Lipschitz extension of f to a mapping $\tilde{f} : L_\infty \to \ell_2$ should act like the identity ℓ_2^N, so the techniques of [8] should yield a linear projection from L_∞ onto ℓ_2^N whose norm is of order $\|f\|_{\ell ip}$. Since ℓ_2^N is complemented in L_∞ only of order $\sqrt{N}$ and there are ε-nets for S^{N-1} of cardinality $n \equiv [4/\varepsilon]^N$, we should get that

$$L(\infty,n) \geq \sqrt{N} \geq \delta \left(\frac{\operatorname{Log} n}{-\operatorname{Log} \varepsilon}\right)^{1/2} .$$

In Theorem 2 we make this approach work when ε is of order N^{-2}, so we get

$$L(\infty,n) \geq \delta' \left(\frac{\operatorname{Log} n}{\operatorname{Log}\operatorname{Log} n}\right)^{1/2} .$$

That the difficulties we incur with the outlined approach for larger values of ε are not purely technical is the gist of the following extension result.

(*)THEOREM 2. <u>Suppose that</u> X <u>is a metric space</u>, $A \subset X$, $f : A \to \ell_2$ <u>is Lipschitz and</u> $d(x,y) \geq \varepsilon > 0$ <u>for all</u> $x \neq y \in A$. <u>Then there is an extension</u> $\tilde{f} : X \to \ell_2$ <u>of</u> f <u>so that</u>

$$\|\tilde{f}\|_{\ell ip} \leq \frac{6D}{\varepsilon} \|f\|_{\ell ip} ,$$

<u>where</u> D <u>is the diameter of</u> A.

PROOF. We can assume by translating f that there is a point $0 \in A$ so that $f(0) = 0$. Set $B = A \sim \{0\}$ and define

$$F : A \to \ell_1^B \text{ by}$$

$$F(b) = \begin{Bmatrix} \delta_b, & b \neq 0 \\ 0, & b = 0 \end{Bmatrix} .$$

Define

$$G : \ell_1^B \to \ell_2$$

by

$$G\Big(\sum_{b \in B} \alpha_b \delta_b\Big) = \sum_{b \in B} \alpha_b f(b) .$$

(*) See the appendix for a generalization of Theorem 2 proved by Yoav Benyamini.

Then

$$G\,F = f,\quad G \text{ is linear with}$$

$$\|G\| \le D\,\|f\|_{\ell ip},\quad\text{and}\quad \|F\|_{\ell ip} \le 2/\varepsilon .$$

A weakened form of Grothendieck's inequality (see section 2.6 in [9]) yields that G (as any bounded linear operator from an L_1 space into a Hilbert space) factors through an $\ell_\infty(\aleph)$ space:

$$G = H\,J,\ \|J\| = 1,\ \|H\| \le 3\,\|G\|,$$

$$J : \ell_1^B \to \ell_\infty(\aleph),\quad H : \ell_\infty(\aleph) \to \ell_2 .$$

By the non-linear Hahn-Banach Theorem the mapping $J\,F$ has an extension

$$E : X \to \ell_\infty(\aleph) \text{ which satisfies}$$

$$\|E\|_{\ell ip} \le \|J\,F\|_{\ell ip} \le 2/\varepsilon .$$

Then $\tilde{f} \equiv H\,E$ extends f and $\|\tilde{f}\| \le \frac{6D}{\varepsilon}\,\|f\|_{\ell ip}$, as desired. □

For the proof of Theorem 3, we need three well known facts which we state as lemmas.

LEMMA 2. _Suppose that_ Y, X _are normed spaces and_ $f : S(Y) \to X$ _is Lipschitz with_ $f(0) = 0$. _Then the positively homogeneous extension of_ f, _defined for_ $y \in Y$ _by_

$$\tilde{f}(y) = \|y\|\, f\left(\frac{y}{\|y\|}\right),\ (y \ne 0);\quad \tilde{f}(0) = 0$$

is Lipschitz and

$$\|\tilde{f}\|_{\ell ip} \le 2\,\|f\|_{\ell ip} + \|f\|_\infty .$$

PROOF. Given $y_1, y_2 \in Y$ with $0 < \|y_1\| \le \|y_2\|$,

$$\|\tilde{f}(y_1) - \tilde{f}(y_2)\| \le \left\|\, \|y_1\|\, f\left(\frac{y_1}{\|y_1\|}\right) - \|y_2\|\, f\left(\frac{y_1}{\|y_1\|}\right)\right\| + \|y_2\|\,\left\| f\left(\frac{y_1}{\|y_1\|}\right) - f\left(\frac{y_2}{\|y_2\|}\right)\right\|$$

$$\le \left(\|y_2\| - \|y_1\|\right)\left\|f\left(\frac{y_1}{\|y_1\|}\right)\right\| + \|y_2\|\,\|f\|_{\ell ip}\left\|\frac{y_1}{\|y_1\|} - \frac{y_2}{\|y_2\|}\right\|$$

$$\le \|y_1 - y_2\|\,\|f\|_\infty + \|f\|_{\ell ip}\left\|\frac{\|y_2\|}{\|y_1\|}\,y_1 - y_2\right\|$$

$$\leq \|f\|_\infty \|y_1 - y_2\| + \|f\|_{\ell ip} \left[\left(\frac{\|y_2\|}{\|y_1\|} - 1\right) \|y_1\| + \|y_1 - y_2\|\right]$$

$$\leq \left(\|f\|_\infty + 2\|f\|_{\ell ip}\right) \|y_1 - y_2\|. \qquad \square$$

LEMMA 3. _If_ Y _is an n-dimensional Banach space and_ $0 < \varepsilon$, _then_ $S(Y)$ _admits an_ ε_-net of cardinality at most_ $(1 + 4/\varepsilon)^n$.

PROOF. Let M be a subset of $S(Y)$ maximal with respect to "$\|x-y\| \geq \varepsilon$ for all $x \neq y \in M$".

Then the sets

$$b(y, \varepsilon/2) \cap S(Y), \quad (y \in M)$$

are pairwise disjoint hence so are the sets

$$b(y, \varepsilon/4), \quad (y \in M).$$

Since these last sets are all contained in $b(1 + \varepsilon/4)$, we have that

$$\text{card } M \cdot \text{vol } b(\varepsilon/4) \leq \text{vol } b(1 + \varepsilon/4)$$

so that

$$\text{card } M \leq \left[\frac{4}{\varepsilon}(1 + \varepsilon/4)\right]^n. \qquad \square$$

LEMMA 4. _There is a constant_ $\delta > 0$ _so that for each_ $1 \leq p < 2$ _and each_ $N = 1, 2, \ldots$, L_p _contains a subspace_ E _such that_

$$d(E, \ell_2^N) \leq 2$$

and every projection from L_p _onto_ E _has norm at least_

$$\delta N^{1/p - 1/2}.$$

PROOF. Given a finite dimensional Banach space X and $1 \leq p < \infty$, let

$$\gamma_p(x) = \inf \{\|T\| \|S\| : T : X \to L_p, \ S : L_p \to X, \ S T = I_X\}.$$

So $\gamma_\infty(X)$ is the projection constant of X, hence by [4], [12]

$$\gamma_1(\ell_2^N) = \gamma_\infty(\ell_2^N) = \sqrt{2n/\pi}.$$

This gives the $p = 1$ case.

For $1 < p < 2$ we reduce to the case $p = 1$ by using Example 3.1 of [2], which asserts that there is a constant $C < \infty$ so that for $1 \le p < 2$ ℓ_p^{CN} contains a subspace E with $d(E, \ell_2^N) \le 2$. Since, obviously,

$$d(\ell_p^{CN}, \ell_1^{CN}) \le (CN)^{1 - 1/p}$$

we get that if E is K-complemented in ℓ_p^{CN}, then

$$\pi^{-1/2} (2n)^{1/2} = \gamma_1(\ell_2^N) \le d(E, \ell_2^N)\, d(\ell_p^{CN}, \ell_1^{CN})\, K$$

$$\le 2\, (CN)^{1 - 1/p}\, K. \qquad \square$$

The next piece of background information we need for Theorem 3 is a linearization result which is an easy consequence of the results in [8].

PROPOSITION 1. *Suppose* $X \subset Y$ *and* Z *are Banach spaces,* $f : Y \to Z$ *is Lipschitz, and* $U : X \to Z$ *is bounded, linear. Then there is a linear operator* $G : Z^* \to Y^*$ *so that* $\|G\| \le \|f\|_{\ell ip}$ *and*

$$\|R_2 G - U^*\| \le \|f_{|X} - U\|_{\ell ip},$$

where R_2 *is the natural restriction map from* Y^* *onto* X^*.

REMARK. Note that if Z is reflexive, the mapping $F \equiv G^*_{|Y} : Y \to Z$ satisfies $\|F\| \le \|f\|_{\ell ip}$ and $\|F_{|X} - U\| \le \|f_{|X} - U\|_{\ell ip}$.

PROOF. We first recall some notation from [8]. If Y is a Banach space, $Y^{\#}$ denotes the Banach space of all scalar valued Lipschitz functions $y^{\#}$ from Y for which $y^{\#}(0) = 0$, with the norm $\|y^{\#}\|_{\ell ip}$. There is an obvious isometric inclusion from Y^* into $Y^{\#}$. For a Lipschitz mapping $f : Y \to Z$, Z a normed space, we can define a *linear* mapping

$$f^{\#} : Z^* \to Y^{\#} \text{ by}$$

$$f^{\#} z^* = z^* f.$$

Given Banach spaces $X \subset Y$, Theorem 2 of [8] asserts that there are norm one linear projections

$$P_Y : Y^{\#} \to Y^*, \quad P_X : X^{\#} \to X^*$$

so that

$$P_X R_1 = R_2 P_Y,$$

where R_1 is the restriction mapping from $Y^{\#}$ onto $X^{\#}$. Thus if $X \subset Y$, f, U, Z are as in the hypothesis of Proposition 1, the linear mapping $P_Y f^{\#}$ satisfies

$$\|P_Y f^{\#}\| \le \|f\|_{\ell ip}, \quad R_2 P_Y f^{\#} = P_X R_1 f^{\#}.$$

Since $U: X \to Z$ is linear,

$$U^* = P_X U^{\#}$$

so

$$\|R_2 P_Y f^{\#} - U^*\| = \|P_X(R_1 f^{\#} - U^{\#})\|$$

$$\le \|R_1 f^{\#} - U^{\#}\| = \sup_{z^* \in S(Z^*)} \|R_1 f^{\#} z^* - U^{\#} z^*\|$$

$$= \sup_{z^* \in S(Z^*)} \|(z^* f)_{|X} - z^* U\| \le \|f_{|X} - U\|_{\ell ip}. \qquad \square$$

The final lemma we use in the proof of Theorem 3 is a smoothing result for homogeneous Lipschitz functions.

LEMMA 5. *Suppose* $X \subset Y$ *and* Z *are Banach spaces with* $\dim X = k < \infty$, $F: Y \to Z$ *is Lipschitz with* F *positively homogeneous* (i.e. $F(\lambda y) = \lambda F(y)$ *for* $\lambda \ge 0$, $y \in Y$) *and* $U : X \to Z$ *is linear. Then there is a positively homogeneous Lipschitz mapping*

$$\tilde{F} : Y \to Z \quad \textit{which satisfies}$$

(1) $\|\tilde{F}_{|X} - U\|_{\ell ip} \le (8k + 2) \quad \|F_{|S(X)} - U_{|S(X)}\|_\infty$

(2) $\|\tilde{F}\|_{\ell ip} \le 4 \|F\|_{\ell ip}$.

PROOF. For $y \in S(Y)$ define

$$\hat{F}y = \int_{B_X(1)} F(y+x) \, d\mu(x)$$

where $\mu(\cdot)$ is Haar measure on X $(= \mathbb{R}^k)$ normalized so that

$$\mu(B_X(1)) = 1.$$

For $y_1, y_2 \in S(Y)$ we have

$$\|\hat{F}y_1 - \hat{F}y_2\| \le \int_{B_X(1)} \|F(y_1 + x) - F(y_2 + x)\| \, d\mu(x)$$

$$\le \|F\|_{\ell ip} \|y_1 - y_2\|$$

so

$$\|\hat{F}\|_{\ell ip} \le \|F\|_{\ell ip}.$$

For $x_1, x_2 \in S(X)$ with $\|x_1 - x_2\| = \delta > 0$ we have, since U is linear, that

$$\|(\hat{F} - U)x_1 - (\hat{F} - U)x_2\| =$$

$$\|\int_{B_X(1)} F(x_1 + x)d\mu(x) - \int_{B_X(1)} U(x_1 + x)d\mu(x) - \int_{B_X(1)} F(x_2 + x)d\mu(x) +$$

$$\int_{B_X(1)} U(x_2 + x)\, d\mu(x)\| \le$$

$$\le \int_{B_X(x_1;\, 1) \,\Delta\, B_X(x_2;\, 1)} \|Fx - Ux\| \, d\mu(x) \le$$

$$\le \sup_{x \in B_X(2)} \|Fx - Ux\| \, \mu\, [B_X(x_1;\, 1) \,\Delta\, B_X(x_2;\, 1)]$$

$$= 2 \sup_{x \in B_X(1)} \|Fx - Ux\| \, \mu\, [B_X(x_1;\, 1) \,\Delta\, B_X(x_2;\, 1)] \quad \left[\text{since } F \text{ is positively homogeneous}\right].$$

Since

$$B_X(x_1;\, 1) \,\Delta\, B_X(x_2;\, 1) \subset [B_X(x_1;\, 1) \sim B_X(x_1;\, 1\text{-}\delta)] \cup [B_X(x_2;\, 1) \sim B_X(x_2;\, 1\text{-}\delta)]$$

we have if $\delta \le 1$ that

$$\mu[B_X(x_2;\, 1) \,\Delta\, B_X(x_2;\, 1)] \le 2[1 - (1\text{-}\delta)^k]$$

$$\le 2\, k\, \delta$$

and hence for all $x_1, x_2 \in S(X)$ that

$$\|(\hat{F} - U)\, x_1 - (\hat{F} - U)\, x_2\| \le 4k \, \|F_{|S(X)} - U_{|S(X)}\| \, \|x_1 - x_2\|$$

whence

$$\|\hat{F}_{|S(X)} - U_{|S(X)}\|_{\ell ip} \le 4k \, \|F_{|S(X)} - U_{|S(X)}\|_\infty .$$

Finally, note that the positive homogeniety of F implies that

$$\|\hat{F}\|_\infty \le 2 \, \|F\|_{\ell ip} \quad \text{and} \quad \|\hat{F}_{|S(X)} - U_{|S(X)}\|_\infty \le 2 \, \|F_{|S(X)} - U_{|S(X)}\|_\infty .$$

It now follows from Lemma 2 that the positively homogeneous extension $\tilde{F}$ of $\hat{F}$ satisfies the conclusions of Lemma 5. □

THEOREM 3. <u>There is a constant</u> $\tau > 0$ <u>so that for all</u> $n = 2, 3, 4, \ldots$ <u>and all</u> $1 \le p < 2$,

$$L(p,n) \ge \tau \left(\frac{\text{Log } n}{\text{Log Log } n}\right)^{1/p - 1/2} .$$

REMARK. Since $L(\infty,n) \ge L(1,n)$, we get the lower estimate for $L(\infty,n)$ mentioned in the introduction.

PROOF. Given p and n, for a certain value of $N = N(n)$ to be specified later choose a subspace E of L_p with $d(E, \ell_2^N) \le 2$ and E only $\delta \, N^{1/p - 1/2}$-complemented in L_p (Lemma 4). For a value $\varepsilon = \varepsilon(n) > 0$ to be specified later, let A be a minimal ε-net of $S(E)$, so, by Lemma 3,

$$\text{card } A \le (1 + 4/\varepsilon)^N .$$

One relation among n, N, ε we need is

$$(1.4) \qquad (1 + 4/\varepsilon)^N + 1 \le n.$$

Let $f : A \cup \{0\} \to E$ be the identify map. Since $d(E, \ell_2^N) \le 2$, we can by Lemma 2 get a positively homogeneous extension $\tilde{f} : L_p \to E$ of f so that

$$\|\tilde{f}\|_{\ell ip} \le 6 \; L(p,n) .$$

Since $\tilde{f}(a) = f(a) = a$ for $a \in A$ and A is an ε-net for $S(E)$, we get that for $x \in S(E)$,

$$\|\tilde{f}(x) - x\| \le (6 \; L(p,n) + 1) \; \varepsilon .$$

Therefore, from Lemma 5 we get a Lipschitz mapping $\hat{f} : L_p \to E$ which satisfies

$$\|\hat{f}\|_{\ell ip} \le 24 \; L(p,n)$$

$$(1.5) \qquad \|\hat{f}_{|E} - I_E\| \le (8N + 2)(6 \; L(p,n) + 1)\varepsilon .$$

Note that if

$$(8N + 2)(6\,L(p,n) + 1)\varepsilon \le 1/2, \tag{1.6}$$

(1.5) implies that there is a linear projection from L_p onto E with norm at most $48\,L(p,n)$, so we can conclude that

$$L(p,n) > \delta/48\,N^{1/p - 1/2}.$$

Finally, we just need to observe that (1.4) and (1.6) are satisfied (at least for sufficiently large n) if we set

$$\varepsilon = \mathrm{Log}^{-2}\, n, \quad N = \frac{\mathrm{Log}\, n}{2\,\mathrm{Log}\,\mathrm{Log}\, n}. \qquad \square$$

2. OPEN PROBLEMS.

Besides the obvious question left open by the preceding discussion (i.e. whether the estimate for $L(\infty,n)$ given in Theorem 1 is indeed the best possible), there are several other problems which arise naturally in the present context. We mention here only some of them.

PROBLEM 1. Is it true that for $1 < p < 2$, every subset X of $L_p(0,1)$, and every Lipschitz map f from X into ℓ_2^k there is an extension $\tilde{f}$ of f from $L_p(0,1)$ into ℓ_2^k with

$$\|\tilde{f}\|_{\ell ip} \le C(p)\,\|f\|_{\ell ip}\,k^{1/p - 1/2} \tag{2.1}$$

where $C(p)$ depends only on p?

A positive answer to problem 1 combined with Lemma 1 above will of course provide an alternative proof to the result of Marcus and Pisier [10] mentioned in the introduction. The linear version of problem 1 (where X is a subspace and f a linear operator) is known to be true (cf. [7] and [3]).

PROBLEM 2. What happens in the Marcus-Pisier theorem if $2 < p < \infty$? Is the Lipschitz analogue of Maurey's extension theorem [11] (cf. also [3]) true? In other words, is it true that for $2 < p < \infty$ there is a $c(p)$ such that for every Lipschitz map f from a subset X of $L_p(0,1)$ into ℓ_2 there is a Lipschitz extension $\tilde{f}$ from $L_p(0,1)$ into ℓ_2 with

$$\|\tilde{f}\|_{\ell ip} \le c(p)\|f\|_{\ell ip}?$$

PROBLEM 3. What are the analogues of Lemma 1 in the setting of Banach spaces different from Hilbert spaces? The most interesting special case seems to be concerning the spaces ℓ_∞^n. It is well known that every finite metric space $X = \{x_i\}_{i=1}^n$ embeds isometrically into ℓ_∞^n (the point x_i is mapped to the n-tuple $\{d(x_1, x_i), d(x_2, x_i), \ldots, d(x_n, x_i)\}$ in ℓ_∞^n). Hence in view of Lemma 1 it is quite natural to ask the following. Does there exist for all $\varepsilon > 0$ (or alternatively for some $\varepsilon > 0$) a constant $K(\varepsilon)$ so that for every metric space X with cardinality n there is a Banach space Y with $\dim Y \le K(\varepsilon)\log n$ and a map f from X into Y so that $\|f\|_{\ell ip} \|f^{-1}\|_{\ell ip} \le 1 + \varepsilon$?

A weaker version of Problem 3 is

PROBLEM 4. It is true that for every metric space X with cardinality n there is a subset $\tilde{X}$ in ℓ_2 and a Lipschitz map F from X onto $\tilde{X}$ so that

$$\|F\|_{\ell ip} \|F^{-1}\|_{\ell ip} \le K\sqrt{\log n} \tag{2.2}$$

for some absolute constant K?

Since for every Banach space Y with $\dim Y = k$ we have $d(Y, \ell_2^k) \le \sqrt{k}$ (cf. [6]) it is clear that a positive answer to problem 3 implies a positive answer to problem 4. V. Milman pointed out to us that it follows easily from an inequality of Enflo (cf. [1]) that (2.2), if true, gives the best possible estimate. (In the notation of [1], observe that the "m-cube"

$$x_\theta = (\theta_1, \theta_2, \ldots, \theta_m) \quad (\theta \in \{-1, 1\}^m)$$

in ℓ_1^m has all "diagonals" of length $2m$ and all "edges" of length 2, so that if F is any Lipschitz mapping from these 2^m points in ℓ_1^m into a Hilbert space, the corollary in [1] implies that

$$\|F\|_{\ell ip} \|F^{-1}\|_{\ell ip} \ge m^{1/2}.)$$

3. APPENDIX.

After this note was written, Yoav Benyamini discovered that Theorem 2 remains valid if ℓ_2 is replaced with any Banach space. He kindly allowed us to reproduce here his proof. The main lemma Benyamini uses is:

LEMMA 6. Let Γ be an indexing set and let $\{e_\gamma\}_{\gamma \in \Gamma}$ be the unit vector basis for $c_o(\Gamma)$. Set

$$A = \{\alpha e_\gamma : 0 \le \alpha \le 1;\ \gamma \in \Gamma\}$$

$$B = \overline{\text{conv}}\ A\ (= \text{positive part of } B_{\ell_1}(\Gamma)).$$

Then

(i) there is a retraction G from $\ell_\infty(\Gamma)$ onto B which satisfies $\|G\|_{\ell ip} \le 2$

(ii) there is a mapping H from $\ell_\infty(\Gamma)$ into A which satisfies $\|H\|_{\ell ip} \le 4$ and $He_\gamma = e_\gamma$ for all $\gamma \in \Gamma$.

PROOF. Since the mapping $x \to x^+$ is a contractive retraction from $\ell_\infty(\Gamma)$ onto its positive cone, $\ell_\infty(\Gamma)^+$; to prove (i) it is enough to define G only on $\ell_\infty(\Gamma)^+$.

For $y \in \ell_\infty(\Gamma)^+$, let

$$g(y) = \inf \{t : \|(y - te)^+\|_1 \le 1\}$$

where $e \in \ell_\infty(\Gamma)$ is the function identically equal to one and $\|\cdot\|_1$ is the usual norm in $\ell_1(\Gamma)$. Clearly the inf is actually a minimum and $0 \le g(y) \le \|y\|_\infty$. Note that

$$|g(y) - g(z)| \le \|y-z\|_\infty.$$

Indeed, assume that $g(y) \ge g(z)$. Then

$$y - [g(z) + \|y-z\|_\infty e] \le y - g(z)e + z - y \le z - g(z)e$$

and hence

$$\|(y-[g(z) + \|y-z\|_\infty]e)^+\|_1 \le 1;$$

that is

$$g(y) \le g(z) + \|y-z\|_\infty.$$

Now set for $y \in \ell_\infty(\Gamma)^+$

$$G(y) = (y - g(y)e)^+.$$

To prove (ii), it is enough, in view of (i), to define H on B with $\|H_{|B}\|_{\ell ip} \le 2$. For $y \in B$, $y = \{y(\gamma)\}_{\gamma \in \Gamma}$, defined Hy by

$$Hy(\gamma) = (2y(\gamma) - 1)^+.$$

For $y \in B$, there is at most one $\gamma \in \Gamma$ for which $y(\gamma) > \frac{1}{2}$, hence $HB \subset A$. Evidently $He_\gamma = e_\gamma$ for $\gamma \in \Gamma$ and $\|H_{|B}\|_{\ell ip} \le 2$.

THEOREM 2 (Y. Benyamini). *Suppose that* X *is a metric space,* Y *is a subset of* X *with* $d(x,y) \ge \varepsilon > 0$ *for all* $x \ne y \in Y$, Z *is a Banach space, and* $f : Y \to Z$ *is Lipschitz. Then there is an extension* $\tilde{f} : X \to Z$ *of* f *so that*

$$\|\tilde{f}\|_{\ell ip} \le (4D/\varepsilon)\|f\|_{\ell ip}$$

where D *is the diameter of* Y.

PROOF. Represent

$$Y = \{0\} \cup \{y_\gamma : \gamma \in \Gamma\}$$

and assume, by translating f, that $f(0) = 0$. We can factor f through the subset $C = \{0\} \cup \{e_\gamma : \gamma \in \Gamma\}$ of $\ell_\infty(\Gamma)$ by defining $g : Y \to C$, $h : C \to Z$ by

$$g(y_\gamma) = e_\gamma, \; g(0) = 0$$

$$h(e_\gamma) = f(y_\gamma), \; h(0) = 0.$$

Evidently,

$$\|g\|_{\ell ip} \le 1/\varepsilon, \; \|h\|_{\ell ip} \le D\|f\|_{\ell ip}.$$

By the non-linear Hahn-Banach theorem, g has an extension to a function $\tilde{g} : X \to \ell_\infty(\Gamma)$ with $\|\tilde{g}\|_{\ell ip} = \|g\|_{\ell ip}$, so to complete the proof, it suffices to extend h to a function $\tilde{h} : B \to Z$ with $\|\tilde{h}\|_{\ell ip} = \|h\|_{\ell ip}$ and apply Lemma 6(ii).

Define for $0 \le t \le 1$ and $\gamma \in \Gamma$

$$\tilde{h}(te_\gamma) = th(e_\gamma).$$

If $1 \ge t \ge s \ge 0$ and $\gamma \ne \Delta \in \Gamma$ then

$$\|\tilde{h}(te_\gamma) - \tilde{h}(se_\Delta)\| \le (t-s)\|h(e_\gamma)\| + s\,\|h(e_\Delta) - h(e_\gamma)\|$$

$$\le (t-s)\|h\|_{\ell ip} + s\|h\|_{\ell ip} = \|h\|_{\ell ip}\|te_\gamma - se_\Delta\|_\infty,$$

so $\|\tilde{h}\|_{\ell ip} = \|h\|_{\ell ip}$. □

REFERENCES

1. P. Enflo, On the non-existence of uniform homeomorphisms between L_p spaces, Arkiv for Matematik 8 (1969), 195-197.

2. T. Figiel, J. Lindenstrauss and V. Milman, The dimension of almost spherical sections of convex bodies, Acta. Math. 139 (1977), 53-94.

3. T. Figiel and N. Tomczak-Jaegermann, Projections onto Hilbertian subspaces of Banach spaces, Israel J. Math 33 (1979), 155-171.

4. D. J. H. Garling and Y. Gordon, Relations between some constants associated with finite-dimensional Banach spaces, Israel J. Math 9 (1971), 346-361.

5. U. Haagerup, The best constant in the Khintchine inequality, Studia Math 70 (1982), 231-283.

6. F. John, Extremum problems with inequalities as subsidiary conditions, Courant anniversary volume, Interscience N.Y. (1948), 187-204.

7. D. Lewis, Finite dimensional subspaces of L_p, Studia Math. 63 (1978), 207-212.

8. J. Lindenstrauss, On non-linear projections in Banach spaces, Mich. Math. J. 11 (1964), 263-287.

9. J. Lindenstrauss and L. Tzafriri, Classical Banach spaces Vol. I sequence spaces, Ergebnisse n. 92, Springer Verlag, 1977.

10. M. B. Marcus and G. Pisier, Characterizations of almost surely continuous p-stable random Fourier series and strongly stationary processes, to appear.

11. B. Maurey, Un theoreme de prolongment, C. R. Acad. Paris 279 (1974), 329-332.

12. D. Rutovitz, some parameters associated with finite dimensional Banach spaces, J. London Math. Soc. 40 (1965), 241-255.

13. S. J. Szarek, On the best constant in the Khintchine inequality, Studia Math. 58 (1978), 197-208.

14. J. H. Wells and L. R. Williams, Embeddings and Extensions in Analysis, Ergebnisse n. 84 Springer Verlag 1975.

William B. Johnson
The Ohio State University and
Texas A & M University

Joram Lindenstrauss
The Hebrew University of Jerusalem,
Texas A & M University, and
The Ohio State University

Contemporary Mathematics
Volume **26**, 1984

A UNIQUENESS RESULT FOR A CLASS OF COMPACT CONNECTED GROUPS

Robert R. Kallman*

ABSTRACT: Let G be a compact connected metric group with totally disconnected center. Then G has a unique topology in which it is a complete separable metric group.

The purpose of this paper is to prove the following theorem.

THEOREM 1. Let G be a compact connected metric group with totally disconnected center, let H be a complete separable metric group, and let $\psi: H \to G$ be an abstract group isomorphism. Then ψ is a topological isomorphism.

Note that this theorem really says something, for the analogous theorem for the circle is false. Stewart ([5], Theorem 3.1) showed that G has a unique topology in the class of compact connected groups with totally disconnected center. The proof of the present more general theorem proceeds along quite different lines. The only point of contact is the following proposition, of independent interest (Stewart [5], Theorem 1.1). The following proof uses group representations. It may be of some pedagogical interest.

PROPOSITION 2. Let G be a compact connected group. Then $G = (\prod_{\gamma\varepsilon\Gamma} S_\gamma \times A)/C$, where each S_γ is a compact connected simply connected simple Lie group, where A is a compact connected abelian group, and C is a compact totally disconnected central subgroup of $\prod_{\gamma\varepsilon\Gamma} S_\gamma \times A$.

PROOF: Recall that each strongly continuous irreducible unitary representation of G is finite dimensional. Let $\pi_\beta (\beta \,\varepsilon\, \beta)$ be a family of strongly continuous irreducible unitary representations of G, no two of which are unitarily equivalent, such that each strongly continuous irreducibly unitary representation of G is unitarily equivalent to one of the π_β's. Let $\pi = \bigoplus_{\beta\varepsilon\beta} \pi_\beta$. The mapping $g \to \pi(g)$, $G \to U(\underline{H}(\pi))$, is strongly continuous and is one-to-one by the Peter-Weyl theorem. Hence, it is a topological isomorphism onto its range, for G is compact.

Each $\pi_\beta(G)$ is a compact, and therefore closed, connected subgroup of the finite dimensional unitary group $U(\underline{H}(\pi_\beta))$. So each $\pi_\beta(G)$ is a connected

*Supported in part by an NSF Grant and a North Texas State University Faculty Research Grant.

compact Lie group. Since $\pi_\beta(G)$ acts irreducibly on $\underline{H}(\pi_\beta)$, well known facts from Lie groups (Hochschild [3], Theorem 1.3, p. 144) imply that $\pi_\beta(G) = P_\beta T_\beta$, where P_β is a compact connected semisimple Lie group or the identity and T_β is either the identity or a circle group which lies in the center of $\pi_\beta(G)$.

Define $N_\beta = \pi_\beta^{-1}(\text{center}(\pi_\beta(G)))$. Note that $G/N_\beta = P_\beta/(\text{center}(P_\beta))$ is either the identity or a compact connected centerless semisimple Lie group. Define an equivalence relation on the set of β's such that G/N_β is a simple Lie group by defining $\beta_1 \sim \beta_2$ if and only if $N_{\beta_1} = N_{\beta_2}$. Call the set of such equivalence classes Γ and define $N_\gamma (\gamma \in \Gamma)$ by setting $N_\gamma = N_\beta$ for any $\beta \in \gamma$. It is clear that $\text{center}(G) = \bigcap_{\gamma \in \Gamma} N_\gamma$. Before proceeding any further, we need the following lemmas.

LEMMA 3. <u>Let</u> $\gamma_1, \ldots, \gamma_n$ <u>be distinct elements of</u> Γ. <u>Then</u>:

(1) $(N_{\gamma_1} \cap \ldots \cap N_{\gamma_{n-1}})/(N_{\gamma_1} \cap \ldots \cap N_{\gamma_n}) = G/N_{\gamma_n}$ for $n \geq 2$;

(2) $G/(N_{\gamma_1} \cap \ldots \cap N_{\gamma_n}) = G/N_{\gamma_1} \times \ldots \times G/N_{\gamma_n}$;

(3) <u>if</u> ρ <u>is a strongly continuous unitary representation of</u> G, <u>then</u> $\rho(N_{\gamma_1} \cap \ldots \cap N_{\gamma_n})$ <u>contains center</u>$(\rho(G))$;

(4) if $[\beta] \in \Gamma$ and $\beta \notin \gamma_i$ $(1 \leq i \leq n)$, then $\pi_\beta(G) = \pi_\beta(N_{\gamma_1} \cap \ldots \cap N_{\gamma_n})$.

PROOF: The proof proceeds by induction on n. Let $n = 1$ and $\gamma_1 = \gamma$. (1) is vacuously satisfied. (2) is trivially satisfied for $n = 1$. Let ρ be a strongly continuous unitary representation of G. $N_\gamma \cdot \text{kernel}(\rho)$ is a closed normal subgroup of G which contains N_γ. Since G/N_γ is a connected centerless simple Lie group, either $N_\gamma \cdot \text{kernel}(\rho) = N_\gamma$ or $N_\gamma \cdot \text{kernel}(\rho) = G$. In the first case, $\rho(G)/\rho(N_\gamma) = G/N_\gamma$, a centerless group, and so $\rho(N_\gamma)$ certainly contains $\text{center}(\rho(G))$. In the second case, $\rho(N_\gamma) = \rho(G)$. Hence, (3) is satisfied. Let $[\beta] \in \Gamma$ and $\beta \notin \gamma$. $N_\gamma \cdot N_\beta$ is a closed normal subgroup of G which contains N_γ. Since G/N_γ is a connected centerless simple Lie group, $N_\gamma \cdot N_\beta = G$ or $N_\gamma \cdot N_\beta = N_\gamma$. In the first case, $\pi_\beta(G) = \pi_\beta(N_\gamma)\pi_\beta(N_\beta) = \pi_\beta(N_\gamma)$ since $\pi_\beta(N_\gamma)$ contains $\pi_\beta(N_\gamma)$ by (3). In the second case, N_β is contained in N_γ. But G/N_β is either the identity or a connected centerless simple Lie group, and G/N_γ is a connected centerless simple Lie group. Hence, $N_\beta = N_\gamma$ and $\beta \in \gamma$. Contradiction. So the second case never holds, and $\pi_\beta(N_\gamma) = \pi_\beta(G)$. Thus (4) is true and the inductive step for $n = 1$ is verified.

Assume that the lemma is true for $n - 1$, and let us prove it true for n. To prove (1), choose $\alpha \in \gamma_n$ so that $\text{kernel}(\pi_\alpha) =$

$N_{\gamma_n} \cdot \pi_\alpha(N_{\gamma_1} \cap \ldots \cap N_{\gamma_{n-1}}) = \pi_\alpha(G)$ by (4). Hence, $G/N_{\gamma_n} = \pi_\alpha(G) =$

$(N_{\gamma_1} \cap \ldots \cap N_{\gamma_{n-1}})/(N_{\gamma_1} \cap \ldots \cap N_{\gamma_{n-1}} \cap \text{kernel}(\pi_\alpha)) =$

$(N_{\gamma_1} \cap \ldots \cap N_{\gamma_{n-1}})/(N_{\gamma_1} \cap \ldots \cap N_{\gamma_n})$.

To prove (2), consider the natural homomorphism $\psi\colon G \to G/N_{\gamma_1} \times \ldots \times G/N_{\gamma_n}$. $\psi(G)$ is a closed connected subgroup of a Lie group and hence is a connected Lie group. Kernel$(\psi) = N_{\gamma_1} \cap \ldots \cap N_{\gamma_n}$ and $\dim(G/\text{kernel}(\psi)) = \dim(G/(N_{\gamma_1} \cap \ldots \cap N_{\gamma_{n-1}}) +$ $\dim((N_{\gamma_1} \cap \ldots \cap N_{\gamma_{n-1}})/\text{kernel}(\psi)) = \dim(G/N_{\gamma_1}) + \ldots + \dim(G/N_{\gamma_{n-1}}) +$ $\dim(G/N_{\gamma_n}) = \dim(G/N_{\gamma_1} \times \ldots \times G/N_{\gamma_n})$ by (2) plus induction and (1). This proves (2).

Let ρ be a strongly continuous unitary representation of G. $\rho(G)/\rho(N_{\gamma_1} \cap \ldots \cap N_{\gamma_n}) = G/(N_{\gamma_1} \cap \ldots \cap N_{\gamma_n})\cdot\text{kernel}(\rho)$ is a homomorphic image of $G/(N_{\gamma_1} \cap \ldots \cap N_{\gamma_n})$, a finite product of connected centerless simple Lie groups. Hence, $\rho(N_{\gamma_1} \cap \ldots \cap N_{\gamma_n})$ contains center$(\rho(G))$. This proves (3).

Since (3) now holds for n, it suffices to prove (4) for the special case kernel$(\pi_\beta) = N_\beta$. Suppose $\pi_\beta(N_{\gamma_1} \cap \ldots \cap N_{\gamma_n}) \neq \pi_\beta(G)$. Since $\pi_\beta(G)$ must be a connected centerless simple Lie group, $N_{\gamma_1} \cap \ldots \cap N_{\gamma_n}$ is contained in N_β, and π_β will factor through $G/(N_{\gamma_1} \cap \ldots \cap N_{\gamma_n})$. But π_β will then factor through one of the simple factors G/N_{γ_i}, and N_{γ_i} will be contained in N_β. As before, $N_\beta = N_{\gamma_i}$, and $\beta \in \gamma_i$. Contradiction. Hence, (4) is proved. This completes the proof of lemma 3.

LEMMA 4. <u>Let ρ be a strongly continuous unitary representation of G. Then</u> $\rho(\text{center}(G)) = \text{center}(\rho(G))$ <u>and</u> $\rho((\text{center}(G))^0) = (\text{center}(\rho(G)))^0$.

PROOF: The second equality follows from the first, for $\rho((\text{center}(G))^0)$ certainly is contained in $(\text{center}(\rho(G)))^0$, and the quotient $\text{center}(\rho(G))/\rho((\text{center}(G))^0)$ is isomorphic to $\text{center}(G)\cdot\text{kernel}(\rho)/(\text{center}(G))^0\cdot\text{kernel}(\rho)$, a homomorphic image of $\text{center}(G)/(\text{center}(G))^0$, a totally disconnected group.

$\rho(\bigcap_{\gamma\in\Gamma} N_\gamma) = \text{center}(\rho(G))$ by (3) of lemma 3 and a compactness argument. But $\text{center}(G) = \bigcap_{\gamma\in\Gamma} N_\gamma$. This proves lemma 4.

LEMMA 5. <u>It suffices to prove proposition</u> 2 <u>for the special case</u> $(\text{center}(G))^0 = (e)$.

PROOF: Let $H = \overline{[G,G]}$, a compact connected normal subgroup of G. For any β, a compactness argument implies that $\pi_\beta(H) = \pi_\beta(\overline{[G,G]}) = \overline{[\pi_\beta(G), \pi_\beta(G)]} = \overline{[P_\beta T_\beta, P_\beta T_\beta]} = P_\beta$, a group with finite center. So $\pi_\beta((\text{center}(H))^0)$ is the identity for all β. Hence, $(\text{center}(H))^0)$ is the identity for all . Hence, $(\text{center}(H))^0$ is the identity.

Suppose proposition 2 is true for H. Then H is the homomorphic image of a product $\prod_{\gamma\varepsilon\Gamma} S_\gamma$, where each S_γ is a compact connected simply connected simple Lie group, and the kernel of this homomorphism is a compact totally disconnected subgroup of the center of the product. $H\cdot(\text{center}(G))^0$ is a compact normal subgroup of G. For every β, $\pi_\beta(H\cdot(\text{center}(G))^0) = \pi_\beta(H)\pi_\beta((\text{center}(G))^0) = P_\beta T_\beta$ by lemma 4. Hence, $G = H\cdot(\text{center}(G))^0$. The natural homomorphism $\prod_{\gamma\varepsilon\Gamma} S_\gamma \times (\text{center}(G))^0 \to H \times (\text{center}(G))^0 \to H\cdot(\text{center}(G))^0 = G$ is an onto homomorphism with kernel a compact totally disconnected subgroup of the center of the product. This proves lemma 5.

We therefore can, and do, assume for the rest of the proof of proposition 2 that $(\text{center}(G))^0 = (e)$.

For every γ in Γ, define $Q_\gamma = \bigcap_{\substack{\gamma'\varepsilon\Gamma \\ \gamma'\neq\gamma}} N_{\gamma'}$. Q_γ^0 is a compact connected normal subgroup of G.

LEMMA 6. (1) <u>If</u> $\beta \varepsilon \gamma$, <u>then</u> $\pi_\beta(Q_\gamma^0) = \pi_\beta(G)$; (2) <u>if</u> $\beta \varepsilon \gamma'$ <u>and</u> $\gamma' \neq \gamma$, <u>then</u> $\pi_\beta(Q_\gamma^0)$ <u>is the identity</u>; (3) <u>the</u> Q_γ^0 <u>commute with each other</u>; (4) <u>each</u> Q_γ^0 <u>is a simple Lie group</u>; (5) <u>if</u> U <u>is any neighborhood of</u> e <u>in</u> G, <u>then all except finitely many of the</u> Q_γ^0 <u>are contained in</u> U.

PROOF: A compactness argument plus lemma 3, (4), imply that $\pi_\beta(Q_\gamma) = \pi_\beta(G)$. The argument used in lemma 4 now shows that $\pi_\beta(Q^0) = \pi_\beta(G)$. This proves (1).

If $\beta \varepsilon \gamma'$ and $\gamma' \neq \gamma$, then Q_γ^0 is contained in $N_{\gamma'}$, and $\pi_\beta(N_{\gamma'})$ is the identity. This proves (2).

Let $\gamma, \gamma' \varepsilon \Gamma$, $\gamma \neq \gamma'$. The set of commutators $aba^{-1}b^{-1}$, $a \varepsilon Q_\gamma^0$, $b \varepsilon Q_{\gamma'}^0$, is connected and contained in center(G) by (2). Hence, each commutator is the identity since center(G) is totally disconnected. This proves (3).

Let ρ be a strongly continuous finite dimensional unitary representation of G. Note that (2) implies that $\text{center}(Q_\gamma^0)$ is contained in $N_{\gamma'}$ for every $\gamma' \varepsilon \Gamma$. Hence, $\text{center}(Q_\gamma^0)$ is contained in center(G), and so $\text{center}(Q_\gamma^0)$ is totally disconnected. Therefore, $\rho(Q_\gamma^0)$ is a compact connected semisimple Lie group or the identity by Lemma 4. If $\rho(Q_\gamma^0)$ is not simple, there exist $\gamma_1, \gamma_2 \varepsilon \Gamma$, $\gamma_1 \neq \gamma_2$, and continuous homomorphisms of $\rho(Q_\gamma^0)$ onto G/N_{γ_1} and G/N_{γ_2}. Reasoning as before, we conclude that $G = Q_\gamma^0\cdot N_{\gamma_1} = Q_\gamma^0\cdot N_{\gamma_2}$.

But at least one of γ_1, γ_2 is not equal to γ. Then either $Q^0_\gamma \cdot N_{\gamma_1} = N^0_{\gamma_1}$ or $Q^0_\gamma \cdot N_{\gamma_2} = N_{\gamma_2}$. This shows that at least one of the homomorphisms is not onto. Contradiction. Hence, $\rho(Q^0_\gamma)$ must be a simple Lie group or the identity. Let ρ_1 and ρ_2 be strongly continuous finite dimensional unitary representations of Q^0_γ such that both $\rho_1(Q^0_\gamma)$ and $\rho_2(Q^0_\gamma)$ are simple Lie groups and such that ρ_1 is a direct summand of ρ_2. The natural mapping of $\rho_2(Q^0_\gamma)$ onto $\rho_1(Q^0_\gamma)$ is a covering mapping. Weyl's lemma (Helgason [2], Theorem 6.9, p. 123) implies that there is an upper bound on the size of the centers of any $\rho(Q^0_\gamma)$, where ρ is a strongly continuous finite dimensional unitary representation. Hence, $Q^0_\gamma = \pi(Q^0_\gamma)$ is a simple Lie group. This proves (4).

For each finite subset F of Γ, let $C(F)$ be the closure of the group generated by the set of Q^0_γ, where $\gamma \in \Gamma - F$. $C(F)$ is a compact connected normal subgroup of G. The intersection of the set of $C(F)$'s is a compact connected normal subgroup of G. Every element of this intersection commutes with every Q^0_γ by (3) and hence lies in center(G). But center(G) is totally disconnected, so this intersection reduces to the identity. Hence, if U is any neighborhood of e in G, then there exists a finite subset F of Γ such that $C(F)$ lies in U. This proves (5) and lemma 6.

We now complete the proof of proposition 2. For each $\gamma \in \Gamma$, let S_γ be the simply connected covering group of Q^0_γ. S_γ is a compact connected simple Lie group by Weyl's lemma and lemma 6, (4). Let $G' = \prod_{\gamma\in\Gamma} S_\gamma$. G' is compact. There is a natural homomorphism of the subgroup of G' algebraically generated by the S_γ onto the subgroup of G algebraically generated by the Q^0_γ. Lemma 6, (5) implies that this homomorphism is continuous and extends to be a continuous homomorphism of G' onto a closed subgroup of G. Note that the closure of the subgroup generated by the Q^0_γ is G itself. Hence, this homomorphism is onto. It is easy to check that the kernel is a closed subgroup of center(G'), a totally disconnected compact abelian group. This proves proposition 2.

We now give the proof of theorem 1. Let F be a finite subset of Γ. Let $Q(F)$ be the subgroup of G generated by the Q^0_γ, where $\gamma \in F$. $Q(F)$ is a compact connected semisimple Lie group. For any $\gamma \in \Gamma$, each element of S_γ is a commutator by a theorem of Goto [1]. Hence, the same is true for any homomorphic image of any product of S_γ's. In particular, each element of $C(F)$ is a commutator. Let $Q(F)'$ be the centralizer of $Q(F)$ in G. $Q(F)' = \text{center}(G)\cdot C(F)$. Hence, $C(F) = [xyx^{-1}y^{-1} \mid x, y \in Q(F)']$.

Let V be open in $Q(F)$. Then $V\cdot C(F)$ is open in G. To see this, note that $G = Q(F)\cdot C(F)$. Hence, the mapping $Q(F) \to G/C(F)$ is onto and so is open. In particular, $V\cdot C(F)$ is open in $G/C(F)$, and so $V\cdot C(F)$ is open in G.

Let n_γ be the dimension of Q_γ^0. Fix a noncentral element a in Q_γ^0. Then the set of products of the form $\prod_{1\le i\le n_\gamma} c_i b_i a b_i^{-1} c_i^{-1}$, where b_i and c_i are arbitrary in G, is a neighborhood $N_\gamma(a)$ of e in Q_γ^0, by the structure of G and by van der Waerden [6]. Furthermore, the family of such $N_\gamma(a)$ forms a neighborhood basis at e for the topology of Q_γ^0. Hence, sets of the following form are a neighborhood basis at e for the topology of G: let F be a finite subset of Γ, let $a_\gamma \varepsilon Q_\gamma^0$ be a noncentral element, and form $\prod_{\gamma\varepsilon F} N(a_\gamma)\cdot C(F)$.

Let ψ: $H \to G$ be an algebraic isomorphism. $\psi^{-1}(N(a_\gamma))$ is clearly an analytic set in H. Furthermore, $\psi^{-1}(Q(F)') = \psi^{-1}(Q(F))'$ is a closed set in H, and so $\psi^{-1}(C(F))$, the set of all commutators of pairs of elements from $\psi^{-1}(Q(F))'$, is an analytic set. Hence, $\psi^{-1}(\prod_{\gamma\varepsilon F} N(a_\gamma)\cdot C(F))$ is an analytic set in H. Since every analytic set has the Baire property, ψ has the Baire property, and there exists a residual set H' contained in H so that $\psi|H'$ is continuous (Kuratowski [4]). It follows that ψ actually is continuous on all of H. To see this, let a_n $(n \ge 1)$ and a be elements of H so that $a_n \to a$. The union of $a^{-1}\cdot(H - H')$ and $a_n^{-1}\cdot(H - H')$ $(n \ge 1)$ is a set of first category. Hence, there is an element b in the complement. Then ab is in H' and a_nb is in H' $(n \ge 1)$. But $a_nb \to ab$ and so $\psi(a_nb) \to \psi(ab)$. Hence, $\psi(a_n) \to \psi(a)$. Hence, ψ is continuous. Souslin's theorem says that ψ^{-1} is a Borel mapping, and hence ψ^{-1} has the Baire property. We conclude as before that ψ^{-1} is continuous. This proves theorem 1.

BIBLIOGRAPHY

[1] Goto, M., A theorem on compact semi-simple groups. Journal of the Mathematical Society of Japan, 1 (1949-50), pp. 270-272.

[2] Helgason, S., Differential Geometry and Symmetric Spaces, Academic Press, New York, 1962.

[3] Hochschild, G., The Structure of Lie Groups, Holden-Day, San Francisco, 1965.

[4] Kuratowski, K., Topology, Volume I, Academic Press, New York, 1966.

[5] Stewart, T. E., Uniqueness of the topology in certain compact groups, Transactions of the American Mathematical Society, 97, 1960, pp. 487-494.

[6] van der Waerden, B. L., Stetigkeitssaetze fuer halbeinfache Liesche Gruppen, Mathematische Zeitschrift, 36, 1933, pp. 780-786.

NORTH TEXAS STATE UNIVERSITY
DENTON, TEXAS 76203

Contemporary Mathematics
Volume 26, 1984

TWO EXTREMAL PROPERTIES OF FUNCTIONS

L. A. Karlovitz

1. Extension of the Markov-Duffin-Schaeffer inequality. The classical Markov inequality deals with the maximization problem

$$\text{(P)} \qquad \max\{p'(x)\colon -1\le x\le 1,\ p\in P_n, |p(x)|\le 1,\ -1\le x\le 1\},$$

where prime denotes differentiation w.r.t. x and P_n is the space of real polynomials of degree n or less. The inequality states that the n-th Chebyshev polynomial, $T_n(x) = \cos n\,(\text{arc}\cos x)$, with its derivative evaluated at $x = 1$, solves problem (P). Now (P) can be viewed as the semi-infinite programming problem

$$\text{(P)} \qquad \max\{\psi(p)\colon p\in P_n, |p(x)|\le 1,\ -1\le x\le 1\},$$

where $\psi(p) = \max\{p'(x)\colon -1\le x\le 1\}$, which is a convex functional. Thus we have the unusal difficulty of maximizing a convex functional, of a finite number of variables, subject to an infinite number of linear inequality constraints. Existing techniques apply to minimization problems. Duffin and Schaeffer [1] derived an interesting extension of the Markov inequality which can be expressed from this point of view by considering the finite programming problem

$$(P_0) \qquad \max\{\psi(p)\colon p\in P_n, |p(\cos k\pi/n)|\le 1,\ k = 0,\ldots,n\}.$$

They show that the same Chebyshev polynomial, T_n, solves (P_0). Thus the sub-program, with finite constraints, has the same solution as the original program. Geometrically, the result asserts that the polynomial which attains the maximum of (P_0) also satisfies $|p(x)|\le 1$, $-1\le x\le 1$. Their proof is one of hard, specific analysis which does not lend itself to generalization.

Our purpose is to provide an approach which yields an extension to quite general Chebyshev systems. (The continuous functions $\{g_0,\ldots,g_n\}$ are said to constitute a Chebyshev system on the interval $[a,b]$ if $\Sigma_0^n c_i g_i$ has at most n zeroes in $[a,b]$ for any coefficients $\{c_i\}$ which are not all zero.)

0271-4132/84 $1.00 + $.25 per page

Consider

(C) $\max\{\psi(g)\colon g\in M,\ |g(x)|\le 1,\ a\le x\le b\}$,

where M is an $(n+1)$-dimensional linear subspace of $C^1[a,b]$ and $\psi(g) = \max\{g'(x)\colon a\le x\le b\}$. Also consider

$(C(x_0,\ldots,x_n))$ $\max\{\psi(g)\colon g\in M,\ |g(x_i)|\le 1,\ i = 0,\ldots,n\}$,

where $a\le x_0\le\ldots\le x_n\le b$. Finally consider

(D) $\min\{\max\{\psi(g)\colon g\in M, |g(x_i)|\le 1,\ i = 0,\ldots,n\}\ a\le x_0\le\ldots\le x_n\le b\}$.

THEOREM. Let M = linear span $\{g_0,\ldots,g_n\}$ be a subspace of $C^4[a,b]$ with $g_0(x) = 1$, $g_1(x) = x$, $g_2(x) = x^2$, $g_3(x) = x^3$. Suppose $\{g_0,\ldots,g_n\}$ and $\{g_1',\ldots,g_n'\}$ and $\{g_2'',\ldots,g_n''\}$ constitute Chebyshev systems on $[a,b]$. Let $x_0,\ldots,x_n$ be the points which yield the minimum of (D) and g the corresponding function in M which achieves the maximum of $(C(x_0,\ldots,x_n))$. Then $|g(x)|\le 1$, $a\le x\le b$, and g also achieves the maximum of (C). Moreover, $x_0 = a$, $x_n = b$, and g equi-oscillates, i.e., $g(x_i)g(x_{i+1}) = -1$, $i = 0,\ldots,n-1$.

PROOF. The program $(C(x_0,\ldots,x_n))$ has a finite maximum if and only if the x_k are distinct. We can choose functions $h_k\in M$ so that

$$h_k(x_j) = \delta_{k,j},\ \text{Kronecker delta},\ k,j = 0,\ldots,n.$$

Choose y so that $\Sigma_0^n|h_j'(y)|$ is maximized on $[a,b]$. Let $g = \Sigma_0^n(\operatorname{sgn} h_j'(y))h_j$. Then the function g achieves the maximum of the program $(C(x_0,\ldots,x_n))$ at the point y, i.e., $\psi(g) = g'(y)$. We note that $h_j'(y)\ne 0$, $j = 0,\ldots,n$. For if $h_k'(y) = 0$ then $g+\lambda h_k$, $|\lambda|\le 1$, satisfies the constraints and also achieves the maximum. Hence if y is interior, $h_k''(y) = 0$. By virtue of the definition of h_k, the latter is readily seen to imply that h_k'' has $n-1$ zeroes in $[a,b]$ contradicting the hypothesis. Similarly, if y is an end point, $h_k'(y) = 0$ together with the definition of h_k implies that h_k' has n zeroes, also a contradiction.

Assume that $x_0,\ldots,x_n$ are the points yielding the minimum of (D). Let $g\in M$ be the function which maximizes the program $(C(x_0,\ldots,x_n))$ at the point y, i.e., $\psi(g) = g'(y)$. We will now use a perturbation argument to prove the theorem.

Let x_k be an interior point. Consider the perturbation $\{x_i(\varepsilon) = x_i+\delta_{ik}\varepsilon,\ i = 0,\ldots,n\}$. Choose $A_i(\varepsilon)$ so that the functions $h_i(\varepsilon,x) = h_i(x)+A_i(\varepsilon)h_k(x)$ satisfy $h_i(\varepsilon,x_j) = \delta_{i,j}$, $i,j = 0,\ldots,n$. Clearly $dA_i/d\varepsilon = h_i'(x_k)$ at $\varepsilon = 0$. The function which achieves the maximum of $(C(x_0(\varepsilon),\ldots,x_n(\varepsilon)))$ is $g(\varepsilon,x) = \Sigma_0^n(\mathrm{sgn}(h_i'(\varepsilon,y(\varepsilon)))h_i(\varepsilon,x)$, where $y(\varepsilon)$ is the maximizing point, i.e., $\psi(g(\varepsilon,x)) = g'(\varepsilon,y(\varepsilon))$. We may assume $y(\varepsilon_j) \to y$ for some sequence $e_j \to 0$. Hence $\mathrm{sgn}(h_i'(\varepsilon,y)(\varepsilon)) = \mathrm{sgn}\ h_i'(y) = \alpha_i$. Thus

$$g(\varepsilon,x) = g(x) + (\Sigma_0^n\alpha_iA_i(\varepsilon))h_i(x).$$

If y is an interior point, then

$$g''(\varepsilon,y(\varepsilon)) = g''(y(\varepsilon))+ (\Sigma_0^n\alpha_iA_i(\varepsilon))h_k''(y(\varepsilon)) = 0, \tag{1}$$

for $\varepsilon =\varepsilon_j$. If, moreover, $g'''(y) \neq 0$ then (1) allows $y(\varepsilon_j)$ to be interpreted as values of a differentiable function of ε. Since the points $x_0,\ldots,x_n$ minimize (D) we readily find

$$0 = \left.\frac{dg'(\varepsilon,y(\varepsilon))}{d\varepsilon}\right|_{\varepsilon=0} = (-\Sigma_0^n\alpha_ih_i'(x_k))h_k'(y) = -g'(x_k)h_k'(y).$$

Hence $g'(x_k) = 0$ at each interior x_k. Since g' can have at most $n-1$ zeroes, it follows by standard argument that $x_0 = a$, $x_n = b$, $|g(x)|\le 1$, $a\le x\le b$, and $g(x_i)g(x_{i+1}) = -1$, $i = 0,\ldots,n-1$. Clearly if g maximizes $(C(x_0,\ldots,x_n))$ and also satisfies the constraints of (C) then it also maximizes (C).

We next then show that if y is interior then $g'''(y) \neq 0$. We first note that a standard argument shows that g has at least $n-2$ alternations of sign; i.e., either $g(x_i)g(x_{i+1}) = -1$, $i = 0,\ldots,n-1$ or there is an index j so that $g(x_i)g(x_{i+1}) = -1$, $i = 0,\ldots,j-1, j+1,\ldots,n-1$. Now (1) yields

$$\lim \frac{g''(y(\varepsilon_j))}{\varepsilon_j} = -g'(x_k)h_k''(y) = -\alpha_kh_k''(y) . \tag{2}$$

Since $g''(y) = \Sigma_0^n\alpha_ih_i''(y) = 0$ we can find indices p and q so that $(\alpha_ph_p''(y))(\alpha_qh_q''(y)) < 0$. Hence (2) implies that $g''(z)g''(w) < 0$ for $z<y<w$, $|w-z|$ small. It follows that from this and $g''(y) = 0$ that if $g'''(y) = 0$ then we also have $g''''(y) = 0$. From the fact that g has at least $n-2$ alternation of signs and from $g''(y) = g'''(y) = g''''(y) = 0$ one can deduce that g'' has at least $n-2$ zeroes in $[a,b]$, contradicting the hypothesis.

Now suppose that y is an end point. If $y(\varepsilon_j)$ is interior for an infinite sequence, the proof proceeds as above. If $y(\varepsilon_j) = y$, $j\ge N$, we can certainly interpret $y(\varepsilon_j)$ as values a differentiable function with derivative zero. Hence we again find

$$0 = \frac{dg'(\varepsilon, y(\varepsilon))}{d\varepsilon}\bigg|_{\varepsilon=0} = -g'(x_k)h_k'(y),$$

and the proof is completed as above. This finishes the proof.

2. A general equi-oscillation result. We now state a simple and surprising consequence of the Gohberg-Krein Theorem. (Let U and V be linear subspaces or a normed linear space with $\dim U < \dim V < \infty$.) Then there exists $v \in V$, $v \neq 0$, so that $\min\{\|v-u\| : u \in U\} = \|v\|$). Namely, we show that every finite dimensional linear subspace of $C(K)$, K compact, contains a function which equi-oscillates like a Chebyshev polynomial.

THEOREM. *Let* M *be a linear subspace of* $C(K)$, K *compact, with* $\dim M = n+1$. *Then there exists a function* $f \in M$ *and points* $x_0 < x_1 < \ldots < x_n$ *in* K *so that* $|f(x)| \leq 1$, $x \in K$, *and* $f(x_i)f(x_{i+1}) = -1, i = 0,\ldots,n-1$.

PROOF. Let $U = \mathcal{P}_{n-1}$ the space of polynomials of degree $n-1$ or less. Let $V = M$. Then $\dim U = n < \dim V$. Hence, by the Gohberg-Krein Theorem, there exists $f \in M$, $f \neq 0$, so that $\min\{\|f-p\| : p \in \mathcal{P}_{n-1}\} = \|f\|$. We may assume $\|f\| = 1$. Now a standard argument from approximation theory asserts that the error function $f-0 = f$ must equi-oscillate according to the conclusion of the theorem. This finishes the proof.

REFERENCES

1. R. J. Duffin and A. C. Shaeffer, "A refinement of an inequality of the brothers Markoff", Trans. Amer. Math. Soc., 50 (1941), 517-528.

School of Mathematics
Georgia Institute of Technology
Atlanta, GA 30332

Contemporary Mathematics
Volume **26**, 1984

ON BERNOULLI CONVOLUTIONS

We few, we happy few - Henry V

Robert Kaufman

We present two kinds of theorems concerning Bernoulli convolutions and their transformation by differentiable functions: the first studies the space C^1 from the topological point of view, and the second uses a difficult (and somewhat neglected) method of Erdos to study C^2.

1. C^1 FUNCTIONS

Let (a_n) be a sequence of positive numbers with summable squares and let $\sigma_k^2 = \sum_{k+1}^{\infty} a_n^2 (k = 0,2,3,4,\ldots)$. Then as is known from probability theory (or Fourier analysis, if one prefers), there is a probability measure μ defined by $\hat{\mu}(u) = \prod_1^{\infty} \cos(2\pi u a_n)$. Because it is possible that $\Sigma a_n = +\infty$, we consider $C^1(R)$, made into a Fréchet space in the usual way.

THEOREM 1. <u>Suppose that</u> $\sigma_{k+1}^2 \geq c^2\sigma_k^2$ <u>for all</u> $k \geq 0$ <u>and a certain</u> $c > 0$. <u>Then for all</u> f <u>in</u> $C^1(R)$, <u>except a set of first category</u>,

$$\lim \int e^{-2\pi i u f(x)}\mu(dx) = 0, \quad \text{as} \quad u \to +\infty.$$

PROOF. Let $(r_n)_1^{\infty}$ be the sequence of Rademacher functions, so that μ is simply the distribution of the random variable $X = \sum_1^{\infty} a_n r_n$. The integral is then $E(e(-uf(X))$, with $e(t) \equiv e^{2\pi i t}$. Let $0 < c \leq 1/2$, fix an integer N and $u > N\sigma_0^{-1}$. Then there is an index $k = k(u)$ at which $N \geq u\sigma_k > cN$. We write $S_k = \sum_1^k a_n r_n$, and $T_k = X_k - S_k$, so that S_k has variance $\sigma_0^2 - \sigma_k^2$, and $T_k\sigma_k^2$; S_k and T_k are independent. By the mean-value theorem

$$f(X) = f(S_k) + T_k f'(S_k + \theta_k T_k), \qquad 0 < \theta_k < 1$$

for a certain random variable θ_k. Now $f'(S_k) \to f'(X)$, $f'(S_k + \theta_k T_k) - f'(S_k) \to 0$ almost everywhere, so that

$$uf(X) - uf(S_k) - uT_k f'(S_k) \to 0 \quad \text{in measure},$$

0271-4132/84 $1.00 + $.25 per page

as $u \to +\infty$ (since $E(|uT_k|) < N$). Thus we can replace $e(-uf(X))$ by $e(-uf(S_k))e(-uT_k f'(S_k))$ with error $o(1)$. Integrating first with respect to T_k, we see that it is enough to estimate (for large u)

$$E\left(\prod_{k+1}^{\infty} |\cos 2\pi u a_n f'(S_k)|\right).$$

We intend to replace $f'(S_k)$ by $f'(X)$ in this formula; to justify this we note that the function $\prod_{k+1}^{\infty} \cos(2\pi u a_n y)$ has derivative at most $2\pi N$ in modulus, because $E(|uT_k|) < N$. Thus for every integer $N \geq 1$ we have the inequality

$$\limsup \left|\int e(-uf(x)\mu(dx)\right|$$

$$\leq \sup E(|\hat{\lambda}(f'(X))|), \quad \lambda \in P_N.$$

Here P_N is the family of probability measure whose Fourier-Stieltjes transforms are given by $\hat{\lambda}(y) = \prod^{\infty} \cos(2\pi b_n y)$ with $c^2N^2 \leq \sum_1^{\infty} b_n^2 \leq N^2$, $\sum_{K+2}^{\infty} b_n^2 \geq c^2 \sum_{k+1}^{\infty} b_n^2$, $k \geq 0$. The second property of the measures in P_N is inherited from the basic measure μ. The supremum decreases with N, for if λ is the element of P_N written above, then an element of P_{N+1} is obtained by augmenting $(b_n)_1^{\infty}$ by a sequence of small numbers $\delta < (1 - c^2)c$, until the variance is between $c^2(N + 1)^2$ and $(N + 1)^2$.

Let now $\Gamma_N(f)$ be the supremum written above, so that $\Gamma_1 \geq \Gamma_2 \geq \cdots \geq \Gamma_N \geq \cdots$. Because $|\hat{\lambda}'| \leq 2\pi N$ for λ in P_N, $\Gamma_N(f)$ is continuous on $C^1(R)$ with its Fréchet metric. The set defined by $\{\lim \Gamma_N(f) = 0\}$ is therefore a G_δ, and in a moment we shall prove that it is dense, by means of two lemmas, the first of which is almost trivial.

LEMMA 1. <u>In the space</u> $C^1(R)$, <u>the set of functions</u> f, <u>such that</u> $f'(X)$ <u>has an absolutely continuous distribution with respect to</u> μ <u>is dense (but is not a</u> G_δ).

LEMMA 2. <u>Whenever the distribution</u> ν <u>of the random variable</u> $f'(X)$ <u>is absolutely continuous, we have</u> $\lim \Gamma_N(f) = 0$. <u>More generally</u>, $\lim \Gamma_N(f) = 0$ <u>if</u> $\hat{\nu} = 0$ <u>at</u> ∞.

PROOF OF LEMMA 1. This is valid for any diffuse measure μ. Suppose that f in $C^1(R)$ is known, and $\varepsilon > 0$. Choose on the real line a double sequence $(c_n)_{-\infty}^{\infty}$ with $c_n < c_{n+1}$, $\lim|c_n| = \infty$, and f' oscillates at most $\varepsilon > 0$ on $[c_n, c_{n+1}]$. If $[c_n, c_{n+1}]$ has positive μ-measure, then g' is a quadratic function of the primitive F of μ on $[c_n, c_{n+1}]$; if $[c_n, c_{n+1}]$ has μ-measure 0 then g' is linear. Also $g'(c_n) = f'(c_n)$ for every n and $g(0) = f(0)$. Clearly $g'(X)$ has an a.c. distribution with respect to μ.

PROOF OF LEMMA 2. We majorize the integrals $E(|\hat{\lambda}(f'(X))|^2)$, $\lambda \in P_N$. Now ν is the distribution of $f'(X)$, so these integrals can be written as

$$\int \hat{\nu}(y)\rho(dy), \qquad \rho = \lambda*\lambda, \quad \lambda \in P_N,$$

or

$$\int \hat{\nu}(Ny)\rho(dy), \qquad \rho = \lambda*\lambda, \quad \lambda \in P_1.$$

By assumption $\hat{\nu}(\infty) = 0$ so that it is sufficient to verify that

$$\lim \rho([-\delta,\delta]) = 0 \quad \text{as} \quad \delta \to 0+$$

uniformly for $\lambda \in P_1$. If this fails, then some weak limit of the measures ρ must have a jump at 0, a circumstance that will be excluded in a moment.

Let (b_n) be the sequence determining λ in P_1, let $\sigma^2 = \sum_1^\infty b_n^2$, and let $1 \le n_1 < n_2 < \cdots$. Then $a^2(n_1) \le (1 - c^2)\sigma^2$, $a^2(n_2) \le (1 - c^2)\sigma^2 - (1 - c^2)a_2^2(n_1), \ldots, a^2(n_{p+1}) \le (1 - c^2)(\sigma^2 - a^2(n_1) - \cdots a^2(n_p))$. Thus $a^2(n_1) + \cdots + a^2(n_p) \le (1 - c^{2p})\sigma^2$, $p = 1,2,3,\ldots$. The last inequality can be used when $a(n_1),\ldots,a(n_p)$ are the p largest among the numbers b_n. From the inequality $E(X^4) \le 3$, we see that every weak limit point of P_1 is a probability measure with $c^2 \le \sigma^2 \le 1$, and in fact is a convolution $\mu_1*\mu_2$ of a Gaussian measure μ_1 and a Bernoulli product μ_2. The product is a continuous measure unless μ_2 is a finite product of, say, p terms. In that case μ_1 has variance at least $c^{2p}\sigma^2 > 0$ and the product is again continuous, and so is its square. This proves Lemma 2 and with it Theorem 2.

For measures μ of a special form a more definite result is true. Let $\lim b_{n+1}b_n^{-1} = \theta$, $0 < \theta < 1$. When θ^{-1} is not a PV-number (see [6]) then we need only require that $\mu(f' = 0) = 0$, a result found in [4]. When $\theta^{-1} \in PV$, the situation is almost as favorable.

THEOREM 2. _Suppose that_ $\lim a_{n+1}a_n^{-1} = \theta$, $0 < \theta < 1$, _and that_ $\mu(f' = s) = 0$ _for every real number_ s. _Then_

$$\lim \int e(-uf(x))\mu(dx) = 0, \quad \text{as} \quad u \to +\infty.$$

The proof is based on the fact that $\prod_1^\infty \cos^2(2\pi\tau\theta^{-n}) = 0$, unless θ _and_ τ are algebraic numbers of a very special kind ([6, p. 148]), and in particular for all except a countable set of numbers τ. (See the remark on p. 151, 1.21 of [6]).

Theorem 1 admits a converse in a special but interesting case. If $a_{n+1} \le a_n/2$, the condition on the variances is nothing but $a_{n+1} \ge c'a_n$ (for some c'). When this condition fails, the conclusion also fails (and much more is known) [2]. If $a_{n+1} \le a_n/2$, $\liminf a_{n+1}/a_n = 0$ and $a_n \ge 4^{-n}$ for example, the set of functions f in $C^1(R)$, defined in Theorem 1, is dense but of first category.

Looking more closely at the case $a_n = r^{-n}$, with $r = 5,6,7,\ldots$, we observe that the condition on f' is natural! In the subset of C^1 defined by ($f' = 1$ a.e. $d\mu$) we have the opposite behavior on a residual set ([3]). When f' has a continuous distribution, $f \circ \mu$ can still have some exceptional properties. Let ψ be positive, concave, and increasing on R^+ and $\psi(0+) = 0$, and let E be the support of μ.

THEOREM 3. *There is an element* f *of* $C^\infty(R)$, *such that* $\mu(f' = s) = 0$ *for each* s; *and each function* g *on* R, *such that* $|g(x) - g(y)| \le \psi(|x - y|)$, *is represented on* $f(E)$ *by a series* $g(t) = \Sigma c_j e(jt)$, *with* $\Sigma |c_j| < +\infty$.

The proof, too involved to be presented here, is based on two observations: (i) all sets $E, rE, r^2E, \ldots$ coincide (modulo 1). (ii) $xE + yE$ has Lebesgue measure 0 for real x and y because $r \ge 5$. Sets of multiplicity (e.g. $f(E)$) with the property claimed were produced first by McGehee [5].

2. C^2 FUNCTIONS

Here $\hat{\mu}(u) = \prod_1^\infty \cos(2\pi u b_n)$ and $1 \le c_1 < 2$, $cb_n \le b_{n+1} \le b_n/2$ $(n = 1,2,3,\ldots)$ for a certain $c > 0$.

THEOREM 3. *Let* $F \in C^2(R)$ *and* $F'' > 0$. *Then* $\int e(-2\pi uF)d\mu = 0(|u|^{-\delta})$ *for a certain* $\delta > 0$ *depending only on* c.

PROOF. Let $u > 2$ and define $N = N(u)$ so that $2 \le b_N u^{2/3} < 2c^{-1}$, and factor the measure μ into $\mu_N * \lambda_N$, where μ_N contains the first N factors and λ_N the remainder. The integral is then

$$\iint e(-uF(x + y))\mu_N(dx)\lambda_N(dy)$$

$$= \iint e(-uF(x) - uF'(x)y)\mu_N(dx)\lambda_N(dy) + 0(uc_N^2).$$

Hence we obtain an upper bound

$$\iint |\hat{\lambda}_N(uf'(x))| \mu_N(dx) + 0(u^{-1/3}).$$

Examining $\hat{\lambda}_N(ut)$ more closely, we write it as $\prod_1^\infty \cos(2\pi t d_k)$ with $d_k = uc_{N+k}$, so that $c^{-1}u^{1/3} \ge d_1 \ge 2cu^{1/3}$, and $d_k/2 \ge d_{k+1} \ge cd_k$. We transform this further by omitting all factors in which $d_k < 1$, and reversing the order, to obtain $\prod_1^M \cos 2\pi t e_k$; so that $1 \le e_1 < c^{-1}$, $2e_k \le e_{k+1} < c^{-1}e_k$ for $1 \le k < M$, and $e_M \ge 2c \cdot u^{1/3}$. Hence M is bounded below by $\log u / -3 \log c + 0(1)$. Now let $\varepsilon > 0$ be a small positive number and let $S(M,\varepsilon)$ be defined by this condition: $|\cos 2\pi t e_k| > \cos \pi c/2$ for at least $(1 - \varepsilon)M$ numbers $k = 1,2,3,\ldots M$. When $f'(x) \notin S(M,\varepsilon)$ then $|\hat{\lambda}_N(uf'(x))| < (\cos \pi c/2)^{\varepsilon M} < u^{-\delta}$ for a certain $\delta = \delta(c,\varepsilon)$. To conclude we prove that for small $\varepsilon > 0$

$\mu_N\{f'(x) \in S(M,\varepsilon)\} \ll u^{-\delta}$ for another $\delta > 0$ (cf. [1]). We write A for the supremum of $|f'|$ over E.

To each x on (0,A) we define integers $r_k(x)$ and remainders $\varepsilon_k(x)$ by

$$2e_k(x) = r_k(x) + \varepsilon_k(x), \qquad -\frac{1}{2} < \varepsilon_k \le \frac{1}{2},$$

so that $x \in S(M,\varepsilon)$ precisely when $|\varepsilon_k(x)| < c/2$ at least $(1 - \varepsilon)M$ times among indices $k = 1,2,\ldots M$. To each x of $S(M,\varepsilon)$ we select exactly $M_1 = [(1 - \varepsilon)M]$ inequalities; this can be done in at most $\binom{M}{M_1}$ ways.

Now $0 \le r_1 < 2Ae_1 + 1$ and

$$|r_{k+1} - e_{k+1}e_k^{-1}r_k| \le \varepsilon_{k+1} + c^{-1}\varepsilon_k$$

so that each value of r_k can be followed by at most $c^{-1} + 2$ choices of r_k; when $2|\varepsilon_k| < c$, $2|\varepsilon_{k+1}| < c$ the number on the right is < 1 and <u>at most one</u> value of r_{k+1} is possible. The number of sequences $r_1,\ldots,r_M$ (with a specified selection of M_1 indices k) can be estimated by $(2Ac^{-1} + 2)(c^{-1} + 2)^{2(M-M_1)}$. Estimating $\binom{M}{M_1}$ by Stirling's formula, we see that to each $\eta > 0$ we can find an $\varepsilon > 0$ so that $S(M,\varepsilon) \cap (0,A)$ is covered by $\exp \eta M$ intervals of length $e_M^{-1} < 2^{-M}$, because $|2e_M x - r_M| < 1/2$. We denote these intervals, and their reflections through $x = 0$, by $J_1,\ldots,J_\ell$, $1 \le \ell \le 2 \exp \eta M$.

The set defined by $\{F'(x) \in J_i\}$ is an interval J_i' of length $0(2^{-M})$, because $F' > 0$, so we look for an upper bound for $\mu_N(J_i')$. Recall that μ_N is the standard measure on the set of sums $\sum_1^N \varepsilon_n b_n$, and that $N \ge \frac{2}{3} \log u/{-\log c} + 0(1)$. In case $2^{-M} < b_N$, $\mu_N(J_i') = 0(2^{-N})$, that is $0(u^{-s})$, with $s = -2/3 \log c > 0$. If $2^{-M} \ge b_N$, the measure is $0(2^{-p})$, where $b_p \simeq 2^{-M}$, and again the measure is $0(2^{-s})$, where s depends only on c. Hence we have only to choose η sufficiently small, and then $\varepsilon > 0$, to obtain $\mu_N\{f'(x) \in S(M,\varepsilon)\} = 0(u^{-\delta})$.

REFERENCES

[1] P. Erdős, On the smoothness properties of a family of symmetric Bernoulli convolutions, Amer. J. Math. 62(1940), 180-186.

[2] R. Kaufman, A functional method for linear sets, Israel J. Math 5(1967), 185-187.

[3] R. Kaufman, A functional method for linear sets, II, Israel J. Math 7 (1969), 293-298. (Theorem 3 should be proved like Theorem 2, not as presented.)

[4] R. Kaufman, Bernoulli convolutions and differentiable functions, Trans. Amer. Math. Soc. 200(1974), 427-435.

[5] O. C. McGehee, Lipschitz classes and restrictions of Fourier transforms, Math. Annalen 239(1979), 223-227.

[6] A. Zygmund, Trigonometric Series, II, Cambridge, 1959 and 1968.

Department of Mathematics
University of Illinois
Urbana, Illinois 61801

Contemporary Mathematics
Volume 26, 1984

GENERIC THEOREMS FOR LIFTING DYNAMICAL PROPERTIES BY CONTINUOUS AFFINE COCYCLES

Harvey B. Keynes and Mahesh G. Nerurkar*

§1. INTRODUCTION

In this note we announce several results regarding the lifting of various dynamical properties in a dynamical system by using continuous affine cocycles. The spaces considered will be built essentially by having a fibre group G over a base space, and the action on the fibres being controlled by a transformation group of group automorphisms. As with group extensions, we can oerturb the action with cocycles incorporating these automorphisms, called affine cocycles. We obtain results in the smallest possible class to modify the action, namely, the closures of the affine continuous coboundaries, when G is an n-torus. Using different techniques, similar results are obtained in the class of all continuous affine cocycles for any compact, connected abelian fibre group. In particular we derive a topological version of a result of Rudolph, Moore and Feldman on lifting the Bernoullian property in affine extensions [8]. We also extend a result of Glasner and Weiss in [4] on lifting unique ergodicity to affine extensions.

In contrast to [5], where minimality was lifted generically in the class of all continuous affine cocycles for toral groups and distal automorphisms, the results announced here are based on a different technique. The basic component of this technique is an affine modification of the idea of 'orbit averaging' introduced in [4].

For the sake of completeness, we include the basic definitions and the facts that are basic to the proofs of these results. To give an indication of the techniques involved, we will sketch a proof of the basic ergodicity result in the toral case (Theorem 3.1), and a similar result for total minimality (Theorem 3.7). Most results will simply be stated without proof.

*Research of both authors supported by NSF Grant MCS 81-02034.

§2. DEFINITIONS AND BACKGROUND MATERIAL

By a <u>dynamical system (d.s.)</u> we mean a pair (Y,T), where Y is a compact metric space and T is a locally-compact, separable group acting continuously on the right of Y with the action $(y,t) \to y\cdot t$. We usually will also have a T-invariant Borel probability μ, and then denote the system by (Y,T,μ). For simplicity assume T is abelian.

Now let (X,T) be a d.s. on which a compact metric group G acts continuously and freely on the left with action $(g,x) \to gx$. If in addition T acts continuously on right of G as a group of automorphisms by $(g,t) \to \sigma_t(g)$, and $(gx)\cdot t = [\sigma_t(g)](x\cdot t)$, $\forall g,x,t$, then (G,X,T) is called an <u>affine bitransformation group</u>. In this situation, we can form $Y = G\backslash X$, the space of G orbits with the quotient topology, and we get a natural T action on Y. The quotient map $\pi : X \to Y$ defines an extension with fibre G, and X is called an <u>affine extension of</u> Y. If μ is a T-invariant Borel probability on Y and η is the normalized Haar measure on G, we define the <u>Haar lift</u> $\tilde{\mu}$ of μ by

$$\tilde{\mu}(f) = \int_X f d\tilde{\mu} = \int_Y \left(\int_{\pi^{-1}(y)} f(gx)d\eta(g)\right)d\mu(y), \quad \forall f \in C(X), \quad (\text{where } \pi x = y).$$

Clearly, $\tilde{\mu}$ is T-invariant. The dynamics of $\tilde{\mu}$ is the focal point of this note.

The simplest example is the case of a homeomorphism T acting on Y (here $T = \mathbb{Z}$, an integer action, and we use the same letter for the generator) and an automorphism σ of G. In this case, we set $X = G \times Y$, let G act on X by $g(g_1,y) = (gg_1,y)$ and use the homeomorphism $\hat{T}(g,y) = (\sigma(g),Ty)$.

It is not difficult to see that if the transformation group (G,T) is distal (equicontinuous), so is the corresponding affine extension. This observation hints that to study the dynamical properties of affine extensions it is important to know the dynamics of the automorphism action of T on G. If G is abelian and $T = \mathbb{Z}$ then a Furstenberg-type tower desceibing the dynamics of this action can be explicitly constructed, cf. [5]. For non-abelian G, one can explicitly describe the maximal equicontinuous factor of (G,T) as a homogenous space $(G/G_F,T)$ for a certain closed T-invariant subgroup G_F of G, cf. [6]. More importantly we can reduce the lifting problem to the case when (G,T) is equicontinuous.

THEOREM (2.1) ([6]). <u>Let</u> $\pi:(X,T,\tilde{\mu}) \to (Y,T,\mu)$ <u>be an affine extension with fibre group</u> G.

(1) <u>Let</u> (Y,T,μ) <u>be ergodic. Then</u> $(X,T,\tilde{\mu})$ <u>is ergodic (weakly-mixing) iff the affine bitransformation group</u> $(G/G_F,G_F\backslash X,T,\bar{\mu})$ <u>is ergodic (weakly-mixing). Here</u> $\bar{\mu} = (\pi_1)_*\tilde{\mu}$, <u>where</u> $\pi_1 : X \to G_F\backslash X$ <u>is the quotient map</u>.

(2) <u>Let</u> (Y,T,μ) <u>be uniquely ergodic. Further assume that either</u> (a) (G,T)

is equicontinuous, or (b) $T = \mathbb{Z}^n$, G *is abelian and* (G,T) *is distal. Then* $(X,T,\tilde{\mu})$ *is uniquely ergodic iff it is ergodic.*

The first conclusion uses Fourier analysis on L_2 eigenfunctions, while the conclusion about unique ergodicity is based on the result that when (G,T) is equicontinuous, any two ergodic measures projecting onto μ are G-translates of each other.

The next result reduces the problem of lifting many dynamical properties to the problem of lifting either ergodicity or weak-mixing in the case of integer actions.

THEOREM (2.2). *Let* $\pi:(X,T,\tilde{\mu}) \to (Y,T,\mu)$ *be an affine extension, and* $T = \mathbb{Z}$.

(a) (*Thomas, see* [1]). *Let* (Y,T,μ) *be a K-automorphism. Then* $(X,T,\tilde{\mu})$ *is a K-automorphism iff it is weakly-mixing.*

(b) (Rudolph, see [7]). *Let* $(Y,T, \;)$ *be Bernoullian. Then* $(X,T,\tilde{\mu})$ *is Bernoullian iff it is weakly-mixing.*

(c) (Walters, see [9]). *Let* (Y,T,μ) *be intrinsically ergodic. Then* $(X,T,\tilde{\mu})$ *is intrinsically ergodic iff it is ergodic.*

We finally need to formally define affine cocycles and the resulting perturbed actions called *skew product flows*.

DEFINITION (2.3). Let $\pi:(X,T) \to (Y,T)$ be an affine extension with fibre group G. (a) A *continuous affine cocycle* α for π is a continuous map $\alpha:Y \times T \to G$ satisfying the cocycle condition:

$$\alpha(y,t_1t_2) = \alpha(y,t_1)[\sigma_{t_1^{-1}}(\alpha(y\cdot t_1,t_2))] \ \forall y,t_1,t_2.$$

Let $Z(Y,T,G)$ be the set of all such cocycles.

(b) Given $f \in C(Y,G)$, the space of continuous maps from Y to G, set $1^f(y,t) = f(y)^{-1}[\sigma_{t^{-1}}(f(y\cdot t))]$. It is easy to verify that $1^f \in Z(Y,T,G)$. Cocycles of this form will be called *affine-coboundaries*, and the set of all affine coboundaries will be denoted by $B(Y,T,G)$.

(c) Given $\alpha \in Z(Y,T,G)$ and $1^f \in B(Y,T,G)$, we define $\alpha\cdot 1^f:Y \times T \to G$ by setting,

$$\alpha\cdot 1^f(y,t) = f(y)^{-1}\alpha(y,t)[\sigma_{t^{-1}}(f(y\cdot t))], \quad \forall y,t.$$

Note that when G is abelian, $\alpha\cdot 1^f$ is the pointwise product. It is direct to verify that $\alpha\cdot 1^f \in Z(Y,T,G)$. Two affine cocycles α_1 and α_2 are *cohomologous* if $\alpha_2 = \alpha_1\cdot 1^f$ for some $f \in C(Y,G)$.

(d) As a subset of $C(Y \times T,G)$, we can restrict the compact - open topology to $Z(Y,T,G)$. This topology can be shown to be induced by a complete separable metric, and thus $Z(Y,T,G)$ is a Baire space.

(e) Given $\alpha \in Z(Y,T,G)$, we can define a new T-action on X by setting $x\circ t = (\alpha(y,t)^{-1}x)\cdot t$, if $\pi x = y$. The new d.s. so obtained will be denoted by (X,T_μ) and is called the *skew product extension* of Y by α. Note that (Y,T) is still a factor of (X,T_α), although it need not be an affine extension, and that the Haar lift $\tilde{\mu}$ is invariant under the new action. When $T = \mathbb{Z}$ and $X = G \times Y$ is dynamically trivial with action $(g,y) \to (\sigma(g),Ty)$, then α can be regarded as in $C(Y,G)$ and the new action is

$$(g,y) \to (\sigma(\alpha(y)^{-1}g),Ty).$$

With these preliminaries, we can now give the main results.

§3. STATEMENTS OF THE RESULTS; SKETCHES OF PROOFS

THEOREM (3.1). *Let* $T = \mathbb{Z}$ *and* $\pi:(X,T,\tilde{\mu}) \to (Y,T,\mu)$ *be an affine extension with fibre* G. *Further assume that*,

(i) μ *is non atomic and supported*,

(ii) $G = K^n$, *the n-torus, and the generating automorphism* σ *is equicontinuous (hence periodic)*,

(iii) (Y,T^p,μ) *is ergodic, where* p *is the period of* σ.

Then the set $\{\alpha|\alpha \in \overline{B(Y,T,G)}$ *such that* $(X,T_\alpha,\tilde{\mu})$ *is ergodic\} is residual in the Baire space* $\overline{B(Y,T,G)}$.

We will later state a more general theorem with fewer assumptions. However the above setting is sufficient to indicate the basic ideas behind the method of proof.

SKETCH OF PROOF: For simplicity, let $X = G \times Y$. Given $f \in L^2(X,\tilde{\mu})$, (and assuming WLOG that $\int_X f d\tilde{\mu} = 0$), $\varepsilon > 0$ and $m \in \mathbb{N}$, the crucial sets to analyze are

$$W(f,\varepsilon,m) = \{\alpha|\alpha \in \overline{B(Y,T,G)} \text{ and } \exists M,\ M > m \text{ with } |<\frac{1}{M}\sum_{i=0}^{M-1} f\circ T_\alpha^{pi},f>| < \varepsilon\},$$

where $<\cdot,\cdot>$ is the inner product. It is straightforward to show that the theorem reduces to proving that each $W(f,\varepsilon,m)$ is open and dense in $\overline{B(Y,T,G)}$. Openess is routine; the problem of proving denseness is then reduced to proving that given any $f \in L^2(X,\tilde{\mu})$, with $\int_X f d\tilde{\mu} = 0$, $\varepsilon > 0$, and $\delta > 0$, there exists a continuous map $\psi:Y \to G$ such that

(a) in the uniform metric, $d(1^\psi,\hat{e}) < \delta$, where $\hat{e}(y) = e$, the identity element of G,

(b) in the L_2-norm, $\|\int_Y f(g\psi(y),y)d\mu(y)\| < \varepsilon$.

This last condition reflects that analysis takes place after projecting into the T-invariant functions. Letting $h:\mathbb{R}^n \to K^n$ be the exponential map, we need to construct $\theta:Y \to \mathbb{R}^n$ such that the map $h\circ\theta$ will satisfy the required conditions. Condition (a) will be satisfied if θ is chosen so that $A(\theta(y))$ and $\theta(Ty)$ are uniformly close, where $A:\mathbb{R}^n \to \mathbb{R}^n$ is a linear map inducing the automorphism σ on K^n. Condition (b) demands that θ should at least "approximately" take the measure μ onto Lebesgue measure. Such a θ is constructed by the use of Rokhlin's lemma, and an orbit averaging process which involves taking average of values of θ with "suitable weights" over orbits of points in Y. Here, the periodicity of σ and (iii) play important roles. □

REMARKS (3.2). (1) Theorem (3.1) also holds for $\mathbb{Z}^r$ actions as well, with essentially technical extensions required.

(2) A more substantial generalization is that we can replace assumption (iii) in the above theorem by assuming just ergodicity of (Y,T,μ). This generalization is based on the idea of viewing σ as a cocycle from Y into the automorphism group of G. Then applying a result of R. Zimmer, and replacing σ by a suitable cohomologous cocycle, we reduce the problem again to the construction of a suitable continuous map from Y to G.

The above results required the periodicity of σ in order to obtain results in the closure of the coboundaries. If we are willing to deal with the larger class of all affine cocycles, we can relax this condition. In fact, using the previously mentioned Furstenberg-type tower for the transformation group (K^n,σ), we can reduce this problem again to the case of periodic automorphism and the previous analysis. This gives the following theorem:

THEOREM (3.3). Let $T = \mathbb{Z}^r$ and $\pi:(X,T,\tilde{\mu}) \to (Y,T,\mu)$ be an affine extension with fibre G. Further, assume that

(i) μ is non atomic and supported,

(ii) $G = K^n$ the n-torus, and

(iii) (Y,T,μ) is ergodic.

Then the set $\{\alpha \mid \alpha \in Z(Y,T,G)$ such that $(X,T_\alpha,\tilde{\mu})$ is ergodic$\}$ is residual in $Z(Y,T,G)$.

Theorem (3.3) permits generalizations in several directions.

COROLLARY (3.4). (I). Ergodicity in Theorem (3.3) can be replaced by:

(a) weak-mixing,

(b) k-automorphism,

(c) Bernoullian.

(II). If σ is distal, then ergodicity can also be replaced by:

(d) unique ergodicity,

(e) strict ergodicity.

Using Theorem (2.2) and known minimality results, the proof reduces to lifting weak-mixing. This case is not an immediate corollary of Theorem (3.3) when the bundle $\pi:X \to Y$ is not trivial. However, techniques used to prove Theorem (3.3) after some modifications enable us to also handle this case.

EXAMPLE (3.5). In [3] H. Furstenberg gave some conditions on a cocycle into K^n which would make the corresponding skew-product action strictly ergodic. Our theorem enables us to extend certain examples of Furstenberg. To see this, let $(Y,T) \equiv (K,R_\alpha)$ be an irrational rotation on the circle by α, $G = K^{r-1}$ - the $(r-1)$ torus and σ be the automorphism on G given by an $(r-1) \times (r-1)$ lower triangular unimodular matrix $A = (a_{ij})$ with integer entries, with 1's down the diagonal and 0's down the lower subdiagonal, [i.e. $a_{ij} = 0$ if $i < j$, $a_{i,i} = 1$, $a_{i+1,i} = 0$]. Set $X = K^r$ and define $T(x_0,x_1,\ldots,x_{r-1}) = (\alpha x_0,\sigma(x_1,\ldots x_{r-1}))$. Then for almost all $\emptyset = (\emptyset_1,\ldots\emptyset_{r-1}) \in C(K,K^{r-1})$, the map

$$T_\emptyset(x_0,x_1,\ldots x_{r-1}) = (\alpha x_0,\ldots,\emptyset_j(x_0)x_1^{a_{j1}}\ldots x_{j-2}^{a_{j,j-2}}x_j,\ldots)$$

is strictly ergodic. These maps are in the class of transformations of K^r considered by H. Furstenberg for which his condition does not apply.

Finally, we consider the situation when G is not necessarily a torus. We can obtain another set of similar results, valid for any compact, connected abelian and in some cases, non-abelian fibre groups. The proofs here are based on a different technique and require the assumption of total ergodicity.

THEOREM (3.6). *Let* $T = Z$, *and* $\pi:(X,T,\tilde{\mu}) \to (Y,T,\mu)$ *be an affine extension with fibre group* G. *Assume that*:

(i) μ *is non atomic and supported*,

(ii) G *is connected, abelian, and*

(iii) (Y,T,μ) *is totally ergodic*.

Then the set $\{\alpha | \alpha \in Z(Y,T,G)$ *such that* $(X,T_\alpha,\tilde{\mu})$ *is totally ergodic*$\}$ *is residual in* $Z(Y,T,G)$.

A parallel result holds for weak-mixing, k - and Bernoulli automorphisms. We also have an analogous result for lifting total minimality in a fairly general setting.

THEOREM (3.7). *Let* $T = Z$ *and* $\pi:(X,T) \to (Y,T)$ *be an affine extension with fibre* G. *Assume that*:

(i) (Y,T) *is totally minimal with* Y *infinite, and*

(ii) G *is connected abelian and* σ *is distal*.

Then the set $\{\alpha | \alpha \in Z(Y,T,G)$ *such that* (X,T_α) *is totally minimal*$\}$ *is residual in* $Z(Y,T,G)$.

SKETCH OF PROOF: Standard arguments and the distality of σ reduces the theorem to the situation when σ is periodic, say $\sigma^p = I$. Moreover, we can assume $X = G \times Y$. Given the skew action T_α induced by $\alpha \in Z(Y,T,G)$, we can define a new action $T_{\tilde{\alpha}}$ on $X' = \underbrace{G \times G \cdots \times G}_{p \text{ times}} \times Y$ by

$$T_{\tilde{\alpha}}(g_1,\ldots g_p,y) = (g_1\alpha(y),g_2\alpha(Ty),\ldots g_p\alpha(T^{p-1}y),T^p y).$$

One then shows that minimality of $T_{\tilde{\alpha}}$ implies minimality of T_α. The proof is then finished by applying techniques similar to those in [4] with suitable modifications to ensure total minimality of $T_{\tilde{\alpha}}$. □

As a last item, we give a couple of examples to further illustrate these results.

EXAMPLE (3.8). Let $(Y,T) \equiv (K,R_\alpha)$ be the irrational rotation on circle. Let $G = K^2$ and σ be the automorphism given by matrix $\begin{matrix} 1 & 1 \\ 0 & 1 \end{matrix}$. Then σ is distal. Put $X = K^3$; now Corollary (3.4) implies that for almost all $\emptyset \in C(K,K^2)$, $\phi = (\phi_1,\phi_2)$, the transformation $T_\emptyset(x_1,x_2,x_3) = (\alpha x_1,\emptyset_1(x_1)x_2x_3,\emptyset_2(x_1)x_3)$ is strictly ergodic.

Now let τ be the automorphism given by matrix $\begin{pmatrix} 0 & 1 \\ 1 & 0 \end{pmatrix}$. Since τ is equicontinuous, Theorem (3.1) gives us many continuous maps $\emptyset \equiv (\emptyset_1,\emptyset_2) \in C(K,K^2)$ such that $T_\emptyset(x_1,x_2,x_3) = (\alpha x_1,\emptyset_1(x_1)x_3,\emptyset_2(x_1)x_2)$, is strictly ergodic and $\emptyset$ is in the closure of the maps of the form $y \to (\psi_1(y)^{-1}\psi_2(y\alpha), \psi_2(y)^{-1}\psi_1(y\alpha))$ where $\psi_1,\psi_2 \in C(K,K)$.

EXAMPLE (3.9). Let (Y,T) be either $(\{0,1\}^{Z},\tau)$ - the Bernoulli-shift, or (Λ,τ) a topologically transitive subshift of finite type. Put $X = K^2 \times Y$, and the automorphism on K^2 be again given by $\begin{pmatrix} 1 & 1 \\ 0 & 1 \end{pmatrix}$. Then Theorem (3.3) and Theorem (2.2) imply that for almost all $\emptyset \equiv (\emptyset_1,\emptyset_2) \in C(Y,K^2)$, the map $(y,x_2,x_3) \to (\tau y,\emptyset_1(y)x_2x_3,\emptyset_2(y)x_3)$ is Bernoullian for (Y,T) the Bernoulli shift and if $(Y,T) \equiv (\Lambda,\tau)$ it is intrinsically ergodic.

REFERENCES

[1] J. P. Conze, Extensions de systemes dynamiques par des endomorphismes de groupes compacts, Ann Inst. Henri Poincaré, Vol. VIII No. 1, p. 33-66.

[2] M. Denker, C. Grillenberger, K. Sigmund, Ergodic theory on compact spaces, Springer Verlag L.N. #527.

[3] H. Furstenberg, Strict ergodicity and transformations of the torus, Amer. J. of Math. 83, (1961), p. 573-601.

[4] S. Glasner and B. Weiss, On the construction of minimal skew products, Israel J. of Math. Vol. 34, No. 4, (1979).

[5] H. Keynes and D. Newton, Minimal (G,σ) extensions, Pacific J. of Math. Vol. 77, (1978), p. 145-163.

[6] H. Keynes and D. Newton, Ergodicity in (G,σ) extensions, Proc. Inter. Conf. on Dyn-systems. Springer-Verlag LN #819, p. 265.

[7] D. Rudolph, Classifying the isometric extensions of a Bernoulli shift, J. Analyse Math. 34, (1978), p. 36-60.

[8] D. Rudolph, J. Feldman, C. Moore, Affine extensions of a Bernoulli shift, Trans. A.M.S. 257, (1980), p. 171-191.

[9] R. Walters, Transformations with unique measure with maximal entropy, Journal London Math. Soc. 43, (1974), p. 500-510.

National Science Foundation

and

University of Minnesota
Minneapolis, MN 55455

Contemporary Mathematics
Volume **26**, 1984

LINEAR ALGEBRA AND SUBSHIFTS OF FINITE TYPE

Bruce Kitchens

ABSTRACT. Continuous factor maps between topological Markov shifts are examined using elementary counting and linear algebra techniques. Many of the known structural properties and invariants are discussed.

INTRODUCTION

A subshift of finite type is a dynamical system defined by a square 0-1 matrix. The space is a compact, symbolic, metric space, Σ_A. The transformation is a homeomorphism $\sigma\colon \Sigma_A \to \Sigma_A$ called the shift. The space also supports a collection of Markov measures which are shift invariant and defined by stochastic matrices.

A map between two subshifts of finite type that commutes with the shifts is called a factor map. The problem of understanding continuous, onto, factor maps has received a good deal of attention lately. The problem is to understand the structure of these maps and to find necessary and sufficient conditions for their existence. There has been some progress, but the main questions are still unanswered. This discussion addresses these questions. It is a summary of many of the results that have been obtained. The approach is different from those found elsewhere. The proofs depend primarily on elementary counting and linear algebra arguments.

Continuous factor maps between subshifts of finite type fall naturally into two categories; infinite-to-one and boundedly finite-to-one. Section 2 deals with this difference and then examines in detail some of the structure of finite-to-one factor maps. A continuous factor map may also carry one Markov measure to another. Some of the relationships that hold when this happens are also discussed.

Recently, a number of invariants of continuous finite-to-one factor maps between subshifts of finite-type have been discovered. Some of these are strictly topological in nature and some involve Markov measures. Many of the invariants of [d.J+], [K], [P-T,1], [P-T,2], and [T] are discussed in Section 3. They are derived using elementary linear algebra considerations. Also, some of the consequences of these results are explored.

Section 4 is put in for completeness. It explains a very nice idea of

W. Parry and S. Tuncel. The idea is to view the Markov measures supported by a subshift of finite type as a vector space. Then, by applying some of the results of Sections 2 and 3, it can be seen that a continuous, onto, finite-to-one factor map between two subshifts of finite type induces a vector space map between their vector spaces of Markov measures.

I would like to thank Roy Adler, Brian Marcus, Bill Parry and Selim Tuncel for their discussions on these topics.

1. BACKGROUND

An $n \times n$ matrix of 0's and 1's determines a subshift of finite type (Σ_A, σ). The space is the closed subset of $\{1,\ldots,n\}^Z$ consisting of all $x = \ldots x_{-1}x_0x_1\ldots$ such that $A_{x_i x_{i+1}} = 1$ for all i. The transformation σ, is the shift transformation, $(\sigma x)_i = x_{i+1}$ [P]. A metric can be defined on this by $d(x,y) = 2^{-k}$ where $k = \sup \{r \in Z^+ : x_i = y_i \text{ all } |i| < r\}$. A Markov measure is defined on the space by a pair (p,P), where P is a stochastic matrix compatible with A (i.e., non-negative, row-sum 1, and $P_{ij} > 0$ exactly when $A_{ij} = 1$) and p is a probability vector with $pP = p$ [P].

One linear algebra theorem that will play a key role in the following discussion is the Perron-Frobenius Theorem [G]. It applies to non-negative irreducible square matrices. A matrix is irreducible if, for each pair i,j there is an n such that $(A^n)_{ij} > 0$. The theorem says that for such a matrix, there is a unique largest positive real eigenvalue. It is greater than or equal to the modulus of all other eigenvalues and to it corresponds a strictly positive eigenvector. This is the only eigenvector lying in the positive cone.

This theorem is useful because a subshift of finite type is topologically transitive if and only if it is ergodic with respect to any Markov measure. This happens exactly when its transition matrix, A, is irreducible [P]. The following dynamical properties will also be used. The zeta function of a dynamical system is $\zeta(s) = \exp[\sum_{n=1}^{\infty} (N_n)x^n/n]$, where N_n is the number of points fixed under the n-th power of the transformation. For a subshift of finite type, Σ_A, this has the form $\zeta_A(s) = [s^n C_A(1/s)]^{-1} = [\det(I-sA)]^{-1}$[B-L], where $C_A(x)$ is the characteristic polynomial of A. The topological entropy of an irreducible subshift of finite type is $\log \lambda$, where λ is the largest real eigenvalue [P], and the measure-theoretic entropy with respect to a Markov measure (p,P) is $-\sum_{i,j} p_i P_{ij} \log P_{ij}$ [P]. An irreducible subshift of finite type has a unique measure of maximal entropy [P] which is Markov. To define it, let λ be the maximal eigenvalue, ℓ a left eigenvector for it and r a right one. Then, $p = (\ell_1 r_1,\ldots,\ell_n r_n)$ normalized to have sum one, and $P_{ij} = A_{ij}\, r_j/\lambda r_i$.

A k-block map φ from Σ_A to Σ_B is defined by a map $\hat{\varphi}$ from the k-blocks of Σ_A to the 1-blocks of Σ_B. Then, $(\varphi x)_n = \hat{\varphi}([x_n,\ldots,x_{n+k-1}])$. We will usually use φ for both the map on the space and the map on the blocks. The Curtis-Hedlund-Lyndon theorem [H] asserts that any continuous map between two subshifts of finite type that commutes with the shift is a block map composed with some power of the shift.

2. STRUCTURE OF FINITE-TO-ONE FACTOR MAPS

It is often useful to "rename" a subshift of finite type by going to its higher block presentation. If Σ_A is a subshift of finite type with alphabet L_A and Markov measure (p,P), then its k-block presentation has alphabet $L_A^{[k]} = \{[i_1,\ldots,i_k]:i_r \in L_A,\ 1 \le r \le k$ with $A_{i_t i_{t+1}} = 1,\ 1 \le t \le k-1\}$ and transition matrix $A^{[k]}$ defined by saying $[i_1,\ldots,i_k] \to [j_1,\ldots,j_k]$ when $i_r = j_{r-1},\ 2 \le r \le k$. The Markov measure is carried along in the obvious manner, $(p^{[k]})_{[i_1,\ldots,i_k]} = p_{i_1} P_{i_1 i_2} \cdots P_{i_{k-1} i_k}$ and $(P^{[k]})_{[i_1,\ldots,i_k][j_1,\ldots,j_k]} = P_{i_k j_k}$. A symbolic system is said to be a k-step subshift of finite type if its k-block presentation is a (1-step) subshift of finite type. If $\varphi:\Sigma_A \to \Sigma_B$ is a k-block map, then when it is expressed as a map from $\Sigma_{A^{[k]}}$ to Σ_B, it is a 1-block map. It is often very useful to express maps in this manner.

Continuous shift-commuting maps between subshifts of finite type divide naturally into two different types; infinite-to-one and boundedly finite-to-one. The difference is exhibited by the following:

PROPOSITION: <u>A continuous shift-commuting map,</u> $\varphi:\Sigma_A \to \Sigma_B$ <u>where</u> Σ_A <u>and</u> Σ_B <u>are topologically transitive, is boundedly finite-to-one if and only if</u> $x,y \in \Sigma_A$ <u>with</u> $\varphi(x) = \varphi(y)$ <u>and</u> $d(\sigma^n x,\sigma^n y) \to 0$ <u>as</u> $n \to \pm\infty$ <u>implies</u> $x = y$.

PROOF. By the previous discussion, we may assume φ is a 1-block map. Suppose $\varphi(x) = \varphi(y)$ and $d(\sigma^n x,\sigma^n y) \to 0$ as $n \to \pm\infty$ implies $x = y$. The condition $d(\sigma^n x, \sigma^n y) \to 0$ as $n \to \pm\infty$ is the same as saying that there is an M such that for all $i \in Z$, $|i| \ge M$, $x_i = y_i$. Then, the supposition means that there is no pair of blocks $[i_1,\ldots,i_r]$, $[j_1,\ldots,j_r]$ with $[i_1,\ldots,i_r] \ne [j_1,\ldots,j_r]$, $i_1 = j_1$, $i_r = j_r$ and $\varphi[i_1,\ldots,i_r] = \varphi[j_1,\ldots,j_r]$. Usually, it is said that the map has no "diamonds". If this is the case, any block $[a_1,\ldots,a_r]$ in Σ_B has less than $|L_A|^2$ pre-images, where L_A is the alphabet of Σ_A. This means φ is boundedly finite-to-one.

Conversely, suppose the map is boundedly finite-to-one. Then, the map can have no diamonds. If it did, say $[i_1,\ldots,i_r]$ and $[j_1,\ldots,j_r]$ form one.

Then, the periodic point y in Σ_B, that is the image of the periodic point in Σ_A, obtained by finding a connecting block $[i_1', i_2', \ldots, i_s']$ where $i_1' = i_r$ and $i_s' = i_1$ and then repeating $[i_1, \ldots, i_r, i_2', \ldots, i_{s-1}']$ over and over would have infinitely many pre-images. This is beacuse if $p = r+s-2$, then the block $[y_0, \ldots, y_{np}]$ would have at least 2^n pre-images. In fact, almost every point in Σ_B would have infinitely many pre-images. □

This fact, that finite-to-one maps are those with no diamonds is the crucial observation. It will be heavily exploited in the following discussion of other properties of finite-to-one maps.

LEMMA 1: If $\varphi: \Sigma_A \to \Sigma_B$ is a continuous, onto, shift-commuting, finite-to-one map that takes Markov measure (p,P) to Markov measure (q,Q) then

(i) If $\varphi(x) = y$ and x is periodic of period p and the "weight" of a block is defined to be $W_P([i_1, \ldots, i_r]) = P_{i_1 i_2} \cdots P_{i_{r-1} i_r}$ then $W_P([x_0, \ldots, x_p]) = W_Q([y_0, \ldots, y_p])$;

(ii) [F-P] The information function of a Markov measure (p,P) is defined by $I_P(x) = -\log P_{x_0 x_1}$ and then $I_P = I_Q \circ \varphi + g \circ \sigma - g$ for some continuous function $g: \Sigma_A \to \mathbb{R}$. When this type of relation holds between I_P and I_Q, they are said to be cohomologous;

(iii) [P-T,1] If φ is also a 1-block map, then $P_{ij} = Q_{\varphi(i)\varphi(j)} w(j)/w(i)$ some $w(i), w(j) > 0$ for all i,j with $A_{ij} > 0$.

PROOF: From (i), (ii) and (iii) follow easily. Let φ be a k-block map. Choose $x \in \Sigma_A$ with least period p. Let $y = \varphi(x)$ with least period $\bar{p}$ so $p = s\bar{p}$ for some $s \in Z^+$. Consider $[y_0, \ldots, y_{\bar{p}}]^N \subseteq \Sigma_B$ the cylinder set obtained by repeating $[y_0, \ldots, y_{\bar{p}}]$ N times and always overlapping y_0 and $y_{\bar{p}}$. This has measure $q_{y_0}(W_Q([y_0, \ldots, y_{\bar{p}}]))^N$. Suppose $x = x^1, x^2, \ldots, x^r$ are the pre-images of y with periods $p = p^1 = s(1)\bar{p}$, $p^2 = s(2)\bar{p}, \ldots, p^r = s(r)\bar{p}$ respectively. Consider $\varphi^{-1}([y_0, \ldots, y_{\bar{p}}]^N) \subseteq \Sigma_A$. It looks something like:

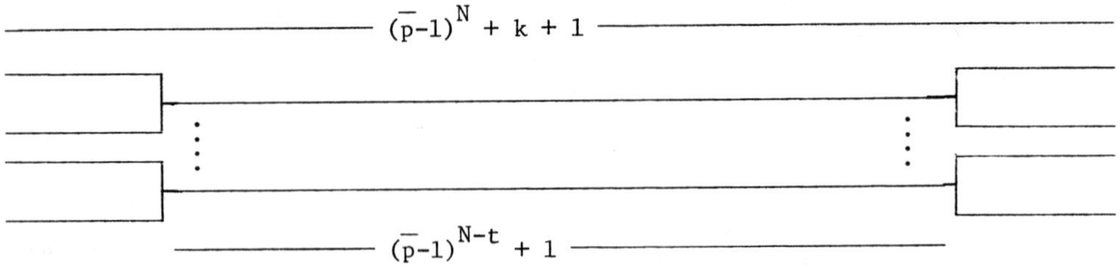

This is because the cylinder sets making up $\varphi^{-1}([y_0, \ldots, y_{\bar{p}}]^N)$ must be converging to the pre-images of y. There will be a t, independent of N, so that each point in $\varphi^{-1}([y_o, \ldots, y_{\bar{p}}]^N)$ will agree with one of x^i on a central $(\bar{p} - 1)^{N-t} + 1$ block. The measure of $\varphi^{-1}([y_0, \ldots, y_{\bar{p}}]^N)$ can be expressed as $M(1)(W_P[x_0^1,\dots,x^1_{p^1}])^{N-t/s(1)} + \dots + M(r)(W_P[x_0^r,\dots,x^r_{p^r}])^{N-t/s(r)}$ some positive $M(i)$. The $s(i)$ are needed because $p^i = s(i)\bar{p}$. The N must be chosen so that the $N-t/s(i)$ are integers. Since φ is measure preserving, we have

$$q_{y_0}(W_Q[y_0,\dots,y_{\bar{p}}])^N = \sum_{i=1}^{r} M(i)((W_P[x_0^i,\dots,x^i_{p^i}])^{-s(i)})^{N-t}$$

for a sequence of N going to infinity. This says $(W_Q[y_0,\dots,y_{\bar{p}}])^{s(i)} = W_P[x_0^i,\dots,x^i_{p^i}]$ for all i. Or more simply $W_Q[y_0,\dots,y_p] = W_P[x_0,\dots,x_p]$.

To see that (ii) follows from (i), pick $x \in \Sigma_A$ with a dense forward and backward orbit. Let $y = \varphi(x)$ and if φ is a k-block map, take the k-block presentation of Σ_A to make it a 1-block map. Define g on the orbit of x by $g(x) = 0$ and

$$g(\sigma^\ell x) = \begin{cases} -\log\left[\dfrac{P_{x_0x_1}\cdots P_{x_{\ell-1}x_\ell}}{Q_{y_0y_1}\cdots Q_{y_{\ell-1}y_\ell}}\right] & \ell \ge 0 \\[2ex] \log\left[\dfrac{P_{x_\ell x_{\ell+1}}\cdots P_{x_{-1}x_0}}{Q_{y_\ell y_{\ell+1}}\cdots Q_{y_{-1}y_0}}\right] & \ell < 0 \end{cases}.$$

Since $I_P = -\sum_{[i,j]} \chi_{[i,j]} \log P_{ij}$ this is just what is needed to satisfy $I_P = I_Q \circ \varphi + g \circ \sigma - g$ on the orbit of x. g is well-defined and depends on only the zero-th coordinate of $\sigma^\ell x$ because of the condition that $W_P[x_0,\dots,x_p] = W_Q[\varphi(x_0),\dots,\varphi(x_p)]$, for any block with $x_0 = x_p$. This means g will extend to a 1-block map from Σ_A to $\mathbb{R}$. If φ is originally a k-block map, then this procedure produces a k-block map g on Σ_A.

To obtain (iii) from (ii), simply exponentiate both sides of $I_P = I_Q \circ \varphi + g \circ \sigma - g$. Since

$$\log P_{ij} = \log Q_{\varphi(i)\varphi(j)} - g(j) + g(i)$$

we have

$$P_{ij} = Q_{\varphi(i)\varphi(j)}\, w(j)/w(i)$$

where $w(i) = e^{-g(i)} > 0$. □

If $\varphi:\Sigma_A \to \Sigma_B$ is a continuous finite-to-one map, it is natural to ask about the relationship of the Markov measures supported by Σ_A and those supported by Σ_B. The image of a Markov measure need not be Markov, but every Markov measure supported by Σ_B is the image of a Markov measure from Σ_A.

LEMMA 2 [T]: _If_ Σ_A, Σ_B _are transitive and if_ $\varphi:\Sigma_A \to \Sigma_B$ _is continuous,_ onto, shift-commuting, finite-to-one map, then every Markov measure supported by Σ_B is the image (under φ) of a unique Markov measure on Σ_A. If φ is a k-block map then the measure on Σ_A that goes to a 1-step Markov measure on Σ_B is at most k-step.

PROOF: We may assume that φ is a 1-block map. Let (q,Q) be a Markov measure on Σ_B. We will construct a Markov measure (p,P) on Σ_A that projects to it. The uniqueness of this Markov measure follows from Lemma 1 (iii). To see this, suppose it exists and define P' by

$$(P')_{ij} = \begin{cases} Q_{\varphi(i)\varphi(j)} & \text{if } A_{ij} = 1 \\ 0 & \text{otherwise} \end{cases} .$$

The previous Lemma says that $P = W^{-1}P'W$ for a diagonal matrix W, whose diagonal entries are strictly positive. Since P must have row sum 1, P' must have maximal eigenvalue 1 and W must be the diagonal matrix with $W_{ii} = w(i)$, $w = (w(1),\ldots,w(m))$ and $P'w = w$.

We need to show that such a P exists, and that φ takes (p,P) to (q,Q). First, define P' as above. We must show that $\lambda(P')$, the maximal eigenvalue of P', is 1. Suppose φ is at most M-to-one. Look at $(P')^n$. Consider a block $[i_1,\ldots,i_n,i_1]$ that is represented on the trace of P'. It goes under φ to an element represented on the trace of Q^n. Several other blocks represented on $\mathrm{tr}\,(P')^n$ may go to the same block on $\mathrm{tr}\,Q^n$, but not more than M of them. Notice that not all blocks represented on $\mathrm{tr}\,Q^n$ have a pre-image on $\mathrm{tr}\,(P')^n$. But we do have $\mathrm{tr}\,(P')^n \le M\,\mathrm{tr}\,Q^n$. Since P' is irreducible, for large n, $\mathrm{tr}\,(P')^n$ acts like $\lambda(P')^n$. The same is true for $\mathrm{tr}\,Q^n$, but $\lambda(Q) = 1$ so $\lambda(P') \le 1$. Next, notice that the sum of all the elements of Q^n is less than or equal to the sum of the elements of $(P')^n$. This is because each block of length $n+1$ in Σ_B has at least one pre-image of length $n+1$ in Σ_A. $\sum_{[a,b]} (Q^n)_{ab} = |L_B|$ for all n since Q is stochastic. Because P' is irreducible, $(P')^n_{ab}$ acts like a constant times $\lambda(P')^n$. If $\lambda(P') < 1$, then the sum of the entries of $(P')^n$ would be going to zero, but it does not. Therefore, $\lambda(P') = 1$. P' has largest eigenvalue 1 so we can stochasticize it to get P by choosing $w > 0$, $P'w = w$, letting W be the diagonal matrix with w down the diagonal and taking $P = W^{-1}P'W$. Take p to be the unique probability vector such that $p = pP$ and we have a Markov measure (p,P) on Σ_A.

The final step is to show that φ takes (p,P) to (q,Q). This is done by showing that the image of (p,P) is absolutely continuous with respect to (q,Q). Let $\mu = \varphi(p,P)$ and $\nu = (q,Q)$. We need two constants, m,M such that $m\nu(E) \le \mu(E) \le M\nu(E)$ for all measurable E. In this case, it is enough to do it for all cylinder sets. Let m be the smallest element and M the largest element of the finite set $\{\Sigma p_i w(j)/w(i)q_a$: any sum where $\varphi(i) = a$ and all j in the sum have the same image$\}$. Then, because

$$\mu[a_1,\ldots,a_n] = \Sigma p_{i_1} P_{i_1 i_2} \ldots P_{i_{n-1} i_n} = \Sigma p_i Q_{a_1 a_2} \frac{w(i_2)}{w(i_1)} \ldots Q_{a_{n-1} a_n} \frac{w(i_n)}{w(i_{n-1})}$$

$$= \left(\Sigma p_i \frac{w(i_2)}{w(i_1)} \right) Q_{a_1 a_2} \ldots Q_{a_{n-1} a_n}$$

where the sum is over all $[i_1,\ldots,i_n] \subseteq \Sigma_A$ that go to $[a_1,\ldots,a_n]$, we have the desired result. Since μ is absolutely continuous with respect to ν, apply the Radon-Nikodym Theorem to find a measurable function $[d\mu/d\nu]$, so that $\mu(E) = \int_E [d\mu/d\nu] d\nu$ for all measurable E. Since μ and ν are shift-invariant measures, $[d\mu/d\nu]$ must be a shift-invariant function. But, (Σ_B, σ) is ergodic so $[d\mu/d\nu]$ is constant and identically one. This shows that φ takes (p,P) to (q,Q). □

One final note, we have shown that (p,P) is unique among all Markov measures. Actually, it is unique among all invariant probability measures. S. Tuncel's proof [T] also includes this fact.

The proofs of the previous two Lemmas contain almost all the techniques needed for the following characterization of finite-to-one maps.

THEOREM 1: <u>If $\varphi:\Sigma_A \to \Sigma_B$ is continuous, onto, shift-commuting map, then the following are equivalent.</u>

(i) <u>φ is finite-to-one;</u>

(ii) [C-P,1] <u>φ takes the measure of maximal entropy of Σ_A to the measure of maximal entropy of Σ_B, and Σ_A and Σ_B have equal topological entropy;</u>

(iii) [C-P,1] <u>Σ_A and Σ_B have equal topological entropy;</u>

(iv) <u>φ takes some (p,P) to some (q,Q) and $I_P = I_Q \circ \varphi + g \circ \sigma - g$ for some continuous g;</u>

(v) <u>φ takes some (p,P) to some (q,Q) and for every periodic point $x \in \Sigma_A$, $W_P[x_0,\ldots,x_t] = W_Q[y_0,\ldots,y_t]$ where $y = \varphi(x)$ and x has some period t;</u>

(vi) <u>φ takes some (p,P) to some (q,Q) and they have the same measure-theoretic entropy.</u>

PROOF: We will assume φ is a 1-block map. To see that (i) implies (ii) we must first see that $\lambda(A) = \lambda(B)$. By exactly the same reasoning as in the proof of Lemma 2, $\operatorname{tr} A^n \leq M \operatorname{tr} B^n$ for some M. So, $\lambda(A) \leq \lambda(B)$. Also, $\sum_{[a,b]} B^n_{ab} \leq \sum_{[i,j]} A^n_{ij}$ so $\lambda(B) \leq \lambda(A)$ and $\lambda(A) = \lambda(B) = \lambda$. This means they have the same topological entropy. Now recall the construction of the measure of maximal entropy. Here, let $P = 1/\lambda\ R^{-1}AR$ and $Q = 1/\lambda\ S^{-1}BS$ where R is the diagonal matrix obtained by putting a right eigenvector for A with respect to λ down the diagonal, and S is a corresponding matrix for B. Again, as in the proof of Lemma 2, we can see that the image of P is absolutely continuous with respect to Q and so φ takes (p,P) to (q,Q).

To see that (ii) implies (i). Suppose it does not, then φ must have a diamond. Let it be $[i_1,\ldots,i_r]$ and $[j_1,\ldots,j_r]$ with $[i_1,\ldots,i_r] \neq [j_1,\ldots,j_r]$ but $i_1 = j_1$, $i_r = j_r$ and $\varphi[i_1,\ldots,i_r] = \varphi[j_1,\ldots,j_r]$. This is just as in the proof of Lemma 1. Now find a connecting block $[i'_1,\ldots,i'_s]$ so that $[i_1,\ldots,i_r,i'_2,\ldots,i'_s]$ and $[j_1,\ldots,j_r,i'_2,\ldots,i'_s]$ are periodic cylinders, $i_1 = i'_s = j_1$, that form a diamond. Now, notice that if this were possible, φ could not take (p,P) to (q,Q). If x is the periodic point formed from $[i_1,\ldots,i_r,i'_2,\ldots,i'_s]$ then its image $y = \varphi(x)$ has uncountably many preimages, because taking $r+s-2 = t$; the cylinder set $[y_0,\ldots,y_{nt}]$ has at least 2^n pre-images. Notice also that the measure of $[y_0,\ldots,y_{nt}]$ is

$$q_{y_0}(Q_{y_0y_1}\cdots Q_{y_{t-1}y_t})^n = q_{y_0}(1/\lambda^t)^n,$$

because

$$Q_{y_0y_1}\cdots Q_{y_{t-1}y_t} = (1/\lambda\ \frac{s(y_1)}{s(y_0)})\cdots(1/\lambda\ \frac{s(y_t)}{s(y_{t-1})}) = 1/\lambda^t$$

since $y_0 = y_t$. The measure of $[y_0,\ldots,y_{nt}]$ must be greater than or equal to 2^n times p_{x_0}, times the weight of $[i_1,\ldots,i_r,i'_2,\ldots,i'_s]^n$ = the weight of $[j_1,\ldots,j_r,i'_2,\ldots,i'_s]^n = (1/\lambda^t)^n$. This says

$$q_{y_0}(1/\lambda^t)^n \geq 2^n \cdot p_{x_0} \cdot (1/\lambda^t)^n,$$

for all n. This is clearly impossible so φ has no diamonds and is finite-to-one.

To see that (ii) and (iii) are equivalent, it is only necessary to notice that if $\lambda(A) = \lambda(B)$, φ must take the measure of maximal entropy of Σ_A to the measure of maximal entropy of Σ_B. This is again the same as the proof of Lemma 2. Construct, the measures of maximal entropy and see that the image of (p,P) is absolutely continuous with respect to (q,Q) and you are done.

Next we see that (iv) and (v) are equivalent. This is contained in the proof of Lemma 1 where it is shown how to go back and forth between the condition on the periodic cylinders and the cohomology equation.

The fact that (iv) and (v) are implied by (i) is part of the conclusion of Lemma 1.

To see that (v) implies (i) notice that the same argument which showed that (ii) implied (i) works here. Simply replace $1/\lambda^t$ by $W_Q[y_0,\dots,y_t]$.

We will now establish the equivalence of (iv), (v), and (vi). (iv) implies (vi) because h(P), the measure-theoretic entropy of (p,P), is

$$-\sum_{[i,j]} p_i P_{ij} \log P_{ij} = \int_{\Sigma_A} I_P \, dP$$

we have

$$h(P) = \int_{\Sigma_A} I_P \, dP = \int_{\Sigma_A} (I_Q \circ \varphi + g \circ \sigma - g) dP$$

$$= \int_{\Sigma_B} I_Q \, dQ + \int_{\Sigma_A} g \circ \sigma \, dP - \int_{\Sigma_A} g \, dP = h(Q).$$

To see that (vi) implies (v) is slightly more difficult. This argument can be found in [d.J+]. First, notice that if φ satisfies the hypotheses of the theorem and takes (p,P) to (q,Q), $W_P[i_0,\dots,i_{t-1},i_0] \le W_Q[a_0,\dots,a_{t-1},a_0]$ for any periodic cylinder $[i_0,\dots,i_0] \subseteq \Sigma_A$ and its image $[a_0,\dots,a_0]$. This is true even if the map is infinite-to-one. Let x be the point of period t defined by $[i_0,\dots,i_0]$ and y its image. Consider the cylinder set $[y_0,\dots,y_{nt}] \subseteq \Sigma_B$ and its inverse image $\varphi^{-1}[y_0,\dots,y_{nt}] \subseteq \Sigma_A$.

Since $[x_0,\dots,x_{nt}] \subseteq \varphi^{-1}[y_0,\dots,y_{nt}]$ and φ is measure preserving:

$$p_{x_0} W_P[x_0,\dots,x_{nt}] \le q_{y_0} W_Q[y_0,\dots,y_{nt}]$$

or

$$(p_{x_0})^{1/n} W_P[i_0,\dots,i_0] \le (q_{y_0})^{1/n} W_Q[a_0,\dots,a_0]$$

for all n, and we have $W_P[i_0,\dots,i_0] \le W_Q[a_0,\dots,a_0]$.

We now need to show that $h(P) = h(Q)$ forces equality. The ergodic theorem states that for almost every $z \in \Sigma_A$ and any $[i_1,\dots,i_k] \subseteq \Sigma_A$, the orbit of z under σ spends $p_{i_1} W_P[i_1,\dots,i_k]$ (the measure of $[i_1,\dots,i_k]$) of its time in $[i_1,\dots,i_k]$. Such a z is said to be "normal". For a normal z it follows that

$$\lim_{k\to\infty} 1/k \log p_{z_0} W_P[z_0,\dots,z_k] = -h(P).$$

We have $x \in \Sigma_A$ of period t, $y = \varphi(x)$, $W_P[x_0,\dots,x_t] \le W_Q[y_0,\dots,y_t]$ and we are trying to prove equality. Choose a $z \in \Sigma_A$ with $z_0 = x_0$. Make sure it is normal and $u = \varphi(z)$ is normal in Σ_B. Let $k_0 = 0$ and $k_1 < k_2 < k_3 \dots$ be the sequence of positive integers such that $[z_{k_i},\dots,z_{k_i+t}] = [x_0,\dots,x_t]$. Then

$$
\begin{aligned}
0 &= h(P) - h(Q) \\
&= \lim_{k\to\infty} 1/k(\log q_{u_0} W_Q[u_0,\dots,u_k] - \log p_{z_0} W_P[z_0,\dots,z_k]) \\
&= \lim_{k\to\infty} 1/k(\log(q_{u_0}/p_{z_0}) + \log(W_Q[u_0,\dots,u_k]/W_P[z_0,\dots,z_k])) \\
&= \lim_{\ell\to\infty} 1/k_\ell \sum_{i=1}^{\ell} \log(W_Q[u_{k_{i-1}},\dots,u_{k_i}]/W_P[z_{k_{i-1}},\dots,z_{k_i}]) \\
&\ge \lim_{\ell\to\infty} 1/k_\ell \sum_{j=0}^{\ell-1} \log(W_Q[u_{k_j},\dots,u_{k_j+t}]/W_P[z_{k_j},\dots,z_{k_j+t}])
\end{aligned}
$$

using the fact that $W_P[z_{k_{j+1}},\dots,z_{k_{j+1}}] \le W_Q[u_{k_j+t},\dots, u_{k_j+1}]$ for all j

$$= \delta \log (W_Q[y_0,\dots,y_t]/W_P[x_0,\dots,x_t]) = 0$$

where δ is the frequency that the k_i occur. Since z is normal, this is positive so $\log(W_Q[y_0,\dots,y_t]/W_P[x_o,\dots,x_t]) = 0$ and we have the desired result. □

3. INVARIANTS

For a matrix P, let J_P be the Jordan form of P and G_P be the block of the Jordan form obtained by discarding the rows and columns where a 0 occurs on the diagonal. If Σ_A is a subshift of finite type with Markov measure (p,P), then $G_{A^{[k]}} = G_A$ and $G_{P^{[k]}} = G_P$, so these are invariant under going to higher block presentations. To see this, define an $|L_A^{[k]}| \times |L_A|$ matrix R by

$$R_{[i_1,\dots,i_k]j} = \begin{cases} 1 & \text{if } i_k = j \\ 0 & \text{otherwise} \end{cases}$$

then $P^{[k]}R = RP$. Since rank $R = |L_A|$ and $\ker R = \ker P^{[k]}$, we have the desired result. The same fact follows about A when (p,P) is the measure of maximal entropy since $P = 1/\lambda\, S^{-1}AS$, S the diagonal matrix with A's right eigenvalue down the diagonal and λ is A's maximal eigenvalue.

THEOREM 2 [K]: <u>If Σ_A and Σ_B are subshifts of finite type and if there is a continuous, onto, shift-commuting map between them that takes</u> (p,P) <u>to</u> (q,Q) <u>then $G_Q \subseteq G_P$. Where $G_Q \subseteq G_P$ means that G_Q can be obtained from G_P by choosing some diagonal entries and discarding their rows and columns, or simply, G_Q is a principal submatrix of G_P.</u>

PROOF: Since this only concerns G_P and G_Q we may assume φ is a 1-block map. For $i \in L_A$ and $a \in L_B$ define R an $|L_A| \times |L_B|$ 0-1 matrix by:

$$R_{ia} = \begin{cases} 1 & \text{if} \quad \varphi(i) = a \\ 0 & \text{otherwise} \end{cases} .$$

A time zero cylinder set is defined by $[a_1,\ldots,a_k] = \{x: x_0 = a_1,\ldots,x_{k-1}=a_k\}$. The inverse image of a time zero cylinder set in Σ_B is a finite union of time zero cylinder sets in Σ_A, of the same length. For any $[a_1,\ldots,a_k] \subseteq \Sigma_B$ define $u^{[a_1,\ldots,a_k]} \in \mathbb{R}^{|L_A|}$ and $v^{[a_1,\ldots,a_k]} \in \mathbb{R}^{|L_B|}$ by:

$(u^{[a_1,\ldots,a_k]})_i = \Sigma\, p_{i_1} P_{i_1 i_2} \cdots P_{i_{k-1} i_k}$ where the sum is taken over all $[i_1,\ldots,i_k] \in \varphi^{-1}[a_1,\ldots,a_k]$ with $i_k = i$, and

$$(v^{[a_1,\ldots,a_k]})_a = \begin{cases} q_{a_1} Q_{a_1 q_2} \cdots Q_{a_{k-1} a_k} & \text{if} \quad a_k = a \\ 0 & \text{otherwise} \end{cases} .$$

Notice that $(u^{[a_1,\ldots,a_k]})R = v^{[a_1,\ldots,a_k]}$, because R adds up all the entries of $u^{[a_1,\ldots,a_k]}$ and puts them in the a_k-th entry of the image vector, φ is measure preserving so this is $v^{[a_1,\ldots,a_k]}$. Let U be the collection of all such $u^{[a_1,\ldots,a_k]}$. Notice now that

$(u^{[a_1,\ldots,a_k]})P = \Sigma\, u^{[a_1,\ldots,a_k,b]}$ where the sum is taken over all b that can follow a_k. This means $\mathcal{U}$, the subspace of $\mathbb{R}^{|L_A|}$ generated by U, is invariant under P and that the diagram

$$\begin{array}{ccc} \mathcal{U} & \xrightarrow{\quad P \quad} & \mathcal{U} \\ R \big\downarrow & & \big\downarrow R \\ \mathbb{R}^{|L_B|} & \xrightarrow[\quad Q \quad]{} & \mathbb{R}^{|L_B|} \end{array}$$

commutes. R is clearly onto so we have the desired result. This linear algebra situation is equivalent to the existence of such a map. □

The previous theorem applies to any continuous, onto, shift-commuting map. In the case where φ is finite-to-one, the proof can be altered to provide much more information. The idea [T], [P-T,1] is to look at the matrix P^t defined by $(P^t)_{ij} = (P_{ij})^t$, for all $t \in \mathbb{R}$. The first necessary observation is that for higher block presentations we have $(P^{[k]})^t R = RP^t$ by the same computation as for $t = 1$ and $G_{(P^{[k]})^t} = G_{P^t}$ for all $t \in \mathbb{R}$.

THEOREM 3: <u>If</u> $\varphi: \Sigma_A \to \Sigma_B$ <u>is a continuous, onto, finite-to-one shift-commuting map that is measure preserving with respect to Markov measures</u> (p,P) <u>and</u> (q,Q) <u>then</u> $G_{Q^t} \subseteq G_{P^t}$ <u>for all</u> $t \in \mathbb{R}$.

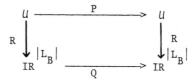

PROOF: This is a slight generalization of the proof of Theorem 2. Assume φ is a 1-block map so by condition (iii) of Lemma 1, $P_{ij}\frac{w(j)}{w(i)} = Q_{\varphi(i)\varphi(j)}$.

Define

$$[R(t)]_{ia} = \begin{cases} [w(i)]^{t-1} & \text{if } \varphi(i) = a \\ 0 & \text{otherwise} \end{cases}$$

leave the v's unchanged and let

$$[u^{[a_1,\ldots,a_k]}(t)]_i = (\Sigma\, p_{i_1} P_{i_1 i_2} \cdots P_{i_{k-1} i_k})[w(i)]^{1-t}$$

where the sum, as before, is over all $[i_1,\ldots,i_k] \in \varphi^{-1}[a_1,\ldots,a_k]$ with $i_k = i$. Here again, we have

$$[u^{[a_1,\ldots,a_k]}(t)][R(t)] = v^{[a_1,\ldots,a_k]}$$

and almost as before

$$([u^{[a_1,\ldots,a_k]}(t)]P^t)_j = \Sigma_i (u^{[a_1,\ldots,a_k]}(t))_i (P^t)_{ij}$$

$$= \Sigma\, p_{i_1} P_{i_1 i_2} \cdots P_{i_{k-1} i_k} P^t_{i_k j}[w(i_k)]^{1-t}$$

$$= \Sigma\, p_{i_1} P_{i_1 i_2} \cdots P_{i_{k-j} i_k} P_{i_k j}[w(j)]^{1-t}\, [P_{i_k j}\frac{w(j)}{w(i_k)}]^{t-1}$$

$$= [\Sigma\, p_{i_1} P_{i_1 i_2} \cdots P_{i_k j}\, w(j)^{1-t}][Q_{\varphi(i_k)\varphi(j)}]^{t-1}.$$

So

$$[u^{[a_1,\ldots,a_k]}(t)]P^t = \Sigma[Q_{a_k b}]^{t-1}[u^{[a_1,\ldots,a_k,b]}(t)]$$

where the sum is over all b that can follow a_k. This means that the subspace $\mathcal{U}(t)$ generated by the u's is invariant under P^t. Then,

$$[u^{[a_1,\ldots,a_k]}]P^t[R(t)] = \Sigma[Q_{a_k b}]^{t-1}\, v^{[a_1,\ldots,a_k b]}$$

$$= [v^{[a_1,\ldots,a_k]}]Q^t = [u^{[a_1,\ldots,a_k]}][R(t)]Q^t$$

and the diagram

$$\begin{array}{ccc} \mathcal{U}(t) & \xrightarrow{\;P^t\;} & \mathcal{U}(t) \\ R(t)\downarrow |L_B| & & \downarrow |L_B|\, R(t) \\ \mathbb{R} & \xrightarrow[\;Q^t\;]{} & \mathbb{R} \end{array}$$

commutes. As before this gives the desired result. □

Notice that when $t = 1$, we have the situation of Theorem 2 and when $t = 0$, we have $P^o = A$, $Q^o = B$, so $G_B \subseteq G_A$. There are many corollaries to this. Some of them follow.

COROLLARY 1: If Σ_A and Σ_B are conjugate by a map that preserves Markov measures (p,P) and (q,Q) then $G_{P^t} = G_{Q^t}$. The generalized zeta function of W. Parry and S. Tuncel [P-T,2] is defined to be $\zeta_P(t,s) = [\det(I-sP^t)]^{-1} = [\det(I-sG_Pt)]^{-1}$ then $\zeta_P(t,s) = \zeta_Q(t,s)$. If Σ_A and Σ_B are topologically conjugate then $G_A = G_B$ and $\zeta_A(t) = \zeta_B(t)$ since the map must preserve maximal measures [W].

COROLLARY 2: If there is a continuous, onto, finite-to-one shift commuting map from Σ_A to Σ_B, then since it preserves measures of maximal entropy $G_B \subseteq G_A$, $C_B(x)$ the characteristic polynomial of B divides $C_A(x)$, and $\zeta_B(s)/\zeta_A(s)$ is a polynomial [K][N]. If the map also preserves Markov measures (p,P) and (q,Q) then $C_{Q^t}(x)$ divides $C_{P^t}(x)$ for all t, and $\zeta_Q(t,s)/\zeta_P(t,s)$ is a polynomial for all t[P-T,2].

COROLLARY 3: There exist equal entropy mixing [P] subshifts of finite type that have no common equal entropy continuous factor. This is in contrast to the Adler-Marcus Theorem [A-M], which asserts that any such pair have a common equal entropy extension.

PROOF: Take

$$A = \begin{bmatrix} 0 & 0 & 1 \\ 1 & 0 & 1 \\ 0 & 1 & 0 \end{bmatrix} \qquad B = \begin{bmatrix} 0 & 0 & 0 & 0 & 1 \\ 1 & 0 & 0 & 0 & 0 \\ 0 & 1 & 0 & 0 & 0 \\ 0 & 0 & 1 & 0 & 0 \\ 0 & 0 & 0 & 1 & 1 \end{bmatrix} .$$

Then $C_A(x) = x^3-x-1$ and $C_B(x) = x^5-x^4-1 = (x^3-x-1)(x^2-x+1)$ so they have the same entropy. Since $C_A(x)$ is irreducible over Z, any continuous, finite-to-one factor of Σ_A must have the same characteristic polynomial. This means the same number of points fixed under σ^k for all k. Σ_B has a fixed point, so any factor of it must also have a fixed point. There is no subshift of finite type that meets both of these conditions. □

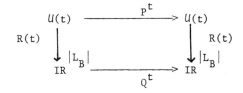

COROLLARY 4 [K]: *There exists a mixing, strictly sofic system* [C-P,2] *that has no equal entropy subshift of finite type as a continuous factor. This should be compared to the fact that any sofic system is a continuous equal entropy factor of a subshift of finite type* [C-P,2].

PROOF: Begin with Σ_A as in the previous corollary. Obtain a strictly sofic system of the same entropy by identifying the pair of two-blocks [2,3] and [3,2]. This sofic system has a fixed point. Any subshift of finite type that is an equal-entropy continuous factor of this system is also one for Σ_A. We already know there is no such shift. □

COROLLARY 5 [M]: *Any equal-entropy continuous factor of the full shift that is a subshift of finite type is shift-equivalent (in the sense of Williams* [W]) *to the same full shift*.

PROOF: Any equal-entropy continuous factor of the full n-shift that is a subshift of finite type has the zeta function $(1-nt)^{-1}$. R. Williams [W] has shown that any subshift of finite type with this zeta function is shift-equivalent to the full n-shift. □

COROLLARY 6 [P-T,1]: *If* $\beta_P(t)$ *is defined to be the largest positive eigenvalue for* P^t *(irreducible for all* t) *and if there is a continuous, onto, finite-to-one, shift-commuting map from* Σ_A *to* Σ_B *that preserves Markov measures* (p,P) *and* (q,Q) *then* $\beta_P(t) = \beta_Q(t)$.

PROOF: We know $G_{Q^t} \subseteq G_{P^t}$, so we need only make sure that $\mathcal{U}(t)$ contains an eigenvector corresponding to the maximal eigenvalue of P^t. $\mathcal{U}(t)$ by construction contains some strictly positive vectors. Choose one and call it x. Since P^t is irreducible $x(P^t)^n/\lambda^n$ converges exponentially fast to an eigenvector. Since $\mathcal{U}(t)$ is closed and P^t invariant, the eigenvector is in $\mathcal{U}(t)$. □

The beta function, $\beta_P(t)$, was defined by S. Tuncel in [T] and examined by W. Parry and S. Tuncel in [P-T,1]. It has many interesting properties. It is a real analytic function, $\beta(0) = h$ the topological entropy of Σ_A, $\beta_P'(1) = -h(P)$ the negative of the measure theoretical entropy, and $\beta_P''(1) = \sigma^2(P) + h(P)^2$ the variance plus the square of the m.t. entropy. It is conjectured to be a complete invariant of finite equivalence [P-T,1].

COROLLARY 7 [d,J+],[T]: *If* $B(p)$ *and* $B(q)$ *are Bernoulli shifts with the same measure-theoretic entropy and if one is a continuous factor of the other, then* q *is a permutation of* p *and they are then isomorphic*.

PROOF: The beta function of $B(p)$ is $\beta_p(t) = p_1^t + \ldots + p_n^t$ since the two have the same beta function $p_1^t + \ldots + p_n^t = q_1^t + \ldots + q_n^t$ for all t and the result follows. □

Given two matrices, P, P' that define Markov measures on Σ_A we can define the matrix $P \circ P'$ by $(P \circ P')_{ij} = P_{ij}P'_{ij}$.

THEOREM 4: <u>If</u> $\varphi:\Sigma_A \to \Sigma_B$ <u>is a continuous, onto, finite-to-one, shift commuting map that takes</u> (p,P) <u>and</u> (q,Q) <u>and</u> (p,P') <u>to</u> (q',Q') <u>then</u> $G_{Q\circ Q'} \subseteq G_{P\circ P'}$.

PROOF: This proof uses the same idea that the two previous proofs use. The point is to produce a subspace $\mathcal{U}(P\circ P') \subseteq \mathbb{R}^{|L_A|}$ and a $|L_A| \times |L_B|$ matrix R such that the following diagram commutes:

$$\begin{array}{ccc} \mathcal{U}(P\circ P') & \xrightarrow{\;P\circ P'\;} & \mathcal{U}(P\circ P') \\ \Big\downarrow R & & \Big\downarrow \\ \mathbb{R}^{|L_B|} & \xrightarrow[\;Q\circ Q'\;]{} & \mathbb{R}^{|L_B|} \end{array}.$$

To do this, first find $w' \in (\mathbb{R}^+)^{|L_A|}$ such that $P'_{ij}\,w'(j)/w'(i) = Q'_{\varphi(i)\varphi(j)}$ as in the proof of Theorem 3. Then define $u^{[a_1,\ldots,a_k]} \in \mathbb{R}^{|L_A|}$ by $(u^{[a_1,\ldots,a_k]})_i = (\Sigma\, p_{i_1}, P_{i_1 i_2} \cdots P_{i_{k-1} i_k})[w'(i)]^{-1}$ where the sum, as before, is taken over all $[i_1,\ldots,i_k] \in \varphi^{-1}[a_1,\ldots,a_k]$ with $i_k = i$. Next, define $v^{[a_1,\ldots,a_k]}$ as before and the matrix R by

$$R_{ia} = \begin{cases} w'(i) & \text{if } \varphi(i) = a \\ 0 & \text{otherwise} \end{cases}.$$

Now see that

$$u^{[a_1,\ldots,a_k]}R = v^{[a_1,\ldots,a_k]},$$

$$u^{[a_1,\ldots,a_k]}(P\circ P') = \Sigma\, Q'_{a_k b}\, u^{[a_1,\ldots,a_k,b]}$$

where the sum is over all b that can follow a_k. This makes use of $P'_{ij}w'(j)/w'(i) = Q'_{\varphi(i)\varphi(j)}$. Also,

$$v^{[a_1,\ldots,a_k]}(Q\circ Q') = \Sigma\; Q'_{a_k b}\, v^{[a_1,\ldots,a_k,b]}$$

where the sum is again over all b that can follow a_k. These are exactly the same type of computations as in the proof of Theorem 3. They give the desired result □

COROLLARY 1 (Parry-Tuncel): *If* $\varphi:\Sigma_A \to \Sigma_B$ *is a topological conjugacy that takes* (p,P) *to* (q,Q) *and* (p',P') *to* (q',Q') *then* $G_{P\circ P'} = G_{Q\circ Q'}$, *and* $C_{P\circ P'}(x) = C_{Q\circ Q'}(x)$.

COROLLARY 2 (Parry-Tuncel): *If there is a continuous, onto, finite-to-one, shift commuting map that takes* (p,P) *to* (q,Q) *and* (p',P') *to* (q',Q') *then* $C_{Q\circ Q'}(x)$ *divides* $C_{P\circ P'}(x)$, *and* $\beta(P\circ P')$ *the maximal eigenvalue of* $P\circ P'$ *is equal to* $\beta(Q\circ Q')$ *the one for* $Q\circ Q'$.

PROOFS: The proofs are the same as for the corresponding corollaries of Theorem 3.

4. A VECTOR SPACE OF MARKOV MEASURES

There is a very nice interpretation of the relationship between the matrices P^t, Q^t, $P\circ P'$, and $Q\circ Q'$. It is due to W. Parry and S. Tuncel.

Let P_t be the stochastic version of P^t, $(P_t)_{ij} = P^t_{ij}u^t(j)/\beta_P(t)u^t(i)$, where u^t is the right eigenvector corresponding to $\beta_P(t)$ for P^t. This means each Markov measure, P, on Σ_A generates a line, $\{P_t\}$, of Markov measures on Σ_A.

The thing to notice is that if $\varphi:\Sigma_A \to \Sigma_B$ is finite-to-one and takes P to Q, then it will also take P_t to Q_t for each t. To see this, consider a periodic point $x \in \Sigma_A$ of period p. Take $y = \varphi(x)$ and consider $W_{P_t}[x_0,\ldots,x_p]$ and $W_{Q_t}[y_0,\ldots,y_p]$.

$$W_{P_t}[x_0,\ldots,x_p] = (P_t)_{x_0x_1}\cdots(P_t)_{x_{p-1}x_p} = [\beta_P(t)]^{-p}(Q_{y_0y_1}\cdots Q_{y_{p-1}y_p})^t$$

and

$$W_{Q_t}[y_0,\ldots,y_1] = (Q_t)_{y_0y_1}\cdots(Q_t)_{y_{p-1}y_p} = [\beta_Q(t)]^{-p}(Q_{y_0y_1}\cdots Q_{y_{p-1}y_p})^t .$$

Since $\beta_P(t) = \beta_Q(t)$, they are equal. Then, by the methods of Theorem 1, φ takes P_t to Q_t.

Define $P*P'$ to be the stochastic version of $P\circ P'$. Then, by exactly the same reasoning we just applied, we get that if $\varphi:\Sigma_A \to \Sigma_B$ is finite-to-one and takes P to Q and P' to Q', it will also take $P*P'$ to $Q*Q'$.

This gives the Markov measures supported by a subshift of finite type a vector space structure. Addition is $P,P' \to P*P'$, scalar multiplication is $P \to P_t$, scalar multiplication distributes over addition, and the zero is the measure of maximal entropy. Lemma 2 says that a finite-to-one map pulls back every Markov measure on the image shift to one on the domain shift. This together with the two previous observations means that if $\varphi:\Sigma_A \to \Sigma_B$ is an onto, finite-to-one, 1-block map and if $M(\Sigma_A)$ is the set of 1-step Markov measures supported by Σ_A then φ induces a map $\varphi^*:M(\Sigma_B) \to M(\Sigma_A)$ which is one-to-one and preserves the vector space operations.

REFERENCES

[A-M] R. Adler and B. Marcus, "Topological Entropy and the Equivalence of Dynamical Systems", Memoirs of the AMS, no. 219. (1979).

[B-L] R. Bowen and O. Lanford, "Zeta Functions of Restrictions of the Shift Transformation", Proc. Symp. Pure Math. Vol. 14, pp. 43-50, Providence, Rhode Island, Amer. Math. Soc. (1970).

[C-P,1] E. Coven and M. Paul, "Endomorphism of Irreducible Subshifts of Finite Type," Math. Syst. Th. 8, 167-175 (1974).

[C-P,2] E. Coven and M. Paul, "Sofic Systems," Israel J. Math. 20, 165-177 (1974).

[d,J+] A. delJunco, M. Keane, B. Kitchens, B. Marcus, and L. Swanson, "Continuous Homomorphisms of Bernoulli Schemes," Ergodic Theory and Dynamical Systems I (Proc. Special Year, Maryland, 1979-80, Ed. A. Katok) Progress in Math. Vol. 10, Birkhauser, Boston, Basel, Stuttgart (1981).

[F-P] R. Fellgett and W. Parry, "Endomorphisms of a Lebesgue Space II," Bull. London Math. Soc. 7, 151-158 (1975).

[G] F. Gantmacher, "Theory of Matrices," Vol. 2, New York, Chelsea (1959).

[H] G. A. Hedlund, "Endomorphisms and Automorphisms of the Shift Dynamical System," Math. Syst. Th. 3, 320-375 (1969).

[K] B. Kitchens, "An Invariant for Continuous Factors of Markov Shifts," Proc. AMS 83, 825-828 (1981).

[M] B. Marcus, "Factors and Extensions of Full Shifts," Mh. Math. 77, 462-474 (1973).

[N] M. Nasu, "Uniformly Finite-to-One and Onto Extensions of Homomorphisms Between Strongly Connected Graphs," Preprint, Research Institute of Electrical Communication, Tohoku University, Sendai, Japan.

[P] W. Parry, "Intrinsic Markov Chains," Trans. AMS 112, 55-66 (1964).

[P-T,1] W. Parry and S. Tuncel, "On the Classification of Markov Chains by Finite Equivalence," (to appear in Journal of Ergodic Theory and Dynamical Systems).

[P-T,2] W. Parry and S. Tuncel, "On the Stochastic and Topological Structure of Markov Chains," Bull. L.M.S.,14, 16-27 (1982).

[T] S. Tuncel, "Conditional Pressure and Coding," Israel J. Math. 39, 101-112 (1981).

[W] R. F. Williams, "Classification of Subshifts of Finite Type," Ann. Math. 98, 120-153 (1973). Errata, Ann. Math. 99, 380-381 (1974).

IBM WATSON RESEARCH LABORATORY
BOX 218
YORKTOWN HEIGHTS, N.Y. 10598

Contemporary Mathematics
Volume 26, 1984

THE ETYMOLOGY OF THE WORD ERGODIC

Anthony Lo Bello

The word ergodic was concocted by Boltzmann as a name for his hypothesis that each surface of constant energy consists of a single trajectory.[1] The correct etymology was given by P. and T. Ehrenfest, who observed that it arose from the combination ἔργον (work) + ὁδός (path).[2] It cannot be maintained that ergodic is produced from ἔργον + ο + εἶδος, i.e., that it means "work-like", since this is forbidden by the laws of philology, for such a combination would require ergoidic, just as σφαῖρα + ο + εἶδος = spheroid, μόνος + ο + εἶδος = monoid, etc. etc.

It cannot be objected to the derivation proposed by P. and T. Ehrenfest that the word should then be erchodic, for the form ἐρχοδικός is impossible; cf. Λεύκιππος and Γλαύκιππος.

DEPARTMENT OF MATHEMATICS
ALLEGHENY COLLEGE
MEADVILLE, PENNSYLVANIA 16335

1. Boltzmann, L., 1887, Über die mechanischen Analogien des zweiten Hauptsatzes der Thermodynamik, J. Reine Angew. Math., 100, p. 201.

2. Ehrenfest, P. and T., 1911, Begriffliche Grundlagen der Statistischen Auffassung in der Mechanik, in Encyklopädie der Mathematischen Wissenschaften, Band IV, Art. 32, Leipzig, Teubner, p. 30, note 88.

Contemporary Mathematics
Volume **26**, 1984

A FUNCTIONAL APPROACH TO NONSTANDARD MEASURE THEORY[1]

Peter A. Loeb

ABSTRACT: In 1973, the author used the Carathéodory extension theorem to construct structurally rich, standard measure spaces on nonstandard point sets. This construction has since been used by a number of researchers to obtain new results in many areas including measure and probability theory, potential theory, mathematical physics and mathematical economics. The author's measure theory is simplified and extended in this paper by starting from first principles with an internal functional and constructing a standard representing measure.

1. INTRODUCTION

Throughout this paper, X will be an internal set in an $\aleph_1$-saturated enlargement of a structure containing the real numbers R. The reader may consult [10] or [16] for terminology and notation.[2] Here, L will denote an internal vector lattice of ${}^*R$-valued functions on X and I will denote an internal (not necessarily order continuous) positive linear functional on L. We use throughout this paper the pointwise ordering and lattice operations. For example, $\varphi \le \psi$ if $\varphi(x) \le \psi(x)$ for all $x \in X$. We assume that for each $\varphi \in L$, $\varphi \wedge 1 \in L$. The class of null functions L_0 is the set of all internal and external ${}^*R$-valued functions h on X such that for any $\varepsilon > 0$ in R there is a $\varphi \in L$ with $|h| \le \varphi$ and $I(\varphi) < \varepsilon$.

To start, we indicate the results that will follow for the special case that $1 \in L$ and ${}^\circ I(1) < +\infty$. For this case, we let L_1 denote the real vector lattice of R-valued functions f on X such that $f = \varphi + h$ where $\varphi \in L$ and $h \in L_0$; we let J denote the well-defined, positive linear functional on L_1 obtained by setting $J(f) = {}^\circ I(\varphi)$. If $\{f_n : n \in N\}$ is an increasing sequence in L_1 with real upper envelope F and $\sup_n J(f_n) < +\infty$, then $F \in L_1$ and $J(F) = \lim_n J(f_n)$. Setting $M = \{A \subset X : \chi_A \in L_1\}$ and $\mu(A) = J(\chi_A)$ for each $A \in M$, we obtain a complete standard measure space (X,M,μ) on X such that $J(f) = \int_X f\,d\mu$ for each $f \in L_1$. If (X,A,ν) is an internal measure space on X with ${}^\circ\nu(X) < +\infty$ and L is the class of internal A-*simple functions with $I(\varphi) = \int_X \varphi\,d\nu$ for each $\varphi \in L$, then (X,M,μ) is the completion of the standard measure space given by [7].

0271-4132/84 $1.00 + $.25 per page

On the other hand, suppose Y is a compact Hausdorff space, $X = {}^*Y$ and $L = {}^*C(Y)$. An example of the functional I can be obtained by integration with respect to an internal Baire measure ν with ${}^\circ\nu(X) < +\infty$, or I might equal ${}^*T$ for some standard positive linear functional T on $C(Y)$. Given "adequate" saturation of our enlargement, for each Borel set B in Y, $\widetilde{B} = st^{-1}[B] \in M$. By setting $\mu_Y(B) = \mu(\widetilde{B})$ one obtains a regular Borel measure on Y such that for each $f \in C(Y)$, $\int f d\mu_Y = {}^\circ I({}^*f)$; in particular, $\int f d\mu_Y = T(f)$ if $I = {}^*T$. Thus the Riesz representation theorem follows.

As an example, one may take $Y = [0,1]$ and let T be the Riemann integral on $C(Y)$. For each $g:[0,1] \to R$ and $x \in {}^*[0,1]$, set $\widetilde{g}(x) = g({}^\circ x)$. A real-valued g is Lebesgue integrable on $[0,1]$ if and only if $\widetilde{g} = \varphi + h$ where $\varphi \in {}^*C(Y)$ and $h \in L_0$. In this case, the Lebesgue integral of g equals ${}^\circ({}^*T(\varphi)) = {}^\circ\int_0^1 \varphi(x)dx$. One can show that a bounded g is Riemann integrable if and only if ${}^\circ({}^*g) \in L_1$ (result with A. Cornea).

It is easy to see that the above results also hold when $I(\varphi)$ is in an internal Banach lattice E and $J(f)$ is in the nonstandard hull $\hat{E}$ [11, 4,5], provided $\hat{E}$ satisfies any condition of Theorem 4.3 of [4]. (A simple, stronger condition is that when $v \geq u \geq 0$ in E and $\|v\| \simeq \|u\|$, then $\|v - u\| \simeq 0$.) This viewpoint was suggested by L. C. Moore, Jr., W. A. J. Luxemburg and C. W. Henson as a modification of work by Lester Helms and the author on vector integrals. We now proceed without the assumption that $1 \in L$.

§2. THE SPACE L_1

In what follows, L_1 will denote the set of all real-valued functions f on X such that $f = \varphi + h$ where $\varphi \in L$, ${}^\circ I(|\varphi|) < +\infty$, and $h \in L_0$.

PROPOSITION 2.1. <u>The sets</u> L_0 <u>and</u> L_1 <u>are vector lattices over</u> R. <u>If</u> f_1 <u>and</u> f_2 <u>are in</u> L_1 <u>and</u> $f_i = \varphi_i + h_i$ <u>where</u> $\varphi_i \in L$, $h_i \in L_0$ <u>for</u> $i = 1,2$, <u>then each integral</u> $I(|\varphi_i|)$ <u>is finite</u>, $(f_1 \vee f_2) - (\varphi_1 \vee \varphi_2) \in L_0$ <u>and</u> $(f_1 \wedge f_2) - (\varphi_1 \wedge \varphi_2) \in L_0$. <u>Moreover, if</u> $f_1 = f_2$, <u>then</u> $\varphi_1 - \varphi_2 \in L_0$ <u>whence</u> ${}^\circ I(\varphi_1) = {}^\circ I(\varphi_2) < +\infty$.

PROOF. Assume first that $f_1 = f_2$. Then $\varphi_1 - \varphi_2 = h_2 - h_1 \in L_0$, so $I(|\varphi_1| - |\varphi_2|) \simeq 0$ and $I(\varphi_1) \simeq I(\varphi_2)$. Moreover, since $f_1 = \widetilde{\varphi} + \widetilde{h}$ for some $\widetilde{h} \in L_0$ and $\widetilde{\varphi} \in L$ with ${}^\circ I(|\widetilde{\varphi}|) < +\infty$, it follows that ${}^\circ I(|\varphi_1|) < +\infty$. Now even when $f_1 \neq f_2$, ${}^\circ I(|\varphi_i|) < +\infty$, $i = 1,2$, and for any $\varepsilon > 0$ in R there is a $\psi \in L$ with $|h_i| \leq \psi$, $i = 1,2$, and $I(\psi) < \varepsilon$. From the inequality

$$(\varphi_1 \vee \varphi_2) - \psi = (\varphi_1 - \psi) \vee (\varphi_2 - \psi) \leq (\varphi_1 + h_1) \vee (\varphi_2 + h_2)$$
$$= f_1 \vee f_2 \leq (\varphi_1 \vee \varphi_2) + \psi$$

and the arbitrary choice of $\varepsilon \in R^+$, it follows that $(f_1 \vee f_2) - (\varphi_1 \vee \varphi_2) \in L_0$. Similarly, $(f_1 \wedge f_2) - (\varphi_1 \wedge \varphi_2) \in L_0$. □

DEFINITION 2.2. For each $f \in L_1$, $\varphi \in L$ and $h \in L_0$ with $f = \varphi + h$, set $J(f) = {}^\circ I(\varphi)$.

PROPOSITION 2.3. <u>The well-defined mapping</u> $J: L_1 \to R$ <u>is a positive linear functional on</u> L_1.

The proof of Proposition 2.3 is left to the reader. We show now that even without assuming continuity for I and without extending the lattice L_1, the monotone convergence property for the functional J on L_1 follows from $\aleph_1$-saturation.

THEOREM 2.4. <u>Let</u> $\{f_n : n \in N\}$ <u>be an increasing sequence in</u> L_1 <u>and let</u> $F = \sup_n f_n$. <u>Assume that</u> $F(x) < +\infty$ <u>for each</u> $x \in X$ <u>and</u> $\sup J(f_n) < +\infty$. <u>Then</u> $F \in L_1$ <u>and</u> $J(F) = \lim_n J(f_n)$.

PROOF. By replacing f_n with $f_n - f_1$, we may assume that each $f_n \geq 0$. By Proposition 2.1, we may fix $\varphi_n \in L$ and $h_n \in L_0$ for each $n \in N$ so that $f_n = \varphi_n + h_n$ and $0 \leq \varphi_n \leq \varphi_{n+1}$. By the $\aleph_1$-saturation of our enlargement, there is a $\varphi_\omega \in L$ with $\varphi_\omega \geq \varphi_n$ for each $n \in N$ and ${}^\circ I(\varphi_\omega) = \lim_{n \in N} {}^\circ I(\varphi_n)$. We need only show that $F - \varphi_\omega \in L_0$. Fix $\varepsilon > 0$ in R. Choose for each $n \in N$ a $\psi_n \in L$ with $|h_n| \leq \psi_n$ and $I(\psi_n) < \frac{\varepsilon}{2^n}$. By $\aleph_1$-saturation, we may extend the sequence $\{\psi_n : n \in N\}$ to an internal sequence $\{\psi_n : n \in {}^*N\} \subset L$. We may choose $\gamma \in {}^*N - N$ so that $\psi_n \geq 0$ and $I(\psi_n) < \frac{\varepsilon}{2^n}$ when $1 \leq n \leq \gamma$ in ${}^*N$. Setting $\psi = \sum_{n=1}^{\gamma} \psi_n$, we have $I(\psi) < \varepsilon$. Now for each $n \in N$,

$$\varphi_n - \psi \leq \varphi_n - \psi_n \leq \varphi_n + h_n \leq F \leq (1+\varepsilon)(\varphi_\omega + \psi),$$

so

$$(\varphi_n - \varphi_\omega) - \psi \leq F - \varphi_\omega \leq \varepsilon\, \varphi_\omega + (1+\varepsilon)\psi.$$

The rest is clear. □

We next sketch further the close relationship between the internal lattice L and the external lattice L_1. Proposition 2.7 corresponds to "pushing down" and the "S-integrability" of a bounded function on a finite measure space in [7] and [10]. Recall that for $\varphi \in L$, ${}^\circ\varphi(x) = {}^\circ(\varphi(x))$ for each $x \in X$.

PROPOSITION 2.5. <u>A real-valued function</u> f <u>on</u> X <u>is in</u> L_1 <u>if and only if for each</u> $\varepsilon > 0$ <u>in</u> R <u>there exist functions</u> ψ_1 <u>and</u> ψ_2 <u>in</u> L <u>with</u> $\psi_1 \leq f \leq \psi_2$, ${}^\circ I(|\psi_1|) < +\infty$, <u>and</u> $I(\psi_2 - \psi_1) < \varepsilon$, <u>in which case</u>, ${}^\circ I(\psi_1) \leq J(f) \leq {}^\circ I(\psi_1) + \varepsilon$.

PROOF. First assume $f = \varphi + h \in L_1$ with $\varphi \in L$, $h \in L_0$. Fix $\varepsilon > 0$ in R. For each $n \in N$, choose $\varphi_n \in L$ with $|h| \le \varphi_n$ and $I(\varphi_n) < \frac{\varepsilon}{n}$. Setting $\psi_1 = \varphi - \varphi_2$ and $\psi_2 = \varphi + \varphi_2$, we have $\psi_1 \le f \le \psi_2$, $°I(|\psi_1|) < +\infty$, and $I(\psi_2 - \psi_1) < \varepsilon$. Moreover, given any ψ_1 and ψ_2 in L satisfying these conditions, we have $\psi_1 - \varphi_n \le \varphi \le \psi_2 + \varphi_n$ for each $n \in N$, whence

$$°I(\psi_1) - \frac{\varepsilon}{n} \le °I(\varphi) = J(f) \le °I(\psi_2) + \frac{\varepsilon}{n} \le °I(\psi_1) + \varepsilon + \frac{\varepsilon}{n}.$$

It follows that $°I(\psi_1) \le J(f) \le °I(\psi_1) + \varepsilon$.

Assume now that f is an arbitrary real-valued function on X for which there exists an increasing sequence $\{\psi_n : n \in N\} \subset L$ and a decreasing sequence $\{\tilde{\psi}_n : n \in N\} \subset L$ with $\psi_n \le f \le \tilde{\psi}_n$, $°I(|\psi_n|) < +\infty$ and $I(\tilde{\psi}_n - \psi_n) < \frac{1}{n}$ for each $n \in N$. By $\aleph_1$-saturation, we may extend both sequences to *N and choose a $\psi_\omega \in L$ with $\psi_n \le \psi_\omega \le \tilde{\psi}_n$, whence $\psi_n - \tilde{\psi}_n \le f - \psi_\omega \le \tilde{\psi}_n - \psi_n$ for each $n \in N$. It follows that $f - \psi_\omega \in L_0$ and thus $f \in L_1$. □

COROLLARY 2.6. *For each* $f \in L_1$, $f \wedge 1 \in L_1$.

PROPOSITION 2.7. *Fix* $\varphi \in L$ *with* $\varphi(x)$ *finite in* *R *for each* $x \in X$. *Assume either that* $1 \in L$ *and* $°I(1) < +\infty$ *or more generally that there is a* $\psi \ge 0$ *in* L *with* $°I(\psi) < +\infty$ *and* $\{x \in X : \varphi(x) \ne 0\} \subset \{x \in X : \psi(x) = 1\}$. *Then* $°\varphi \in L_1$, $\varphi - °\varphi \in L_0$, *and* $J(°\varphi) = °I(\varphi)$.

PROOF. For each $\varepsilon > 0$ in R, $|\varphi - °\varphi| \le \varepsilon\psi$, so $\varphi - °\varphi \in L_0$. For some $n \in N$, $|\varphi| \le n\psi$, so $°I(|\varphi|) < +\infty$. Thus $°\varphi = \varphi + (°\varphi - \varphi) \in L_1$ and $J(°\varphi) = °I(\varphi)$. □

PROPOSITION 2.8. *The function* 1 *is in* L_1 *if and only if* $1 \in L$ *and* $°I(1) < +\infty$. *In this case, if* $f = \varphi + h$ *is a real-valued function on* X *with* $\varphi \in L$ *and* $h \in L_0$, *then* $°I(|\varphi|) < +\infty$, *i.e.*, $f \in L_1$.

PROOF. If $1 \in L_1$, then there is a $\psi \ge 1$ in L with $°I(\psi) < +\infty$, whence $1 = \psi \wedge 1 \in L$. If $1 \in L$ and $°I(1) < +\infty$, then by Proposition 2.7, $1 = °1 \in L_1$. Assuming $1 \in L_1$ and given $f = \varphi + h$ as above, we fix $\psi \in L$ with $|h| \le \psi$ and $I(\psi) < 1$. Then $\varphi - \psi \le f \le \varphi + \psi$. Since f is real-valued and both $\varphi - \psi$ and $\varphi + \psi$ are internal, there is an $n \in N$ with $\varphi - \psi \le n$ and $-n \le \varphi + \psi$. It follows that $°I(|\varphi|) < +\infty$. □

We close this section by showing that L_1 contains "sufficiently" many characteristic functions. We will often denote subsets of X with notation such as $\{f > \alpha\}$.

PROPOSITION 2.9. *Fix* $f \ge 0$ *in* L_1 *and* $\alpha > 0$ *in* R. *Let* $A = \{f > \alpha\}$. *Then the characteristic function of* A, χ_A, *is in* L_1.

PROOF. We may assume that $\alpha = 1$ since $A = \{\alpha^{-1}f > 1\}$. Let $f_1 = f - (f \wedge 1)$.

Then $f_1 \in L_1$ and for each $n \in N$, $J(1 \wedge nf_1) \le J(f)$. By Theorem 2.4, $\chi_A = \lim_{n\in N} (1 \wedge nf_1) \in L_1$. □

§3. THE SPACE M

DEFINITION 3.1. Let M^+ denote the set of nonnegative, extended real-valued functions g on X such that for each $f \ge 0$ in L_1, $g \wedge f \in L_1$. For each $g \in M^+$ set $J(g) = \sup_{f\in L_1} J(g \wedge f)$. Let $M = \{g : g \vee 0 \in M^+$ and $-g \vee 0 \in M^+\}$. For each $g \in M$ set $J(g) = J(g \vee 0) - J(-g \vee 0)$ if at least one of the values $J(g \vee 0)$ and $J(-g \vee 0)$ is finite. Let $\mathcal{L}_1 = \{A \subset X : \chi_A \in L_1\}$ and let $\mathcal{M} = \{A \subset X : \chi_A \in M^+\}$. For each $A \in \mathcal{M}$, set $\mu(A) = J(\chi_A)$.

We will show that $(X, \mathcal{M}, \mu)$ is a standard measure space on X and $J(g) = \int_X g d\mu$ for each $g \in M^+$. First we note that $M \supset L_1$ and the restriction of J to L_1 is the functional defined in Section 2. By Corollary 2.6, $1 \in M^+$. It need not be the case if $1 \in M^+ - L_1$ that $J(1) = +\infty$. For example, L might consist of functions which are zero except at a single point $x_0 \in X$ and $I(\varphi) = \varphi(x_0)$ for each $\varphi \in L$. We also note that for $g \in M^+$, $J(g) = \sup\{J(f) : f \in L_1,\ 0 \le f \le g\}$. On the other hand, $J(g)$ may be less than the supremum of the values ${}^\circ I(\varphi)$ for $\varphi \in L$ with $0 \le \varphi \le g$. For example, if $X = \{x,y\}$, $L = {}^*R^X$, and for some $\omega \in {}^*N - N$, $I(\varphi) = \varphi(x) + \omega\varphi(y)$ for all $\varphi \in L$, then $f(y) = 0$ for each $f \in L_1$. Here $1 \in M - L_1$, $J(1) = 1$, but $\sup\{{}^\circ I(\varphi) : \varphi \in L,\ 0 \le \varphi \le 1\} = +\infty$. We consider next the following useful criterion for membership in M^+.

PROPOSITION 3.2. <u>A nonnegative extended real-valued function</u> g <u>on</u> X <u>is in</u> M^+ <u>if and only if for each</u> $A \in \mathcal{L}_1$ <u>and</u> $n \in N$, $g \wedge n\chi_A \in L_1$. <u>In particular,</u> $B \subset X$ <u>is in</u> $\mathcal{M}$ <u>if and only if for each</u> $A \in \mathcal{L}_1$, $A \cap B \in \mathcal{L}_1$. <u>If</u> $g \in M^+$, $J(g) = \sup_{n\in N, A\in\mathcal{L}_1} J(g \wedge n\chi_A)$.

PROOF. The proof follows from the fact that for each $f \ge 0$ in L_1,

$$g \wedge f = \sup_{n\in N} g \wedge f \wedge n\chi_{\{f > \frac{1}{n}\}} . \qquad \square$$

PROPOSITION 3.3. <u>Fix a sequence</u> $\{g_n : n \in N\} \subset M^+$ <u>and an</u> $\alpha \ge 0$ <u>in</u> R. <u>Then</u> $g_1 + g_2$, αg_1, $g_1 \vee g_2$ <u>and</u> $g_1 \wedge g_2$ <u>are in</u> M^+. <u>Moreover,</u> $J(g_1 + g_2) = J(g_1) + J(g_2)$, $J(\alpha g_1) = \alpha J(g_1)$, <u>and if</u> $g_1 \le g_2$ <u>then</u> $J(g_1) \le J(g_2)$. <u>If</u> $g_n \nearrow G$ <u>then</u> $G \in M^+$ <u>and</u> $J(G) = \sup_{n\in N} J(g_n)$.

PROOF. Given $f \ge 0$ in L_1, $(g_1 + g_2) \wedge f = [(g_1 \wedge f) + (g_2 \wedge f)] \wedge f \in L_1$, and for $\alpha > 0$, $(\alpha g_1) \wedge f = \alpha(g_1 \wedge \frac{1}{\alpha} f) \in L_1$. If, moreover, $f \le g_1 + g_2$, then $f \wedge g_1 \le g_1$ and $f - (f \wedge g_1) \le g_2$, so

$$J(f) = J(f \wedge g_1) + J(f - (f \wedge g_1)) \le J(g_1) + J(g_2) \le J(g_1 + g_2).$$

The rest is clear. □

THEOREM 3.4. The collection M is a σ-algebra in X and μ is a complete countably additive extended real-valued measure on (X, M). If $1 \in L_1$, then $M = L_1$ and $\mu(X) < +\infty$.

PROOF. If $B \in M$ and $A \in L_1$, then $A \cap (X-B) = A - (A \cap B) \in L_1$, whence $X - B \in M$. If, moreover, $\mu(B) = 0$ and $C \subset B$, then $\chi_{A \cap B}$ and thus $\chi_{A \cap C}$ are in $L_1 \cap L_0$, so $C \in M$. The rest follows from Proposition 3.3. □

THEOREM 3.5. A nonnegative extended real-valued function g on X is M-measurable if and only if $g \in M^+$ in which case $J(g) = \int_X g d\mu$.

PROOF. Fix $g \in M^+$, $\alpha > 0$ in R and $A \in L_1$. Then $A \cap \{g > \alpha\} = \{g \wedge 2\alpha\chi_A > \alpha\} \in L_1$, so $\{g > \alpha\} \in M$ and by Theorem 3.4, $\{g > 0\} \in M$. Thus g is M-measurable. The converse and the equality $J(g) = \int g d\mu$ follow from the corresponding facts for finite sums $\Sigma \alpha_i \chi_{A_i}, A_i \in M$. □

It is easy to see that if $g \in M^+$ with $J(g) < +\infty$, then there is an $f \in L_1$ with $0 \le f \le g$ and $J(g-f) = 0$. We next show that one can "push down" an element of L to an element of M and "lift" an element of M to one in L.

PROPOSITION 3.6. If $\varphi \in L$, then $°\varphi \in M$.

PROOF. We need only show that for $\varphi \ge 0$ in L and $A \in L_1$, $°\varphi \wedge \chi_A \in L_1$. (The general result follows by rescaling $\varphi \vee 0$ and $-\varphi \vee 0$.) Given $\varepsilon > 0$ in R, we may choose ψ_1 and ψ_2 in L so that $0 \le \psi_1 \le \chi_A \le \psi_2 \le 1$, $I(\psi_2) < \mu(A) + 1$, and $I(\psi_2 - \psi_1) < \varepsilon$. We then have

$$-\varepsilon\psi_2 + (\varphi \wedge \psi_1) \le °\varphi \wedge \chi_A \le (\varphi \wedge \psi_2) + \varepsilon\psi_2.$$

Since

$$I((\varphi \wedge \psi_2) - (\varphi \wedge \psi_1) + 2\varepsilon\psi_2) \le I(\psi_2 - \psi_1) + 2\varepsilon I(\psi_2) \le 3\varepsilon + 2\varepsilon\mu(A)$$

and ε is arbitrary, it follows from Proposition 2.5 that $°\varphi \wedge \chi_A \in L_1$. □

PROPOSITION 3.7. Assume $1 \in L_1$ and fix $g \ge 0$ in M. There is a $\varphi \ge 0$ in L such that for each $n \in N$, $(g \wedge n) - (\varphi \wedge n) \in L_0$, whence

$$J(g) = \sup_{n \in N} J(g \wedge n) = \sup_{n \in N} °I(\varphi \wedge n) = °I(\varphi \wedge \omega)$$

for some $\omega \in {}^*N - N$.

PROOF. By Proposition 2.5, we may choose sequences $\{\varphi_n : n \in N\}$ and $\{\psi_n : n \in N\}$ in L so that $0 \le \varphi_n \le g \wedge n \le \psi_n$, $\varphi_n \le \varphi_{n+1}$ and $I(\psi_n - \varphi_n) < \frac{1}{n}$ for each $n \in N$. Given $k \ge m \ge n$ in N,

$$\psi_m \wedge n \ge g \wedge n \ge \varphi_k \wedge n \ge \varphi_m \wedge n.$$

By $\aleph_1$-saturation we may choose a $\varphi \in L$ so that for every m and n with $m \geq n$ in N, $\psi_m \wedge n \geq \varphi \wedge n \geq \varphi_m \wedge n$. Clearly, $g \wedge n - \varphi \wedge n \in L_0$ for each $n \in N$. □

A result analogous to Proposition 3.7 can be established with the assumption (more general than $1 \in L_1$) that there is an increasing sequence $\{\chi_{A_n}\} \subset L \cap L_1$ with the set $\{g > \alpha\}$ contained in $\cup A_n$; details are left to the reader. We have shown that for each $\varphi \in L$, ${}^\circ\varphi \in M$. Following Anderson [1], we call φ S-integrable if ${}^\circ I(|\varphi|) < +\infty$ and $J({}^\circ\varphi) = {}^\circ I(\varphi)$. Here, as in [10], S-integrability is a straightforward interpretation of the definition of the integral and the S-integrability of bounded functions when $1 \in L_1$. Note that when $1 \notin L$, we still have $(|\varphi| - \frac{1}{n}) \vee 0 = (|\varphi| - (|\varphi| \wedge \frac{1}{n})) \in L$.

PROPOSITION 3.8. <u>Assume</u> $1 \in L_1$. <u>Then</u> $\varphi \in L$ <u>is</u> S-<u>integrable if and only if for each</u> $\omega \in {}^*N - N$, $I(|\varphi| - |\varphi| \wedge \omega) \simeq 0$.

PROOF. We may assume $\varphi \geq 0$. By the definition of the integral and Proposition 2.7,

$$J({}^\circ\varphi) = \sup_{n\in N} J({}^\circ\varphi \wedge n) = \sup_{n\in N} {}^\circ I(\varphi \wedge n).$$

Since ${}^\circ I(\varphi) = {}^\circ I(\varphi - \varphi \wedge n) + {}^\circ I(\varphi \wedge n)$ for each $n \in {}^*N$, the proposition follows. □

PROPOSITION 3.9. <u>Assume</u> $1 \notin L_1$ <u>and fix</u> $\varphi \in L$. <u>Then</u> φ <u>is</u> S-<u>integrable if and only if for each</u> $\omega \in {}^*N - N$, $I(|\varphi| - |\varphi| \wedge \omega) \simeq 0$ <u>and</u> $I(|\varphi| \wedge \frac{1}{\omega}) \simeq 0$. <u>If</u> φ <u>is finite valued, then</u> ${}^\circ\varphi \in L_1$ <u>if and only if</u> $\sup_{n\in N} {}^\circ I((|\varphi| - \frac{1}{n}) \vee 0) < +\infty$.

EXAMPLE 3.10. Let A be an internal algebra (or σ-algebra) in X and let ν be an internal finitely additive (or σ-additive) measure on (X,A). Let I be the ν-integral on the class L of internal A-*simple functions. For each $A \in A$, let $\nu_o(A) = {}^\circ(\nu(A))$. Then (X,M,μ) is an extension of the measure space given by [7] provided that for each $A \in A$, $\nu_o(A) = \sup\{\nu_o(B): B \in A, B \subset A, \nu_o(B) < +\infty\}$. If $B \in L_1$ and $\varepsilon > 0$ is in R, then from the existence of A-*simple functions ψ_1, ψ_2 with $\psi_1 \leq \chi_B \leq \chi_2$ and $I(\psi_2 - \psi_1) < \varepsilon$, we obtain sets $A_1 = \{\psi_1 > 0\}$ and $A_2 = \{\psi_2 \geq 1\}$ in A. Clearly $A_1 \subset B \subset A_2$ and $\nu(A_2 - A_1) \leq I(\psi_2 - \psi_1) < \varepsilon$. When $\nu_o(X) = +\infty$, M is the class considered by Stroyan and Bayod in [15]. Examples in [15] show that M may be strictly larger than the completion of the smallest σ-algebra generated by A.

§4. INTERNAL FUNCTIONALS ON CONTINUOUS FUNCTIONS

In this section we consider a set Y with a locally compact Hausdorff topology T and an $\aleph_1$-saturated enlargement of a structure containing Y and

R; we assume $\aleph \geq \kappa_1$ and $\kappa \geq \mathrm{card}(\mathcal{T})$. We let C_c or $C_c(Y)$ denote the space of continuous real-valued functions with compact support in Y; $\mathcal{K}$ denotes the compact subsets of Y. If $(Y,\mathcal{T})$ is not compact and P_∞ is the point at infinity for the one point compactification, then we let the monad $m(P_\infty)$ consist of just the points that are not near-standard in $^*Y$. If $(Y,\mathcal{T})$ is compact, we set $m(P_\infty) = \phi$.

As indicated in Footnote 2, the standard part map can be used to transfer sets from Y to $^*Y$; a similar use to transfer functions appears in Anderson [2]. For this section, we associate with each extended real-valued function g on Y the function $\tilde{g}$ on $^*Y$ where $\tilde{g}(x) = g(\mathrm{st}(x))$ for $x \in {}^*Y - m(P_\infty)$ and $\tilde{g}(x) = 0$ for $x \in m(P_\infty)$. With each set $A \subset Y$, we associate the set $\tilde{A} = \bigcup_{y \in A} m(y) \subset {}^*Y - m(P_\infty)$. Clearly, $\chi_{\tilde{A}} = \tilde{\chi}_A$ on $^*Y$.

We now fix an internal positive linear functional I on $^*C_c(Y)$ such that for each standard $f \in C_c(Y)$, $^\circ I(^*f) < +\infty$. We apply our previous results to the case that $X = {}^*Y$ and $L = {}^*C_c(Y)$ keeping the same notation as in Sections 1-3. The next result, needed only when Y is not compact, extends measurability results in [3] and [9]; it may be replaced with Proposition 4.2 when Y is compact.

PROPOSITION 4.1. <u>For each</u> $U \in \mathcal{T}$, $\tilde{U} \in M$ <u>and</u> $\mu(\tilde{U}) = \sup\{^\circ I(^*f) : f \in C_c,$ $0 \leq f \leq 1,$ <u>support of</u> $f \subset U\}$. <u>In particular</u>, $\tilde{Y} \in M$.

PROOF. Fix $U \in \mathcal{T}$ and $A \in L_1$. We will show that $\chi_{\tilde{U}} \wedge \chi_A \in L_1$. Let $\mathcal{K}_U$ denote the set of compact subsets of U and let $\mathcal{F}_U = \{f \in C_c : 0 \leq f \leq 1,$ support of $f \subset U\}$. Given $\varepsilon > 0$ in R, there are functions ψ_1 and ψ_2 in L with $0 \leq \psi_1 \leq \chi_A \leq \psi_2 \leq 1$ and $I(\psi_2 - \psi_1) < \frac{\varepsilon}{3}$. For each $K \in \mathcal{K}_U$, let $\alpha_K = \inf\{^\circ I(\psi_1 \wedge {}^*f) : f \in \mathcal{F}_U, \chi_K \leq f\}$ and let $\beta_K = \inf\{^\circ I(\psi_2 \wedge {}^*f) :$ $f \in \mathcal{F}_U, \chi_K \leq f\}$. Let $\alpha = \sup_{K \in \mathcal{K}_U} \alpha_K$ and $\beta = \sup_{K \in \mathcal{K}_U} \beta_K$. For each $K \in \mathcal{K}_U$, $\beta_K - \alpha_K \leq \frac{\varepsilon}{3}$, so $\beta - \alpha \leq \frac{\varepsilon}{3}$. We now choose $f \in \mathcal{F}_U$ so that $I(\psi_1 \wedge {}^*f) > \alpha - \frac{\varepsilon}{3}$. By κ-saturation, we may choose a set $K' \in {}^*\mathcal{K}_U$ and a $\varphi \in L = {}^*C_c$ so that $K' \supset {}^*K$ for each $K \in \mathcal{K}_U$, $\chi_{K'} \leq \varphi \leq \chi_{^*U}$, and $I(\psi_2 \wedge \varphi) < \beta + \frac{\varepsilon}{3}$. It follows that $\psi_1 \wedge {}^*f \leq \chi_A \wedge \chi_{\tilde{U}} \leq \psi_2 \wedge \varphi$ and $I(\psi_2 \wedge \varphi) - I(\psi_1 \wedge {}^*f) < \beta - \alpha + \frac{2\varepsilon}{3} \leq \varepsilon$. By Proposition 2.5, $\kappa_A \wedge \chi_{\tilde{U}} \in L_1$ and $^\circ I(\psi_1 \wedge {}^*f) \leq \mu(A \cap \tilde{U}) \leq$ $^\circ I(\psi_1 \wedge {}^*f) + \varepsilon$. By Proposition 2.7, $^\circ(^*f) \in L_1$ and $^\circ I(\psi_1 \wedge {}^*f) \leq {}^\circ I(^*f)$ $= J(^\circ(^*f)) \leq J(\chi_{\tilde{U}})$ since $^\circ(^*f) \leq \chi_{\tilde{U}}$. The rest is clear. □

PROPOSITION 4.2. <u>For each</u> $K \in \mathcal{K}$, $\tilde{K} \in L_1$ <u>and if</u> $\alpha_K = \inf\{^\circ I(^*f) : f \in C_c,$ $\chi_K \leq f \leq 1\}$ <u>then</u> $\mu(\tilde{K}) = \alpha_K$.

PROOF. There is a $\varphi \in {}^*C_c = L$ with $\chi_{^*K} \leq \varphi \leq \chi_{\tilde{K}}$ and $^\circ I(\varphi) = \alpha_K$. Given any $f \in C_c$ with $\chi_K \leq f \leq 1$ and given $\varepsilon > 0$ in R, we have $\varphi \leq \chi_{\tilde{K}} \leq$ $(1+\varepsilon)^*f$. It follows that $\chi_{\tilde{K}} - \varphi \in L_0$, $\chi_{\tilde{K}} \in L_1$, and $\mu(\tilde{K}) = J(\chi_{\tilde{K}}) = {}^\circ I(\varphi) = \alpha_K$. □

We now establish the main result of this section.

THEOREM 4.3. Let $\mathcal{M}_Y = \{B \subset Y : \tilde{B} \in \mathcal{M}\}$, and let $\mu_Y(B) = \mu(\tilde{B})$ for each $B \in \mathcal{M}_Y$. Then $\mathcal{M}_Y$ is a σ-algebra in Y containing the Borel σ-algebra. A function g on Y is $\mathcal{M}_Y$-measurable if and only if $\tilde{g}$ is $\mathcal{M}$-measurable. The set function μ_Y is a complete, inner-regular measure on $(Y, \mathcal{M}_Y)$ such that for each $K \in \mathcal{K}$, $\mu_Y(K) = \inf_{\substack{U \in \mathcal{T} \\ K \subset U}} \mu_Y(U)$. For each nonnegative $\mathcal{M}_Y$-measurable g, $\int_Y g\,d\mu_Y = \int_{*Y} \tilde{g}\,d\mu$, and for each $f \in C_c(Y)$, $\int_Y f\,d\mu_Y = {}^\circ I(*f)$.

PROOF. To show that μ_Y is inner-regular, choose $B \in \mathcal{M}_Y$, $\varepsilon > 0$ in R and $n \in N$. Fix $f \geq 0$ in L_1 so that if $J(\chi_{\tilde{B}}) < +\infty$ then $J(f \wedge \chi_{\tilde{B}}) > J(\chi_{\tilde{B}}) - \varepsilon$, and if $J(\chi_{\tilde{B}}) = +\infty$ then $J(f \wedge \chi_{\tilde{B}}) > n$. Now choose $\psi \in L$ with $0 \leq \psi \leq f \wedge \chi_{\tilde{B}}$ and ${}^\circ I(\psi) \geq J(f \wedge \chi_{\tilde{B}}) - \varepsilon$. Let K be the set of standard parts of points in the internal set $\{x \in *Y : \psi(x) > 0\} \subset \tilde{B}$. By Luxemburg's Theorem 3.6.1 of [11], K is compact, and of course $K \subset B$. Since $\mu_Y(K) = \mu(\tilde{K}) \geq {}^\circ I(\psi)$, it follows that μ_Y is inner-regular. If $f \in C_c$, then $*f \in L$, $\tilde{f} = {}^\circ(*f) \in L_1$, and $\tilde{f} - *f \in L_0$, whence $\int_Y f\,d\mu_Y = \int_{*Y} \tilde{f}\,d\mu = {}^\circ I(*f)$. The rest is clear. □

Note that by the Transfer Principle [13,10], a bounded standard function g on $[0,1]$ is Riemann integrable if and only if for any $\varepsilon > 0$ in R there exist internal continuous functions ψ_1 and ψ_2 on $*[0,1]$ with $\psi_1 \leq *g \leq \psi_2$ and $*\int_0^1 \psi_2(x) - \psi_1(x)\,dx < \varepsilon$. By Theorem 4.3, g is Lebesgue integrable if and only if the same is true for $\tilde{g}$.

Finally, assume that $(Y, \mathcal{T})$ is a completely regular Hausdorff space and Z is the Stone-Čech compactification of Y. Anderson and Rashid's results [3] on weak convergence can be formulated in terms of internal positive linear functionals on $*C(Z)$. Tightness corresponds to having $\chi_{\tilde{Z}-\tilde{Y}} \in L_0$ in which case, $\tilde{Y} \in L_1$ so $Y \in \mathcal{M}_Z$ and $\mu_Z(Y) = \sup_{\substack{K \text{ compact} \\ K \subset Y}} \mu_Z(K) = \mu_Z(Z)$.

FOOTNOTES

1) This work was supported in part by U.S. National Science Foundation Grants MCS76-07471 and MCS82-00494. Results for the case $1 \in L_1$ were presented at the July 1981 meeting on nonstandard analysis at Oberwolfach.

2) Background: In [7], the author used the Carathéodory extension theorem to show that an internal measure space built on X can be transformed into a standard measure space, also on X. Thus, for example, one can construct a standard probability space on an internal, hyperfinite, cognitive experiment such as coin tossing of length γ for γ an infinite element of the nonstandard natural numbers $*N$. Along with the construction in [7], a measurable map, the standard part map, was used to produce standard measures on standard spaces in a potential theoretic setting. A connection with weak convergence of measures was then established by the works of Anderson [1], [3], Rashid [12], [3], and the author [8], [9]. In brief, if X is the

nonstandard extension of a compact Hausdorff space Y and ν is an internal Baire measure on X with $^\circ(\nu(Y)) < +\infty$, then the standard part of ν, ν_Y, in the weak* topology can be obtained by standardizing ν on X using [7] and transferring the resulting measure to Y via the standard part map. Clearly, the measure ν_Y can also be obtained by applying the Riesz representation theorem to the functional $f \to {}^\circ\int{}^*f d\nu$, $f \in C(Y)$.

The coincidence of these two methods for obtaining ν_Y suggested the functional approach of this paper for obtaining and extending the theory in [7]. In [7], $\aleph_1$-saturation is used to show that a set of finite measure can be approximated from both inside and outside by internal sets from the generating algebra. This fact has made possible a development of the results from [7] in Stroyan and Bayod [15] and Hrbacek [6] that makes no direct use of the Carathéodory theorem. Similarly, the development here starts from first principles with the Riesz representation theorem following as a simple application.

REFERENCES

1. R. M. Anderson, A nonstandard representation for Brownian motion and Itô integration, Israel J. Math. 25 (1976), 15-46.

2. ______________, Star-finite representations of measure spaces, Trans. Amer. Math. Soc. 271 (1982), 667-687.

3. R. M. Anderson and S. Rashid, A nonstandard characterization of weak convergence, Proc. Amer. Math. Soc. 69 (1978), 327-332.

4. D. Cozart and L. C. Moore, Jr., The nonstandard hull of a normed Riesz space, Duke Math. J. 41 (1974), 263-275.

5. C. W. Henson and L. C. Moore, Jr., Nonstandard analysis and the theory of Banach spaces. To appear.

6. K. Hrbacek, Nonstandard set theory, Amer. Math. Monthly, 86 (1979), 659-677.

7. P. A. Loeb, Conversion from nonstandard to standard measure spaces and applications in probability theory, Trans. Amer. Math. Soc. 211 (1975), 113-122.

8. __________, Applications of nonstandard analysis to ideal boundaries in potential theory, Israel J. Math. 25 (1976), 154-187.

9. __________, Weak limits of measures and the standard part map, Proc. Amer. Math. Soc. 77 (1979), 128-135.

10. __________, An introduction to nonstandard analysis and hyperfinite probability theory, Probabalistic Analysis and Related Topics, A. T. Barucha-Reid, ed., Academic Press, New York, 1979.

11. W. A. J. Luxemburg, A general theory of monads, Applications of Model Theory to Algebra, Analysis, and Probability, W. A. J. Luxemburg, ed., Hold, Rinehart, and Winston, New York, 1969, pp. 18-86.

12. S. Rashid, Economies with infinitely many traders, Ph.D. Dissertation, Yale University, May 1976.

13. A. Robinson, Non-standard Analysis, North-Holland, Amsterdam, 1966.

14. H. L. Royden, Real Analysis, Macmillan, New York, 1968.

15. K. D. Stroyan and J. M. Bayod, Foundations of Infinitesimal Stochastic Analysis, to appear in North-Holland Studies in Logic series.

16. K. D. Stroyan and W. A. J. Luxemburg, Introduction to the Theory of Infinitesimals, Series on Pure and Applied Mathematics, vol. 72, Academic Press, New York, 1976.

University of Illinois
Champaign-Urbana

Contemporary Mathematics
Volume 26, 1984

ON POSITIVE OPERATORS

Dorothy Maharam

1. INTRODUCTION.

The subject of this survey is some old work on (linear) operators between spaces of functions (modulo "negligible" functions). (See items [3]-[7] in the bibliography.) The operators considered are, in the first instance, positive and between spaces of real-valued functions, but the results extend (with some trouble) to certain operators between spaces of complex-valued functions. This survey includes a few applications, one or two new results, some open questions. Neither the domain nor the range space need consist of measurable functions in the ordinary sense; but if the latter does, then so does the former (see §3 below), and we comment on that special case from time to time.

The main feature of the treatment of positive linear operators considered here, is that it emphasizes the structure induced on the space X on which the functions in the domain of the operator are defined. Roughly speaking, under very weak hypotheses (and in particular no measure-theoretic assumptions), X can be embedded in a product $R \times (S,\mu)$, where R is the underlying space of the range functions and (S,μ) is an ordinary numerical measure space, in such a way that the given operator becomes a restriction of the "slice integral", $f \longmapsto \int_S f(r,s)\,d\mu(s)$, $r \in R$, $s \in S$, defined for the "measurable" functions on $R \times (S,\mu)$.

Another difference between the work surveyed here and most other treatments of operator theory is the heavy use here of techniques from Boolean algebra. To a first approximation, such techniques were not being used at all by functional analysts when this work was being done, though they have become more popular since. Because a positive operator must live on a partially ordered linear space, which in turn is represented by a space of continuous functions on the Stone space of a Boolean algebra, the methods of Boolean algebra seem natural

0271-4132/84 $1.00 + $.25 per page

in the present context. For other instances of Boolean algebra techniques (used for different purposes) see for example J. D. M. Wright ([12], [13] and elsewhere). For the study of positive operators in which the setting of Boolean spaces is replaced by the equally natural one of Riesz spaces, see for example [1] and [2].

2. PRELIMINARY DEFINITIONS AND NOTATION

The Boolean algebras considered here are all complete and "ccc" (satisfying the countable chain condition). As a general reference for the aspects of Boolean algebra used here, see [11] and the introductory sections of [3]-[7], especially [3], The letters E, E' always denote such Boolean algebras.

A realization of E is a triple $(X,\mathcal{B},\mathcal{N})$ where $\mathcal{B}$ is a σ-field of subsets of X, $\mathcal{N}$ a σ-ideal in $\mathcal{B}$, and $\mathcal{B}/\mathcal{N}$ is isomorphic to E. (This isomorphism will generally be suppressed.) The standard realization of E is $(R_E, \mathcal{B}_E, \mathcal{N}_E)$, where R_E is the Stone space of E, $\mathcal{B}_E = \mathcal{B}(R_E)$ is the family of Borel sets in R_E, and $\mathcal{N}_E$ is the σ-ideal of first category sets in $\mathcal{B}_E$. For an arbitrary realization space X, we denote the set of extended-real $\mathcal{B}(X)$-measurable functions by $\mathcal{F}(X)$, the ideal of functions with supports in $\mathcal{N}(X)$ by $\mathcal{Z}(X)$, and the quotient ring $\mathcal{F}(X)/\mathcal{Z}(X)$ by $F(X)$. If is easy to show that if X' is any other realization space for E (that is, if $\mathcal{B}(X)/\mathcal{N}(X)$ and $\mathcal{B}(X')/\mathcal{N}(X')$ are isomorphic) then $F(X)$ and $F(X')$ are isomorphic. Hence we use the symbol $F(E)$ for any such space $F(X)$ (or, more properly, $F(X,\mathcal{B},\mathcal{N})$), and move from one realization to another as convenient.

Usually the most convenient realization is the standard one in R_E. We call such a space (i.e., the Stone space of a ccc complete Boolean algebra) a Boolean space. Such spaces some with compact Hausdorff topologies, satisfy the ccc, and are extremally disconnected (the closure of each open set is open). Moreover, each extended-real Borel measurable function on R_E agrees with some continuous function except on a first category set. It follows that $F(R_E)$ is isomorphic (by inclusion) to $C^e(R_E)$, the set of extended-real continuous functions on R. (Note, however, that in $C^e(R_E)$ $\lim_n f_n$ means not the pointwise limit but the corresponding continuous function.) Thus one advantage of working in R_E is that we can work with single (continuous) functions rather than with residue classes of functions.

We write $F^+(E)$ for the positive cone of $F(E)$; thus if $f_n \in F^+(E)$ for all $n = 1,2,\ldots$, then $\Sigma_n f_n$ always exists (possibly infinite). Using the standard representation $C^{e+}(R_E)$ for $F^+(E)$, we define $[f]$ to be the (closed) support of f (an open-closed set), and write $f \ll g$ to mean $[g - f] = R_E$, i.e., $g > f$ except on a first category set. We use e for the unit of E (but in C^e the "e" stands for "extended"), and 1_E for the unit function. If $a \in E$, the principal ideal $\{a' \in E : a' \leq a\}$ is written $E(a)$, and 1_a is the characteristic function (in $C^e(R_E)$) of the open-closed set in R_E that corresponds to a, or, more generally, is the corresponding element of $F(E)$.

3. DEFINITION OF "POSITIVE OPERATOR"

Throughout, a "positive operator" from $F(E)$ to $F(E')$ (where E' is another ccc complete Boolean algebra) is a map ϕ, defined in the first instance from $F^+(E)$ to $F^+(E')$, and satisfying conditions 3.1 and 3.2 below. We then extend ϕ by linearity as far as possible without encountering $\infty - \infty$, mapping a subset of $F(E)$ to $F(E')$. We write (inaccurately but conveniently $\phi : F(E) \to F(E')$.

(3.1) If $f_n \in F^+(E)$ and α_n is a non-negative real number for each $n = 1,2,\ldots,$ then

$$\phi(\Sigma_n \alpha_n f_n) = \Sigma_n \alpha_n \phi(f_n) .$$

(3.2) For each n there exists $g_n \in F^+(E)$ such that

$$\phi(g_n) \ll \infty \quad \text{and} \quad V_n g_n \gg 0.$$

REMARK. The condition (3.2) cannot be strengthened, either by requiring the g_n's to be bounded (rather than essentially finite), or by replacing the sequence of g_n's by a single function, without altering its content. For example, take $E = E' = P(N)$, the power set of the positive integers; then $F^+(E) = F^+(E')$ = the space of all non-negative real sequences. A positive operator $\phi : F(E) \to F(E')$ can here be defined by specifying $\phi(e_n)$ for all n, where e_n is the sequence whose k^{th} term is δ_{nk}. For our example we specify $\phi(e_n)$ = sequence whose k^{th} term is k^n. Then (3.1) and (3.2) are satisfied, but not the altered forms of (3.2).

We shall also assume, when needed, the harmless normalizing conditions:

(3.3) $\phi(f) = 0_{E'} \Longleftrightarrow f = 0_E$.

(3.4) $\phi(1_E) \gg 0_{E'}$ (though it may be infinite).

In any case these conditions can be achieved by replacing E and E' by suitable principal ideals.

The measure-algebraic case is described by:

PROPOSITION 1. If ϕ: $F(E) \to F(E')$ satisfies (3.1)-(3.4), and if E' is a measure algebra, then so is E.

The proof (unpublished) is straightforward except that a little care is needed to ensure that the measure constructed is σ-finite. The converse fails trivially.

4. SITUATIONS PRODUCING OPERATORS

If Φ: $\mathcal{F}^+(X) \to \mathcal{F}^+(X')$ is a positive linear operator, countably additive modulo null functions, satisfying the analog of (3.2), and such that $\Phi(\mathcal{Z}(X)) \subseteq \mathcal{Z}(X')$, it defines a positive operator ϕ: $F(E) \to F(E')$ by

$$\phi(f + \mathcal{Z}(X)) = \Phi(f) + \mathcal{Z}(X').$$

This is, of course, analogous to the standard method of defining operators between L^p spaces, where the distinction between functions and equivalence classes of functions modulo null functions is usually ignored in the notation. I use a similar licence, so that f can denote either a function or its residue class in $\mathcal{F}(X)/\mathcal{Z}(X)$; and similarly A can denote either a "measurable" subset of X or a residue class in $\mathcal{B}(X)/\mathcal{N}(X)$.

Conversely, given ϕ, we may try to realize it by a suitable Φ. This can always be done if X' is not too pathological--for instance, if $\mathcal{B}(X')/\mathcal{N}(X')$ admits a lifting. This is always the case if $X' = R_{E'}$, where the continuous functions provide a lifting. (Note that no condition is needed on X.) Thus every ϕ has a realization Φ if we realize $F(E')$ suitably; and it may perhaps be the case that every ϕ has a realization for the given $\mathcal{F}(X')$.

Another way in which positive operators arise is from measurable maps of the underlying spaces. Let $A' \in \mathcal{B}(X')$, and suppose T is a measurable map of A' onto X, so that $T^{-1}(\mathcal{B}(X')) \subseteq \mathcal{B}(A')$ and $T^{-1}(\mathcal{N}(X)) \subseteq \mathcal{N}(A')$. Then a corresponding Φ : $\mathcal{F}^+(X) \to \mathcal{F}^+(X')$ is defined by

$$(\Phi f)(x') = f(T(x')), \quad x' \in A'.$$

As above, Φ determines an operator ϕ : $F(E) \to F(E')$.

Note that T^{-1} here induces an "embedding" τ of E into E'--that is, τ is an isomorphism of E onto a subalgebra of the principal ideal $E'(a')$ of E' corresponding to A'. We say that the pair (X,X') is "good" if each embedding E into E' is generated by such a map T. In particular,

$(R_E, R_{E'})$ is a good pair; each embedding of E into E' is always generated by a suitable continuous map T from an open-closed subset $R_{E'}(a')$ of $R_{E'}$ onto R_E.

5. ISOMORPHISMS

An isomorphism of $F(E)$ into $F(E')$ is, of course, a special kind of positive operator. Such isomorphic operators are characterized by the following well known theorem:

A positive operator ϕ of $F(E)$ onto $F(E')$ is a (linear) isomorphism of $F(E)$ onto $F(E')$ (or, equivalently, of $C^e(R_E)$ onto $C^e(R_{E'})$) if, and only if, there are : a homeomorphism T of $R_{E'}$ onto R_E, and a fixed $k' \in C^e(R_{E'})$ such that $0 << k' << \infty$, satisfying

$$(\phi f)(\alpha') = f(T^{-1}\alpha')k'(\alpha')$$

for all $f \in C^e(R_E)$ and $\alpha' \in R_{E'}$. (In fact, k' is $\phi(1_E)$.)

More generally, if $C^e(R_E)$ is isomorphic to $C^e(R_{E''})$, where $R_{E''}$ is the Stone space of a σ-subalgebra E'' of E, then the isomorphism is induced, as above, by a continuous map T of $R_{E'}$ onto R_E; and conversely. We do not use this more general result, since we can replace E' by E''.

6. STANDARD INTEGRALS

We now define the "standard integral", in terms of which we shall express an arbitrary positive operator.

Let $(X', \mathcal{B}', \mathcal{N}')$ be any realization of E', and let (J,μ) be a σ-finite measure algebra, realized by a measure space $(S,\mathcal{B},\mu)$. Put $X^* = X' \times S$, and define $\mathcal{B}(X^*) = \mathcal{B}(X') \times \mathcal{B}(S)$. For $H^* \in \mathcal{B}(X^*)$ and $x' \in X'$, set $\mu(H^*,x') = \mu\{s \in S : (x',s) \in H^*\}$. The "slice-measure function" $x' \mapsto \mu(H^*,x')$ is written μ_{H^*}; it is $\mathcal{B}(X')$-measurable, and we define $\mathcal{N}(X^*) = \{H^* : \mu_{H^*} \in \mathcal{Z}(X')\}$. (Thus if there is a measure μ' on $\mathcal{B}(X')$ with $\mathcal{N}(X')$ as its family of null sets, $(X^*, \mathcal{B}(X^*), \mathcal{N}(X^*))$ is just the product measure space, the product measure being given by integrating μ_{H^*} with respect to μ'.)

Define $E^* = \mathcal{B}(X^*)/\mathcal{N}(X^*)$. It is easy to verify that E^* is ccc and complete, and that it is independent of the choice of realizations X' of E', S of (J,μ); and accordingly we denote it by $E' \times (J,\mu)$.

Now define $M : \mathcal{F}(X^*) \to \mathcal{F}(X')$ by setting, for $f^* \in \mathcal{B}(X^*)$ and $x' \in X'$,

$$(6.1)\qquad (Mf^*)(x') = \int_S f^*(x',s)\,d\mu(s).$$

Then M induces a positive operator from $F(E^*)$ to $F(E)$, still denoted by M; (6.1) continues to hold when f^* and Mf^* denote functions modulo negligible functions. Again we note that, if X' is a measure space, Mf^* is the ordinary "slice integral" of f^*.

We shall later (§9) have to consider more general integral operators from $F(E' \times (J,\mu))$ to $F(E')$, of the form

$$(\psi f)(x') = \int_S k(x',s)f(x',s)\,d\mu(s),$$

where k is fixed, and k and f are $\mathcal{B}(X^*)$-measurable and may in fact be complex-valued. We shall see that every sufficiently "good" operator is expressible in this form.

7. THE MAIN THEOREM

The following version of the structure theorem, though not the strongest, is the simplest to state. (See [5, Th. 8].)

THEOREM 1. *If ϕ is a positive operator from $F(E)$ to $F(E')$, then there are: a σ-finite measure algebra (J,μ), and an embedding τ of E into $E' \times (J,\mu)$, such that, for each $f \in F^+(E)$, $\phi f = M(\tau f)$.*

More concretely, by choosing suitably realization spaces X, X' and S for E, E' and (J,μ), there are (a) a measurable subset A^* of $X' \times S$, and (b) a measurable map $T : A^* \to X$ such that ϕ is induced by Φ, where (for all $f \in \mathcal{F}^+(X)$ and $x' \in X'$)

$$(\Phi f)(x') = \int_{A^*} (f \circ T)(x',s)\,d\mu(s).$$

8. SKETCH OF PROOF

In order to throw more light on the theorem and its applications, I outline a sketch of its proof, beginning of necessity rather far afield.

Let E be a Boolean algebra on which an equivalence relation $\sim$ is given, subject to:

(8.1) $\sim$ is countably additive: that is, if x is partitioned into $x_1, x_2, \ldots,$ and y into $y_1, y_2, \ldots,$ and $x_n \sim y_n (n = 1.2.,,,)$, then $x \sim y$.

(8.2) $\sim$ is a σ-bounded: there exist elements $e_n \in E$ $(n = 1,2,\ldots)$ such that $V_n e_n = e$ and, for each n, if $x \sim e_n \geq x$ then $x = e_n$.

(8.3) $\sim$ is homogeneous: if $x \sim y \geq y'$, then there is $x' \leq x$ such that $x' \sim y'$.

Clearly, a positive operator $\phi : F(E) \to F(E')$ induces an equivalence relation on E satisfying (8.1) and (8.2), though not (8.3) in general. Nevertheless the homogeneity condition (8.3), which arose in [3], will be crucial to the proof of Theorem 1. It was motivated by decomposition equivalence of measure space under a group of measure-preserving transformations, and we shall apply the next theorem to that situation in §11.

The first step ([3,§12]), is:

THEOREM 2. _If_ (8.1), (8.2) _and_ (8.3) _hold, the set of equivalence classes of_ E, _with the natural definitions of addition, scalar multiplication and order, can be naturally embedded in_ $F^+(U)$, _where_ $U = \{u \in E: x \sim u \Rightarrow x \leq u\}$ _is the "invariant subalgebra"_.

REMARK. The idea of arithmetizing the equivalence classes goes back to Murray and von Neumann [8, Ch. VI] in the special case in which they are totally ordered.

One next deduces ([3,§19]):

THEOREM 3. _Let_ $\psi : F^+(E) \to F^+(U)$ _be the map induced by the correspondence_ $x \to [x]$, _where_ $x \in E$ _and_ $[x]$ _is its equivalence class. Then Theorem_ 1 _holds for_ ψ _(in much stronger form; more can be said about_ τ _and_ μ).

In deducing Theorem 1 from Theorems 2 and 3, we exploit the following strengthening of homogeneity (which can be obtained by a suitable embedding). (Recall that 1_x is the characteristic function of $x \in E$.) We say that E is _full-valued_ (with respect to an operator $\phi : F(E) \to F(E')$) if, for each $x \in E$ and $g' \in F^+(E')$ such that $g' \leqq \phi(1_x)$, there is $y \leq x$ such that $\phi(1_y) = g'$. And we call $F^+(E)$ _full-valued_ if, for all $f \in F^+(E)$ and $g' \in F^+(E')$ such that $g' \leq \phi(f)$, there is $g \in F^+(E)$ such that $\phi(g) = g'$.

Associated with ϕ is an equivalence relation:

$$x \sim y \Leftrightarrow \phi(1_x) = \phi(1_y);$$

and if E is full-valued then this equivalence relation is homogeneous, and also $F^+(E)$ is full-valued. Both converses fail, and the full-valuedness of $F^{|}(E)$ does not imply homogeneity. (In fact we do not use the full-valuedness of $F^+(E)$ at all in this survey; I mention it because it is a more natural and more general condition than full-valuedness of E, and it is also useful in producing results). Examples and counterexamples are to be found in [6, pp. 242, 247].

Now if E is full-valued (with respect to ϕ), Theorem 1 follows for E. For Theorem 2 applies to the equivalence relation induced by ϕ; here U can be shown to be isomorphic to E'. Now apply Theorem 3 to the corresponding ψ, and note that (as is not hard to see) $\phi(f) = \psi(f)k$, where k is a fixed element of $F(E')$ and $0 \ll k \ll \infty$.

To complete the proof of Theorem 1, we show that ϕ can be extended, by suitably extending E, so that E becomes full-valued. This is done in the next theorem, from [5, Th, 6].

THEOREM 4. If ϕ is a positive operator from $F(E)$ to $F(E')$, then there is an algebra $\tilde{E} \supseteq E$, and there is a positive operator $\tilde{\phi} : F^+(\tilde{E}) \to F^+(E')$, such that $\tilde{\phi}$ extends ϕ and $\tilde{E}$ is full-valued (with respect to $\tilde{\phi}$).

The essential step in defining $\tilde{E}$ is to embed E in an intermediate extension $E_1 = (E,\phi) \times E'$, and then one takes $\tilde{E} = E_1 \times (I,\lambda)$, where (I,λ) is the (Lebesgue) measure algebra of the unit interval. E_1 itself is defined as a suitable quotient field of $\mathcal{B}(R_E \times R_{E'})$. It is convenient here to treat ϕ as a map from $C^e(R_E)$ to $C^e(R_{E'})$. For each continuous function h on $R_E \times R_{E'}$ of the form $h(\alpha,\alpha') = f(\alpha)g'(\alpha')$, we define

$$(\phi_1(h))(\alpha') = ((\phi f)(\alpha'))g'(\alpha') .$$

We extend ϕ_1 to $F(R_E \times R_{E'})$ by linearity and continuity (this is the main difficulty of the argument) and, of course, take quotients modulo a suitable null ideal. The effect of this construction is that disjoint open-closed sets in $R_{E'}$ can be separated by members of the range of ϕ_1; and this is an essential step on the way to getting a full-valued extension.

As a corollary we obtain (see [5, Th. 4.3]):

COROLLARY. If E is full-valued with respect to ϕ, then the measure algebra (J,μ) and embedding τ of Theorem 1 can be chosen so that $\tau(E)$ is a principal ideal of $E' \times (J,\mu)$ (rather than merely a subalgebra of such an ideal).

9. EXTENSION TO THE COMPLEX CASE

Now let $F_c(E)$, $F_c(E')$ (c for "complex") denote the spaces of complex-valued functions corresponding to E and E', and suppose ξ is a linear operator from $F_c(E)$ to $F_c(E')$. We wish to deduce an integral representation for ξ. Now in "nice" cases we can write

$$(9.1) \qquad \xi = (\phi_1 - \phi_2) + i(\phi_3 - \phi_4) ,$$

where the ϕ's are positive operators. When this is possible we say that ξ is of bounded variation, and define $|\xi|$ to be the positive operator $\phi = \phi_1 + \phi_2 + \phi_3 + \phi_4$, extended by linearity to complex-valued functions. (This makes $|\xi|$ depend on the choice of the representation (9.1) rather than on ξ alone. A canonical representation can be given, but is not needed here.) The desired integral representation would be something like

$$(9.2) \qquad \xi(f) = \phi(kf), \ f \in F_c(E) ,$$

for some fixed $k \in F_c(E)$. (Conversely, given k, 9.2 defines a ξ of the type we are considering.) But in this form (9.2) turns out to be too restrictive (it eliminates some desirable examples), so we extend it by allowing k to be drawn from a larger space. This follows a model familiar (for example) for operators on $L^p(\mathbb{R})$ of the form $(Tf)(x) = \int_{\mathbb{R}} k(x,y)f(y)dy$.

It is, of course, enough to represent each ϕ_j $(j = 1,2,3,4)$ in the desired form; thus the problem returns to the real case, and we seek representations of the form

$$(9.3) \qquad \phi_j(f) = \tilde{\phi}(\tilde{k}_j f), \ f \in F(E), \ j = 1,2,3,4,$$

where $\tilde{k}_j$ is a fixed element of $F(\tilde{E})$, $\tilde{E}$ being the full-valued extension of E described in §8, and $\tilde{\phi}$ denoting the extension of ϕ to $F(\tilde{E})$. Note that $\tilde{\phi}$ "dominates" ϕ_j, in a sense to be made precise in §10.

We regard E as being isomorphically embedded in $E' \times (J,\mu)$ (by Theorem 1), which is realized as $X^* = X' \times S$. Then $\tilde{E} = \mathcal{B}(A^*)/\mathcal{N}(A^*)$ for a suitable $A^* \in \mathcal{B}(X^*)$, and ϕ is realized in the form

$$(\phi f)(x') = \int_{A^*} f(x',s)d\mu(s) .$$

We are going to represent each ϕ_j as

$$(\phi_j f)(x') = \int_{A^*} \tilde{k}_j(x',s)f(x',s)d\mu(s)$$

for some $\tilde{k}_j \in \mathcal{F}(A^*)$.

10. DOMINANCE

Let ϕ and ψ be positive operators from $F(E)$ to $F(E')$. We say that ϕ <u>dominates</u> ψ, and write $\psi \prec \phi$, if there is $k \in F(\widetilde{E})$ such that $\psi f = \widetilde{\phi}(kf)$ for all $f \in F^+(E)$. We realize ϕ and ψ as positive maps from $C^e(R_E)$ to $C^e(R_{E'})$, and recall that, for $f \in C^{e+}(R_E)$, the support $[f]$ of f is an open-closed set. One easily sees:

(10.1) For each positive operator $\sigma : F(E) \to F(E')$,

$$[f] = [g] \text{ implies } [\sigma f] = [\sigma g].$$

Hence, for each σ, we can define a map $\tau_\sigma : E \to E'$ by

$$\tau_\sigma(a) = [\sigma f] \text{ whenever } [f] = a.$$

If $\psi \prec \phi$, then clearly $\tau_\psi(a) \le \tau_\phi(a)$ for all $a \in E$; but an example in [6, p. 247] shows that the converse fails in general. The facts are (see [6, p. 245]):

(10.2) $\psi \prec \phi$ if and only if, for each $n = 1,2,\ldots$, a positive operator $\psi_n : F(E) \to F(E')$ and a non-negative constant ρ_n exist such that $\rho_n\psi_n \le \phi$ and $\Sigma_n\psi_n = \psi$.

(10.3) If E is full-valued with respect to ϕ, then the following are equivalent:

(a) $\psi \prec \phi$,

(b) $\tau_\psi(a) \le \tau_\phi(a)$ for all $a \in E$,

(c) ψ has the Hahn decomposition property with respect to ϕ (that is, for each non-negative real number r, there is $a_r \in E$ such that $\psi(f) \le r\phi(f)$ if $[f] \le [a_r]$, and $\psi(f) > r\phi(f)$ if $o \ne [f] \le -a_r$).

From (10.2) it follows that, if $\xi = (\phi_1 - \phi_2) + i(\phi_3 - \phi_4)$ is a complex-valued operator of bounded variation and $\phi = \phi_1 + \phi_2 + \phi_3 + \phi_4$, then ϕ dominates each ϕ_j, and from this follows the desired representation of ξ as a kernel operator, $\xi(f) = \widetilde{\phi}(\widetilde{k}f)$; that is,

$$(\xi f)(x') = \int_{A^*} \widetilde{k}(x',s)f(x',s)d\mu(s) .$$

For an extension of the Hahn decomposition theorem (10.3(c)), in stronger form, to Riesz spaces, see [2]. A variant of the theorem is in [13].

11. APPLICATIONS

A few typical applications follow.

(11.1) Let (E,λ) be a measure algebra, E' a σ-subalgebra of E, and ν a σ-finite reduced measure on E'. Then there are: a measure algebra (J,μ) and a measure-preserving isomorphism τ of E onto a principal ideal in $(E' \times J, \nu \times \mu)$, such that, if $a' \in E'$, $\tau(a')$ is the cylinder over a' in $E' \times J$. (This is [4, Th. 2b].)

In particular ([4, Th. 6]):

(11.2) If E' is the subalgebra of E consisting of the elements that are invariant under a group G of measure-preserving transformations, then there is a partition $e = V_n u_n$, where $u_n \in E'$ $(n = 1,2,\ldots)$, and for each n there is a homogeneous measure algebra (J_n,λ_n), such that $E(u_n) = E'(u_n) \times J_n$ (to within a measure-preserving isomorphism).

A sharpened generalization of (11.1) is as follows. In stating it, we use the notation $E(R)$ for the measure algebra of a measure space $(R,\mathcal{B},m)$. The isomorphisms referred to need not be measure-preserving.

(11.3) Let $(X,\mathcal{B},m)$ and $(X',\mathcal{B}',m')$ be measure spaces, and suppose $T\colon X \to X'$ is measurable. Then there are pairwise disjoint measurable sets $X_n (n = 1,2,\ldots)$ partitioning X, and measurable sets $X'_n \subseteq X'$, such that, for each n, either T induces an isomorphism of $E(X_n)$ onto $E(X'_n)$, or there is an infinite cardinal k_n such that T induces an isomorphism σ_n of $E(X_n)$ onto $E(X'_n \times 2^{k_n}, m' \times \lambda_{k_n})$, where λ_{k_n} is Lebesgue product measure and (modulo null sets) $\sigma_n(T^{-1}Y') = Y' \times 2^{k_n}$ for all $Y' \in \mathcal{B}(X'_n)$. (See [4, Th. 2].)

If X and X' here are sufficiently "good", which can be ensured if we replace them by other realizations of $E(X)$ and $E(X')$, the isomorphisms σ_n can be replaced by measurable point-maps S_n, each defined a.e. on the corresponding set X_n. This leads to the following version of the theorem in (11.3).

(11.4) If X and X' are "good" in the sense just described, we can suppose in (11.3) that $X'_n = TX_n$ $(n = 1,2,\ldots)$; and that, for each n, either $T|X_n$ in invertible (from almost all of X_n to almost all of X'_n), or there is an invertible point-map S_n from almost all of X_n to almost all of $(X'_n \times 2^{k_n}, m' \times \lambda_{k_n})$ such that

$$T(x) = \pi_n(S_n(x)) \qquad (x \in X_n),$$

where π_n is the projection from $X'_n \times 2^{k_n}$ to X'_n.

Again, the construction used in §8 to produce full-valuedness has the following analog. Let $R \times R'$ denote the tensor product. Then define $T_1 : R \times R' \to R'$ by $T_1(r \times r') = T(r)r'$. Is $R \times R'$ full-valued with respect to T_1?

Finally, the analog of "dominance" (§9) is: define $T' \measuredangle T$ to mean $T'(r) = T(kr)$ for some fixed $k \in R$.

It would be nice if the results for function spaces could be extended to this more general setting, in which the requirement of positivity is no longer needed.

(12.4) A considerable amount of attention has been given by functional analysts to finding reasonable conditions on a Banach space X for X to contain a subspace isomorphic to ℓ^1, or to various other L^p ot ℓ^p spaces. (See [10], for example.) Similarly one can ask for conditions on a Boolean algebra E (still complete and ccc) for there to exist a non-atomic measure algebra embedded in E. (The atomic case is trivial.) The work considered here provides some information. For instance, if there exist E' and a positive operator ϕ: F(E) → F(E') such that E is full-valued with respect to ϕ, one can prove that E contains a non-trivial measure algebra. On the other hand, if ϕ : F(E) → F(E') is an isomorphism (hence E is far from full-valued) we get no information (unless E' is itself a measure algebra). I do not know what happens when E' and ϕ exist with the property that ϕ is not an isomorphism on any non-zero principal ideal of E. But I conjecture that in this case E will support many <u>finitely</u> additive non-atomic measures. If we regard ϕ as going from $C^e(R_E)$ to $C^e(R_{E'})$, some finitely additive measures are provided by the maps

$$a \mapsto \phi(1_a)(\alpha'), \; a \in E, \; \alpha' \in R_E' \;;$$

but they may be atomic.

REFERENCES

[1] W. A. J. Luxemburg and A. C. Zaanen, Riesz spaces I, North Holland, Amsterdam 1971.

[2] W. A. J. Luxemburg and A. R. Schep, A Radon-Nikodym type theorem for positive operators and a dual, Indag. Math. 81 (1978) 357-375.

[3] Dorothy Maharam, The representation of abstract measure functions, Trans. Amer. Math. Soc. 65 (1949) 279-330.

[4] ______________, Decompositions of measure algebras and spaces, Trans. Amer. Math. Soc. 69 (1950) 142-160.

[5] ______________, The representation of abstract integrals, Trans. Amer. Math. Soc. 75 (1953) 154-184.

[6] ______________, On kernel representation of linear operators, Trans. Amer. Math. Soc. 79 (1955) 229-255.

[7] ______________, Homogeneous extensions of positive linear operators, Trans. Amer. Math. Soc. 99 (1961) 62-82.

[8] F. J. Murray and J. von Neumann, On rings of operators, Ann. of Math. 37 (1936) 116-229.

[9] V. A. Rokhlin, On the fundamental ideas in measure theory, Math. Sbornik 25 (1949) 107-150: American Math. Soc. Translations 71 (1952) 1-54.

[10] H. P. Rosenthal, A characterization of Banach spaces containing ℓ^1, Proc. Nat. Acad. Sci. 71 (1974) 2411-2413.

[11] R. Sikorski, Boolean algebras, 2nd. ed., Ergebnisse der Math. u. i. Grenzgebiete, N.S. 25 (1964).

[12] J. D. M. Wright, The measure extension problem for vector lattices, Ann. Inst. Fourier (Grenoble) 21 (1971) 65-85.

[13] ______________, A Radon-Nikodym theorem for Stone algebra valued measures, Trans. Amer. Math. Soc. 139 (1969), 75-94.

University of Rochester
Rochester, NY

Contemporary Mathematics
Volume 26, 1984

TRANSMISSION RATES AND FACTORS OF MARKOV CHAINS[1]

Brian Marcus[2], Karl Petersen[3], and Susan Williams

0. INTRODUCTION

A sofic system is a symbolic system which is a continuous factor of a topological Markov chain. A metrically sofic measure is a continuous factor of a measure-theoretic Markov chain. Metrically sofic measures were called sub-markov processes by Furstenberg (1960) and functions of Markov chains by Blackwell (1957). In this paper we study maps from sofic systems to sofic systems, and we look at what they do to metrically sofic measures.

For motivation we first describe (Section 1) the information-theoretic problems from which systems of this type arose. Basically, we have studied a very special type of channel, and we hope to apply our ideas to more general channels.

In Section 2, we discuss the problem of computing the entropy of a metrically sofic measure. We exhibit an infinite series formula for special cases similar to Blackwell's formulas (1957). We also give an example of a metrically sofic measure which is not a factor of any measure-theoretic Markov chain with the same entropy (in contrast to the topological case (Coven and Paul, 1977 and Weiss, 1973)).

By definition, the image of a metrically sofic measure via a continuous factor map is still metrically sofic. In Section 3, we consider the problem of lifting measures via inverse images. We show that any metrically sofic measure can be lifted to a metrically sofic measure of the same entropy - by constructing a sofic subsystem S_0 of the domain such that the restriction of the map to S_0 is finite-to-one. We show that for Markov chains this is the best that can be done: namely, we give an example of two topological Markov chains

---

[1]This paper was presented by Karl Petersen at the Yale University Conference on Modern Analysis and Probability, June, 1982.

[2]Partially supported by NSF Grant MCS-8001796.

[3]Partially supported by NSF Grant MCS-8001590.

and a continuous factor map from one onto the other so that no Markov measure of any order in the image can be lifted to a Markov measure of any order in the domain.

BACKGROUND AND NOTATION

The full shift on n symbols, denoted Σ_n, is $\{1,\ldots,n\}^Z$ with the left shift map:

$$\sigma:\Sigma_n \to \Sigma_n$$

$$\sigma(\underline{x})_i = x_{i+1} .$$

A subshift S is a closed σ-invariant subset of Σ_n. The set of symbols of S, $\{1,\ldots,n\}$, is denoted $A(S)$. A block of S is simply a finite word in some point of S. If u is a block of S, then u^n means the block u repeated n times. If μ is a σ-invariant measure on S and $b_0\ldots b_n$ is a block in S, then $\mu(b_0\ldots b_n)$ is an abbreviation for

$$\mu(\{\underline{x} \in S: x_0 = b_0,\ldots,x_n = b_n\}).$$

And if $b_0\ldots b_n$ is a block of S and $m \leq n$, then $\mu(b_0\ldots b_n|b_0\ldots b_m)$ means the conditional probability

$$\frac{\mu(b_0\ldots b_n)}{\mu(b_0\ldots b_m)} .$$

A topological Markov chain is a subshift which is determined by excluding a finite collection of blocks. We use the abbreviation SFT (subshift of finite type) to mean a topological Markov chain. We often denote a SFT by Σ_A, where A is an $(n\times n)$ 0-1 matrix and

$$\Sigma_A = \{\underline{x} \in \{1,\ldots,n\}^Z: A_{x_i,x_{i+1}} = 1 \ \forall i \in Z\} .$$

Σ_A often means both the set Σ_A and the shift map σ. When presented in this way a SFT has only one step of memory. However, if it had more steps of memory, it could be recoded to a one-step SFT.

By a factor map we mean a continuous shift-commuting map from one subshift onto another. Such a map is always generated by a block map from blocks of the domain (of a fixed length) to symbols of the range. For more background on SFT's, sofic systems and factor maps, see Adler and Marcus (1979), Hedlund (1969), Parry (1964), and Weiss (1973).

1. INFORMATION THEORY CONNECTION

A basic notion in information theory is that of a channel. For us, a channel consists of input and output spaces $\mathcal{I}$ and $\mathcal{O}$, both of which are sofic subshifts, and a collection of probability measures $\{\nu_x : x \in \mathcal{I}\}$ on $\mathcal{O}$. The measure ν_x describes the probability distribution on the output given that the input is x. (See Billingsley, 1965). We assume that the channel is stationary, i.e.

$$\sigma\nu_x = \nu_{\sigma x},$$

where σ is the left shift map.

Now, any shift-invariant measure μ on $\mathcal{I}$ defines a shift-invariant measure λ on $\mathcal{I} \times \mathcal{O}$:

$$\lambda = \int \nu_x d\mu.$$

Letting α and β denote the canonical 0th coordinate partitions on $\mathcal{I}$ and $\mathcal{O}$, one defines the transmission rate of μ as follows:

$$\begin{aligned} R(\mu) &= \lim_{n\to\infty} \frac{1}{n}[H(\alpha_0^n) - H(\alpha_0^n | \beta_0^n)] \\ &= \lim_{n\to\infty} \frac{1}{n}[H(\beta_0^n) - H(\beta_0^n | \alpha_0^n)]. \end{aligned}$$

(Here H denotes the entropy with respect to the measure λ on $\mathcal{I} \times \mathcal{O}$, $\alpha_0^n = \bigvee_{i=0}^{n-1} \sigma^{-i}\alpha$, and α_0^n, β_0^n are considered as partitions of $\mathcal{I} \times \mathcal{O}$.) The capacity of the channel is defined by

$$C = \sup_{\mu} R(\mu).$$

The basic problems are:

(1) Compute $R(\mu)$ for nice μ's.

(2) Compute C.

(3) Describe the measures of maximal transmission rate, namely those for which $R(\mu) = C$.

These problems are important because C represents the maximal amount of information that can be transmitted through the channel, and a maximal measure μ describes the optimal input statistics.

In the case where

(a) I is the full shift on n symbols (i.e., all sequences on $1,2,\ldots,n$ are allowed) and

(b) the channel is memoryless (i.e., $\nu_x(y_0)$ depends only on x_0, and $\nu_x(y_0\ldots y_m) = \prod_{i=0}^{m} \nu_{\sigma^i x}(y_i))$,

C can be explicitly computed and a maximal μ exists, is unique, and is memoryless itself (i.e., μ is i.i.d.).

However, the case with memory is quite different. If we allow even one step of memory in the channel, or in the input space I, it may happen that no n-step Markov measure is maximal for any n. Channels with memory arise naturally: first, for example, the errors may tend to come in bursts; second, we may want to restrict the input sequences in order to reduce the chance of error or to modulate the signal (e.g., for recording data on a magnetic medium (Adler, Coppersmith and Hassner, 1982)).

Consider the case of "deterministic noise". The idea here is that information is being distorted in a very definite way. Thus, information is lost, even though it is lost in a completely deterministic way. This kind of channel is sometimes called "noiseless" in the literature. An example is as follows: imagine the transcript of a congressional hearing. The input here is the speech and the output is the written transcript. While the output does contain every uttered word, it ignores the tone and emotion of the speaker's voice, which often carry considerable information. Deterministic channels can also be used to approximate honest noisy channels.

To be precise, by a <u>deterministic channel</u> we mean one where I is sofic and each ν_x is a point mass on O. Then, since the channel is stationary, it is described by a factor map $\varphi: I \to O$ with ν_x being the point mass on $\varphi(x)$. We then have $R(\mu) = h(\varphi\mu)$. This paper deals with this case. The problem of computing $R(\mu)$ is then the problem of computing the entropy of a metrically sofic measure (discussed in Section 2). As for computing capacity in the deterministic case, since $R(\mu) = h(\varphi\mu)$ and every measure on O lifts to I (by Hahn-Banach), it follows that

$$C = \sup_{\nu \text{ on } B} h_\nu(O) .$$

But this is just the topological entropy (see Walters, 1975) of O, which can be computed effectively (see Coven and Paul, 1977). As a corollary of Section 3, we will see that C is achieved by a metrically sofic measure (Corollary 3.3).

In the deterministic case, $H(\beta_0^n|\alpha_0^n)$ is always 0. With real noise it may still happen that $H(\beta_0^n|\alpha_0^n)$ is a constant independent of μ, so that maximizing the transmission rate is still equivalent to maximizing the entropy of the output measure. For example, let I be a SFT on two symbols 0 and 1, O the full shift on two symbols and the noise measures Bernoulli:

$$\nu_x = \begin{cases} B(1-\eta_0,\eta_0) & \text{if } x_0 = 0 \\ B(\eta_0,1-\eta_0) & \text{if } x_0 = 1. \end{cases}$$

Here η_0 is simply the probability of receiving the wrong symbol, regardless of what is sent. The problem can be reformulated as follows: letting Σ_2 denote the full shift on 0 and 1, define

$$\varphi: I \times \Sigma_2 \to O \quad \text{by}$$

$$\varphi(x_n,y_n) = x_n + y_n \mod 2.$$

Then $C = \sup_{\mu} h(\varphi(\mu \times B(1-\eta_0,\eta_0)))$. One can show by a standard semicontinuity argument (Breiman, 1960) that the sup is achieved. Computer evidence with Markov μ suggests that the max is unique. Any maximum is probably not Markov of any order.

2. ENTROPY OF A METRICALLY SOFIC MEASURE

Let Σ_A be an irreducible SFT and μ a one-step Markov measure (chain) on Σ_A. Let S be a sofic system and $\varphi:\Sigma_A \to S$ an (onto) factor map. By recoding, we may assume that φ is a one-block map.

We want to compute $h(\varphi\mu)$. At first glance, it would seem reasonable to expect a simple formula, in closed form, for $h(\varphi\mu)$ in terms of the one-block map φ (which is just an equivalence relation on a finite set) and the transition probabilities which define μ (especially in light of the fact that $h(\mu)$ has such a simple formula). However, Blackwell (1957) showed that $h(\varphi\mu)$ is an intrinsically complicated function. He did give an infinite series formula, using conditional entropy, which works in a special case. We have a related idea.

Suppose that there is a subset B of S such that the first return (induced) map σ_B of σ to B is nice. "Nice" here means that σ_B has a natural countable Bernoulli or Markov partition. Then one can use Abramov's formula (1959) to "compute" $h(\varphi\mu)$.

To be more concrete, take Blackwell's example. Let μ be the Markov measure defined by the transitions

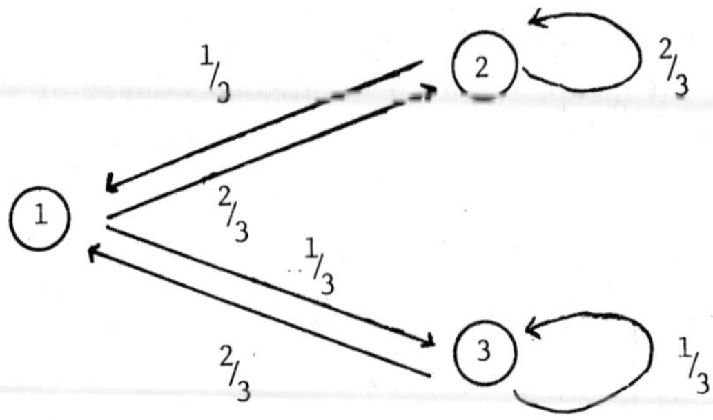

let Σ_A be the support of μ, let S be the image of Σ_A via the map $\varphi(1) = b$, $\varphi(2) = \varphi(3) = a$. Let $B = \{\underline{x} \in S: x_0 = b)$. Then, from the fact that $\varphi^{-1}(b)$ is a single symbol, one easily sees that the partition $\{ba^k b\}_{k\geq 1}$ is a countable Bernoulli partition for σ_B. An easy computation gives

$$\mu(B) = \frac{2}{7} \quad \text{and}$$

$$\mu_B(ba^k b) = \left(\frac{2}{9}\right)\left(\left(\frac{2}{3}\right)^{k-1} + \left(\frac{1}{3}\right)^{k-1}\right).$$

Thus, by Abramov's formula,

$$h(\varphi\mu) = \frac{4}{63}\sum_{k=0}^{\infty}\left(\left(\frac{2}{3}\right)^k + \left(\frac{1}{3}\right)^k\right)\log\left(\left(\frac{2}{9}\right)\left[\left(\frac{2}{3}\right)^k + \left(\frac{1}{3}\right)^k\right]\right)$$

(cf. Blackwell, 1957).

As another example, let μ be the Markov measure defined by the transitions

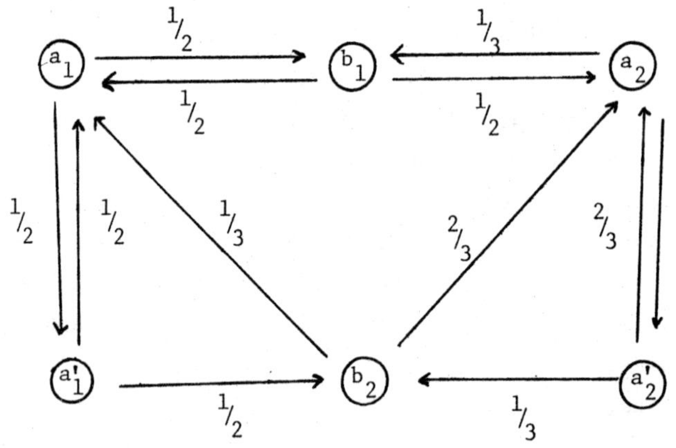

Define $\varphi(b_1) = \varphi(b_2) = b$, $\varphi(a_1) = \varphi(a_1') = \varphi(a_2) = \varphi(a_2') = a$. Here, Σ_A is the support of μ and S is the image of Σ_A via φ. Then for the same set $B = \{x \in S: x_0 = b\}$, one can show that $\{ba^k b\}$ is a countable Markov partition for the first return map σ_B. The idea is that if you've just seen the atom $ba^k b$ with k odd, then you know you are in b_1, and so your transition probabilities to each $a^j b$ are determined (and different from what they would be if you had just seen the atom ba^kb with k even, in which case you would be in b_2). Thus we may find Markov transition probabilities among the atoms ba^kb and write down a formula for $h(\varphi\mu)$.

One can also construct examples where the inducing set B is more complicated than a simple cylinder set. Moreover, the entropy of the output is a continuous function of the transition probabilities of the input Markov measure, and indeed of the input measure itself; thus transmission rates can be approximated in certain circumstances.

These infinite series formulas are not entirely satisfying. Notice, however, that if the map φ were finite-to-one, then $h(\varphi\mu) = h(\mu)$, and so in this case there is an explicit closed form. To the optimist, then, the following question leaps to mind.

Given Σ_A, μ, S and φ (as in the beginning of this section), does there exist another SFT Σ_B, another Markov ν on B, and another factor map $\varphi':\Sigma_B \to S$ such that $\varphi'\nu = \varphi\mu$ and <u>φ' is finite-to-one</u>?

If there were, and if the procedure were explicit, then it would yield, in principle, a closed form. For some metrically sofic measures, it is possible to construct Σ_B, ν, and φ' as above. In particular, if $\varphi\mu$ were the measure of maximal entropy on S, then this is possible: for one can explicitly find a Σ_B and $\varphi':\Sigma_B \to S$ such that φ' is finite-to-one; then $\varphi'(\nu) = \varphi\mu$, where ν is the (Markov) measure of maximal entropy for Σ_B. (See Coven and Paul, 1975). However, in general the answer to this question is NO. We outline a proof that <u>$\varphi\mu$ = Blackwell's example above cannot be the finite-to-one image of any Markov chain.</u> (This is equivalent (see the proof of Theorem 2 in Parry, 1977) to saying that it is not the image of any Markov chain of the same entropy.)

To understand the defect of this example, first observe that for a Markov chain, of any order with symbols a and b, the sequence $\mu(ba^kb|b)$ must be geometric (i.e., of the form γq^k for some γ, some q, and all k sufficiently large). We basically show that the same must hold for any metrically sofic measure which is the finite-to-one image of a Markov chain and satisfies the following independence condition (as the example does): for all k and ℓ sufficiently large,

$$\mu((ba^k)^\ell b|b) = (\mu(ba^kb|b))^\ell.$$

To be more precise:

Suppose that Σ_C is a SFT, $\varphi':\Sigma_C \to S$ is a finite-to-one factor map, and ν is a Markov measure with $\varphi'\nu = \varphi\mu$. We will produce a contradiction. Notice that we may as well assume that Σ_C and ν are one-step and φ' is a one-block map.

Let $k > 0$. One sees by direct computation that for all $\ell > 0$

$$\varphi\mu((ba^{k-1})^{\ell}b|b) = \left(\frac{1}{3}\frac{2}{3}^{k-1} + \frac{2}{3}\frac{1}{3}^{k-1}\right)^{\ell} .$$

By arguments similar to those of del Junco,et.al. (1981), it follows that if $\underline{x} \in \Sigma_C$ is any periodic point with period nk and $\varphi'(\underline{x}) = (ba^{k-1})^{\infty}$, then

$$(*) \qquad \nu(x_0 \dots x_{nk}|x_0)^{\frac{1}{n}} = \frac{1}{3}\frac{2}{3}^{k-1} + \frac{2}{3}\frac{1}{3}^{k-1} .$$

Now, fix $k \geq \# A(\Sigma_C) + 2$ and $\underline{x}$ as above (for some n). Observe that for each $i = 0,\dots,n-1$, the block

$$E_i = x_{ik+1} \dots x_{(i+1)k-1}$$

contains a repeated symbol whose image is a. Let f_i be the first appearance of such a symbol in E_i and $f_i + d_i$ the next appearance of the same symbol. Let $d = \text{l.c.m.}(d_0,\dots,d_{n-1})$. For each positive integer r, we construct a periodic point $\underline{y}^r$ from $\underline{x}$ as follows: replace E_i by the block

$$x_{ik+1} \dots x_{f_i - 1}(x_{f_i} \dots x_{f_i + d_i - 1})^{1+(rd)/d_i} x_{f_i + d_i} \dots x_{(i+1)k-1} .$$

Then $\underline{y}^r$ has the following properties:

(1) $\underline{y}^r$ has period $n(k+rd)$

(2) $\varphi'(\underline{y}^r) = (ba^{k-1+rd})^{\infty}$

(3) $\nu(y_0 \dots y_{n(k+rd)}|y_0)$ is geometric as a function of r.

But then, applying (*) above to the point $\underline{y}^r$ and its image under φ', we get

$$\frac{1}{3}\frac{2}{3}^{k-1+rd} + \frac{2}{3}\frac{1}{3}^{k-1+rd}$$

is geometric as a function of r. Thus, there exist α,β,γ and q such that for all r

$$\alpha\left(\left(\frac{2}{3}\right)^d\right)^r + \beta\left(\left(\frac{1}{3}\right)^d\right)^r = \gamma q^r .$$

But this is impossible: dividing through by q^r and letting $r \to \infty$ one sees that $q = \left(\frac{2}{3}\right)^d$ and $q = \left(\frac{1}{3}\right)^d$ also.

This example gives some necessary conditions for a metrically sofic measure to be the finite-to-one image of a Markov chain.

PROBLEM. Find sufficient conditions. A good answer to this problem would help understand which metrically sofic measures have "computable" entropies. As mentioned in Section 1, this would be important for computing capacities of channels.

3. LIFTING MEASURES

Let Σ_C and Σ_D be SFT's. If $\varphi:\Sigma_C \to \Sigma_D$ is a finite-to-one onto factor map, then every measure on Σ_D can be lifted uniquely to a Markov measure on Σ_C (Tuncel, 1981). If φ were *not* finite-to-one, one might guess that the lifting is still possible, even though the uniqueness doesn't even have a chance. The following is a sobering example: *There exist SFT's $\Sigma_C \to \Sigma_D$ such that the image of every Markov measure (of every order) on Σ_C is not both Markov (of any order) and fully supported on Σ_D.*

REMARK. Thus while every measure on Σ_D can be lifted to a measure on Σ_C (by Hahn-Banach), no fully supported Markov measure on Σ_D can be lifted to a Markov measure on Σ_C (fully supported or not).

The example is as follows: Let Σ_D be defined by the graph

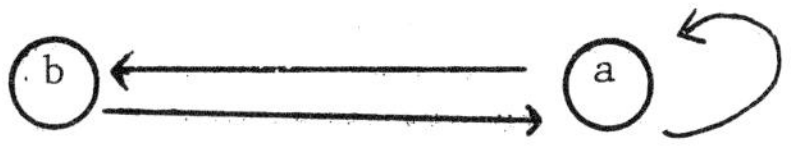

Let Σ_C be defined by the graph

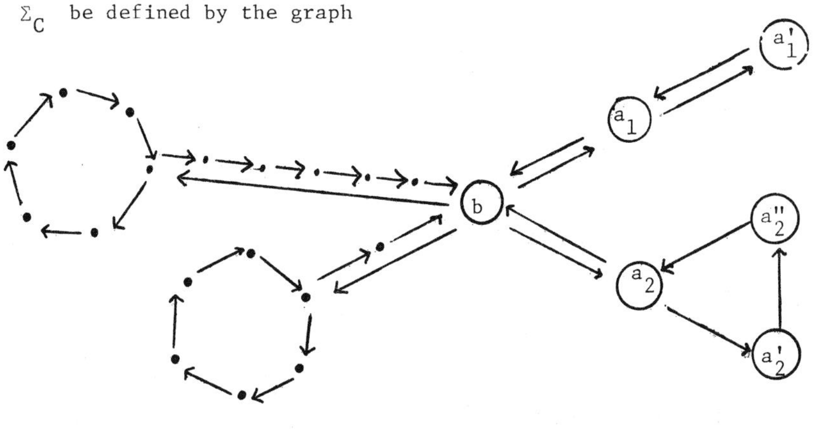

The states of Σ_C are $b, a_1, a_1', a_2, a_2', a_2''$ and the black dots above, who will remain anonymous. The map $\varphi:\Sigma_C \to \Sigma_D$ is defined by setting $\varphi(b) = b$ and for all other states α in Σ_C, $\varphi(\alpha) = a$.

To prove that images of Markov measures are not Markov, we compute measures of the blocks ba^ib (and their inverse images) for $i \equiv 3 \bmod 6$, then for $i \equiv 4 \bmod 6$, and finally for $i \equiv 1 \bmod 6$. For the first congruence class, $\varphi^{-1}(ba^ib)$ involves only the cycle $a_1'a_1$; for the second congruence class, $\varphi^{-1}(ba^ib)$ involves only the cycle $a_2'a_2''a_2$; for the third congruence class, $\varphi^{-1}(ba^ib)$ involves both cycles $a_1'a_1$ and $a_2'a_2''a_2$. This interference produces a contradiction.

We will use the following notations: Fix ν and μ Markov measures on Σ_C, Σ_D (respectively) so that $\varphi(\nu) = \mu$, ν is m-step and μ is n-step. Choose k with $6k > \max(n,m)$. Let

$$\Pi = \mu(ba^{6k+1})$$
$$\Pi_1 = \nu(ba_1(a_1'a_1)^{3k})$$
$$\Pi_2 = \nu(ba_2(a_2'a_2''a_2)^{2k}) .$$

We also set

$$Q = \mu(ba^{n+\ell+1} | ba^{n+\ell}),$$

which is constant for $\ell \geq 0$. Observe that

$$1 - Q = \mu(ba^{n+\ell}b | ba^{n+\ell}) \quad \text{for} \quad \ell \geq 0.$$

Likewise, define

$$p = \nu(ba_1(a_1'a_1)^{3k+\ell+1} | ba_1(a_1'a_1)^{3k+\ell}).$$

Then

$$1-p = \nu(ba_1(a_1'a_1)^{3k+\ell}b | ba_1(a_1'a_1)^{3k+\ell}) .$$

Also let

$$q = \nu(ba_2(a_2'a_2''a_2)^{2k+\ell+1} | ba_2(a_2'a_2''a_2)^{2k+\ell}) ,$$

so that

$$1 - q = \nu(ba_2(a_2'a_2''a_2)^{2k+\ell}b | ba_2(a_2'a_2''a_2)^{2k+\ell}).$$

We first show that

$$\Pi(1-Q) = \Pi_1(1-p) . \tag{1}$$

To see this, first observe that for all $i \geq 1$

$$\mu(ba^{6k+3+6i}b) = \Pi Q^{2+6i}(1-Q)$$

and

$$\nu(ba_1(a_1'a_1)^{3k+1+3i}b) = \Pi_1 p^{1+3i}(1-p) .$$

But since $\varphi(\mu) = \nu$, these two expressions are equal for all $i \geq 1$. Since the first expression grows geometrically with Q^6 and the second with p^3, it follows that $p = Q^2$, and so $\Pi(1-Q) = \Pi_1(1-p)$. Next, we show that

(2) $$\Pi(1-Q) = \Pi_2(1-q) \quad .$$

For this, note that

$$\mu(ba^{6k+4+6i}) = \Pi Q^{3+6i}(1-Q)$$

and

$$\nu(ba_2(a_2'a_2''a_2)^{2k+1+2i}b) = \Pi_2 q^{1+2i}(1-q) \quad .$$

Again, since $\varphi(\mu) = \nu$, these two expressions are equal for all $i \geq 1$; thus $q = Q^3$ and so $\Pi(1-Q) = \Pi_2(1-q)$.

Finally, we show that

(3) $$\Pi(1-Q) = \Pi_1(1-p) + \Pi_2(1-q) \quad .$$

Notice that (1), (2), and (3) imply that $\Pi(1-Q) = 2\Pi(1-Q)$, which is impossible if μ is fully supported. To prove (3), observe

$$\mu(ba^{6k+1}b) = \Pi(1-Q)$$

and

$$\nu(ba_1(a_1'a_1)^{3k}b) + \nu(ba_2(a_2'a_2''a_2)^{2k}b = \Pi_1(1-p) + \Pi_2(1-q) \quad .$$

Again, since $\varphi(\nu) = \mu$ and

$$\varphi^{-1}(ba^{6k+1}b) = ba_1(a_1'a_1)^{3k}b \cup ba_2(a_2'a_2''a_2)^{2k}b \ ,$$

we have the desired equality (3).

REMARK. There is no SFT $\Sigma_E \subset \Sigma_C$ such that $\varphi|\Sigma_E$ is finite-to-one and onto. (Otherwise, by Tuncel, 1981, one could do the lifting.)

However, we do have the following positive result in this direction.

PROPOSITION 3.1. Let S_1 and S_2 be two transitive sofic systems and $\varphi:S_1 \to S_2$ an (onto) factor map. Then there exists a transitive sofic system $S_0 \subset S_1$ such that $\varphi|_{S_0}$ is finite-to-one and onto.

PROOF. First, we show that we may assume that S_1 is a one-step SFT. To see this, recall that there is a transitive SFT Σ_A and a finite-to-one onto factor map $\theta:\Sigma_A \to S_1$ (Weiss 1973, Coven and Paul 1975). So, if we can produce a sofic $S_0 \subset \Sigma_A$ with $\varphi o\theta|_{S_0}$ finite-to-one and onto S_2, then $\varphi|_{\theta(S_0)}$ will be finite-to-one and onto, with $\theta(S_0)$ sofic as well. (The continuous image of any sofic system is, of course, still sofic.)

So, assume that S_1 is a one-step SFT and also that φ is a one-block map. For each symbol $i \in A(S_2)$, fix a linear ordering on the elements of $\varphi^{-1}(i)$. These orderings then determine a natural lexicographic ordering on the set $\varphi^{-1}(w)$ for each block w in S_2. Now, fix an n-block w in S_2 and a pair of symbols $a^*,\bar{a} \in A(S_1)$. Let $U(w,a^*,\bar{a})$ be the n-block $a_1 \ldots a_n$ in $\varphi^{-1}(w)$ which begins with a^*, ends with $\bar{a}$ and is smallest in the ordering on $\varphi^{-1}(w)$ (if such a word exists).

The set of blocks $\mathcal{U} = \{U(w,a^*,\bar{a})\}$ determines a subshift S_0 of S_1. S_0 consists of all bisequences all of whose subblocks are in $\mathcal{U}$. Notice that some blocks of $\mathcal{U}$ may not occur in any point of S_0.

For each block w in S_2, there is certainly some block s in S_0 for which $\varphi(s) = w$. Then, by the usual compactness argument, one sees that $\varphi|_{S_0}$ is onto. But also by construction $\varphi|_{S_0}$ has no "diamonds" (a diamond is a pair of different n-blocks in the domain which agree in the first and last coordinates and map via φ to the same block). This is sufficient for any factor map to be finite-to-one. (The number of points in any inverse image is then at most $(\#A(S_1))^2$.)

We must now show that S_0 is sofic. For each block s in $\mathcal{U}$, let

$$F(s) = \{t \text{ in } \mathcal{U} : st \text{ is in } \mathcal{U}\}.$$

$F(s)$ is not quite the set of all successors of s, of all lengths, in S_0. Nevertheless, to show that S_0 is sofic, it still suffices, as in Weiss (1973), to prove that $\{F(s) : s \text{ in } \mathcal{U}\}$ is finite.

Let $s = s_1 \ldots s_n$ and $t = t_1 \ldots t_m$ be in $\mathcal{U}$. Then st is <u>not</u> in $\mathcal{U}$ if and only if

(1) st is not in S_1

or

(2) there is a block r in S_1 of length $n+m$ with $r_1 = s_1$, $r_{n+m} = t_m$, $\varphi(r) = \varphi(st)$ and $r < st$.

The question is: how do (1) and (2) depend on the block s? Well, since S_1 is a one-step SFT, (1) only depends on s_n, and not on $s_1 \ldots s_{n-1}$. And (2) above depends only on the set

$$E(s) = \{i \in A(S_1) : \text{there exists a block } s' = s'_1 \ldots s'_n \text{ with } s'_1 = s_1, s'_n = i, \varphi(s'_1 \ldots s'_n) = \varphi(s_1 \ldots s_n) \text{ and } s' < s\}.$$

Thus, $F(s)$ depends only on s_n and $E(s)$. Since there are only finitely many choices for s_n and only finitely many choices for $E(s)$ (at most the number of subsets of $A(S_1)$), there are only finitely many $F(s)$. Therefore, S_0 is sofic.

Finally, observe that if S_0 is not transitive, then we can find a transitive component of maximal entropy (just find an irreducible component of maximal entropy for a SFT which factors finite-to-one onto S_0). This component is still sofic, the restriction of φ to this component is still obviously finite-to-one, and the restriction is still onto since its image has full entropy (see Coven and Paul 1974 for the latter).

COROLLARY 3.2. <u>Let S_1 and S_2 be transitive sofic systems and $\varphi: S_1 \to S_2$ a factor map. Let μ be a metrically sofic measure on S_2. Then there exists a metrically sofic measure ν on S_1 such that</u>

(1) $\varphi\nu = \mu$

and

(2) $h(\nu) = h(\mu)$.

PROOF. Let $S_0 \subset S_1$ as in Proposition 3.1. So, $\varphi: S_0 \to S_2$ is a finite-to-one factor map. Let Σ_A be a SFT with $\theta: \Sigma_A \to S_0$ a finite-to-one factor map. Since μ is metrically sofic, there is a one-step Markov measure m on a SFT Σ_B and a factor map $\alpha: \Sigma_B \to S_2$ with $\alpha(m) = \mu$. Now let

$$T = \{(x,y) \in \Sigma_A \times \Sigma_B : \varphi \circ \theta(x) = \alpha(y)\} .$$

It is easy to see that T is a SFT imbedded in $\Sigma_A \times \Sigma_B$.

Let ρ_1 and ρ_2 be the projections $\rho_1(x,y) = x$, $\rho_2(x,y) = y$. We have the commutative diagram

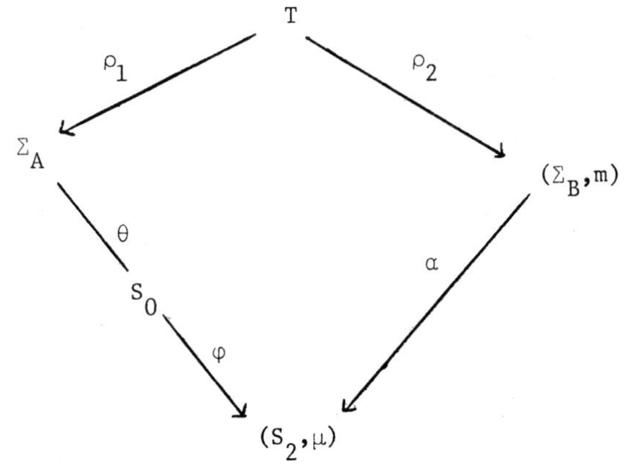

Since $\varphi o \theta$ is finite-to-one and onto, so is ρ_2. Thus, by Tuncel (1981), the measure m can be lifted to a Markov measure $\bar{m}$ on T:

$$\rho_2(\bar{m}) = m.$$

Thus,

$$\varphi\theta\rho_1(\bar{m}) = \alpha\rho_2(\bar{m}) = \alpha(m) = \mu,$$

and then $\theta\rho_1(\bar{m})$ is our desired measure ν. It is metrically sofic, since $\bar{m}$ is Markov. It has the same entropy as μ since $\varphi|_{S_0}$ is finite-to-one.

COROLLARY 3.3. Any channel with sofic input and deterministic noise has a metrically sofic measure of maximal transmission rate.

PROOF. Let μ be the measure of maximal entropy on S_2. Apply the preceding corollary to μ. (See Section 1.)

REFERENCES

1. L. M. Abramov, Entropy of induced automorphisms, Dokl. Akad. Nauk. SSSR 128 (1959), 647-650.

2. R. Adler, D. Coppersmith, M. Hassner, Algorithms for sliding block codes, to appear in IEEE Trans. - Information Theory.

3. R. Adler, B. Marcus, Topological entropy and equivalence of dynamical systems, Memoirs AMS 219 (1979).

4. P. Billingsley, Ergodic Theory and Information, J. Wiley, New York (1965).

5. J. Birch, Approximations for the entropy for functions of Markov chains, Ann. Math. Stat. 33 (1962), 930-938.

6. D. Blackwell, The entropy of functions of finite state Markov chains, Trans. 1st Prague Conf. (1957), 13-20.

7. L. Breiman, On achieving capacity in finite-memory channels, Illinois J. Math. 4 (1960), 246-252.

8. E. Coven, M. Paul, Finite procedures for sofic systems, Monats. Math. 83 (1977), 265-278.

9. E. Coven, M. Paul, Sofic systems, Israel J. Math. 20 (1975), 165-177.

10. E. Coven, M. Paul, Endomorphisms of irreducible SSFT, Math. Systems Theory 8 (1974), 167-175.

11. H. Furstenberg, Stationary Processes and Prediction Theory, Annals of Math. Studies No. 44, Princeton University Press, Princeton (1960).

12. G. A. Hedlund, Endomorphisms and automorphisms of the shift dynamical system, Math. Systems Theory 3 (1969), 320-375.

13. A. del Junco, M. Keane, B. Kitchens, B. Marcus, L. Swanson, Continuous homomorphisms of Bernoulli schemes, Ergodic Theory and Dynamical Systems II, Proc. Spec. Yr., Univ. of Maryland (1981), 91-111, Birkhäuser, Boston.

14. W. Parry, Intrinsic Markov chains, Trans. Amer. Math. Soc. 112 (1964), 55-66.

15. W. Parry, A finite classification of topological Markov chains and sofic systems, Bull. London Math. Soc. 9 (1977), 86-92.

16. C. Shannon, Mathematical theory of communication, Bell Syst. Tech. J. 27 (1948), 379-423; 623-656.

17. S. Tuncel, Conditional pressure and coding, Israel J. Math. 39 (1981), 101-112.

18. P. Walters, Ergodic Theory - Introductory Lectures, Lecture Notes in Math. 458, Springer-Verlag, Berlin (1975).

19. B. Weiss, Subshifts of finite type and sofic systems, Monats. Math. 77 (1973), 462-474.

DEPARTMENT OF MATHEMATICS
UNIVERSITY OF NORTH CAROLINA
CHAPEL HILL, NC 27514

Contemporary Mathematics
Volume 26, 1984

ELLIS GROUPS AND COMPACT RIGHT TOPOLOGICAL GROUPS

I. Namioka

ABSTRACT. We give a concrete representation of the Ellis group of a certain distal flow on the 2-torus. We also describe examples of compact Hausdorff right topological groups that are not topological groups.

A compact flow (S,X) consists of a compact Hausdorff space X and a semigroup (under the composition) S of continuous maps $X \to X$. The semigroup S is clearly a subsemigroup of the semigroup X^X of all maps (not necessarily continuous) of X into itself. It is easy to see that the closure Σ of S in X^X with respect to the product topology is again a semigroup, which is called the enveloping semigroup of the flow (S,X). We shall always assume that the enveloping semigroup Σ is provided with the relativization of the product topology. Then Σ is a compact Hausdorff space, and the map $\sigma \mapsto \sigma\tau$: $\Sigma \to \Sigma$ is continuous for each τ in Σ and the map $\sigma \mapsto \tau\sigma$: $\Sigma \to \Sigma$ is continuous for each τ in S.

A compact flow (S,X) is said to be distal if $x = y$ whenever $x,y \in A$ and $\lim_\alpha s_\alpha(x) = \lim_\alpha s_\alpha(y)$ for some net s_α in S. Ellis [2] proved that a compact flow (S,X) is distal if and only if the enveloping semigroup Σ of the flow is a group in which the identity element is the identity map. The enveloping semigroup of a distal flow is called the Ellis group of the flow. One of the aims of the present note is to compute explicitly the Ellis group of one specific distal flow.

The flow we are interested in is described by Furstenberg [3]. Let $\mathbb{T} = \mathbb{R}/\mathbb{Z}$ be the 1-torus, and let ξ be an arbitrary member of $\mathbb{T}$. The group operation in $\mathbb{T}$ will be written additively. Define a continuous map $T_\xi : \mathbb{T}^2 \to \mathbb{T}^2$ by $T_\xi(x,y) = (\xi + x, x + y)$, and let S_ξ denote the semigroup of iterates of T_ξ, i.e. $S_\xi = \{T_\xi^n : n = 1,2,\ldots\}$. Then $(S_\xi,\mathbb{T}^2)$ is a distal flow (cf. [3]). In what follows, we determine the Ellis group of $(S_\xi,\mathbb{T}^2)$ in case ξ is an irrational member of $\mathbb{T}$.

Let G be a compact Abelian topological group with the normalized Haar measure μ. A sequence $\{x_n\}$ in G is said to be uniformly distributed (in G) if for each complex valued continuous function f on G,

$$\lim_{N\to\infty} \frac{1}{N} \Sigma_{n=1}^{N} f(x_n) = \int_G f\, d\mu.$$

It is clear that if a sequence $\{x_n\}$ is uniformly distributed in G then it is dense in G. We need the following slightly generalized version of a celebrated theorem of Weyl [7].

THEOREM 1. *Let $\{x_n\}$ be a sequence in a compact Abelian topological group G such that, for each continuous character χ on G not identically equal to 1, $\lim_{N\to\infty} \frac{1}{N} \Sigma_{n=1}^{N} \chi(x_n) = 0$. Then $\{x_n\}$ is uniformly distributed, and hence dense, in G.*

For an Abelian group $(G,+)$, let $E(G)$ denote the set of all endomorphisms $G \to G$. For $f,g \in E(G)$, define $f + g: G \to G$ by $(f + g)(x) = f(x) + g(x)$. Then $f + g \in E(G)$ and $(E(G),+)$ is an Abelian group. If G is a compact Abelian group, then $E(G)$ is a closed (hence compact) subset of the product space G^G, and under this topology $E(G)$ is a compact Abelian group.

LEMMA 2. *Let η be an irrational member of $\mathbb{T}$, let $f \in E(\mathbb{T})$ and let $u \in \mathbb{T}$. Then there is a net $\{n_\gamma\}$ of positive integers such that*

i) $\lim_\gamma n_\gamma x = f(x)$ *for each* x *in* $\mathbb{T}$, *and*

ii) $\lim_\gamma n_\gamma^2 \eta = u$.

PROOF. Let $G = E(\mathbb{T}) \times \mathbb{T}$. Then, as noted above, G is a compact Abelian group. For each positive integer n, let g_n denote the member of $E(\mathbb{T})$ given by $g_n(x) = nx$ for each x in $\mathbb{T}$. Then the conclusion of the lemma is equivalent to the fact that the sequence $\{(g_n, n^2\eta): n = 1,2,3,\ldots\}$ is dense in G. We use Weyl's theorem to prove the latter.

The character group $\hat{G}$ can be identified (algebraically) with the product group $\mathbb{T} \times \mathbb{Z}$. Under this identification, the pairing of G and $\hat{G}$ can be expressed as follows: for $(f,x) \in G = E(\mathbb{T}) \times \mathbb{T}$ and $(y,m) \in \mathbb{T} \times \mathbb{Z}$,

$$<(f,x),(y,m)> = e(f(y) + mx),$$

where $e: \mathbb{T} \to \mathbb{C}$ is the map induced by the map $t \mapsto \exp(2\pi i t): \mathbb{R} \to \mathbb{C}$. In view of Theorem 1, it is sufficient to prove that: whenever $(y,m) \in \mathbb{T} \times \mathbb{Z}$ and $(y,m) \neq (0,0)$,

$$(*) \qquad \lim_{N\to\infty} \frac{1}{N} \Sigma_{n=1}^{N} <(g_n,n^2\eta),(y,m)> = \lim_{N\to\infty} \frac{1}{N} \Sigma_{n=1}^{N} e(ny + mn^2\eta) = 0.$$

Case (1). $m \neq 0$. In this case, Satz 9 of Weyl [7] shows that (*) is valid because $m\eta$ is irrational.

Case (2). $m = 0$. In this case, $y \neq 0$ or equivalently $e(y) \neq 1$. Hence $|\Sigma_{n=1}^{N} e(ny + mn^2\eta)| = |\Sigma_{n=0}^{N-1} e(ny)| = \frac{|1 - e(Ny)|}{|1 - e(y)|} \leq \frac{2}{|1 - e(y)|}$, and consequently (*) holds in this case too. This completes the proof of the lemma.

Let us now fix an irrational member ξ in $\mathbb{T}$, and we choose one irrational member η of $\mathbb{T}$ such that $2\eta = \xi$. Then for each positive integer n,

$$T_\xi^n(x,y) = (n\xi + x, \tfrac{1}{2}n(n-1)\xi + nx + y) = (n\xi + x, (n^2 - n) \quad + nx + y)$$
$$= (n\xi + x, n^2\eta + n(x - \eta) + y).$$

Let σ be an arbitrary element of the Ellis group Σ_ξ of the distal flow $(S_\xi, \mathbb{T}^2)$. Then there is a net $\{n_\eta\}$ of positive integers such that

$T_\xi^{n_\gamma}(x,y) = (n_\gamma\xi + x, n_\gamma^2\eta + n_\gamma(x - \eta) + y) \to \sigma(x,y)$ for each (x,y) in $\mathbb{T}^2$. Since $E(\mathbb{T})$ and $\mathbb{T}$ are compact, by taking a subnet if necessary, we may assume that there exists a pair (f,u) in $E(\mathbb{T}) \times \mathbb{T}$ such that

(a) $\lim_\gamma n_\gamma z = f(z)$ for each z in $\mathbb{T}$, and

(b) $\lim_\gamma n_\gamma^2 \eta = u$.

Then σ can be written as follows:

$$\sigma(x,y) = (f(\xi) + x, u + f(x - \eta) + y) = (f(\xi) + x, u - f(\eta) + f(x) + y).$$

This formula prompts us to define a map $\rho\colon E(\mathbb{T}) \times \mathbb{T} \to (\mathbb{T}^2)^{\mathbb{T}^2}$ as follows: for (f,v) in $E(\mathbb{T}) \times \mathbb{T}$ and (x,y) in $\mathbb{T}^2$,

$$\rho(f,v)(x,y) = (f(\xi) + x, v + f(x) + y).$$

Then the above computation shows that $\Sigma_\xi \subset \operatorname{Im}\rho$. We note in passing that $T_\xi = \rho(\mathrm{id}, 0)$. On the other hand, by lemma 2, given an arbitrary pair (f,v) in $E(\mathbb{T}) \times \mathbb{T}$, there is a net $\{n_\gamma\}$ of positive integers that satisfies (a) and (b) above with $u = v + f(\eta)$. Then $T_\xi^{n_\gamma}(x,y) \to \rho(f,v)(x,y)$ for all (x,y) in $\mathbb{T}^2$. We have thus shown that $\Sigma_\xi = \operatorname{Im}\rho$. The map ρ is obviously continuous, and it is easily seen that ρ is one-to-one. Hence ρ is a homeomorphism.

To see what happens to the group operation of Σ_ξ under ρ, let $(f,v),(g,w) \in E(\mathbb{T}) \times \mathbb{T}$. Then for each (x,y) in $\mathbb{T}^2$,

$$[\rho(f,v)\rho(g,w)](x,y) = \rho(f,v)(g(\xi) + x, w + g(x) + y)$$
$$= (f(\xi) + g(\xi) + x, v + f(g(\xi) + x) + w + g(x) + y)$$
$$= \rho(f + g, v + w + f \circ g(\xi))(x,y).$$

The following theorem is now proved.

THEOREM 3. *Let* ξ *be an irrational member of* $\mathbb{T}$. *Then the Ellis group* Σ_ξ *of the distal flow* $(S_\xi, \mathbb{T}^2)$ *is homeomorphically isomorphic to the compact Hausdorff space* $E(\mathbb{T}) \times \mathbb{T}$ *provided with the group operation given by the following formula: for* $(f,v),(g,w)$ *in* $E(\mathbb{T}) \times \mathbb{T}$,

$$(f,v) \cdot (g,w) = (f + g, v + w + f \circ g(\xi)).$$

Under this isomorphism T_ξ corresponds to (id,0) in $E(\mathbb{T}) \times \mathbb{T}$.

REMARKS. (1) The computation given above contradicts a remark (stated without a proof) in p. 486 of Furstenberg [3], according to which the Ellis group Σ_ξ consists of all transformations $\sigma: \mathbb{T}^2 \to \mathbb{T}^2$ of the form $\sigma(x,y) = (\alpha + x, \varphi(x) + y)$ where $\alpha \in \mathbb{T}$ and $\varphi: \mathbb{T} \to \mathbb{T}$ (not necessarily continuous) are such that $\phi(z + \xi) = \phi(z) + \alpha$ for all z in $\mathbb{T}$. However, for such σ to be in the image of ρ (and hence in Σ_ξ), it is necessary and sufficient that the map φ is affine, i.e. $\varphi(z) = v + f(z)$ for some (f,v) in $E(\mathbb{T}) \times \mathbb{T}$. The set of all maps $\varphi:\mathbb{T} \to \mathbb{T}$ for which $z \mapsto \varphi(z + \xi) - \varphi(z)$ is constant is considerably larger than the set of all affine maps: $\mathbb{T} \to \mathbb{T}$.

(2) On account of Lemma 2, one might conjecture (in a moment of weakness) that given $(f,g) \in E(\mathbb{T}) \times E(\mathbb{T})$ there is a net $\{n_\gamma\}$ of positive integers such that $\lim_\gamma n_\gamma x = f(x)$ and $\lim_\gamma n_\gamma^2 x = g(x)$ for each x in $\mathbb{T}$. This conjecture is false. For suppose $f(x) = g(x) = 2x$ and $\{n_\gamma\}$ is a net such that $n_\gamma \frac{1}{3} \to f(\frac{1}{3}) = \frac{2}{3}$. Then eventually $n_\gamma \equiv 2 \bmod 3$. But then $n_\gamma^2 \equiv 1 \bmod 3$ eventually, and therefore $n_\gamma^2 \frac{1}{3} \to \frac{1}{3} \neq g(\frac{1}{3})$. This fact prevents us from determining the Ellis group of the distal flow on $\mathbb{T}^3$ generated by the map $(x,y,z) \to (\xi + x, x + y, y + z)$ by the method similar to the one for $(S_\xi, \mathbb{T}^2)$.

A <u>compact Hausdorff right topological group</u> [5] is a group G provided with a compact Hausdorff topology such that $x \mapsto xy: G \to G$ is continuous for each y in G. Following Ruppert [6], we let $\Lambda(G)$ be the set of all y in G such that the map $x \mapsto yx: G \to G$ is continuous. As noted earlier, the Ellis group Σ of a distal flow (S,X) is a compact Hausdorff right topological group such that $\Lambda(\Sigma)$ is dense in Σ. Conversely, if G is a compact Hausdorff right topological group, then $\Lambda(G)$ is a subgroup of G, $\Lambda(G)$ acts on G by the left multiplications and the closure $\overline{\Lambda(G)}$ of $\Lambda(G)$ in G is homeomorphically isomorphic to the Ellis group of the distal flow $(\Lambda(G),G)$.

Let ξ be an irrational member of $\mathbb{T}$, and let Σ_ξ be the Ellis group of the flow $(S_\xi, \mathbb{T}^2)$. In [4], Milne proved that $\Lambda(\Sigma_\xi)$ is properly larger than the cyclic subgroup of Σ_ξ generated by T_ξ. As a corollary to Theorem 3, we can determine $\Lambda(\Sigma_\xi)$ more precisely.

COROLLARY 4. <u>Let ξ and Σ_ξ be as above. Then the group $\Lambda(\Sigma_\xi)$ is isomorphic to $\mathbb{Z} \times \mathbb{T}$ provided with the multiplication rule:</u>

$$(m,v) \cdot (n,w) = (m + n, v + w + mn\xi).$$

<u>Under this isomorphism the map T_ξ corresponds to (1,0) in $\mathbb{Z} \times \mathbb{T}$.</u>

PROOF. Via the isomorphism of Theorem 3, the discussion of Σ_ξ can be transferred to that of $E(\mathbb{T}) \times \mathbb{T}$ with the multiplication rule given in

Theorem 3. An element (f,v) of $E(\mathbb{T}) \times \mathbb{T}$ belongs to $\Lambda(E(\mathbb{T}) \times \mathbb{T})$ if and only if $(g,w) \mapsto (f + g, v + w + f \circ g(\xi))$ is continuous, and this is the case if and only if $g \mapsto f \circ g(\xi): E(\mathbb{T}) \to T$ is continuous. Since ξ is irrational, the map $g \mapsto g(\xi): E(\mathbb{T}) \to T$ is onto, and hence the above condition is equivalent to f's being continuous, i.e. for some integer n, $f(x) = nx$ for all x in $\mathbb{T}$. The corollary follows from these remarks and Theorem 3.

Compact Hausdorff right topological groups G that are not topological groups (i.e. $\Lambda(G) \neq G$) arise naturally as the Ellis groups of non-equicontinuous distal flows. However only a few explicit descriptions of such groups are recorded in the literature (see e.g. [1]). By generalizing the representation of Σ_ξ in Theorem 3, we can give further examples.

EXAMPLES OF COMPACT HAUSDORFF RIGHT TOPOLOGICAL GROUPS.

Let G be a compact Hausdorff Abelian topological group, and let $R = E(G)$. Then as observed above, $(R,+)$ is a compact Hausdorff Abelian topological group. Furthermore, R is a ring with a unit, where the ring multiplication is the composition of maps, and the map $x \mapsto xy: R \to R$ is continuous for each y in R. Let $\Lambda(R)$ denote the set of all x in R such that $y \mapsto xy: R \to R$ is continuous.

For each positive integer n, let R^n denote the n-fold cartesian product of the R's, and let R^∞ denote the countable product of the R's. One can then define group operations on R^n and R^∞ as follows:

$$(x \cdot y)_k = x_k + x_{k-1}y_1 + x_{k-2}y_2 + \dots + x_1 y_{k-1} + y_k,$$

whenever $x,y \in R^n$ and $1 \le k \le n$, or whenever $x,y \in R^\infty$ and $k = 1,2,\dots$. It is easy to see that R^n $(n = 1,2,\dots)$ and R^∞ are compact Hausdorff right topological groups with respect to the product topology, and

$\Lambda(R^1) = R^1$ (in fact, R^1 is the topological group $(R,+)$),
$\Lambda(R^n) = (\Lambda(R))^{n-1} \times R$ $(n \ge 2)$, and
$\Lambda(R\ ) = (\Lambda(R))^\infty$.

For $n \ge 2$, we have the exact sequence:

$$0 \to R \overset{i}{\to} R^n \overset{p}{\to} R^{n-1} \to 1,$$

where $i(x) = (0,0,\dots,x)$ and $p((x_1,\dots, x_{n-1},x_n)) = (x_1,\dots,x_{n-1})$, i.e. R^n is an extension*) of R by R^{n-1}. It is straightforward to check that this extension is given by the trivial action of R^{n-1} on R and a cocycle $c: R^{n-1} \times R^{n-1} \to R$, where $c(x,y) = \Sigma_{i=1}^{n-1} x_i y_{n-i}$. Let $(A,+)$ be a compact

---

*We acknowledge gratefully the enlightening discussions with Prof. R. Warfield on the matter of group extensions.

Hausdorff Abelian topological group, and let $\varphi: (R,+) \to (A,+)$ be a continuous homomorphism. Then $\varphi\circ c: R^{n-1} \times R^{n-1} \to A$ is also a cocylce with the trivial action of R^{n-1} on A, and consequently it gives rise to an extension Γ of A by R^{n-1}. As a (topological) space $\Gamma = R^{n-1} \times A$, and the group operation in Γ is given by the formula:

$$(x,a) \cdot (y,b) = (x \cdot y, a + b + \Sigma_{i=1}^{n-1} \varphi(x_i y_{n-i})).$$

Clearly Γ is another example of compact Hausdorff right topological space.

As a special case, let $R = E(\mathbb{T})$ and let ξ be a fixed element in $\mathbb{T}$. Then we can define a homomorphism $\varphi: R \to \mathbb{T}$ by $\varphi(f) = f(\xi)$. Then the multiplication rule given above turns $E(\mathbb{T})^n \times \mathbb{T}$ into a compact Hausdorff right topological group. For $n = 1$ and 2, the explicit multiplication rules are respectively:

$$(f,u) \cdot (g,v) = (f + g, u + v + f \circ g(\xi)), \quad \text{and}$$

$$(f_1,f_2,u)\cdot(g_1,g_2,v) = (f_1 + g_1, f_2 + f_1\circ g_1 + g_2,\ u + v + f_1\circ g_2(\xi) + g_2\circ f_1(\xi))$$

The first formula is, of course, the one given in Theorem 3.

References

1. J. F. Berglund, H. D. Junghenn and P. Milnes, Compact right topological semigroups and generalizations of almost periodicity, Lecture Notes in Math., vol. 663, Springer-Verlag, Berlin-Heidelberg-New York, 1978.

2. R. Ellis, Distal transformation groups, Pacific J. Math., 8(1958), 401-405.

3. H. Furstenberg, The structure of distal flows, Amer. J. Math., 85(1963), 477-515.

4. P. Milnes, Continuity properties of compact right topological groups, Math. Proc. Cambridge Philos. Soc., 86(1979), 427-435.

5. I. Namioka, Right topological groups, distal flows and a fixed point theorem, Math. Systems Theory, 6 (1972), 193-209.

6. W. Ruppert, Rechtstopologische Halbgruppen, J. reine angew. Math., 261 (1973), 123-133.

7. H. Weyl, Uber die Gleichverteilung von Zahlen mod. Eins., Math. Ann., 77(1916), 313-352.

Department of Mathematics
University of Washington
Seattle, Washington 98195 U.S.A.

Contemporary Mathematics
Volume **26**, 1984

INVARIANTS OF FINITARY ISOMORPHISMS WITH FINITE EXPECTED CODE-LENGTHS

William Parry* and Klaus Schmidt

1. INTRODUCTION

Throughout this talk we shall concentrate on stationary Markov shifts with a finite number of states.

Let $X = \prod_{n=-\infty}^{\infty} \{1,2,\ldots k\}$ be the compact zero-dimensional space of doubly infinite sequences of symbols chosen from $1,2,\ldots,k$, and let P be a $k \times k$ irreducible non-negative matrix with maximum eigenvalue equal to 1. Let $\bar{p}$, p be corresponding left and right invariant vectors ($\bar{p}P = \bar{p}$, $Pp = p$) such that $\bar{p} > 0$, $p > 0$ and $\sum_{i=1}^{k} \bar{p}_i p_i = 1$.

Define $X_P = \{x \in X : P(x_n, x_{n+1}) > 0, n \in Z\}$ then $T_P X_P = X_P$ where $(T_P x)_n = x_{n+1}$ is the shift. The Markov probability

$$\begin{aligned} m_P[i_{-m},\ldots,i_o,\ldots,i_n]_{-m} &= m_P\{x \in X_P : x_{-m} = i_{-m},\ldots,x_n = i_n\} \\ &= \bar{p}_{i_{-m}} P(i_{-m}, i_{-m+1}) \ldots P(i_{n-1}, i_n) p_{i_n} \end{aligned}$$

is T_P invariant. T_P is called a Markov shift defined on the probability space (X_P, m_P).

Two Markov shifts T_P, T_Q are measure-theoretically isomorphic if there exists an invertible measure-preserving transformation ϕ of almost all X_P onto almost all of X_Q such that $\phi T_P = T_Q \phi$ a.e. The isomorphism ϕ is called finitary if for each state j of Q $\phi^{-1}[j]_o$ is a.e. a countable union of P cylinders and for each state i of P $\phi[i]_o$ is a.e. a countable union of Q cylinders.

The main results concerning isomorphism are:

FRIEDMAN AND ORNSTEIN [2]. *If P,Q have the same period and if T_P, T_Q have the same entropy then T_P and T_Q are isomorphic.*

KEANE AND SMORODINSKY [3]. *If P,Q have the same period and if T_P, T_Q have the same entropy then T_P and T_Q are finitarily isomorphic.*

*Presented by the first author.

If ϕ is a finitary isomorphism then for almost all $x \in X_P$ there exist positive integral valued functions a, m such that $\phi(x_o)$ is known once $x_{-m}\ldots x_o \ldots x_a$ are known. In other words $\phi(x)_o = \phi(y)_o$ if $x_i = y_i$, $-m \le i \le a$. Of course ϕ^{-1} shares a similar property. We define $a_\phi(x) = a$ (anticipation) and $m_\phi(x) = m$ (memory). We say that ϕ has *finite expected code-length* if $\int (a_\phi + m_\phi)dm < \infty$.

We shall consider a finitary isomorphism between T_P and T_Q satisfying the following *Hypothesis F.E.*: ϕ and ϕ^{-1} have finite expected code-lengths.

In [7] the first author showed that there are a number of obstructions to this hypothesis and in [5] Krieger constructed a group invariant which he denoted by Δ_P.

Our main aim is to give an alternative proof that Krieger's group is an invariant (under the above hypothesis) and construct other related invariants as well.

In order to specify $\phi(x)_{-1} = (\phi T_P^{-1}x)_o$ we have to look $m_\phi(T_P^{-1}x)$ units to the right of x_{-1} and $m_\phi(T_P^{-1}x)$ units to the left of x_{-1} i.e. we look at $a_\phi(T_P^{-1}x) - 1$ units to the right of x_o and $m_{1\phi}(T_P^{-1}x) + 1$ to the left of x_o. Hence to specify $\phi(x)_{-1}, \phi(x)_o$ we look $\bigvee_{n=0}^{1} a_\phi(T_P^{-n}x) - n$ to the right of x_o and $\bigvee_{n=0}^{1} m_\phi(T_P^{-n}x) + n$ to the left of x_o.

Continuing in this way we see that $a_\phi^*(x) = \bigvee_{n=0}^{\infty} a_\phi(T^{-n}x) - n$ units to the right of x_o and the entire left sequence are needed in order to specify

$$(\ldots.\phi(x)_{-2}, \phi(x)_{-1}, \phi(x)_o).$$

In a similar way $m_\phi^*(x) = \bigvee_{n=0}^{\infty} m(T_x^n) - n$ units to the left of x_o and the entire right sequence are needed in order to specify

$$(\phi(x)_o, \phi(x)_1, \ldots).$$

It is important to note that our hypothesis ensures that $a_\phi^*, m_\phi^*, a_{\phi-1}^*, m_{\phi-1}^*$ are finite a.e.

Further analysis along these lines reveals that there exists a set of measure zero E such that if $x, y \in X_P - E$ and $x_n = y_n$ for $|n| \ge M$ then $\phi(x)_n = \phi(y)_n$ for $|n| \ge N$ (where N depends on x,y,M). In other words if we define the equivalence relation $x \sim y$ when $x_n = y_n$ for all but finitely many n then

PROPOSITION. Under the hypothesis that ϕ and ϕ^{-1} have finite expected code-lengths we have (neglecting sets of measure zero) $x \sim y$ if and only if $\phi(x) \sim \phi(y)$.

This observation is due to Krieger who defined Δ_P to be the multiplicative group consisting of all ratios

$$\frac{P(i_o,i_1)\cdots P(i_{n-1}i_o)}{P(i_o,j_1)\cdots P(j_{n-1}i_o)}$$

(where the $P(i_r,i_{r+1}) \neq 0$)

$$\text{i.e.}\quad \Delta_P = \left\langle \frac{P(i_o,j_1)\cdots P(i_{n-1},i_o)}{P(i_o,j_1)\cdots P(j_{n-1},i_o)} \right\rangle .$$

He used the above proposition to show that $\Delta_P = \Delta_Q$ when ϕ, ϕ^{-1} have finite expected code-lengths.

2. THE INFORMATION COCYCLE

We propose to extend Krieger's result by examining the information cocycles $I_P(A_P|T_P^{-1}A_P) = I_P$ and $I_Q(A_Q|T_Q^{-1}A_Q) = I_Q$. These were used previously in [7] where it was proved (under the finite expected code-length hypothesis) that

$$* \qquad I_P = I_Q \circ \phi + f \circ T_P - f.$$

Here f is the real valued function

$$f = I(A_P|\phi^{-1}A_Q) - I(\phi^{-1}A_Q|A_P).$$

For reasons which will become clear later, we need to show that f assumes (almost surely) a countable number of values. Since T_P is ergodic and I_P, I_Q assume only finitely many values, it is enough to show that f is constant on some set of positive measure. The proof that f has this property is indirect and involves extending * to an equation involving a larger group than $\{T_P^n\}$. An information cocycle recently introduced by Butles and Schmidt in [1] fits our requirements exactly.

Quite generally let G be a discrete group of measure preserving transformations of the probability space $(\Omega,\mathcal{B},\mu)$ and let A be a sub-σ-algebra such that $I(A|g^{-1}A) < \infty$ a.e. and $I(g^{-1}A|A) < \infty$ (a.e.) i.e $gA \underset{I}{\sim} A$ for all $g \in G$.

Define $J_\mu(A,g) = I(A|g^{-1}A) - I(g^{-1}A|A)$ then

(1) $J_\mu(A,\cdot)$ is a cocycle i.e.

$$J_\mu(A,gh) = J_\mu(A,g) \circ h + J_\mu(A,h).$$

(2) If A' is another sub-σ-algebra such that $A' \underset{I}{\sim} A$ then $J_\mu(A,\cdot)$ and $J_\mu(A',\cdot)$ are cohomologous i.e. $J_\mu(A,g) - J_\mu(A',g) = f \circ g - f$ for some real valued f. In fact

$$f = I(A|A') - I(A'|A).$$

In the Butler-Schmidt theory one need only assume that G is a group of non-singular transformations in which case

$$-\log E\left(\frac{dmg^{-1}}{dm}\Big|\mathcal{A}\right) \circ g$$

has to be added in the definition of $J_\mu(\mathcal{A},g)$. However, we shall only need the restricted version.

The groups we shall consider are G_P, G_Q where G_P is the group generated by G'_P, T_P and G'_P is the group of all locally finite dimensional measure preserving transformations. A measure preserving transformation S of (X_P,m_P) is said to be locally finite dimensional if (almost surely) $(Sx)_n = x_n$ for $|n| \geq N$ depending on x i.e. if $Sx \sim x$.

It is not difficult to see that for almost all x the G'_P orbit of x is the $\sim$ equivalence class of x. This being the case Proposition 1 shows that $\phi^{-1}G'_Q\phi = G'_P$ and $\phi^{-1}G_Q\phi = G_P$. And since $\phi^{-1}\mathcal{A}_Q \underset{I}{\sim} \mathcal{A}_P$ we have

PROPOSITION 2. <u>If</u> ϕ, ϕ^{-1} <u>have finite expected code-lengths then</u>

$$J(\mathcal{A}_P,g) = J(\phi^{-1}\mathcal{A}_P,g) + f \circ g - f$$

<u>for all</u> $g \in G_P$, <u>where</u>

$$f = I(\mathcal{A}_P|\phi^{-1}\mathcal{A}_Q) - I(\phi^{-1}\mathcal{A}_Q|\mathcal{A}_P).$$

We conclude this section with a sketch of the proof of our main result:

THEOREM. <u>The function</u> f <u>assumes a countable number of values on a set of measure one. Equivalently,</u> f <u>is constant on some set of positive measure.</u>

SKETCH OF PROOF.

For any cylinder $C = [i_{-N}\cdots i_N]_{-N}$ let H_C^-, H_C^+ be the following groups of *uniformly* finite dimensional measure preserving homeomorphisms: $S \in H_C^-$ means that S keeps all co-ordinates x_n fixed if $-N \leq n$ or $n \leq -M$ where $M > N$ depends on S but not on x. $S \in H_C^+$ means that S keeps all co-ordinates x_n fixed if $n \leq N$ or $n \geq M$ where $M > N$ again depends on S but not on x.

We note that C is $H_C^- \cdot H_C^+$ invariant and the first step is to prove that this group acts ergodically on C. The proof is very similar to the proof of the Hewitt and Savage theorem [4] on exchangeable events.

Choose $C = [i_{-N}\cdots i_0\cdots i_N]_{-N}$ such that $D = C \cap \{x:a^*(x) \leq N, m^*(x) \leq N\}$ has positive measure and note that for $x,gx \in D$, $g \in H_C^- \cdot H_C^+$ one has $J(\mathcal{A}_P,g)(x) = 0$ and $J(\phi^{-1}\mathcal{A}_Q,g) = 0$ so that $f(gx) = f(x)$.

For $\varepsilon > 0$ let $A = \{x \in D: f(x) \in [\alpha-\varepsilon,\alpha+\varepsilon]\}$ have positive measure, and let $B = D-A$. By ergodicity if $m_P(B) > 0$ then for almost all $x \in A$ there exists $g \in H_C^- \cdot H_C^+$ such that $gx \in B$. Hence $f(gx) = f(x)$ contradicting the definition of A. In other words on D the values of f are within ε of α for some α. Hence f is a.e. constant on D, and the sketch of the proof is complete.

3. THE GROUP INVARIANTS

As usual we assume that ϕ is a finitary isomorphism with ϕ and ϕ^{-1} having finite expected code-lengths. In view of Theorem 1 and the equation *, we have

$$I_P = I_Q \circ \phi + f \circ T_P - f \quad \text{a.e.}$$

where f assumes a countable number of values. Since T_P and T_Q are Markov shifts we know that

$$\begin{aligned} I_P(x) &= I(A_P | T_P^{-1} A_P)(x) \\ &= \log (\bar{p}_{x_1} p_{x_1} / \bar{p}_{x_o} P(x_o, x_1) p_{x_1}) \end{aligned}$$

with a similar expression for I_Q. The equation * can therefore be written as

$$** \qquad P(x_o, x_1) = Q(\phi(x)_o, \phi(x)_1) \cdot g(T_P x)/g(x) \qquad \text{a.e.}$$

where g assumes a countable number of values.

Define $\Gamma_P = \langle P(i_o, i_1) \ldots P(i_{n-1}, i_o) \rangle$ (the multiplicative group generated by such non-zero products), then clearly $\Gamma_P \supset \Delta_P$.

THEOREM 2. $\Gamma_P = \Gamma_Q$, $\Delta_P = \Delta_Q$ and (if P,Q are aperiodic) there exist canonical cosets $C_P \Delta_P = C_Q \Delta_Q$ such that $C_P \Delta_P$ is a generator for the cyclic group Δ_P / Δ_P.

PROOF. There exists a positive vector r such that $P(i,j) r_j / r_i \in \Gamma_P$. To see this, for i,j, choose sequences so that $P(j,x_1)P(x_1 x_2) \ldots P(x_a, 1) > 0$ and $P(i,y_1)P(y_1,y_2) \ldots P(y_b, 1) > 0$. Let r_j, r_i be these respective products. Then

$$P(i,j) r_j / r_i = P(i,j) \frac{P(j,x_1) \ldots P(x_a, 1)}{P(i,y_1) \ldots P(x_b, 1)} .$$

Now premultiply top and bottom by the same product involving a path from 1 to i. We see that $P(i,j) r_j / r_i \in \Gamma_P$. Consequently there is no loss in generality in assuming that $P(i,j) \in \Gamma_P$ and similarly $Q(i,j) \in \Gamma_Q$, and this we do.

Now let Γ denote the multiplicative group generated by Γ_P, Γ_Q and the values of g. Let χ be a character of Γ which annihilates Γ_Q then

$$\chi(P(x_o, x_1)) = \chi(g) \circ T_P / \chi(g) \quad (g)\text{a.e.}$$

From this (cf. [6]) it follows that

$$\chi(P(x_o, x_1)) = h(x_1)/h(x_o) \quad \text{(everywhere)}$$

for some function h of the state space. Multiplying over a 'cycle' $(i, x_1, \ldots, x_{n-1}, i)$ we obtain

$$\chi(P(i,x_1)P(x_1,x_2)\cdots P(x_{n-1},i)) = 1$$

and hence χ annihilates Γ_P. Thus $\Gamma_P \subset \Gamma_Q$ and similarly $\Gamma_Q \subset \Gamma_P$ i.e. $\Gamma_P = \Gamma_Q$.

The proof that $\Delta_P = \Delta_Q$ is entirely similar and makes use of the fact that there exists $C_P \in \Gamma_P$ and $s = (s_1,\ldots,s_k)$ such that $s_i \in \Gamma_P$ and $P(i,j)s_j/c_P s_i \in \Delta_P$.

From ** one concludes that

$$*** \qquad \frac{C_P P'(x_o,x_1)}{C_Q} = Q'(\phi(x)_o,\phi(x)_1)\frac{g' \circ T_P}{g'}$$

where $P'(i,j) \in \Delta_P$, $Q'(i,j) \in \Delta_Q$. If χ is a character of $\Gamma_P = \Gamma_Q$ and annihilates $\Delta_P = \Delta_Q$ we see that

$$\chi(C_P/C_Q) = 1.$$

Hence $C_P\Delta_P = C_Q\Delta_Q$. Finally $C_P\Delta_P$ generates Γ_P/Δ_P since

$$P(i,j)s_j/s_i \in C_P\Delta_P$$

and therefore

$$P(i,x_1)\ldots P(x_{n-1},i) \in C_P^n\Delta_P.$$

4. FURTHER INVARIANTS

Whenever we have a cocycle-coboundary equation

$$u = v + w \circ T_P - w$$

where $u,v \in L^2(X_P)$ and w is real-valued and measurable we can assert

$$\int u\,dm_P = \int v\,dm_P$$

and if

$$\sigma_P^2(u) = \lim_{n\to\infty} \frac{1}{n}\int(u + .. + u\,T_P^{n-1} - n\int u\,dm_P)^2 \quad \text{exists,}$$
$$\sigma_P^2(u) = \sigma_Q^2(v).$$

This being the case we can establish the following invariants:
For each homomorphism J of Γ_P into the additive group of $\mathbb{C}$ we define

$$\mu_{J,P} = \int J(P(x_o,x_1))dm_P$$

$$\sigma_{J,P}^2 = \sigma_P^2(J(P(x_o,x_1))),$$

where P has been adjusted so that $P(i,j) \in \Gamma_P$ when $P(i,j) \neq 0$. In view of ** we have

THEOREM 3. *If ϕ is a finitary isomorphism such that ϕ, ϕ^{-1} have finite expected code-lengths then for each homomorphism* J *of* $\Gamma_P = \Gamma_Q$ *into the additive group of* $\mathbb{C}$ *we have*

$$\mu_{J,P} = \mu_{J,Q}$$
$$\sigma^2_{J,P} = \sigma^2_{J,Q} .$$

REFERENCES

1. R. Butler and K. Schmidt: An information cocycle for groups of non-singular transformations. (To appear).

2. N. Friedman and D. S. Ornstein: On the isomosphism of weak Bernoulli transformations, Adv. in Math., 5 (1970), 365-394.

3. M. Keane and M. Smorodinsky: Finitary isomorphism of irreducible Markov shifts, Israel J. Math., 34 (1979), 281-286.

4. E. Hewitt and L. J. Savage: Symmetric measures on Cartesian products, Trans. Amer. Math. Soc., 80 (1956), 470-501.

5. W. Krieger: On the finitary isomorphisms of Markov shifts that have finite expected coding time. (To appear).

6. W. Parry: Endomorphisms of a Lebesgue space III, Israel J. Math., 21 (1975), 167-172.

7. W. Parry: Finitary isomorphisms with finite expected code-lengths, Bull. L.M.S., 11 (1979), 170-176.

Mathematics Institute
University of Warwick
Coventry, England

Contemporary Mathematics
Volume **26**, 1984

ERGODIC THEORY IN HYPERBOLIC SPACE

Marina Ratner*)

INTRODUCTION

In this paper we shall exhibit a striking contrast in the ergodic behavior of two classical dynamical systems: the geodesic and the horocycle flows on the unit tangent bundle of a surface of constant negative curvature with finite volume. We shall see that horocycle flows, though very random, are rigidly attached to the conformal class of the underlying surfaces, while geodesic flows are statistically the same for all the surfaces due to their extremal randomness caused by exponential instability of orbits.

The study of statistical properties of the two flows began in the thirties when the ergodic theorems of J. von Neumann and G. D. Birkhoff gave a powerful stimulus to the development of ergodic theory. In 1936 E. Hopf [Ho1] proved that geodesic flows are ergodic. In 1936-39 G. A. Hedlund [He1,2] introduced the horocycle flows and used their association with the geodesic flows to give an elegant proof that horocycle flows are ergodic and geodesic flows are mixing [He3]. In 1939 E. Hopf wrote a beautiful paper [Ho2] where he gave simple geometrical proofs to his and Hedlund's results.

The geometrical and dynamical method of Hedlund and Hopf was based on the study of the global behavior of the orbits. This method turned out to be more powerful than the algebraic method suggested by Gelfand and Fomin in 1952 [GF], based on representation theory and Fourier analysis. The approach of Hedlund and Hopf led to wide generalizations such as Anosov and Axiom A systems [An], [Sma] and the construction of a rich ergodic theory of these systems.

Though some of our theorems below have an algebraic content, they all have purely dynamical and geometrical proofs.

1. ERGODIC THEORY BACKGROUND (See [KSF],[Wa])

Let T and S be two measure preserving transformations (m.p.t.'s) of probability spaces $(X,\mathbb{B}_X,\mu)$ and $(Y,\mathbb{B}_Y,\nu)$ respectively. (In this paper all m.p.t.'s are assumed to be invertible.) T and S are called isomorphic ($T\sim S$) if there is an invertible m.p. ψ: X onto Y called an isomorphism

*) Partially supported by NSF grant MCS 74-19388.

s.t. $\psi T(x) = S\psi(x)$ for μ-a.e. $x \in X$. M.p. flows T_t and S_t are isomorphic if there is ψ as above with $\psi T_t(x) = S_t\psi(x)$ for all $t \in \mathbb{R}$ and μ-a.e. $x \in X$. With each m.p. T we associate the unitary operator U_T on $L_2(X,\mu)$ defined by $(U_T f)(x) = f(Tx)$, $x \in X$, $f \in L_2(X,\mu)$. If $T \sim S$ then U_T is isometric to U_S. By the spectral theorem $(U_T f,f) = \int s d\mu_f(s)$, where μ_f is a finite Borel measure on the unit circle and (g,f) denotes the inner product in $L_2(X,\mu)$. Let L_0 be the subspace of $L_2(X,\mu)$ orthogonal to the constants. T is said to have continuous spectrum if $\mu_f\{s\} = 0$ for every s and every $f \in L_0$. T is said to have Lebesgue spectrum of multiplicity m (m can be infinite) if there are $f_1,\dots,f_m \in L_0$ s.t. each μ_{f_i} is equivalent to Lebesgue measure and L_0 is the orthogonal sum of the subspaces spanned by f_i, $i = 1,\dots,m$. An m.p. flow T_t is said to have Lebesgue spectrum of multiplicity m if each T_t, $t \neq 0$ does.

The central problem of ergodic theory is classifying all m.p.t.'s or flows up to an isomorphism. To approach this problem it is natural to seek properties invariant under isomorphisms. Here are some such properties listed in order of increasing randomness.

1) T is <u>ergodic</u> iff $TA = A$, $A \in \mathcal{B}_X$ implies $\mu(A) = 0$ or 1 iff U_T has no invariant functions different from the constants iff

$$\frac{1}{n}\sum_{i=0}^{n-1} \mu(T^{-i}A \cap B) \to \mu(A)\mu(B)$$

when $n \to \infty$, $A,B \in \mathcal{B}_X$.

2) T is <u>weak mixing</u> iff U_T has no eigenfunctions different from the constants iff T has continuous spectrum iff

$$\frac{1}{n}\sum_{i=0}^{n-1} |\mu(T^{-i}A \cap B) - \mu(A)\mu(B)| \to 0$$

when $n \to \infty$, $A,B \in \mathcal{B}_X$.

3) T is <u>mixing</u> iff $\mu(T^{-n}A \cap B) \to \mu(A)\mu(B)$ when $n \to \infty$, $A,B \in \mathcal{B}_X$. This property can also be stated in spectral terms.

4) T is <u>mixing of degree</u> $r \geq 1$ iff for every $A_0,A_1,\dots,A_r \in \mathcal{B}_X$

$$\mu(A_0 \cap T^{-k_1}A_1 \cap\cdots\cap T^{-(k_1+\cdots+k_r)}A_r) \to \prod_{i=0}^{r} \mu(A_i)$$

when $k_1,\dots,k_r \to \infty$.

5) T is a K-<u>automorphism</u> iff for every $A_0,A_1,\dots,A_r \in \mathcal{B}_X$ and every $r \geq 1$ $\sup|\mu(A_0 \cap B) - \mu(A_0)\mu(B)| \to 0$ when $n \to \infty$ where sup is taken over $B \in \mathcal{A}_n$, and $\mathcal{A}_n = \mathcal{A}_n(A_1,\dots,A_r)$ is the σ-algebra generated by $T^{-k}A_i$ for $k \geq n$, $i = 1,\dots,r$. One can show that K-automorphisms have Lebesgue spectrum of infinite multiplicity.

6) T is a <u>Bernoulli automorphism</u> iff there is a measurable partition

$\alpha = \{A_1,\ldots,A_n,\ldots\}$ of X finite or infinite s.t. $\mathbb{B}_X$ is generated by the sets $T^k A_i$, $-\infty < k < \infty$, $i = 1,2,\ldots$ and

$$\mu(T^{k_1} A_{i_1} \cap \cdots \cap T^{k_m} A_{i_m}) = \prod_{j=1}^{m} \mu(A_{i_j})$$

for every $A_{i_1},\ldots,A_{i_m} \in \alpha$, every $m \geq 1$ and every $k_1,\ldots,k_m$ with $k_i \neq k_j$ if $i \neq j$.

Each of the listed properties implies the preceding one, but not vice versa. Similarly, these properties can be stated for m.p. flows. An m.p. flow T_t is Bernoulli if every T_t, $t \neq 0$ is a Bernoulli automorphism. K-flows are defined similarly.

There is a numerical invariant of isomorphism called the <u>entropy</u>. This was introduced by Kolmogorov and Sinai in 1958-59. Let $\alpha = \{A_1,\ldots,A_k\}$ be a finite measurable partition of X and let

$$E(\alpha) = -\sum_{i=1}^{k} \mu(A_i)\log \mu(A_i).$$

Let α_n be the largest refinement of the partitions α, $T^{-1}\alpha,\ldots,T^{-(n-1)}\alpha$. One can show that the

$$\lim_{n\to\infty} \frac{E(\alpha_n)}{n} = E(T,\alpha)$$

exists and $E(T) = \sup_\alpha E(T,\alpha)$ is called the entropy of T. $E(T) \geq 0$ and $E(T)$ can be infinite. One can show that if T_t is an m.p. flow, then $E(T_t) = |t|E(T_1)$, $t \in R$. The entropy $E(\{T_t\})$ of the flow T_t is defined to be $E(T_1)$.

In 1970 D. Ornstein made a major advance in ergodic theory by proving that the entropy is a complete invariant for the class of Bernoulli automorphisms and flows. This had been conjectured by Kolmogorov and partially proved by Sinai [Si2].

ORNSTEIN'S THEOREM [O1],[O2],[O3]. <u>Any two Bernoulli automorphisms or flows with the same entropy are isomorphic</u>.

2. GEOMETRIC AND ALGEBRAIC DEFINITIONS OF GEODESIC AND HOROCYCLE FLOWS

2.1. The geometric definition [He1,2,3],[Ho1,2],[AGH].

Let $\mathbb{H}$ denote the upper half of the complex plane $\mathrm{Im} z > 0$ equipped with the hyperbolic metric

$$ds^2 = \frac{dx^2 + dy^2}{y^2}, \quad z = x + iy.$$

It is well known that the curvature of $\mathbb{H}$ equals -1 at every point and geodesics in $\mathbb{H}$ are the circles or straight lines orthogonal to the real axis.

Let $\mathbf{V}$ denote the 3-dimensional bundle of unit tangent vectors to $\mathbf{H}$. If v is a unit tangent vector at $z \in \mathbf{H}$ then there is a unique geodesic $g(v)$ through z with v as its tangent. Let $z(t)$ be the point on $g(v)$ s.t. the hyperbolic length of the oriented geodesic segment $(z,z(t))$ is t. The geodesic flow G_t on $\mathbf{V}$ is defined by: $G_t(v)$ is the unit tangent to $g(v)$ at $z(t)$. It follows from the classical Liouville's theorem that the geodesic flow preserves the measure m on $\mathbf{V}$ given by $dm = dVd\omega$, where V is the hyperbolic area on $\mathbf{H}$ and ω is the length measure on the unit circle.

A horocycle in $\mathbf{H}$ is either a circle tangent to the real axis or a straight line parallel to the real axis (tangency at ∞). Let v be a unit tangent at z and $g(v)$ the geodesic determined by v. Let x_- and x_+ be the points of intersection of $g(v)$ with the real axis and let h_- and h_+ be the horocycles through z tangent to the real axis at x_- and x_+ respectively. The vector v is an inward normal to h_+ and an outward normal to h_-. Let v' be the unit tangent vector at z obtained from v by rotating it clockwise through a right angle. The vector v' defines an orientation for each horocycle h_+ and h_-. Let $z_-(t)$ be the point on h_- s.t. the hyperbolic length of the oriented horocycle segment $(z,z_-(t))$ is t. The horocycle flow H_t on $\mathbf{V}$ is defined by: $H_t(v)$ is the outward normal to h_- at $z_-(t)$. Similarly, we can define a flow H_t^* on $\mathbf{V}$ by taking $z_+(t)$ on h_+ and defining $H_t^*(v)$ to be the inward normal to h_+ at $z_+(t)$. One can easily show that the horocycle flows H_t and H_t^* preserve the measure m on $\mathbf{V}$ and

$$G_t \circ H_s = H_{se^t} \circ G_t, \quad G_t \circ H_s^* = H^*_{se^{-t}} \circ G_t \tag{2.1}$$

for all $s,t \in R$. The relation (2.1) exhibits the exponential divergence of geodesic orbits.

Let $\mathbf{G}$ be the group of isometries of $\mathbf{H}$ and let Γ be a subgroup of $\mathbf{G}$ which acts discontinuously on $\mathbf{H}$. Let $\Gamma z = \{\gamma(z) : \gamma \in \Gamma\}$ be the Γ-orbit of $z \in \mathbf{H}$ and let $X = \{\Gamma z : z \in \mathbf{H}\}$ be the space of Γ-orbits. X is covered by $\mathbf{H}$ via the projection $p: \mathbf{H} \to X, p(z) = \Gamma z$ and inherits the Riemannian structure of $\mathbf{H}$. X is a surface of constant negative curvature and Γ is the fundamental group of X. We shall consider only those Γ for which X has finite area.

To give the geometric definition of the geodesic and horocycle flows we take the space Y of unit tangent vectors to X which is covered by $\mathbf{V}$ via the differential dp of p and define the geodesic flow g_t and the two horocycle flows h_t and h_t^* on Y to be the projections via dp of the geodesic flow G_t and the two horocycle flows H_t and H_t^* on $\mathbf{V}$. The flows g_t, h_t, and h_t^* preserve the normalized Riemannian volume on Y derived from m.

2.2 The algebraic definition [AGH],[GF]

It is well known that the group $\mathbb{G}$ of isometries of $\mathbb{H}$ is the group of Möbius transformations $Tz = \frac{az+b}{cz+d}$, $z \in \mathbb{H}$, where a,b,c,d are real and $ad - bc = 1$. There is a natural continuous homomorphism $SL(2,R) \to \mathbb{G}$ given by $\begin{pmatrix} a & b \\ c & d \end{pmatrix} \to \frac{az+b}{cz+d}$, $z \in \mathbb{H}$, where $SL(2,R)$ is the group of two-by-two real matrices of determinant one. The kernel of this homomorphism is $Z_2 = \left\{\begin{pmatrix} 1 & 0 \\ 0 & 1 \end{pmatrix}, \begin{pmatrix} -1 & 0 \\ 0 & -1 \end{pmatrix}\right\}$. This implies that $\mathbb{G}$ is isomorphic and homeomorphic to $SL(2,R)/Z_2$. For simplicity we shall identify $\mathbb{G}$ with $SL(2,R)/Z_2$.

Let v_0 be the unit tangent at i whose direction is the same as the imaginary axis. For each $v \in \mathbb{V}$ there is exactly one isometry $g_v \in \mathbb{G} = SL(2,R)/Z_2$ s.t. $g_v(v_0) = v$. The map $\varphi: g_v \to v$ is a homeomorphism from $\mathbb{G}$ onto the unit tangent bundle $\mathbb{V}$ of $\mathbb{H}$. The geodesic flow G_t and the horocycle flows H_t and H_t^* on $\mathbb{V}$ can be realized on $\mathbb{G}$ via φ as the flows $\tilde{G}_t(u) = \varphi^{-1}G_t\varphi(u)$, $\tilde{H}_t(u) = \varphi^{-1}H_t\varphi(u)$, $\tilde{H}_t^*(u) = \varphi^{-1}H_t^*\varphi(u)$, $u \in \mathbb{G}$. We have $\varphi(Id) = v_0$ and one can easily see that

$$\varphi^{-1}(G_tv_0) = \begin{pmatrix} e^{t/2} & 0 \\ 0 & e^{-t/2} \end{pmatrix} Z_2, \quad \varphi^{-1}(H_tv_0) = \begin{pmatrix} 1 & 0 \\ t & 1 \end{pmatrix} Z_2 \text{ and } \varphi^{-1}(H_t^*v_0) = \begin{pmatrix} 1 & t \\ 0 & 1 \end{pmatrix} Z_2.$$

This implies that

$$\tilde{G}_t(gZ_2) = g\begin{pmatrix} e^{t/2} & 0 \\ 0 & e^{-t/2} \end{pmatrix} Z_2$$

$$\tilde{H}_t(gZ_2) = g\begin{pmatrix} 1 & 0 \\ t & 1 \end{pmatrix} Z_2$$

$$\tilde{H}_t^*(gZ_2) = g\begin{pmatrix} 1 & t \\ 0 & 1 \end{pmatrix} Z_2$$

for all $g \in SL(2,R)$.

Let now Γ be a subgroup of $\mathbb{G}$ which acts discontinuously on $\mathbb{H}$. Then Γ is a discrete subgroup of $\mathbb{G}$ and the space M of cosets Γu, $u \in \mathbb{G}$ is homeomorphic to the unit tangent bundle Y of the surface X of constant negative curvature from section 2.1. This homeomorphism carries the geodesic flow g_t and the horocycle flows h_t and h_t^* on Y to the flows $\tilde{g}_t, \tilde{h}_t, \tilde{h}_t^*$ on M given by

$$\tilde{g}_t(\Gamma u) = \Gamma u\begin{pmatrix} e^{t/2} & 0 \\ 0 & e^{-t/2} \end{pmatrix}, \quad \tilde{h}_t(\Gamma u) = \Gamma u\begin{pmatrix} 1 & 0 \\ t & 1 \end{pmatrix}, \quad \tilde{h}_t^*(\Gamma u) = \Gamma u\begin{pmatrix} 1 & t \\ 0 & 1 \end{pmatrix}, \quad t \in R.$$

This leads us to the following algebraic definition of the flows, which in fact embraces a more general situation than the one described above. In this definition we shall slightly change the above notations.

Let $\mathbb{G}$ denote the group $SL(2,R)$ equipped with a left invariant Riemannian metric and let Γ be a discrete subgroup of $\mathbb{G}$ s.t. the quotient space $M = \Gamma\backslash G = \{\Gamma g : g \in G\}$ has finite volume. The geodesic flow g_t on M and the horocycle flows h_t and h_t^* on M are defined by

$$g_t(\Gamma g) = \Gamma g\begin{pmatrix} e^{t/2} & 0 \\ 0 & e^{-t/2} \end{pmatrix}$$

$$h_t(\Gamma g) = \Gamma g\begin{pmatrix} 1 & 0 \\ t & 1 \end{pmatrix}$$

$$h_t^*(\Gamma g) = \Gamma g\begin{pmatrix} 1 & t \\ 0 & 1 \end{pmatrix}$$

$g \in G$, $t \in R$. It is well known that a left invariant volume V on G is also right invariant. This implies that the flows g_t, h_t and h_t^* preserve the normalized volume μ on M derived from V. One can check that

$$g_t \circ h_s = h_{se^t} \circ g_t,\ g_t \circ h_s^* = h_{se^{-t}}^* \circ g_t \tag{2.2}$$

$t,s \in R$. This is the algebraic version of (2.1). The commutation relation (2.2) plays a crucial role in the ergodic theory of the flows.

3. ANOSOV FLOWS AND ASSOCIATED HOROCYCLE FLOWS (see [An],[An,Si])

The relation (2.2) motivated the introduction of Anosov flows. Let M be a compact connected n-dimensional Riemannian manifold and let T_t be a C^1-flow on M generated by a nonvanishing vector field f on M. Let $\tilde{T}_t$ denote the differential of T_t acting on the tangent bundle TM of M. T_t is called Anosov if TM is the direct sum of $\tilde{T}_t$-invariant continuous subbundles E^k, E^ℓ and F s.t. $\dim E^k = k \neq 0$, $\dim E^\ell = \ell \neq 0$, $k + \ell + 1 = n$, F is generated by f and if $u \in E_x^k$, $v \in E_x^\ell$ then

$$|\tilde{T}_{t,x}u| \geq a\rho^t|u|,\quad |\tilde{T}_{t,x}v| \leq b\rho^{-t}|v|, \tag{3.1}$$

for all $t > 0$ and some $\rho > 1$, $a,b > 0$ independent of x,t, where $\tilde{T}_{t,x}$ denotes the differential of T_t at x. One shows that E^k and E^ℓ are tangent bundles of some T_t-invariant continuous, oriented foliations H^k and H^ℓ, called horocycle foliations. (3.1) is a generalized version of (2.2). It shows that the Anosov flow T_t exponentially expands leaves of H^k and exponentially contracts leaves of H^ℓ if $t > 0$. An Anosov flow T_t is called transitive if there is a leaf of H^k or of H^ℓ which is dense in M. The geodesic flow defined in section 2 is transitive Anosov. More generally, the geodesic flow on the unit tangent bundle of a compact n-dimensional Riemannian manifold of negative curvature is a transitive Anosov flow.

If H^k is one-simensional we can define a flow h_t on M by setting $h_t(x)$ to be the point on $H^k(x)$, $x \in M$ s.t. the length of the oriented segment

$(x,h_t(x)) \subset H^k(x)$ is t. The flow h_t is called the horocycle flow associated with the Anosov T_t. Margulis [Mgl] proved that if T_t is a transitive C^2-Anosov flow then there is a continuous function $u:M \times R \to R$ unique up to a multiplicative constant s.t. if we denote $\mathbf{h}_t(x) = h_{u(x,t)}(x)$, $x \in M$, $t \in R$ then the following commutation relation

$$T_t \circ \mathbf{h}_s = \mathbf{h}_{s\lambda t} \circ T_t$$

holds for all $s,t \in R$ and some $\lambda > 1$. This is an exact analog of (2.2). The flow $\mathbf{h}_t$ is called the uniformly parameterized horocycle flow associated with T_t. One can show [Si3] that $\log \lambda$ is the topological entropy of T_t and that there is a unique Borel probability measure on M preserved by both T_t and $\mathbf{h}_t$. This measure is positive on open sets and is the measure of maximal entropy for T_t.

4. SUMMARY OF ERGODIC PROPERTIES OF GEODESIC AND HOROCYCLE FLOWS

As we mentioned in the introduction Hopf proved in 1936 that geodesic flows are ergodic and Hedlund proved in 1939 that horocycle flows are ergodic and geodesic flows are mixing. Gelfand and Fomin proved in 1952 [GF] using representation theory that geodesic flows have Lebesgue spectrum of infinite multiplicity. A work of Parasyuk [P] combined with [GF] showed that horocycle flows have Lebesgue spectrum of infinite multiplicity. Actually, the last two results follow from a more general theorem of Stepin [St]. Sinai proved [Sil] in 1960 that geodesic flows are K-flows. Finally, Ornstein and Weiss proved [OW] in 1973 that geodesic flows are Bernoulli. Sinai [Si3] generalized his result to transitive C^2-Anosov flows with respect to a wide class of Borel invariant measures called Gibbs measures (see also [Bw]). These were shown [Ral] to be Bernoulli, too. Marcus proved [Ma2] in 1978 that horocycle flows are mixing of all degrees. His theorem also applies to uniformly parameterized horocycle flows associated with transitive C^2-Anosov flows.

One can show using (2.2) that the entropy of the geodesic flow g_t is 1. (2.2) also shows that h_α and h_β are isomorphic if $\alpha \cdot \beta > 0$. This implies that the entropy of the horocycle flow h_t is either 0 or ∞. Gurevich showed [Gu] that it is 0. (This follows also from a theorem of Kushnirenko [Kul].) This is equally true for h_t^* and for horocycle flows associated with Anosov flows.

5. INVARIANT MEASURES

A continuous flow T_t on a compact metric space X is called minimal if every orbit of T_t is dense in X. T_t is called uniquely ergodic if there is only one T_t-invariant Borel probability measure on X. This measure is necessarily ergodic and if it is positive on open sets then T_t is minimal (see [Wa]).

Hedlund [He1] proved in 1936 that horocycle flows h_t and h_t^* on a compact M as defined in section 2.1 are minimal. Anosov [An] extended this to horocycle foliations associated with geodesic flows on the unit tangent bundles of compact manifolds of negative curvature. Furstenberg [Fu2] proved in 1972 that h_t and h_t^* on a compact M are uniquely ergodic. His theorem was extended to general horospherical flows by Veech [V].

Let us note that the geodesic flow on M has uncountably many ergodic invariant measures, even with the additional property of positivity on open sets (see [Si3],[Bw]).

Marcus [Ma1] generalized Furstenberg's theorem using geometrical and dynamical methods. He gave a simple proof that horocycle flows associated with transitive C^2-Anosov flows on compact manifolds are uniquely ergodic. This was generalized to horocycle foliations by Bowen and Marcus [BwM].

A finite volume quotient space $M = \Gamma\backslash G$, $G = SL(2,R)$ (see section 2.2) is noncompact iff Γ contains a parabolic element A (see [Fo],[GGP]). A matrix A is parabolic if $A = g_0\begin{pmatrix}1 & \alpha\\ 0 & 1\end{pmatrix}g_0^{-1}$ for some $\alpha \in R$, $g_0 \in G$. Let $x = \Gamma g_0 \in M$. We have $h_\alpha(x) = \Gamma g_0\begin{pmatrix}1 & \alpha\\ 0 & 1\end{pmatrix} = \Gamma g_0\begin{pmatrix}1 & \alpha\\ 0 & 1\end{pmatrix}g_0^{-1}g_0 = \Gamma g_0 = x$, since $A \in \Gamma$. This says that x lies on a closed horocycle orbit of length $|\alpha|$ (we assume that the length of the horocycle segment $(y,h_t y)$ is $t, y \in M$). (2.2) shows then that $g_t(x)$ lies on a closed horocycle orbit of length $|\alpha|e^t$. Thus h_t has uncountably many closed orbits of arbitrary small and arbitrary large length. The normalized length measure on such an orbit is an ergodic invariant measure for h_t and hence h_t is not uniquely ergodic.

Dani [D] proved that if $M = \Gamma\backslash G$ is noncompact and m is an ergodic h_t-invariant Borel probability measure on M then either m is the normalized volume on M or m is the normalized length measure on a closed orbit of h_t. Recently, he and J. Smillie [DS] showed that if $x \in M$ is not on a closed orbit of h_t then x is a generic point for the normalized volume on M. Sarnak [Sa] proved that closed orbits of h_t are uniformly distributed in M. This means that given an open set $B \subset M$ and $\varepsilon > 0$ there is $r_0 > 0$ s.t. if x lies on a closed h_t-orbit of length $r \geq r_0$ then $|\frac{1}{r}\int_0^r \chi_B(h_t x)dt - \mu(B)| < \infty$, where χ_B denotes the characteristic function of B and μ is the normalized volume on M. Of course, everything said about h_t applies to h_t^* as well. Recently, Ornstein and the author have found a simple proof of Dani's theorem using a geometrical construction from [Ra2]. This proof also shows that every σ-finite h_t-invariant ergodic measure on M which is finite on compact sets is finite. The proof also implies Sarnak's theorem. It will appear elsewhere.

J. Smillie seems to be able to apply our proof to horocycle flows associated with geodesic flows on the unit tangent bundles of noncompact surfaces of variable negative curvature provided the curvature is bounded away from 0 and $-\infty$ and horocycle flows have the uniform parameterization as in section 3.

6. THE ISOMORPHISM PROBLEM FOR GEODESIC AND HOROCYCLE FLOWS

Henceforth we shall use only the algebraic definition of the flows.

Let $\mathbf{T}$ denote the totality of all discrete subgroups Γ of $G = SL(2,R)$ s.t. the quotient space $M = \Gamma\backslash G$ has finite volume. Let $\Gamma_i \in \mathbf{T}$ and let $g_t^{(i)}$, $h_t^{(i)}$, $h_t^{*(i)}$ be the geodesic and horocycle flows on $(M_i = \Gamma_i\backslash G, \mu_i)$, where μ_i is the normalized volume on M_i, $i = 1,2$. We pose the following questions: Is $g_t^{(1)}$ isomorphic to $g_t^{(2)}$? Is $h_t^{(1)}$ isomorphic to $h_t^{(2)}$? The same about $h_t^{*(1)}$ and $h_t^{*(2)}$.

As we mentioned in section 4, for any $\Gamma \in \mathbf{T}$ the geodesic flow on $\Gamma\backslash G$ is a Bernoulli flow with entropy 1. It follows then from Ornstein's theorem that for any $\Gamma_1,\Gamma_2 \in \mathbf{T}$ the geodesic flows $g_t^{(1)}$ and $g_t^{(2)}$ are isomorphic.

To approach the problem for horocycle flows let us ask whether there is an m.p. $\psi:M_1$ onto M_2 not necessarily one-to-one s.t. $\psi h_t^{(1)}(x) = h_t^{(2)}\psi(x)$ for all $t \in R$ and μ_1 - a.c. $x \in M_1$. Such a map ψ is called a conjugacy between $h_t^{(1)}$ and $h_t^{(2)}$.

Suppose that $C\Gamma_1C^{-1} \subset \Gamma_2$ for some $C \in G$ and let $\psi_C:M_1$ onto M_2 be defined by $\psi_C(\Gamma_1 g) = \Gamma_2 Cg$, $g \in G$. ψ_C is an isometry and is a conjugacy between $h_t^{(1)}$ and $h_t^{(2)}$. Actually, ψ_C is a conjugacy between any two flows $f_t^{(1)}$ and $f_t^{(2)}$ defined by a one-parameter subgroup of G, that is $f_t^{(i)}(\Gamma_i g) = \Gamma_i g \exp\{tF\}$, $i = 1,2$, for some element F of the Lie algebra of G. The flow $f_t^{(i)}$ is called the F-flow on M_i. It is clear that $\psi = h_\sigma^{(2)} \circ \psi_C$, $\sigma \in R$ is also a conjugacy between $h_t^{(1)}$ and $h_t^{(2)}$. We shall call such a ψ an algebraic conjugacy.

Surprisingly, it turns out that every conjugacy between $h_t^{(1)}$ and $h_t^{(2)}$ is algebraic.

THEOREM 6.1. (<u>Rigidity of horocycle flows</u>) [Ra5]. <u>Suppose there is a measure preserving</u> $\psi:M_1$ <u>onto</u> M_2 <u>s.t.</u> $\psi h_1^{(1)}(x) = h_1^{(2)}\psi(x)$ <u>for</u> μ_1 - a.e. $x \in M_1$ ($h_1^{(i)}$ <u>is the time one transformation of the flow</u> $h_t^{(i)}$). <u>Then there are</u> $C \in G$ <u>and</u> $\sigma \in R$ s.t. $C\Gamma C^{-1} \subset \Gamma_2$ <u>and</u> $\psi(x) = h_\sigma^{(2)}\psi_C(x)$ <u>for</u> μ_1 - <u>a.e.</u> $x \in M_1$.

COROLLARY 6.1. $h_1^{(1)}$ <u>and</u> $h_1^{(2)}$ <u>are isomorphic iff</u> $\Gamma_2 = C\Gamma_1C^{-1}$ <u>for some</u> $C \in G$. <u>Any isomorphism</u> $\psi:M_1$ <u>onto</u> M_2 <u>between</u> $h_1^{(1)}$ <u>and</u> $h_1^{(2)}$ <u>has the form as in Theorem</u> 6.1.

When $h_t^{(i)}$, $i = 1,2$ are viewed as horocycle flows on the unit tangent bundles of surfaces of constant negative curvature, the condition $C\Gamma_1C^{-1} = \Gamma_2$ for some $C \in G$ means that the underlying surfaces are isometric.

Let $f = \{f_t, t \in R\}$ be an m.p. flow on a probability space (X,μ) and let $\Psi(f)$ denote the set of all measure isomorphisms ψ from X onto itself s.t. $\psi f_t(x) = f_t\psi(x)$ for a.e. $x \in X$ and all $t \in R$ (i.e. ψ commutes with every f_t, $t \in R$). We say that $\psi_1 \in \Psi(f)$ and $\psi_2 \in \Psi(f)$ are equivalent iff there is $\sigma \in R$ s.t. $\psi_2 = f_\sigma \circ \psi_1$ a.e. Let $\varkappa(f)$ denote the set of equivalence classes in Ψ. $\varkappa(f)$ becomes a group if we define $[\psi_1]\cdot[\psi_2] = [\psi_1\circ\psi_2]$ where $[\psi]$ denotes the equivalence class of $\psi \in \Psi(f)$. $\varkappa(f)$ is called the commutant of f.

Let $\Gamma \in \mathbf{T}$ and let $h = \{h_t, t \in R\}$ be the horocycle flow on $M = \Gamma\backslash G$. Let $\tilde{\Gamma} = \{C \in G: C\Gamma C^{-1} = \Gamma\}$ be the normalizer of Γ in G. It is easy to see that $\tilde{\Gamma} \in \mathbf{T}$. This implies that the quotient group $\Gamma\backslash\tilde{\Gamma}$ is finite. It follows from Corollary 6.1 that $\varkappa(h) = \{[\psi_C]: C \in \tilde{\Gamma}\}$. Let $\omega:\Gamma\backslash\tilde{\Gamma} \to \varkappa(h)$ be defined by $\omega(\Gamma C) = [\psi_C]$, $C \in \tilde{\Gamma}$. We show that ω is a group isomorphism and we get

COROLLARY 6.2. <u>The commutant</u> $\varkappa(h)$ <u>of the horocycle flow</u> h_t <u>on</u> $\Gamma\backslash G$ <u>is isomorphic to</u> $\Gamma\backslash\tilde{\Gamma}$. $\varkappa(h)$ <u>is finite and if</u> $\Gamma = \tilde{\Gamma}$ <u>then</u> $\varkappa(h)$ <u>is trivial</u>.

One can show that the commutant of the geodesic flow is uncountable, since it is a Bernoulli flow.

Let $P = \begin{pmatrix} 1 & 0 \\ 0 & -1 \end{pmatrix}$, $\det P = -1$, and suppose that $\beta\Gamma_1\beta^{-1} = \Gamma_2$ for some β with $\det \beta = -1$, $\Gamma_1, \Gamma_2 \in \mathbf{T}$. Let $\psi:M_1 \to M_2$ be $\psi(\Gamma_1 g) = \Gamma_2\beta gP$, $g \in G$. One can see that ψ is an isomorphism between $h_t^{(1)}$ and $h_{-t}^{(2)}$. This implies via Corollary 6.1.

COROLLARY 6.3. $h_1^{(1)}$ <u>and</u> $h_{-1}^{(2)}$ <u>are isomorphic iff</u> $\beta\Gamma_1\beta^{-1} = \Gamma_2$ <u>for some</u> β <u>with</u> $\det \beta = -1$. <u>Every isomorphism</u> ψ <u>between</u> $h_1^{(1)}$ <u>and</u> $h_{-1}^{(2)}$ <u>has the form</u> $\psi(\Gamma_1 g) = h_\sigma^{(2)}(\Gamma_2\beta gP)$ <u>a.e. for some</u> $\sigma \in R$. <u>In particular</u>, $h_1^{(1)} \sim h_{-1}^{(1)}$ <u>iff</u> $\beta\Gamma_1\beta^{-1} = \Gamma_1$ <u>for some</u> β <u>with</u> $\det \beta = -1$.

Let us note that the geodesic flow g_t is isomorphic to g_{-t} by Ornstein's theorem.

Of course, all our theorems hold for h_t^* as well. Let $M = \Gamma\backslash G$, $\Gamma \in \mathbf{T}$ and let $\vartheta:M \to M$ be $\vartheta(\Gamma g) = \Gamma g\begin{pmatrix} 0 & 1 \\ -1 & 0 \end{pmatrix}$, $g \in G$. One can check that ϑ is an isomorphism between h_{-t} and h_t^* on M. Geometrically, ϑ rotates unit tangents by a right angle. So $h_t^* \sim h_{-t}$ and we can substitute $h_1^{*(2)}$ instead of $h_{-1}^{(2)}$ in Corollary 6.3. In particular, $h_1 \sim h_1^*$ on $\Gamma\backslash G$ iff $\beta\Gamma\beta^{-1} = \Gamma$ for some β with $\det \beta = -1$.

The proof of Theorem 6.1 uses (2.2) and the polynomial divergence of horocycle orbits. First we show that if ψ is a conjugacy between $h_1^{(1)}$ and $h_1^{(2)}$ then it is a conjugacy between the flows $h_t^{(1)}$ and $h_t^{(2)}$. Then we show that there is a unique $\sigma \in R$ s.t. $h_\sigma^{(2)} \circ \psi = \varphi$ is a conjugacy between $g_t^{(1)}$ and $g_t^{(2)}$ and between $h_t^{*(1)}$ and $h_t^{*(2)}$. This implies that if $\varphi(\Gamma_1 g) = \Gamma_2 g'$ then $\varphi(\Gamma_1 g\alpha) = \Gamma_2 g'\alpha$ for all $\alpha \in G$ and μ_1 - a.e. $\Gamma_1 g \in M_1$. This may happen only if $C\Gamma_1 C^{-1} \subset \Gamma_2$ for some $C \in G$ and $\varphi(\Gamma_1 g) = \Gamma_2 Cg$ for μ_1- a.e. $\Gamma_1 g \in M_1$.

Recently, Feldman and Ornstein [FeO] were able to apply our method to give a partial generalization of Theorem 6.1. Their result is as follows: Let $\mathbf{h}_t^{(i)}$, $\mathbf{h}_t^{*(i)}$ be the uniformly parametrized horocycle flows associated with the geodesic flow $g_t^{(i)}$ on the unit tangent bundle M_i of a compact surface S_i of variable negative curvature and let μ_i be the unique Borel probability measure on M_i preserved by each of $\mathbf{h}_t^{(i)}$, $g_t^{(i)}$, $\mathbf{h}_t^{*(i)}$, $i = 1,2$ (μ_i is the measure of maximal entropy for $g_t^{(i)}$). Let $\psi:M_1 \to M_2$ be an isomorphism between $\mathbf{h}_t^{(1)}$ and $\mathbf{h}_t^{(2)}$. Then 1) there exists a unique $\sigma \in R$ s.t. $\mathbf{h}_\sigma^{(2)} \circ \psi$ is μ_1- a.e. equal to a homeomorphism φ, s.t. φ is also an isomorphism between $g_t^{(1)}$ and $g_{pt}^{(2)}$ and between $\mathbf{h}_t^{*(1)}$ and $\mathbf{h}_{ct}^{*(2)}$ for some $c > 0$, where p is the topological entropy of $g_t^{(1)}$ divided by the topological entropy of $g_t^{(2)}$. 2) S_1 and S_2 are homeomorphic. Using this theorem they also showed that the commutant of $\mathbf{h}_t$ is finite.

We believe that Theorem 6.1 can be generalized to higher dimensions as follows: let $H^{(1)}$ and $H^{(2)}$ be the expanding horocycle foliations for the geodesic flows on the unit tangent bundles M_1 and M_2 of manifolds S_1 and S_2 of constant negative curvature with finite volume. Suppose that there is a measure preserving invertible $\psi:M_1 \to M_2$ s.t. ψ maps leaves of $H^{(1)}$ to leaves of $H^{(2)}$ and ψ is a C^∞-isometry on every leaf of $H^{(1)}$. Then S_1 and S_2 are isometric. Details will appear elsewhere.

7. FACTORS

Let T be an m.p.t. on (X,μ) and S an m.p.t. on (Y,ν). We say that S is a factor of T if there is a measure preserving $\psi:X \to Y$ not necessarily one-to-one s.t. $\psi T(x) = S\psi(x)$ for μ -a.e. $x \in X$. ψ is called a factor map or a conjugacy between T and S. The conjugacy ψ induces a T-invariant measurable partition ξ of X into sets $\psi^{-1}\{y\}$, $y \in Y$ called ψ-fibers. The m.p.t. T induces an m.p.t. T^ξ on the quotient space

$(X/\xi, \mu_\xi)$, which is isomorphic to S via the map $\psi_\xi : X/\xi \to Y$, $\psi_\xi(\xi(x)) = \psi(x)$. A factor S is called trivial if $\nu\{y\} = 1$ for some $y \in Y$. Henceforth the word "factor" means nontrivial factor.

Let $\Phi(T)$ denote the set of all isomorphisms $\varphi : X \to X$ s.t. $\varphi T(x) = T\varphi(x)$ for μ-a.e. $x \in X$ and let $\Psi = \Psi(T,S)$ denote the set of all conjugacies between T and S. We say that $\psi_1 \in \Psi$ and $\psi_2 \in \Psi$ are equivalent if there are $\varphi_1 \in \Phi(T)$ and $\varphi_2 \in \Phi(S)$ s.t. $\psi_2 = \varphi_2 \circ \psi_1 \circ \varphi_1$ a.e. Let $\varkappa(T,S)$ denote the set of equivalence classes in Ψ. It is clear that if $T \sim T'$ and $S \sim S'$ then there is a natural one-to-one correspondence between $\varkappa(T,S)$ and $\varkappa(T',S')$. So $|\varkappa(T,S)|$ is an invariant of the isomorphism class of (T,S). Similar definitions are applied to m.p. flows.

It is natural to pose the following problems: 1) classifying all factors of an m.p.t. T up to an isomorphism. 2) describing $\varkappa(T,S)$ for a given factor S of T.

A flow S_t is a factor of a Bernoulli flow T_t iff S_t is Bernoulli and the entropy of S_t does not exceed the entropy of T_t. This implies that the number of nonisomorphic factors of a Bernoulli flow T_t is uncountable. One can show that $\varkappa(T_t, S_t)$ is uncountable for every factor S_t of T_t. This solves our problems for the geodesic flow g_t on $(\Gamma\backslash G, \mu)$, $\Gamma \in \mathbf{T}$.

Suppose that $\Gamma_1, \Gamma_2 \in \mathbf{T}$ and $\Gamma_1 \subset \Gamma_2$. The quotient space $\Gamma_1\backslash\Gamma_2$ is finite and $|\Gamma_1\backslash\Gamma_2|$ is called the index of Γ_1 in Γ_2. Let $\psi : M_1 \to M_2$ be defined by $\psi(\Gamma_1 g) = \Gamma_2 g$, $g \in G$. As we have mentioned in section 6 ψ is a conjugacy between any two F-flows $f_t^{(1)}$ and $f_t^{(2)}$ on M_1 and M_2 respectively. The ψ-fiber of $y = \Gamma_2 g$ is $\psi^{-1}\{y\} = \{\Gamma_1\gamma_i g : i = 1,\dots,n\}$ where $n = |\Gamma_1\backslash\Gamma_2|$, $\gamma_i \in \Gamma_2$, $i = 1,\dots,n$ and $\Gamma_1\backslash\Gamma_2 = \{\Gamma_1\gamma_i : i = 1,\dots,n\}$. Thus $f_t^{(2)}$ is a factor of $f_t^{(1)}$. It is called an algebraic factor of $f_t^{(1)}$.

The following theorem shows that every factor of the horocycle flow h_t is algebraic.

THEOREM 7.1. [Ra6]. 1) <u>If</u> S_t <u>is a factor of</u> h_t <u>on</u> $(M = \Gamma\backslash G, \mu)$ <u>then there is</u> $\Gamma' \in \mathbf{T}$, $\Gamma' \supset \Gamma$ <u>s.t.</u> S_t <u>is isomorphic to the horocycle flow</u> h'_t <u>on</u> $(\Gamma'\backslash G, \mu')$. 2) <u>If</u> S <u>on</u> (Y,ν) <u>is a factor of</u> h_1 <u>via a factor map</u> $\psi : M \to Y$, $\psi h_1(x) = S\psi(x)$ <u>for</u> μ_1-a.e. $x \in M$ <u>then there is an</u> m.p. <u>flow</u> S_t <u>on</u> (Y,ν) <u>s.t.</u> $S = S_1$ <u>and</u> $\psi h_t(x) = S_t\psi(x)$ <u>for all</u> $t \in R$ <u>and</u> μ_1-a.e. $x \in M$.

THEOREM 7.2. [Ra7]. <u>Let</u> S_t <u>be an</u> m.p. <u>flow on</u> (Y,ν) <u>and let</u> S_1 <u>be a factor of</u> h_1 <u>on</u> $(\Gamma\backslash G, \mu)$ <u>via a factor map</u> ψ. <u>Then</u> S_t <u>is a factor of</u> h_t <u>via</u> ψ.

Corollary 6.1 says that $h_1^{(1)}$ on $\Gamma_1\backslash G$ is isomorphic to $h_1^{(2)}$ on $\Gamma_2\backslash G$ iff Γ_1 and Γ_2 are conjugate in G. For $\Gamma \in \mathbf{T}$ let $\alpha(\Gamma) = \{\Gamma' \in \mathbf{T} : \Gamma \subset \Gamma'\}$. It is well known that $\alpha(\Gamma)$ is finite. Γ is called maximal if

$\alpha(\Gamma) = \{\Gamma\}$. We get the following

COROLLARY 7.1. <u>The number of nonisomorphic factors of</u> h_1 <u>on</u> $\Gamma\backslash G$, $\Gamma \in \mathbf{T}$ <u>is finite and equal to the number of conjugacy classes in</u> $\alpha(\Gamma)$. <u>If</u> Γ <u>is maximal and</u> S <u>is a factor of</u> h_1 <u>then</u> S <u>is isomorphic to</u> h_1.

Knowing the commutant of h_t from Corollary 6.2 we can describe $\varkappa(h_1,S)$ for every factor S of h_1 and show that it is finite [Ra6].

In order to prove Theorem 7.1 we first observe the following important dynamical property of h_t called an H-property, which is caused by (2.2) and the polynomial divergence of horocycle orbits: given $\varepsilon > 0$, $N > 0$ there are $\alpha = \alpha(\varepsilon) > 0$, $\delta = \delta(\varepsilon,N) > 0$ s.t. if $d(x,y) < \delta$, $x,y \in M$ and y is not on the h_t-orbit of x then there are $L = L(x,y) > 0$ and $M = M(x,y) \geq N$ with $M/L \geq \alpha$ s.t. either $d(h_nx,h_{n+1}y) < \varepsilon$ for all $n \in [L,L+M]$ or $d(h_nx,h_{n-1}y) < \varepsilon$ for all $n \in [L,L+M]$, where d denotes the distance in M.

We use this property to show that if S is a factor of h_1 via a factor map ψ then μ-a.e. fiber of ψ is finite. Then we show that if the partition ξ into ψ-fibers is invariant under h_1 then it is invariant under h_t, g_t and h_t^* for all $t \in R$. This implies that ξ is invariant under the map $\Gamma g \to \Gamma g\alpha$ for every $\alpha \in G$. This enables us to construct a group $\Gamma' \in \mathbf{T}$ s.t. $\Gamma \subset \Gamma'$ and $\xi(\Gamma g) = \Gamma\backslash\Gamma' g$, $g \in G$.

The H-property is generalized and discussed in details in [Ra7]. We believe that Theorem 7.1 can be generalized to uniformly parameterized horocycle flows associated with geodesic flows on the unit tangent bundles of compact surfaces of variable negative curvature.

8. JOININGS

Let I be a countable set (finite or infinite) and let $(X_i,\mathcal{B}_i)$ $i \in I$ be a measurable space. Let $(X^I,\mathcal{B}^I)$ denote the product space $(\times_{i\in I} X_i, \times_{i\in I} \mathcal{B}_i)$. For $I' \subset I$ let $\pi_{I'}:X^I \to X^{I'}$ be the projection $(\pi_{I'}(x))_i = x_i$ $i \in I'$ where $x_i \in X_i$ is the i-coordinate of $x \in X^I$. Let m be a probability measure on $(X^I,\mathcal{B}^I)$ and $I' \subset I$. The probability measure $m_{I'}$ on $(X^{I'},\mathcal{B}^{I'})$ defined by $m_{I'}(A) = m(\pi_{I'}^{-1}(A))$, $A \in \mathcal{B}^{I'}$ is called the I'-marginal of m. If $|I'| = n$ then $m_{I'}$ is called an n-dimensional marginal of m. We say that m is divisible if there is a nonempty proper subset $I' \subset I$ s.t. $m = m_{I'} \times m_{I-I'}$. If m is undivisible it is called prime. If there are disjoint $I_j \subset I$, $\cup I_j = I$ s.t. $m = m_{I_1} \times m_{I_2} \times \ldots = \times_j m_{I_j}$ and each m_{I_j} is prime, then $\times_j m_{I_j}$ is called the prime expansion of m. It is clear that the prime expansion of m is unique.

Let $T^{(i)}$ be an m.p.t. on a probability space $(X_i, \mathcal{B}_i, \mu_i)$, $i \in I$. We shall denote by T^I the cartesian product $\underset{i\in I}{\times} T^{(i)}$ acting on the product space $(X^I, \mathcal{B}^I)$. A T^I-invariant probability measure m on $(X^I, \mathcal{B}^I)$ is called a joining of $T^{(i)}$, $i \in I$ if all one-dimensional marginals of m are μ_i, $i \in I$. Sometimes the word "joining" will also be used for the pair (T^I, m) meaning T^I acting on $(X^I, \mathcal{B}^I, m)$. It is clear that the product measure $\underset{i\in I}{\times} \mu_i = \mu^I$ is a joining of $T^{(i)}$, $i \in I$. We shall denote by $J(T^I)$ the set of all ergodic joinings of $T^{(i)}$, $i \in I$. $J(T^I)$ might be empty. We say that $T^{(i)}$, $i \in I$ are disjoint if $J(T^I) = \{\mu^I\}$.

Let T_t be an m.p. flow on a probability space $(X, \mathcal{B}_X, \mu)$ and $T = T_p$, $p \neq 0$ be ergodic. Set $(X_i, \mathcal{B}_i, \mu_i) = (X, \mathcal{B}_X, \mu)$, $T^{(i)} = T$ for all $i \in I$, $|I| > 1$. For $x \in X$ and $s \in R^I$, let $x^s \in X^I$ be defined by $(x^s)_i = T_{s(i)}x_i$, $i \in I$. Let m_s be the probability measure on $(X^I, \mathcal{B}^I)$ defined by $m_s\{x^s \in X^I : x \in A\} = \mu(A)$ for all $A \in \mathcal{B}_X$. It is clear that $m_s \in J(T^I)$. The measures m_s, $s \in R^I$ are called the off-diagonal self-joinings of $T = T^{(i)}$, $i \in I$. T is said to have trivial countable self-joinings if for every countable I and every $m \in J(T^I)$ there are disjoint subsets $I_j \subset I$, $\cup I_j = I$ s.t. $m = \underset{j}{\times} m_j$, where $m_j \in J(T^{I_j})$ and for each j either $|I_j| = 1$ and $m_j = \mu$ or $|I_j| > 1$ and m_j is off-diagonal on X^{I_j}.

A joining $(T \times S, m)$ of two m.p.t.'s T on (X, μ) and S on (Y, ν) is called a finite extension of S if there is a $T \times S$-invariant subset $\Omega \subset X \times Y$, $m(\Omega) = 1$ s.t. the intersection $\Omega(y) = \Omega \cap (X \times \{y\})$ is finite for ν-a.e. $y \in Y$. If $(T \times S, m)$ is ergodic, the number of points in $\Omega(y)$ is the same for ν-a.e. $y \in Y$.

Using property H stated in section 7 we prove the following

THEOREM 8.1. [Ra7]. <u>Let</u> h_t <u>be the horocycle flow on</u> $(\Gamma\backslash G, \mu)$, $\Gamma \in \mathbf{T}$. <u>Let</u> $h = h_1$ <u>and</u> $(h \times S, m)$ <u>be an ergodic joining of</u> h <u>and some</u> m.p.t. S <u>on</u> (Y, ν). <u>Then either</u> $m = \mu \times \nu$ <u>or</u> $(h \times S, m)$ <u>is a finite extension of</u> S.

THEOREM 8.2. [Ra7]. <u>Let</u> S_t <u>be an</u> m.p. <u>flow and let</u> m <u>be an ergodic joining of</u> S_1 <u>and</u> h_1 <u>on</u> $(\Gamma\backslash G, \mu)$. <u>Then</u> m <u>is an ergodic joining of the flows</u> S_t <u>and</u> h_t.

Theorem 8.1 enables us to classify up to an isomorphism all ergodic joinings of two horocycle flows $h_t^{(1)}$ and $h_t^{(2)}$ on $M_1 = \Gamma_1\backslash G$ and $M_2 = \Gamma_2\backslash G$, $\Gamma_1, \Gamma_2 \in \mathbf{T}$.

Suppose that $\Gamma_1 \cap \alpha\Gamma_2\alpha^{-1} = \Gamma_0$ belongs to $\mathbf{T}$ for some $\alpha \in G$. Let $\Gamma_0\backslash\Gamma_1 = \{\Gamma_0\gamma_i : i = 1, \ldots, n\}$; $\gamma_i \in \Gamma_1$, $n = |\Gamma_0|\Gamma_1|$ and let $\Omega = \{(\Gamma_1 g, \Gamma_2\alpha^{-1}\gamma_i g) : g \in G, i = 1, \ldots, n\}$. The set Ω is well defined and is an $h_t^{(1)} \times h_t^{(2)}$-invariant subset of $M_1 \times M_2$. Define a probability measure m on $M_1 \times M_2$ by

$m(A \times B) = \int m_x(B)d\mu$, for every measurable sets $A \subset M_1$, $B \subset M_2$ where $m_x(B) = |i \in \{1,\ldots,n\}: \Gamma_2\alpha^{-1}\gamma_i g \in B|/n$ if $x = \Gamma_1 g$. It is easy to see that m is a joining of $h_t^{(1)}$ and $h_t^{(2)}$.

Let $h_t^{(0)}$ be the horocycle flow on $(M_0 = \Gamma_0\backslash G,\mu_0)$, $h_t^{(1)}$ is a factor of $h_t^{(0)}$ via $\psi_1: M_0 \to M_1$, $\psi_1(\Gamma_0 g) = \Gamma_1 g$. Similarly, $\tilde{h}_t^{(2)}$ on $(\tilde{M}_2 = \tilde{\Gamma}_2\backslash G,\mu_2)$, $\tilde{\Gamma}_2 = \alpha\Gamma_2\alpha^{-1}$ is a factor of $h_t^{(0)}$ via $\psi_2:\psi_2(\Gamma_0 g) = \tilde{\Gamma}_2 g$. Also, $\tilde{h}_t^{(2)} \sim h_t^{(2)}$ via $\psi_\alpha:\psi_\alpha(\tilde{\Gamma}_2 g) = \Gamma_2\alpha^{-1}g$. Let $A_1(\Gamma_1 g) = \psi_1^{-1}\{\Gamma_1 g\} \subset M_0$ and $A_2(\tilde{\Gamma}_2 z) = \psi_2^{-1}\{\tilde{\Gamma}_2 z\} \subset M_0$. It is easy to show that for any given $g,z \in G$ the intersection $A_1(\Gamma_1 g) \cap A_2(\tilde{\Gamma}_2 z)$ consists of at most one point and is nonempty iff $(\Gamma_1 g,\Gamma_2\alpha^{-1}z) \in \Omega$. Let $\psi:\Omega \to M_0$ be defined by $\psi(\Gamma_1 g,\Gamma_2 z) = A_1(\Gamma_1 g) \cap A_2(\tilde{\Gamma}_2\alpha z)$. The map ψ is an isomorphism between $(h_t^{(1)} \times h_t^{(2)},m)$ and $(h_t^{(0)},\mu_0)$ as well as between $(f_t^{(1)} \times f_t^{(2)},m)$ and $(f_t^{(0)},\mu_0)$ for any F-flows $f_t^{(i)}$ on M_i, $i = 0,1,2$. We shall call $(h_t^{(1)} \times h_t^{(2)},m)$ an algebraic joining of $h_t^{(1)}$ and $h_t^{(2)}$. The following theorem shows that all nontrivial ergodic joinings of $h_t^{(1)}$ and $h_t^{(2)}$ come from algebraic ones.

THEOREM 8.3. [Ra7]. <u>Let</u> $h_t^{(i)}$ <u>be the horocycle flow on</u> $(M_i = \Gamma_i\backslash G,\mu_i)$, $\Gamma_i \in \mathbf{T}$, $i = 1,2$ <u>and let</u> $(h^{(1)} \times h^{(2)},m)$ <u>be an ergodic joining of</u> $h^{(1)} = h_1^{(1)}$ <u>and</u> $h^{(2)} = h_1^{(2)}$. <u>Then</u> $(h_t^{(1)} \times h_t^{(2)},m)$ <u>is an ergodic joining of</u> $h_t^{(1)}$ <u>and</u> $h_t^{(2)}$ <u>and either</u> 1) $m = \mu_1 \times \mu_2$ <u>or</u> 2) <u>there are</u> $\alpha \in G$, $\sigma \in R$ <u>s.t.</u> $\Gamma_1 \cap \alpha\Gamma_2\alpha^{-1} = \Gamma_0 \in \mathbf{T}$, <u>the</u> $h_t^{(1)} \times h_t^{(2)}$<u>-invariant set</u> $\Omega = \{(\Gamma_1 g,h_\sigma^{(2)}(\Gamma_2\alpha^{-1}\gamma_i g)):g \in G,i = 1,\ldots,n\}$ <u>has m-measure</u> 1 <u>and</u> $(h_t^{(1)} \times h_t^{(2)},m)$ <u>is isomorphic to the horocycle flow</u> $h_t^{(0)}$ <u>on</u> $(\Gamma_0\backslash G,\mu_0)$, <u>where</u> $n = |\Gamma_0\backslash\Gamma_1|$ <u>and</u> $\Gamma_0\backslash\Gamma_1 = \{\Gamma_0\gamma_i:i = 1,\ldots,n\}$. <u>The number of nonisomorphic ergodic joinings of</u> $h_t^{(1)}$ <u>and</u> $h_t^{(2)}$ <u>is at most countable</u>.

Let us note that the number of nonisomorphic ergodic joinings of two Bernoulli flows, in particular of two geodesic flows, is uncountable [SmT].

The proof of Theorem 8.3 uses Theorem 8.1 and is similar to the proof of Theorem 7.1. Both the rigidity Theorem 6.1 and the factor Theorem 7.1 follow from Theorem 8.3.

A group $\Gamma \in \mathbf{T}$ is cocompact or arithmetic [Brl] iff so is every subgroup of Γ of finite index. This implies the following

COROLLARY 8.1. 1) <u>If</u> Γ_1 <u>is cocompact and</u> Γ_2 <u>is not, then</u> $h^{(1)}$ <u>and</u> $h^{(2)}$ <u>are disjoint</u>. 2) <u>If</u> Γ_1 <u>is arithmetic and</u> Γ_2 <u>is not, then</u> $h^{(1)}$ <u>and</u> $h^{(2)}$

are disjoint. If, in addition, Γ_1 and Γ_2 are cocompact then $h^{(1)} \times h^{(2)}$ is uniquely ergodic. 3) If Γ_1 and Γ_2 are not conjugate in G, are maximal and nonarithmetic, then $h^{(1)}$ and $h^{(2)}$ are disjoint.

For $\Gamma \in \mathbf{T}$ denote $\bar{\Gamma} = \{\alpha \in G: \alpha\Gamma\alpha^{-1} \cap \Gamma \in \mathbf{T}\} \supset \Gamma$. $\bar{\Gamma}$ is a countable subgroup of G which is either dense or discrete in G. Margulis [Mg2] proved that $\bar{\Gamma}$ is dense in G iff Γ is arithmetic. We get the following

COROLLARY 8.2. Let h_t be the horocycle flow on $(\Gamma\backslash G,\mu)$, $\Gamma \in \mathbf{T}$ and let $h = h_1$. 1) If Γ is arithmetic then the number of nonisomorphic ergodic self-joinings of h is countably infinite. 2) If Γ is not arithmetic the number is finite. 3) If Γ is maximal and not arithmetic then h has only trivial ergodic self-joinings.

Recently, Glasner and Weiss [GlW] have constructed a countable family Q of maximal cocompact arithmetic groups s.t. if $\Gamma_1,\Gamma_2 \in Q$, $\Gamma_1 \neq \Gamma_2$ then Γ_1 and Γ_2 are not conjugate in G, and $\Gamma_1 \cap \Gamma_2$ is of finite index in Γ_1 and in Γ_2. The existence of such a family follows also from a paper of A. Borel [Br2]. The horocycle flows $h_t^{(1)}$ and $h_t^{(2)}$ for such Γ_1, Γ_2 are minimal (see section 5), are not disjoint and have no common factors by Theorems 6.1 and 7.1. This answers in the negative a question raised by Furstenberg in [Fu1].

In the same way as we got Corollary 6.3 from Corollary 6.1 we now get

COROLLARY 8.3. Let $h_t^{(i)}$ be the horocycle flow on $(\Gamma_i\backslash G,\mu_i)$, $\Gamma_i \in \mathbf{T}$, $i = 1,2$ and let $(h_1^{(1)}\times h_{-1}^{(2)},m)$ be an ergodic joining of $h_1^{(1)}$ and $h_{-1}^{(2)}$. Then $(h_t^{(1)} \times h_{-t}^{(2)},m)$ is an ergodic joining of $h_t^{(1)}$ and $h_{-t}^{(2)}$ and either $m = \mu_1 \times \mu_2$ or there is β with $\det \beta = -1$ s.t. $\Gamma_1 \cap \beta\Gamma_2\beta^{-1} = \Gamma_0 \in \mathbf{T}$ and $(h_t^{(1)} \times h_{-t}^{(2)},m)$ is isomorphic to the horocycle flow $h_t^{(0)}$ on $(\Gamma_0\backslash G,\mu_0)$.

Since $h_t^{*(2)} \sim h_{-t}^{(2)}$ we can substitute $h_t^{*(2)}$ instead of $h_{-t}^{(2)}$ in Corollary 8.3.

9. CARTESIAN PRODUCTS

In this section we shall classify up to an isomorphism all ergodic joinings of n horocycle flows $h_t^{(i)}$ on $\Gamma_i\backslash G$, $i = 1,\ldots,n$. Let us note that all ergodic flows with sufficiently large entropy are joinings of at least three Bernoulli flows [SmT], hence of geodesic flows.

Using Theorem 8.3 and the H-property of horocycle flows from section 7 we prove the following

THEOREM 9.1. [Ra8]. Let $h_t^{(i)}$ be the horocycle flow on $(M_i = \Gamma_i\backslash G, \mu_i)$, $\Gamma_i \in \mathbf{T}$, $h^{(i)} = h_1^{(i)}$, $i \in \mathbf{n} = \{1,\dots,n\}$, $h^{\mathbf{n}} = \times_{i\in\mathbf{n}} h^{(i)}$ and let $m \in J(h^{\mathbf{n}})$. Then there are disjoint subsets $I_j \subset \mathbf{n}$, $I_j = \{i_1^{(j)},\dots,i_{n_j}^{(j)}\}$, $j = 1,\dots,q$, $1 \le q \le \mathbf{n}$, $\cup I_j = \mathbf{n}$ and $\alpha_{i_k}^{(j)} \in G$, $k = 1,\dots,n_j$, $j = 1,\dots,q$ s.t. $\bigcap_{k=1}^{n_j} \alpha_{i_k}^{(j)}\Gamma_{i_k}(\alpha_{i_k}^{(j)})^{-1} = \Gamma_0^{(j)} \in \mathbf{T}$ and $m = \times_{j=1}^{q} m_j$ where $m_j \in J(h^{I_j})$ is prime and $(h_t^{I_j}, m_j)$ is isomorphic to the horocycle flow on $(M_0^{(j)} = \Gamma_0^{(j)}\backslash G, \mu_0^{(j)})$, $j = 1,\dots,q$.

In section 5 we described all ergodic invariant measures for h_t on $\Gamma\backslash G$, $\Gamma \in \mathbf{T}$. This and Theorem 9.1 give the description of all ergodic invariant measures for the horocycle flow acting on $\Gamma_1 \times\dots\times \Gamma_n\backslash G^{\mathbf{n}}$ where $\Gamma_i \in \mathbf{T}$, $i = 1,\dots,n$ and $G^{\mathbf{n}}$ denotes the n-fold cartesian product of G with itself.

PROBLEM. Classify all ergodic invarinat measures of the horocycle flow acting on $\Gamma\backslash G^{\mathbf{n}}$, where Γ is a discrete subgroup of $G^{\mathbf{n}}$ with $\Gamma\backslash G^{\mathbf{n}}$ of finite volume.

THEOREM 9.2. [Ra8]. Let I and L be two disjoint countable sets and let $h_t^{(i)}$ be the horocycle flow on $(M_i = \Gamma_i\backslash G, \mu_i)$, $\Gamma_i \in \mathbf{T}$, $h^{(i)} = h_1^{(i)}$, $i \in I \cup L$. Let $S = h^I$ act on (M^I, μ^I) and $U = h^L$ act on (M^L, μ^L). Let $m \in J(S \times U)$. Then there are $\tilde{I} \subset I$, $\alpha_i \in G$, $i \in \tilde{I}$ ($\tilde{I}$ might be empty) and a one-to-one map $p:\tilde{I} \to L$ s.t. $\Gamma_i \cap \alpha_i\Gamma_{p(i)}\alpha_i^{-1} = \Gamma_0^{(i)} \in \mathbf{T}$, $i \in \tilde{I}$ and $m = \times_{i\in\tilde{I}} m_i \times \mu^{I-\tilde{I}} \times \mu^{L-p(\tilde{I})}$, where $m_i \in J(h^{(i)} \times h^{(p(i))})$ is prime and $(h_t^{(i)} \times h_t^{(p(i))}, m_i)$ is isomorphic to the horocycle flow on $(\Gamma_0^{(i)}\backslash G, \mu_0^{(i)})$, $i \in \tilde{I}$.

THEOREM 9.3. [Ra8]. (Rigidity of products). Let I and L be two disjoint countable sets and let $h_t^{(i)}$ be the horocycle flow on $(M_i = \Gamma_i\backslash G, \mu_i)$, $\Gamma_i \in \mathbf{T}$. $h^{(i)} = h_1^{(i)}$, $i \in I \cup L$. Let $S = h^I$ act on (M^I, μ^I) and $U = h^L$ act on (M^L, μ^L). Suppose that there exists an m.p. $\psi: M^I \to M^L$ s.t. $\psi S(x) = U\psi(x)$ for μ^I-a.e. $x \in M^I$. Then there are $\tilde{I} \subset I$, $\sigma_i \in R$, $\alpha_i \in G$, $i \in \tilde{I}$ and a one-to one map p from $\tilde{I}$ onto L s.t. $\Gamma_i \subset \alpha_i^{-1}\Gamma_{p(i)}\alpha_i$, $i \in \tilde{I}$ and $\psi = \tilde{\psi}\circ \pi_{\tilde{I}}$ where $\tilde{\psi}: M^{\tilde{I}} \to M^L$ is $(\tilde{\psi}(x))_{p(i)} = h_{\sigma_i}^{(p(i))}(\Gamma_{p(i)}\alpha_i g)$ if $x_i = \Gamma_i g$, $i \in \tilde{I}$ and $\pi_{\tilde{I}}$ denotes the projection onto $M^{\tilde{I}}$.

COROLLARY 9.1. Suppose that h^I on (M^I, μ^I) is isomorphic to h^L on (M^L, μ^L) via an isomorphism $\psi: M^I \to M^L$. Then there are $\sigma_i \in R$, $\alpha_i \in G$, $i \in I$ and a one-to-one map p from I onto L s.t. $\Gamma_i = \alpha_i^{-1}\Gamma_{p(i)}\alpha_i$, $i \in I$ and ψ has the form $(\psi(x))_{p(i)} = h_{\sigma_i}^{(p(i))}(\Gamma_{p(i)}\alpha_i g)$, if $x_i = \Gamma_i g$, $i \in I$.

By Margulis' theorem [Mg2] $\Gamma = \bar{\Gamma}$ iff Γ is maximal and nonarithmetic.

Using Corollary 8.2 we prove the following

THEOREM 9.4. [Ra8]. *Let* $\Gamma \in \mathbf{T}$ *be maximal and nonarithmetic and let* h_t *be the horocycle flow on* $(\Gamma\backslash G,\mu)$, $h = h_1$. *Then* h *has trivial countable self-joinings.*

Let $\Psi^I = \Psi^I(h^I)$ be the commutant of h^I acting on (M^I,μ^I), i.e. the group of all isomorphisms $\psi:M^I \to M^I$ which commute with h^I. For $s \in R^I$ let $\varphi_s:M^I \to M^I$ be defined by $(\varphi_s(x))_i = h^{(i)}_{s(i)}x_i$, $i \in I$. It is clear that $\varphi_s \in \Psi^I$ and that $\Phi^I = \{\varphi_s:s \in R^I\}$ is a subgroup of Ψ^I isomorphic to R^I. Corollary 1 shows that Φ^I is a normal subgroup of Ψ^I. We shall denote by $k(h^I)$ the quotient group $\Phi^I|\Psi^I$.

Suppose now that $h^{(i)}_t = h_t$ on $(M = \Gamma\backslash G,\mu)$ for all $i \in I$, $h^{(i)} = h_1 = h$. Let $\mathbf{P}(I)$ denote the group of permutations of I and for $p \in \mathbf{P}(I)$ let $\pi_p:M^I \to M^I$ be $(\pi_p(x))_i = x_{p^{-1}(i)}$, $x \in M^I$, $i \in I$. (In fact we shall use the notation $\pi_p(z)$ for any I-tuple z, $(\pi_p(z))_i = z_{p^{-1}(i)}$). It is clear that $\pi_p \in \Psi^I$ and $P^I = \{\pi_p:p \in \mathbf{P}(I)\}$ is a subgroup of Ψ^I isomorphic to $\mathbf{P}(I)$. We say that h has trivial product commutant if for every countable I the commutant $\Psi(h^I)$ is generated by Φ^I and P^I.

It follows from Corollary 9.1 that $\Psi(h^I) = \{\varphi_s \circ \pi_p \circ \psi_C:p \in \mathbf{P}(I), s \in R^I, C \in \tilde{\Gamma}^I\}$ where $\tilde{\Gamma} \supset \Gamma$ denotes the normalizer of Γ in G and $\psi_C:M^I \to M^I$ is defined by $(\psi_C(\Gamma^I g))_i = \Gamma C_i g_i$, $g \in G^I$, $i \in I$. It is clear that $\psi_{\gamma C} = \psi_{C\gamma} = \psi_C$ for all $\gamma \in \Gamma^I$, $C \in \tilde{\Gamma}^I$. Let $\mathbf{K} = \{\pi_p \circ \psi_C:p \in \mathbf{P}(I),C \in \tilde{\Gamma}^I\}$. One can see that $\pi_p \circ \psi_C = \psi_{\pi_p(C)} \circ \pi_p$. This implies that $\mathbf{K}$ is a subgroup of Ψ^I with $(\pi_p \circ \psi_C)\cdot(\pi_q \circ \psi_D) = \pi_{pq} \circ \psi_{\pi_p^{-1}(C)D}$, $p,q \in \mathbf{P}(I)C$, $C,D \in \tilde{\Gamma}^I$. We get the following

COROLLARY 9.2. *Let* $h^{(i)}_t$ *be the horocycle flow on* $(M_i = \Gamma_i\backslash G,\mu_i)$, $\Gamma_i \in \mathbf{T}$, $h^{(i)} = h^{(i)}_1$, $i \in I$ *and let* h^I *act on* (M^I,μ^I). *If* I *is finite then the group* $k(h^I)$ *is finite. If* $h^{(i)} = h$ *on* $(M = \Gamma\backslash G,\mu)$ *for all* $i \in I$ *then* $k(h^I)$ *is isomorphic to* $\mathbf{K}$. *In particular, if* $|I| = n$ *then* $|k(h^I)| = |\Gamma\backslash\tilde{\Gamma}|^n n!$ *If* $\Gamma = \tilde{\Gamma}$ *then* $k(h^I)$ *is isomorphic to the group of permutations of* I, *and* h *has trivial product commutant.*

Let $M = \Gamma\backslash G$, $\Gamma \in \mathbf{T}$, $\alpha \in G$ and $f_\alpha:M \to M$ be defined by $f_\alpha(\Gamma g) = \Gamma g\alpha$ $g \in G$. Let $f^I_\alpha:M^I \to M^I$ be $(f^I_\alpha(x))_i = f_\alpha(x_i)$, $i \in I$. The map $\alpha \to f^I_\alpha$ is a measure preserving action of $G = SL(2,R)$ on (M^I,μ^I). Let $\Psi(G)$ be the commutant of this action, i.e. the group of all isomorphisms $\psi:M^I \to M^I$ which commute with every f^I_α, $\alpha \in G$.

COROLLARY 9.3. $\Psi(G) = \{\pi_p \circ \psi_C : p \in \mathbf{P}(I),\ C \in \widetilde{\Gamma}^I\}$.

Next we shall study factors of h^I on (M^I, μ^I), assuming that $h^{(i)} = h$ on $(M = \Gamma\backslash G, \mu)$ for all $i \in I$. By a factor we shall mean a measurable partition ξ of (M^I, μ^I) invariant under h^I. ξ is called trivial if $\xi = \{M^I, \phi\}$. For $x \in M^I$ let $L(x,\xi)$ be the smallest subset of I s.t. the set $\{z \in M^I : z_i = x_i,\ i \in L(x,\xi)\}$ is contained in $\xi(x)$, where $\xi(x)$ denotes the atom of ξ containing x. We have $L(x,\xi) = L(h^I x, \xi)$ for μ^I-a.e. $x \in M^I$ and since h^I is ergodic $L(x,\xi)$ is the same say $L = L(\xi)$ for a.e. $x \in M^I$. Let $r: M^I \to M^L$ be $(r(x))_i = x_i$, $i \in L$ and let ξ_L be the h^L-invariant partition of M^L defined by $\xi_L(z) = r(\xi(r^{-1}z))$, $z \in M^L$. We shall say that the factor ξ projects to the factor ξ_L.

Now we are going to classify up to an isomorphism all factors of h^I, assuming for simplicity that I is finite, $I = \{1,\dots,n\} = \mathbf{n}$.

Let $\bar{\Gamma} = \{\alpha \in G : \alpha\Gamma\alpha^{-1} \cap \Gamma \in \mathbf{T}\} \supset \widetilde{\Gamma} \supset \Gamma$ be as in section 8. $\alpha \in \bar{\Gamma}$ iff the set $\{\Gamma\alpha\gamma : \gamma \in \Gamma\}$ is finite. Let $\Lambda = \Lambda(h^{\mathbf{n}}, \Gamma) = \{\pi_p \circ \varphi_s \circ \psi_C : p \in \mathbf{P}(\mathbf{n}),\ s \in R^{\mathbf{n}},\ C \in \bar{\Gamma}^{\mathbf{n}}\}$. Λ is a group with the group operation $(\pi_p \circ \varphi_s \circ \psi_C) \cdot (\pi_q \circ \varphi_t \circ \psi_D) = \pi_{pq} \circ \varphi_{\pi^{-1}(s)+t} \circ \psi_{\pi^{-1}(C)D}$.

Let $\mathbf{U} = \mathbf{U}(h^{\mathbf{n}}, \Gamma)$ be the set of all finite subgroups U of Λ s.t. if $\pi_p \circ \varphi_s \circ \psi_C \in U$ for some $p \in \mathbf{P}(\mathbf{n})$, $s \in R^{\mathbf{n}}$, $C \in \bar{\Gamma}^{\mathbf{n}}$ then $\pi_p \circ \varphi_s \circ \psi_{C\gamma} \in U$ for every $\gamma \in \Gamma^{\mathbf{n}}$. For $U \in \mathbf{U}$ and $x \in M^{\mathbf{n}}$ let $\xi_u(x) = \{u(x) : u \in U\}$. ξ_u is well defined as is a factor of $h^{\mathbf{n}}$. Here are some examples of $U \in \mathbf{U}$.

EXAMPLE 1. Let Γ' be a discrete subgroup of $G^{\mathbf{n}}$ s.t. $\Gamma^{\mathbf{n}} \subset \Gamma'$. $\Gamma^{\mathbf{n}}$ has finite index in Γ' and hence $\Gamma' \subset \bar{\Gamma}^{\mathbf{n}}$. This implies that $U = \{\psi_C : C \in \Gamma'\} \in \mathbf{U}$. One can see that $\xi_u(x) = \Gamma^{\mathbf{n}}\,\Gamma' g$ if $x = \Gamma^{\mathbf{n}} g \in M^{\mathbf{n}}$, $g \in G^{\mathbf{n}}$.

EXAMPLE 2. Let U be a finite subgroup of the commutant $\Psi(h^{\mathbf{n}})$. Then $U \in \mathbf{U}$ since $\widetilde{\Gamma} \subset \bar{\Gamma}$ and $\psi_{C\gamma} = \psi_C$ for every $C \in \widetilde{\Gamma}^{\mathbf{n}}$, $\gamma \in \Gamma^{\mathbf{n}}$. Here are examples of $U, V \in \mathbf{U}$ of this kind that differ from the groups described in Example 1.

$U = \{\pi_{p^k} \circ \varphi_{\sum_{j=0}^{k-1} \pi^{-1}_{p^{k-1}}(s)} : k = 1,\dots,m\}$ for some $p \in \mathbf{P}(\mathbf{n})$ of order m and some $s \in R^{\mathbf{n}}$ with $\sum_{k=1}^{m} \pi^{-1}_{p^{k-1}}(s) = 0$. It is clear that $U \in \mathbf{U}$. Now let $n = 2$, $\mathbf{n} = \{1,2\}$, $\Gamma \neq \widetilde{\Gamma} \supset \Gamma$, $\alpha \in \widetilde{\Gamma}-\Gamma$ and let $p \in \mathbf{P}(\mathbf{n})$ be $p(1) = 2$, $p(2) = 1$. Let $U = \{\pi_p \circ \varphi_{(1,-1)} \circ \psi_{(\alpha,\alpha^{-1})}, \mathrm{Id}\}$ where Id denotes the identity element in Λ. One can check that $U \in \mathbf{U}(h^{\mathbf{n}}, \Gamma)$.

THEOREM 9.5. [Ra8]. <u>Let ξ be a factor of $h^{\mathbf{n}}$ on $(M^{\mathbf{n}} = \Gamma^{\mathbf{n}} G^{\mathbf{n}}, \mu^{\mathbf{n}})$ and let ξ project to ξ_L, $L \subset \mathbf{n}$. Then there is $U \in \mathbf{U}(h^L, \Gamma)$ s.t. $\xi_L = \xi_u$ a.e. on (M^L, μ^L).</u>

We say that a factor ξ on $(M^{\mathbf{n}},\mu^{\mathbf{n}})$ is algebraic if ξ is a factor of each $f_\alpha^{\mathbf{n}}$, $\alpha \in G$. Let $\mathbf{U}(G^{\mathbf{n}},\Gamma)$ be the set of all finite groups U consisting of maps of the form $\pi_p \circ \psi_C$ for some $p \in \mathbf{P}(\mathbf{n})$, $C \in \overline{\Gamma}^{\mathbf{n}}$ and s.t. $\pi_p \circ \psi_C \in U$ iff $\pi_p \circ \psi_{C\gamma} \in U$ for every $\gamma \in \Gamma^{\mathbf{n}}$. We get the following:

COROLLARY 9.4. <u>A factor</u> ξ <u>on</u> $(M^{\mathbf{n}},\mu^{\mathbf{n}})$ <u>is algebraic iff it projects to a factor</u> $\xi_{L,U}$, $L \subset \mathbf{n}$ <u>for some</u> $U \in \mathbf{U}(G^L,\Gamma)$.

Let I be countable (finite or infinite), $h^{(i)} = h$ on $(M = \Gamma\backslash G,\mu)$, $i \in I$ and let h^I act on (M^I,μ^I). Let $\mathbf{V}_I$ denote the set of all subgroups $V \subset \Psi(h^I)$ of the commutant $\Psi(h^I)$ which consist of maps of the form $\pi_p \circ \varphi_s$ for some $p \in \mathbf{P}(I)$, $s \in R^I$. For $x \in M^I$, $V \in \mathbf{V}_I$ let $\xi_V(x) = \{v(x):v \in V\}$. It is clear that ξ is a factor of h^I, which might be trivial. h is called product prime if for every countable I every factor ξ of h^I projects to a factor $\xi_{L,V}$ for some $V \in \mathbf{V}_L$.

We have shown in Theorem 9.4 that if Γ is maximal and nonarithmetic then h has trivial countable self-joinings. This implies

COROLLARY 9.5. <u>Let</u> h_t <u>be the horocycle flow on</u> $(M = \Gamma\backslash G,\mu)$ <u>where</u> $\Gamma \in \mathbf{T}$ <u>is maximal and nonarithmetic. Then</u> $h_1 = h$ <u>is product prime</u>.

10. KAKUTANI EQUIVALENCE

There is another kind of equivalence in ergodic theory called Kakutani equivalence (see [Ka],[Fe],[Ktl],[We]) which is weaker than isomorphism.

Let T_t be an m.p. flow on a probability space (X,μ) and let $\tau:X \to R^+$ be integrable, $\int_X \tau d\mu = \alpha$. We say that a flow T_t^τ is obtained from T_t by the time change τ if $T_t^\tau(x) = T_{v(x,t)}(x)$, $x \in X$, where $v(x,t)$ is defined by $\int_0^{v(x,t)} \tau(T_u x)du = t$. The flow T_t^τ preserves the measure $d\mu_\tau = (\tau/\alpha)d\mu$. An m.p. flow S_t on (Y,ν) is called Kakutani equivalent to $T_t (S_t \approx T_t)$ if there is a time change τ s.t. S_t is isomorphic to T_t^τ. Equivalently, $S_t \approx T_t$ iff there is a measurable invertible $\psi:Y \to X$ s.t. ψ carries the measure ν to a measure equivalent to μ and maps orbits of S_t to orbits of T_t, preserving orientation on the orbits. ψ is called an orbit equivalence between S_t and T_t. Ornstein and Smorodinsky [OSm] showed that if T_t is an ergodic C^1-flow on a compact C^1-manifold M, preserving a Borel probability measure and T_t^τ is obtained from T_t by a time change τ then there is a differentiable function $\tilde{\tau}:M \to R^+$, C^∞ on orbits of T_t s.t. $T_t^{\tilde{\tau}}$ is isomorphic to T_t^τ. The derivative of $\tilde{\tau}$ may be unbounded.

One can show that if $S_t \approx T_t$ then the entropy $E(S_t) = 0$ iff $E(T_t) = 0$ and $E(S_t) < \infty$ iff $E(T_t) < \infty$.

It follows from Ornstein's theorem that any two Bernoulli flows are Kakutani equivalent. The same is true for any two irrational flows on the 2-torus [Ktl] (these flows are sometimes called Kronecker flows). According to Feldman [Fe] an m.p. flow with positive entropy is called loosely Bernoulli (LB) if it is Kakutani equivalent to a Bernoulli flow and an m.p. flow with zero entropy is called LB[1] if it is Kakutani equivalent to an irrational flow on the 2-torus.

Let us give another definition of Kakutani equivalence. Let φ be an m.p.t. on a probability space (X,μ) and let $f:X \to R^+$ be integrable, $\int f d\mu = \bar{f}$. Let $Y = \{(x,u): x \in X,\ 0 \le u \le f(x), (x,f(x)) = (\varphi x, 0)\}$ and let T_t on Y be defined by $T_t(x,u) = (x,u+t)$ if $u+t \le f(x)$ and $T_t(x,u) = (\varphi^n x, u+t - \sum_{k=0}^{n-1} f(\varphi^k x))$ if $u+t > f(x)$ where n is the largest positive integer with $\sum_{k=0}^{n-1} f(\varphi^k x) \le u+t$. The flow T_t preserves the measure ν on Y defined by $d\nu = d\mu \times dt/\bar{f}$ and is called the special flow built over φ with f. We write $T_t = (\varphi, f)$. φ is called a cross-section of T_t.

By the Ambrose-Kakutani theorem [AmK] every m.p. flow is isomorphic to a special flow (φ,f) for some m.p.t. φ and some $f > 0$. One can easily show that $T_t \approx S_t$ iff there are an m.p.t. φ on (X,μ) and $f_1, f_2 > 0$ integrable on X s.t. $T_t \sim (\varphi, f_1)$ and $S_t \sim (\varphi, f_2)$. In other words $T_t \approx S_t$ iff the flows have a common cross-section.

One can see that every irrational flow on the 2-torus is a special flow (φ,f) where φ is an irrational rotation of the circle and f is a positive constant. This implies that a zero entropy flow is LB iff it is isomorphic to a special flow built over an irrational rotation of the circle with a positive function f. The theorem of Ornstein and Smorodinsky [OSm] says that f can be assumed differentiable on the circle. A positive entropy flow is LB iff it is isomorphic to a special flow built over a Bernoulli automorphism.

The following theorem shows that the horocycle flow $h_t^{(1)}$ on $\Gamma_1\backslash G$ is Kakutani equivalent to $h_t^{(2)}$ on $\Gamma_2\backslash G$ for any $\Gamma_1, \Gamma_2 \in \mathbf{T}$. Also h_t on $M = \Gamma\backslash G$ is Kakutani equivalent to h_t^* on M.

THEOREM 10.1. [Ra2]. The horocycle flows h_t and h_t^* on $(\Gamma\backslash G,\mu)$, $\Gamma \in \mathbf{T}$ are loosely Bernoulli.

This theorem has been in fact proved [Ra2] for general horocycle flows associated with transitive C^2-Anosov flows. Marcus [Ma3] showed that if there is a homeomorphic orbit equivalence between $h_t^{(1)}$ and $h_t^{(2)}$ then $C\Gamma_1 C^{-1} = \Gamma_2$ for some $C \in G$.

[1]We propose to call these loosely Kronecker flows (LK) instead of loosely Bernoulli.

Kushnirenko [Ku2] was the first to show that h_t is not isomorphic to $h_t \times h_t$. The following theorem combined with Theorem 10.1 shows even more, namely that h_t is not Kakutani equivalent to $h_t \times h_t$.

THEOREM 10.2. [Ra3]. <u>The cartesian product</u> $h_t \times h_t$ <u>on</u> $(M \times M, \mu \times \mu)$ <u>is not loosely Bernoulli</u>.

Let h_t^n denote the n-fold cartesian product of h_t with itself acting on (M^n, μ^n), $M = \Gamma \backslash G$, $\Gamma \in \mathbf{T}$. We use Theorem 10.2 to prove

THEOREM 10.3. [Ra4]. <u>If</u> $m \neq n$ <u>then</u> h^m <u>is not Kakutani equivalent to</u> h^n.

Let us note that the n-fold cartesian product g_t^n of the geodesic flow g_t is Kakutani equivalent to g_t^m, since g_t^n and g_t^m both are Bernoulli flows.

PROBLEMS

1) Let $h_t^{(i)}$ be the horocycle flow on $M_i = \Gamma_i \backslash G$, $\Gamma_i \in \mathbf{T}$, $i = 1,2$. Is $h_t^{(1)} \times h_t^{(1)}$ Kakutani equivalent to $h_t^{(2)} \times h_t^{(2)}$?

2) Is $h_t \times h_{-t}$ Kakutani equivalent to $h_t \times h_t$?

3) Can Theorems 10.2 and 10.3 be extended to the uniformly parameterized horocycle flows associated with transitive C^2-Anosov flows?

4) Let $H^{(i)}$ be the 2-dimensional horocycle expanding foliation associated with the geodesic flow on the unit tangent bundle (M_i, μ_i) of a 3-dimensional manifold S_i of constant negative curvature with finite volume, where μ_i is the normalized volume on M_i, $i = 1,2$. We say that $H^{(1)}$ is Kakutani equivalent to $H^{(2)}$ if there is a measurable invertible $\psi : M_1 \to M_2$ s.t. 1) ψ carries μ_1 to a measure equivalent to μ_2. 2) ψ maps leaves of $H^{(1)}$ to leaves of $H^{(2)}$ preserving orientation on the leaves. 3) ψ restricted on μ_1-a.e. leaf of $H^{(1)}$ is a C^∞-map with μ_1-a.e. bounded Jacobian. Are $H^{(1)}$ and $H^{(2)}$ Kakutani equivalent?

5) Is $H^{(1)}$ from problem 4 Kakutani equivalent to the 2-dimensional Kronecker foliation on the 3-torus? (See [FeN1,2].)

11. EXAMPLES IN ERGODIC THEORY

A technique has recently been developed for producing m.p.t.'s illustrating certain phenomena in ergodic theory. It is based on a method of cutting an abstract probability space and stacking the parts suitably. Using this technique D. Rudolph [Ru2] has recently constructed an m.p.t. with so called minimal self-joinings and used it to develop a mechanism of producing various kinds of examples.

It is desirable, however, to seek "natural" m.p.t.'s illustrating the same phenomena. One can see from this paper that the horocycle flows h_t serve this purpose very well. Though $h = h_1$ never have minimal self-joinings in the sense of Rudolph, they have nevertheless trivial countable self-joinings in our sense for nonarithmetic maximal Γ by Theorem 9.4. This is enough in order to use h in a substantial part of Rudolph's mechanism of producing examples. Besides this, $h \times h$ gave us a natural example of a non-LB mixing m.p.t. An abstract example of this was first given by Feldman [Fe] (see also [Ru1]). In addition, $h_t \times h_t$ was used to give a natural example of a K-automorphism which is not Bernoulli [Kt2]. An abstract example of this was first constructed by Ornstein [O4]. Theorems 8.1 and 10.3 give some additional natural examples. We would like to mention that our method is completely different from the abstract one.

REFERENCES

[AmK] W. Ambrose, S. Kakutani, Structure and continuity of measurable flows. Duke Math. J. 9(1942), 25-42.

[An] D. V. Anosov, Geodesic flows on closed Riemannian manifolds with negative curvature, Proc. Steklov Inst. Math. 90(1967) (A.M.S. Translation, 1969).

[AnSi] D. V. Anosov, Ya.G. Sinai, Some smooth ergodic systems, Russian Math. Surveys 22(1967), 103-167.

[AGH] L. Auslander, L. Green, F. Hahn, Flows on homogeneous spaces, Princeton Univ. Press, 1963.

[Br1] A. Borel, Introduction aux groupes arithmetiques, Paris, Hermann, 1969.

[Br2] A. Borel, Commensurability classes and volumes of hyperbolic 3-manifolds, Annali Scu. Nor. Sup.-Pisa, Serie IV, Vol. VIII (1981), 1-33.

[Bw] R. Bowen, Equilibrium states and the ergodic theory of Anosov diffeomorphisms, Springer-Verlag 1975.

[BwM] R. Bowen, B. Marcus, Unique ergodicity for horocycle foliations, Israel J. Math. 26(1977), 43-67.

[D] S. G. Dani, Dynamics of the horospherical flow, Bull. Amer. Math. Soc. 3(1980), 1037-1039.

[DS] S. G. Dani, J. Smillie, Uniform distribution of horocycle orbits for Fuchsian groups, to appear.

[Fe] J. Feldman, Non-Bernoulli K-automorphisms and a problem of Kakutani, Israel S. Math. 24(1976), 16-37.

[FeN1] J. Feldman, D. Nadler, Reparameterization of N-flows of zero entropy, Trans. AMS, 256(1979), 289-304.

[FeN2] J. Feldman, D. Nadler, Corrections and some additions to "Reparameterization of N-flows of zero entropy", Trans. AMS, 264(1981), 583-585.

[FeO] J. Feldman, D. Ornstein, A partial extension of Ratner's rigidity theorem, to appear.

[Fo] L. Ford, Automorphic functions, McGraw-Hill, 1929.

[Fu1] H. Furstenberg, Disjointness in ergodic theory, minimal sets and a problem in diophantine approximation, Math. Systems Theory 1(1967), 1-49.

[Fu2] H. Furstenberg, The unique ergodicity of the horocycle flow, Recent Advances in topological dynamics, Springer-Verlag lecture notes 318(1973), 95-115.

[GF] I. M. Gelfand, S. V. Fomin, Geodesic flows on manifolds of constant negative curvature, Russian Math. Surveys 7(1952), 118-137.

[GGP] I. M. Gelfand, M. I. Graev, I. I. Piatetskii-Shapiro, Representation theory and automorphic functions, W. B. Saunders (1969).

[GlW] S. Glasner, B. Weiss, Minimal transformations with no common factor need not be disjoint, to appear.

[Gu] B. M. Gurevich, The entropy of horocycle flows, Soviet Math. Dokl. 2(1961), 124-130.

[He1] G. A. Hedlund, Fuchsian groups and transitive horocycles, Duke J. of Math. 2(1936), 530-542.

[He2] G. A. Hedlund, The dynamics of geodesic flows, Bull. AMS 45(1939), 241-260.

[He3] G. A. Hedlund, Fuchsian groups and mixtures, Annals of Math. 40(1939), 370-383.

[Ho1] E. Hopf, Fuchsian groups and ergodic theory, Trans. AMS 39(1936), 299-314.

[Ho2] E. Hopf, Statistik der Geodätischen Linien in Mannigfaltigkeiten Negativer-Krümmung, Ber. Voch. Sachs. Akad. Wiss. Leipzig 91(1939), 261-304.

[Ka] S. Kakutani, Induced measure preserving transformations, Proc. Imp. Acad. Tokyo 19(1943), 635-641.

[Kt1] A. B. Katok, Monotone equivalence in ergodic theory, Izv. Akad. Nauk Math. 41(1977), 104-157.

[Kt2] A. B. Katok, Smooth non-Bernoulli K-automorphisms, Inventiones Math. 61(1980), 291-300.

[KSF] I. P. Kornfeld, Ya.G. Sinai, S. V. Fomin, Ergodic theory, Springer-Verlag, 1982.

[Ku1] A. G. Kushnirenko, An estimate from above for the entropy of a classical system, Soviet Math. Dokl. 6, No. 2(1965), 360-362.

[Ku2] A. G. Kushnirenko, Metric invariants of entropy type, Russian Math. Surveys 22(1967), 53-61.

[Ma1] B. Marcus, Unique ergodicity of the horocycle flow: variable negative curvature case, Israel J. Math. 21(1975), 133-144.

[Ma2] B. Marcus, The horocycle flow is mixing of all degrees, Inventiones Math. 46(1978), 201-209.

[Ma3] B. Marcus, Topological conjugacy of horocycle flows, to appear in Amer. J. Math.

[Mg1] G. A. Margulis, Certain measures associated with U-flows on compact manifolds, Functional Anal. Appl. 4(1)(1970).

[Mg2] G. A. Margulis, Discrete groups of isometries of manifolds of nonpositive curvature, Proc. of the International Congress of Mathematicians, Vancouver, 1974, 21-34.

[O1] D. S. Ornstein, Bernoulli shifts with the same entropy are isomorphic, Advances in Math. 4(1970), 337-352.

[O2] D. S. Ornstein, Two Bernoulli shifts with infinite entropy are isomorphic, Advances in Math. 5(1970), 339-348.

[O3] D. S. Ornstein, The isomorphism theorem for Bernoulli flows, Advances in Math. 10(1973), 124-142.

[O4] D. S. Ornstein, An example of a Kolmogorov automorphism that is not a Bernoulli shift, Advances in Math. 10(1973), 49-69.

[OSm] D. Ornstein, M. Smorodinsky, Continuous speed changes for flows, Israel J. Math. 31(1978), 161-168.

[OW] D. Ornstein, B. Weiss, Geodesic flows are Bernoullian, Israel J. Math. 14(1973), 184-198.

[P] O. Parasyuk, Horocycle flows on surfaces of constant negative curvature, Russian Math. Surveys 8(1953), 125-126.

[Ra1] M. Ratner, Anosov flows with Gibbs measure are also Bernoullian, Israel J. Math. 17(1974), 380-391.

[Ra2] M. Ratner, Horocycle flows are loosely Bernoulli, Israel J. Math. 31(1978), 122-132.

[Ra3] M. Ratner, The cartesian square of the horocycle flow is not loosely Bernoulli, Israel J. Math. 34(1979), 72-96.

[Ra4] M. Ratner, Some invariants of Kakutani equivalence, Israel J. Math. 38(1981), 231-240.

[Ra5] M. Ratner, Rigidity of horocycle flows, Annals of Math. 115(1982), 587-614.

[Ra6] M. Ratner, Factors of horocycle flows, to appear in Ergodic theory and dynamical systems.

[Ra7] M. Ratner, Joinings of horocycle flows, to appear.

[Ra8] M. Ratner, Rigidity of products of horocycle flows, to appear.

[Ru1] D. Rudolph, Non-equivalence of measure preserving transformations.

[Ru2] D. Rudolph, An example of a measure preserving map with minimal self-joinings and applications. Journal d'analyse mathematique, 35(1979).

[Sa] P. Sarnak, Asymptotic behavior of periodic orbits of the horocycle flow and Eisenstein Series, Communications on Pure and Appl. Math. 34(1981), 719-739.

[Si1] Ya.G. Sinai, Geodesic flows on compact surfaces of negative curvature, Sov. Math. Dokl. 2(1961), 106-109.

[Si2] Ya.G. Sinai, Weak isomorphism of measure preserving transformations, Mat. Sbornik 63(1964), 23-42.

[Si3] Ya.G. Sinai, Gibbs measures in ergodic theory, Russian Math. Surveys 27(1972), 21-63.

[Sma] S. Smale, Differentiable dynamical systems, Bull. AMS 73(1967), 747-817.

[SmT] M. Smorodinsky, J.-P. Thouvenot, Bernoulli factors that span a transformation, Israel J. Math. 32(1979), 39-43.

[St] A. M. Stepin, Dynamical systems on homogeneous spaces of semisimple Lie groups, Izv. Acad. Nauk. USSR 37(1973), 1091-1107.

[V] W. Veech, Unique ergodicity of horospherical flows, Amer. J. Math. 99(1977), 827-859.

[Wa] P. Walters, Ergodic Theory, Springer-Verlag, 1982.

[We] B. Weiss, Equivalence of measure preserving transformations, Lecture Notes, the Institute for Advanced Studies, The Hebrew University of Jerusalem, 1976.

Department of Mathematics
University of California
Berkeley, California 94720

Contemporary Mathematics
Volume 26, 1984

EMBEDDINGS OF L^1 IN L^1

by

Haskell P. Rosenthal*

§1. Statements of the main results

L^1 denotes the classical real Banach space of equivalence classes of Lebesgue integrable functions on the unit interval. For $\lambda \geq 1$ and Banach spaces X and Y, X and Y are said to be isomorphic (resp. λ-isomorphic) if there is a one-one onto operator $T: X \to Y$ (resp. with $\|T\| \|T^{-1}\| \leq \lambda$). (Throughout, "operator" means "bounded linear map"). Our main structural result goes as follows:

THEOREM 1.1. <u>Let</u> X <u>be a subspace of</u> L^1 <u>with</u> X <u>isomorphic to</u> L^1 <u>and</u> $\varepsilon > 0$. <u>There exists a subspace</u> Y <u>of</u> X <u>with</u> Y $1+\varepsilon$ -<u>isomorphic to</u> L^1.

By a result of D. Alspach [1], one obtains that isomorphs of L^1 in L^1 contain subspaces which are small perturbations of isometric copies of L^1. (Actually this can be deduced rather easily from our proof of 1.1.) Precisely, we have the equivalent

THEOREM 1.1'. <u>Let</u> X <u>be a subspace of</u> L^1 <u>with</u> X <u>isomorphic to</u> L^1 <u>and</u> $\varepsilon > 0$. <u>There exists a subspace</u> Y <u>of</u> X, <u>a subspace</u> Z <u>of</u> L^1 <u>isometric to</u> L^1 <u>and a surjective operator</u> $T: Y \to Z$ <u>so that</u>

$$\|y-Ty\| \leq \varepsilon\|y\| \quad \underline{\text{for all}} \quad y \in Y.$$

Of course it follows easily that if X and ε are as in Theorem 1.1, then the Y of Theorem 1.1 may be chosen with Y $1+\varepsilon$-complemented in L^1. (I.e. there is an operator $P: L^1 \to Y$ with $P|Y = Id|Y$ and $\|P\| < 1 + \varepsilon$.) This sharpens the previous result of Enflo-Starbird [2] (c.f. also Kalton [5]) that Y may be chosen with Y complemented in L^1 and isomorphic to L^1.

Theorem 1.1 should be contrasted with results of Lindenstrauss and Pełczyński. Indeed, these authors obtain in [6] that for all $\lambda > 1$, there exists a Banach space B with B isomorphic to L^1 such that B contains no

*The research discussed here was supported in part by NSF-MCS-8002393.

λ-isomorph of L^1. Moreover for all $1 < p < \infty$, $p \neq 2$, there exists a $\lambda_p > 1$ and a subspace B of L^p so that B is isomorphic to L^p yet B contains no λ_p-isomorph of L^p. However it also follows from their results that for all such p there is a $K_p > 1$ so that every subspace B of L^p does contain a K_p-isomorph of L^p. Nevertheless our Theorem 1.1 shows the "singular" nature of L^1 among the L^p-spaces, for one has $\lim_{p \to 1} \lambda_p = \infty$.

We prove our structural result by a rather delicate analysis of general operators on L^1, employing a powerful decomposition of operators on L^1 due to N. Kalton [5] (see also [3]). Our analysis yields the following theorem:

THEOREM 1.2. Let B be a Banach space and $T: L^1 \to B$ an operator such that T is an isomorphism on some subspace of L^1 isomorphic to L^1. Then T is an isomorphism on a subspace X of L^1 with X isometric to L^1. Moreover if $B = L^1$ itself, then there is a $\lambda > 0$ so that given $\varepsilon > 0$, X may be chosen so that there is an operator $S:X \to L^1$ with $\|T|X\| \geq \lambda$, $S/\|S\|$ an isometry and $\|S-T|X\| < \varepsilon$.

We now recall some previous results: we choose to formulate these in a rather general setting. Let $(\Omega, \mathcal{S}, \mu)$ and $(X, \mathcal{F}, \nu)$ be fixed finite measure spaces. We denote $L^1(\Omega, \mathcal{S}, \mu)$ simply by $L^1(\mu)$.

DEFINITION. An operator $T:L^1(\mu) \to L^1(\nu)$ is called an atom if T maps disjoint functions to disjoint functions; i.e. if $f,g \in L^1(\mu)$ satisfy $|f| \wedge |g| = 0$ a.e., then $|Tf| \wedge |Tg| = 0$ a.e.

(Here $u \wedge v$ denotes the minimum of the functions u and v.) It is perhaps noteworthy that functions $f,g \in L^1(\mu)$ satisfy $|f| \wedge |g| = 0$ a.e. if and only if $\|f \pm g\| = \|f\| + \|g\|$ (where $\|f\| = \int |f| d\mu$); thus Definition 1 can be given in "pure" Banach space language.

We are of course mainly interested in the case where $\Omega = [0,1]$, $\mathcal{S}$ the Borel subsets of $[0,1]$, and μ Lebesgue measure on $[0,1]$. Indeed, if μ is an atomless measure with $L^1(\mu)$ separable, then $L^1(\mu)$ is isometric to L^1 as is well-known. When convenient, we shall assume that Ω is a separable complete metric space and $\mathcal{S}$ the family of Borel subsets of Ω. Under this assumption, we have the following known structural result:

PROPOSITION 1.3. $T:L^1(\mu) \to L^1(\nu)$ is an atom if and only if there are measurable functions $a: X \to \mathbb{R}$ and $\sigma: X \to \Omega$ with $(Tf)(x) = a(x)f(\sigma x)$ a.e. for all $f \in L^1(\mu)$.

We shall say that a and σ represent T if the above equality holds. Evidently a pair of such functions a and σ represents an atom T if and only if $K = \sup \{\int_{\sigma^{-1}(S)} |a(x)|d\nu(x)/\mu(S) : S \in \mathcal{S}, \mu(S) > 0\} < \infty$, and then $\|T\| = K$.

The proof of Proposition 1.3 may be deduced from classical results as follows: Let $\mathcal{N}$ denote the σ-ring of members of $\mathcal{F}$ of ν-measure 0. Given an atom T, set $a = T1$. For $f \in L^1(\nu)$, let $\text{supp } f = \{x: f(x) \neq 0\}$. (Of course supp f is thus properly defined as a member of $\mathcal{F}/\mathcal{N}$.) Define $F: \mathcal{S} \to \mathcal{F}/\mathcal{N}$ by $F(S) = \text{supp } T\chi_S$ for all $S \in \mathcal{S}$. Now if $X_0 = \text{supp } T1$, $\mathcal{F}_0 = \mathcal{F} \cap X_0$, $\mathcal{N}_0 = \mathcal{N} \cap X_0$, then F is a σ-homomorphism of $\mathcal{S}$ into $\mathcal{F}_0/\mathcal{N}_0$ where $(X_0, \mathcal{F}_0, \mathcal{N}_0)$ is our "measurable space with null sets", whence by a result of Sikorski (c.f. Theorem 6, page 328 of [8]) there exists a $\sigma: X_0 \to \Omega$ inducing F; i.e. $F(S) =$ (the equivalence class of) $\sigma^{-1}(S)$ for all $S \in \mathcal{S}$. We may trivially extend σ on $\sim X_0$ by mapping the latter set to a point; then a and σ represent T.

Let us return now to our general setting. Given an operator $T: L^1(\mu) \to L^1(\nu)$, recall that there exists a unique operator, denoted $|T|: L^1(\mu) \to L^1(\nu)$ so that for all $f \geq 0$, $f \in L^1(\mu)$,

$$|T|(f) = \sup \{ \sum_{i=1}^{n} |T f_i| : n \text{ is arbitrary, and } f_1, \ldots, f_n \geq 0 \text{ in } L^1(\mu) \text{ are such that } f = \sum_{i=1}^{n} f_i \} .$$

For standard facts about $|T|$ (such as $\| |T| \| = \|T\|$), see [9]. Given $A: L^1(\mu) \to L^1(\nu)$, we say $A \leq T$ if $Af \leq Tf$ a.e. for all $f \geq 0$ in $L^1(\mu)$.

DEFINITION. Let $T: L^1(\mu) \to L^1(\nu)$ be a given operator.

(a) Say that T has <u>atomic part</u> if there exists a non-zero atom $A: L^1(\mu) \to L^1(\nu)$ with $0 \leq A \leq |T|$.

(b) Say that T is <u>purely continuous</u> if T has no atomic part.

(c) Say that T is <u>purely atomic</u> if T is an ℓ^1-sum of atoms in the strong operator topology.

((c) means that there are a sequence (A_j) of atom-operators from $L^1(\mu)$ to $L^1(\nu)$ and a $K < \infty$ so that for all $f \in L^1(\mu)$, $\Sigma\|A_j f\| \leq K\|f\|$ and $Tf = \Sigma A_j f$; when this occurs, we shall say that T is the strong ℓ^1-sum of (A_j).)

In [3], purely atomic operators are termed "atomic" and purely continuous ones are termed "diffuse". We employ the terminology "purely" continuous since evidently continuous operators usually refer to bounded linear maps. Our next fundamental structural result is a reformulation of Kalton's decomposition theorem in [5]. This result is proved in [5] by a random measure representation of operators on L^1 (see also [3]). (Via this representation, one obtains that T is purely atomic iff the random representing measure is almost surely purely atomic, while T is purely continuous iff the random measure is almost surely continuous of "diffuse".) We intend to present a more intrinsic and constructive proof of the decomposition theorem elsewhere.

THEOREM 1.4. Let $T: L^1(\mu) \to L^1(\nu)$ be a given operator with $L^1(\mu)$ separable.

(a) There are unique operators T_a and T_c from $L^1(\mu)$ to $L^1(\nu)$ so that T_a is purely atomic, T_c is purely continuous, and $T = T_a + T_c$.

(b) Suppose T is purely atomic and Ω is a complete metric space, $\mathcal{S}$ the Borel subsets of Ω. Then there is a "special" sequence (A_j) of atoms so that T is the strong ℓ^1-sum of (A_j); namely, for each j there are measurable maps $a_j: X \to R$ and $\sigma_j: X \to \Omega$ representing A_j so that for ν-almost all x,

(i) $|a_j(x)| \geq |a_{j+1}(x)|$

(ii) $\sigma_i(x) \neq \sigma_j(x)$ for all $i \neq j$.

We note in passing that the special sequence (A_j) in (b) necessarily satisfies the condition that $\Sigma\|A_j f\| \leq \|T\| \; \|f\|$ for all $f \in L^1(\mu)$ and moreover $|T|$ is the strong ℓ^1-sum of $(|A_j|)$.

We formulate one last result in this section, most of which is known; the result involves non-trivial criteria for an operator to have atomic part. For $L^1(\mu) = L^1$ itself and a given T, define the Enflo-Starbird maximal function $\lambda_T \in L^1(\nu)$ by

$$\lambda_T(x) = \lim_{n \to \infty} \max_{1 \leq j \leq 2^n} |T\chi_{[\frac{j-1}{2^n}, \frac{j}{2^n})}|(x)$$

(It is proved in [2] that this limit exists for ν-almost all x; actually the function λ_T may be defined for any $(\Omega, \mathcal{S}, \mu)$; we discuss this in §3.)

Say that $T: L^1(\mu) \to L^1(\nu)$ is sign-preserving provided there is a set S of positive measure and a $\delta > 0$ so that $\|Tf\| \geq \delta$ for all f with $\int f d\mu = 0$ and $|f| = \chi_S$ a.e. (See [4] and [7] for results related to this concept.)

For $S \in \mathcal{S}$ of positive measure and $T: L^1(\mu) \to L^1(\nu)$, denote by $T|S$ the operator T restricted to the subspace of $L^1(\mu)$ consisting of all functions f with $f(x) = 0$ for almost all $x \notin S$.

THEOREM 1.5. Let $T: L^1 \to L^1$ be a given operator. The following are equivalent:

(a) T <u>has atomic part</u>.

(b) <u>The Enflo-Starbird maximal function</u> λ_T <u>is a non-zero member of</u> L^1.

(c) T <u>is sign-preserving</u>.

(d) <u>There is a set</u> S <u>of positive Lebesgue measure so that</u> $T|S$ <u>is an (into) isomorphism</u>.

The equivalences (a),(b) and (d) are proved in Kalton's paper [5]. The implication (b) $\Rightarrow$ (d) was essentially proved earlier in the fundamental paper of Enflo-Starbird [2]. The assertion (c) $\Rightarrow$ (b) is new; its proof is given in §3 along with a more general treatment of the function λ_T.

The remainder of this article is organized as follows: In the next section, we discuss the structure of purely atomic operators. The new ingredients required for Theorem 1.1 are obtained here. In §3, we complete the proofs of Theorem 1.1 and 1.2, with further complements concerning the work of Enflo-Starbird and Theorem 1.5. Before passing to §2, we now fix certain notations that will be used in the sequel. As mentioned above, $(\Omega, \mathcal{S}, \mu)$ and $(X, \mathcal{F}, \nu)$ denote fixed finite measure spaces. For a measurable subset E of a measure space, $|E|$ denotes the measure of E. We set $\mathcal{S}^+ = \{S \in \mathcal{S}: |S| > 0\}$. For $S \in \mathcal{S}^+$, set $\mathcal{S}^+_S = \mathcal{S}^+ \cap \{E: E \subset S\}$. When convenient, we identify a subset E of a set S with its characteristic function χ_E. Thus for example, given $E, E_1, E_2 \ldots$ in $\mathcal{S}$; the statement $E = \Sigma E_j$ a.e. is equivalent to the assertion that there is an $S \in \mathcal{S}$ with $|\sim S| = 0$ so that $E \cap S = \bigcup_j E_j \cap S$ and $E_i \cap E_j \cap S = \emptyset$ for all $i \neq j$. For $S \in \mathcal{S}^+$, $L^1(\mu|S) = \{f \in L^1(\mu): f \cdot S = f \text{ a.e.}\}$; recall that given $T: L^1(\mu) \to L^1(\nu)$, $T|S$ refers to the operator $T \mid L^1(\mu|S)$. Given $F \in \mathcal{F}$, we denote the restriction operator onto F by $R(F)$; thus $R(F)(f) = f \cdot F$ for all $f \in L^1(\nu)$. Finally, given Banach spaces X and Y and $T: X \to Y$ an <u>into</u> isomorphism, T^{-1} refers to the inverse of T regarded as an operator from X to TX.

§2. The structure of purely atomic operators

The main result of this section goes as follows: (The notations and conventions of §1 are in effect),

THEOREM. <u>Let</u> $T: L^1(\mu) \to L^1(\nu)$ <u>be a non-zero purely atomic operator</u>. <u>For all</u> $\varepsilon > 0$, <u>there is an</u> $S \in \mathcal{S}^+$ <u>with</u>

(a) $\|T|S\| > \|T\| - \varepsilon$

and

(b) $\|T|S\| \, \|(T|S)^{-1}\| < 1 + \varepsilon$.

REMARK. By a result of D. Alspach [1], the conclusion of the Theorem is equivalent to the following assertion: For all $\varepsilon > 0$, there are an $S \in S^+$ and an operator $V: L^1(\mu|S) \to L^1(\nu)$ with (a) holding and (b'): $\|T|S-V\| < \varepsilon$ and $\frac{V}{\|V\|}$ is an into isometry. Actually, we prove our theorem by obtaining an atom V with $\|T|S-V\| < \varepsilon$ (and (a) holding); hence the above assertion may be deduced just from the fact that it holds for atoms themselves.

We first require a number of preliminary results; these are stated in suitable generality (just to reveal the simplicity of the proofs).

LEMMA 2.1. Let B be a Banach space and $T:L^1(\mu) \to B$ a non-zero operator. For all $\varepsilon > 0$ there is an $S \in S^+$ so that

$$\|Tf\| \geq (\|T\| - \varepsilon)\|f\|$$

for all $f \in L^1(\mu|S)$ with $f \geq 0$.

PROOF. Let $0 < \varepsilon < \|T\|$ and set $\delta = \|T\| - \varepsilon$. Since $\|T\| = \|T^*\|$, choose $b^* \in B^*$ with $\|b^*\| = 1$ and $\|T^*b^*\| > \delta$. Identifying $L^1(\mu)^*$ with $L^\infty(\mu)$, we may thus regard T^*b^* as a function $\varphi \in L^\infty(\mu)$ with $\|\varphi\|_\infty > \delta$. We may then choose $S \in S^+$ with $\varphi(w) > \delta$ for all $w \in S$ (or $-\varphi(w) > \delta$ for all $w \in S$; but we may assume the former condition). Thus if

$$f \in L^1(\mu|S),\ f \geq 0,$$

$$\|Tf\| \geq b^*(Tf) = T^*b^*(f) = \int_S \varphi f d\mu \geq \delta \int_S f d\mu = \delta\|f\| .$$

LEMMA 2.2. The conclusion of the Theorem holds provided T is an atom.

PROOF. Let $\varepsilon > 0$. Let $\bar{\varepsilon} > 0$, to be decided, depending on ε, set $\delta = \|T\| - \bar{\varepsilon}$ and choose $S \in S^+$ with $\|Tf\| \geq \delta\|f\|$ for all $f \geq 0$ with $f \in L^1(\mu|S)$, by the preceding lemma. Now let $f \in L^1(\mu|S)$ and let $f^+ = f \vee 0$, $f^- = -f \vee 0$. Thus $\|Tg\| \geq \delta\|g\|$ for $g = f^+$ or $g = f^-$; but since Tf^+ and Tf^- are disjointly supported, $\|Tf\| = \|T(f^+ - f^-)\| = \|Tf^+\| + \|Tf^-\| \geq \delta\|f\|$. Thus $\|T|S\|^{-1} \leq \frac{1}{\delta}$, so

$$\|T|S\| \; \|(T|S)^{-1}\| \leq \frac{\|T\|}{\|T\| - \bar{\varepsilon}} .$$

Thus choose $0 < \bar{\varepsilon} < \varepsilon$, $\bar{\varepsilon} < \|T\|$ so that

$$\|T\| \, (\|T\| - \bar{\varepsilon})^{-1} < 1 + \varepsilon.$$

We now pass to a lemma rather deeper than the above. We here assume (as we may) that Ω is a spearable metric space with metric ρ and μ is a finite measure on the Borel subsets of Ω .

LEMMA 2.3. Let $f,g:X \to \Omega$ be measurable functions with $f(x) \neq g(x)$ a.e. (with respect to ν). Given $S \in S^+$ and $\varepsilon > 0$, there is an $E \in S^+_S$ with $|f^{-1}(E) \cap g^{-1}(E)| < \varepsilon|E|$.

PROOF. For each positive integer n, set

$$X_n = \{x \in X: \rho(f(x),g(x)) > \tfrac{1}{n}\} . \tag{1}$$

Then evidently the X_n's are measurable, $X_1 \subset X_2 \subset \ldots$ and $X \sim \bigcup_{n=1}^{\infty} X_n$ is of ν-measure 0. Consequently we may choose n so that

$$|{\sim}X_n| < \varepsilon|S|. \tag{2}$$

Now choose a (finite or infinite) partition $S_1,S_2,\ldots$ of measurable subsets of S so that for all i, S_i is of positive measure and diameter less than $\frac{1}{n}$. Then it follows that

$$\bigcup_i f^{-1}(S_i) \cap g^{-1}(S_i) \cap X_n \text{ is empty.} \tag{3}$$

We claim that $|f^{-1}(S_i) \cap g^{-1}(S_i)| < \varepsilon|S_i|$ for some i. Were this false, then

$$|\bigcup_i f^{-1}(S_i) \cap g^{-1}(S_i)| = \sum_i |f^{-1}(S_i) \cap g^{-1}(S_i)| \geq \varepsilon \sum_i |S_i| = \varepsilon|S|. \tag{4}$$

But by (2), $|(\bigcup_i f^{-1}(S_i) \cap g^{-1}(S_i)) \cap {\sim}X_n| < \varepsilon|S|$, so by (4), $\bigcup_i f^{-1}(S_i) \cap g^{-1}(S_i) \cap X_n$ is of positive measure, contradicting (3).

LEMMA 2.4. For any operator $T:L^1(\mu) \to L^1(\nu)$ and $S \in S^+$, there is an $E \in S^+_S$, with $|S|\ \|T|E\| \leq \||T|(S)\|$.

PROOF. It obviously suffices to prove this for the case of $S = \Omega$ itself. Now for any $S \in S^+$ we have that

$$\|T|S\| = \sup \{\|T(E)\| / |E| : E \in S^+_S \}. \tag{5}$$

Thus were the conclusion of the lemma false, by measure exhaustion we could choose a partition $S_1, S_2,\ldots$ of Ω with the S_i's in S^+ so that for all i,

$$|\Omega|\,\|T(S_i)\| > C|S_i|, \text{ where } C = \||T|\Omega\| . \tag{6}$$

We then obtain that

$$
\begin{aligned}
|\Omega|\,\||T|(\Omega)\| &= |\Omega|\Sigma\||T|(S_i)\| \\
&\geq |\Omega|\Sigma\|T(S_i)\| \\
&> C\,\Sigma|S_i| = \||T|\Omega\|\,|\Omega| \quad ,
\end{aligned}
$$

a contradiction.

We are finally prepared for the proof of the Theorem. Let T and ε be as in its statement; let $\bar{\varepsilon} > 0$ be a small number, to be decided later. (For now, we shall assume $\bar{\varepsilon} < \min\{\frac{\|T\|}{3}, \varepsilon\}$.) Now by Lemma 2.1, choose $S_1 \in S^+$ so that $\|Tf\| \geq (\|T\| - \bar{\varepsilon})\|f\|$ for all $f \in L^1(\mu|S_1)$. Note that in particular, we obtain

(7) $$\|T|E\| \geq \delta \quad \text{for all } E \in S^+_{S_1}, \quad \text{where} \quad \delta = \|T\| - \bar{\varepsilon} .$$

By Theorem 1.4(b), (assuming Ω and S satisfy the hypotheses of 1.4(b)), we may choose atoms $T_1, T_2, \ldots$ so that for all $f \in L^1(\mu)$,

(8i) $\Sigma\|T_i f\| \leq \|T\|\ \|f\|$,

(8ii) $Tf = \Sigma T_i f$,

(8iii) for all i, there are measurable functions $a_i: X \to R$ and $\sigma_i: X \to \Omega$ with $(T_i f)(x) = a_i(x) f(\sigma_i x)$ a.e. x,

(8iv) for all $i \neq j$, $\sigma_i(x) \neq \sigma_j(x)$ a.e. .

By 8(i), we may choose an n so that

(9) $$\sum_{i=n+1}^{\infty} \|T_i(S_1)\| < \bar{\varepsilon}|S_1| .$$

Now for any atom A and $f \geq 0$ in L^1, $|A|\ (f) = |Af|$, whence for all $f \in L^1$, $\|Af\| = \||A|f\|$. Thus we have that setting $R = \sum_{i=n+1}^{\infty} T_i|S_1$, then $\||R|\ (S_1)\| \leq \|\sum_{i=n+1}^{\infty} |T_i|(S_1)\| < \bar{\varepsilon}|S_1|$, hence by Lemma 2.4, we may choose $S_2 \in S^+_{S_1}$ so that

(10) $$\|R|S_2\| < \bar{\varepsilon} .$$

Evidently we have thus obtained that

(11) $$\|(T - \sum_{i=1}^{n} T_i)|S_2\| < \bar{\varepsilon} .$$

The last step of the proof is to show that the operator $\sum_{i=1}^{n} T_i$ can be closely approximated by an atom on a suitable $S_3 \in \mathcal{S}^+_{S_2}$. To achieve this, we require the following result:

LEMMA 2.5. Let $\eta > 0$, $i \neq j$ and $E \in \mathcal{S}^+$. There exists an $S \in \mathcal{S}^+_E$ with

$$\|R(\sigma_i^{-1}(S) \cap \sigma_j^{-1}(S))T_\ell|S\| < \eta \qquad \text{for } \ell = i \text{ or } j.$$

Let us introduce some notation. For any measurable function a and positive K, set $a^K(x) = a(x)$ if $|a(x)| \le K$; $a^K(x) = 0$ otherwise. Then set $a^{\sim K} = a - a^K$. For any k and measurable f on Ω, define $F_k f$ a measurable function on X by

$$F_k(f)(x) = f(\sigma_k x) \quad \text{a.e.} \tag{13}$$

For $S \in \mathcal{S}$, let $F(S) = \sigma_i^{-1}(S) \cap \sigma_j^{-1}(S)$. Now since a_i is integrable, we may choose a positive K so that $\int_{\sigma_i^{-1}(E)} |a_i^{\sim K}| < \frac{\eta}{2}|E|$. We thus have that

$$\| |a_i^{\sim K} \cdot F_i|(E)\| < \frac{\eta}{2}|E| . \tag{14}$$

Hence by Lemma 2.4, there is an $S^1 \in \mathcal{S}^+_E$ with

$$\|a_i^{\sim K} \cdot F_i|S^1\| < \frac{\eta}{2} . \tag{15}$$

Now by Lemma 2.3, we may choose $S^2 \in \mathcal{S}^+_{S^1}$ with

$$|F(S^2)| < \frac{\eta}{2K}|S^2| . \tag{16}$$

We then have that

$$\| |R(F(S^2))a_i^K \cdot F_i|(S^2)\| = \int_{F(S^2)} |a_i^K| d\nu \le K|F(S^2)| < \frac{\eta}{2}|S^2| .$$

Hence by Lemma 2.4, we may choose $S^3 \in \mathcal{S}^+_{S^2}$ with

$$\|R(F(S^3)\, a_i^K \cdot F_i|S^3\| < \frac{\eta}{2} . \tag{17}$$

Combining (15) and (17), we have that

$$\|RF(S^3)T_i|S^3\| < \eta. \tag{18}$$

By what we have proved, it finally follows that also there is an $S \in \mathcal{S}'_{S^3}$ with $\|RF(S)T_j|S\| < \eta$, whence S satisfies the conclusion of the Lemma.

Returning now to the main thread of the proof of the Theorem, define η by

(19) $$\eta = \bar{\varepsilon}/n(n-1).$$

By applying Lemma 2.5 $n\cdot(n-1)/2$ times, we may choose $S = S_3 \in \mathcal{S}^+_{S_2}$ so that

(20) $$\|R(\sigma_i^{-1}(S) \cap \sigma_j^{-1}(S))T_\ell|S\| < \eta$$

for all $i \neq j$ and $\ell = i$ or j.

At last, for all i, define V_i by

(21) $$V_i = R(F_i(S) \sim \bigcup_{\substack{j\neq i \\ 1\leq j\leq n}} F_j(S))\cdot T_i ,$$

then define V by

(22) $$V = \sum_{i=1}^{n} V_i .$$

Now V is an atom since $|f| \wedge |g| = 0$ a.e., $f,g \in L^1$ implies $|Vf| \wedge |Vg| = 0$ a.e.

For each i, $1 \leq i \leq n$, we have that

$$(V_i - T_i)|S = R(\bigcup_{\substack{j\neq i \\ 1\leq j\leq n}} \sigma_i^{-1}(S) \cap \sigma_j^{-1}(S))T_i(S) ,$$

hence by (20),

(21) $$\|(V_i - T_i)\ |\ S\| < \eta\cdot(n-1) ,$$

so by the definition of V,

$$\|(V - \sum_{i=1}^{n} T_i)|S\| < \eta\cdot(n-1)\cdot n = \bar{\varepsilon} .$$

Thus by (11) and the above inequality,

(22) $$\|(V-T)|S\| < 2\bar{\varepsilon} .$$

Now recall that by (7), $\|T|S\| > \|T\| - \bar{\varepsilon}$; hence $\|V|S\| > \|T\| - 3\bar{\varepsilon}$, so $V|S$ is a non-zero atom. Thus by Lemma 2.2, we may choose $S' = S_4 \varepsilon S^+_{S_3}$ so that

$$\|V|S'\| \, \|(V|S')^{-1}\| < 1 + \bar{\varepsilon} \, . \tag{23}$$

We now show that for $\bar{\varepsilon}$ small enough, depending only on $\|T\|$ and ε, we obtain that $\|T|S'\| \, \|(T|S')^{-1}\| < 1 + \varepsilon$, completing the proof. Let $\lambda = \|V|S'\|$ and let $f \varepsilon L^1(\mu|S')$. (23) may be rephrased as

$$\|Vf\| \geq \frac{\lambda}{1 + \bar{\varepsilon}} \|f\| \, .$$

Thus applying (22) twice,

$$\|Tf\| \geq \left(\frac{\lambda}{1 + \bar{\varepsilon}} - 2\bar{\varepsilon}\right) \|f\|$$

$$\geq \left(\frac{\|T|S'\| - 2\bar{\varepsilon}}{1 + \bar{\varepsilon}} - 2\bar{\varepsilon}\right) \|f\| \, .$$

Hence we have that

$$\begin{aligned} &\|T|S'\| \, \|(T|S')^{-1}\| \\ &< \|T|S'\| \left(\frac{\|T|S'\| - 2\bar{\varepsilon}}{1 + \bar{\varepsilon}} - 2\bar{\varepsilon}\right)^{-1} \\ &\leq \|T\| \left(\frac{\|T\| - 3\bar{\varepsilon}}{1 + \bar{\varepsilon}} - 2\bar{\varepsilon}\right)^{-1} \end{aligned} \tag{24}$$

by (7).

Evidently, the validity of the above inequalities required that

$$\frac{\|T\| - 3\bar{\varepsilon}}{1 + \bar{\varepsilon}} - 2\bar{\varepsilon} > 0 \, ;$$

we simply choose $\bar{\varepsilon} > 0$ so small that this holds, and such that the right side of the inequalities (24) is less than $1 + \varepsilon$, completing the proof.

§3. Proofs of the main results

For any measure space (Ω, S, μ), let $L^1_0(\mu) = \{f \varepsilon L^1(\mu) : \int f d\mu = 0\}$. Let us assume now that (Ω, S, μ) is the standard measure space, $\Omega = [0,1]$, S = Borel subsets $[0,1]$, μ = Lebesgue measure. Suppose $S \varepsilon S^+$ and A is a σ-subalgebra of the measurable subsets of S. If A is atomless (with respect to μ), then note that $L^1(\mu|A)$ is isometric to L^1 and also $L^1_0(\mu|A)$ is isometric to L_0^1. To prove Theorem 1.1, we first require the following result:

LEMMA 3.1. Let B be a Banach space and $T:L^1 \to B$ a non-sign preserving operator. Then for all $S \in \mathcal{S}^+$ and $\varepsilon > 0$, there exists an atomless σ-subalgebra $\mathcal{A}$ of the measurable subsets of S with $\|T|L_0^1(\mu|\mathcal{A})\| < \varepsilon$.

PROOF. Choose by induction sets E_n in $\mathcal{S}_S^+$ with $E_1 = S$ so that for all n, $E_n = E_{2n} + E_{2n+1}$, $|E_{2n}| = \frac{1}{2}|E_n|$ and $\|T\tilde{h}_n\| < \frac{\varepsilon}{2^{n+1}}$ where $\tilde{h}_n = (\chi_{E_{2n}} - \chi_{E_{2n+1}})|E_n|^{-1}$. (To see that this is possible, let k be an odd integer and suppose $E_1, \ldots, E_k$ constructed. Set $n = (k+1)/2$, choose h_n with $\int h_n d\mu = 0$, $|h_n| = \chi_{E_n}$ and $\|Th_n\| < \varepsilon|E_n|/2^{n+1}$; then simply set $E_{2n} = \{\omega: h_n(\omega) = 1\}$, $E_{2n+1} = \{\omega : h_n(\omega) = -1\}$.) It follows that if $\mathcal{A}$ denotes the σ-ring of sets generated by the E_n's, then $\mathcal{A}$ is atomless, $(\tilde{h}_j)_{j=1}^\infty$ is isometrically equivalent to the usual sequence of mean-zero Haar functions and $(\tilde{h}_j)_{j=1}^\infty$ is a monotone basis for $L_0^1(\mu|\mathcal{A})$. Hence $\|T|L_0^1(\mu|\mathcal{A})\| < \varepsilon$.

Our next result yields Theorem 1.1 as a corollary.

THEROEM 3.2. Let $T:L^1 \to L^1$ have atomic part. There is a $\lambda > 0$ so that for all $\varepsilon > 0$, there is an $S \in \mathcal{S}^+$ and an atomless σ-subalgebra $\mathcal{A}$ of the measurable subsets of S so that

(a) $$\|T|L_0^1(\mathcal{A})\| > \lambda - \varepsilon$$

and

(b) $$\|T|L_0^1(\mathcal{A})\| \, \|(T|L_0^1(\mathcal{A}))^{-1}\| < 1 + \varepsilon.$$

Let us first deduce Theorem 1.1 from this result. Let $X \subset L^1$ with X isomorphic to L^1 and $\varepsilon > 0$. Choose $T: L^1 \to X$ an onto isomorphism. By Theorem 1.5, T has atomic part. Choose S and $\mathcal{A}$ as in Theorem 3.2. Since $L_0^1(\mathcal{A})$ is isometric to L_0^1, there is a subspace Z of $L_0^1(\mathcal{A})$ with Z isometric to L^1. Set $Y = TZ$. Since $\|T|Z\| \, \|(T|Z)^{-1}\| < 1 + \varepsilon$ also, Y is $1 + \varepsilon$ -isometric to L^1.

PROOF OF THEOREM 3.2. Let $T = T_a + T_c$ where T_a is purely atomic, T_c is purely continuous (as given by Theorem 1.4). Thus $T_a \neq 0$; let $\lambda = \|T_a\|$ and let $\bar{\varepsilon} > 0$ to be decided, (depending only on λ and ε). By the Theorem of Section 2, choose $S \in \mathcal{S}^+$ with

(25) $$\|T_a|S\| \, \|(T_a|S)^{-1}\| < 1 + \bar{\varepsilon} \quad \text{and} \quad \|T_a|S\| > \lambda - \bar{\varepsilon}.$$

By Theorem 1.5, T_c is non-sign preserving. So by Lemma 3.1, choose A as in its statement with

$$\|T_c | L_0^1(A)\| < \bar{\varepsilon} . \tag{26}$$

Now let $f \in L_0^1(A)$. We have that

$$\|Tf\| \leq \|T_a f\| + \|T_c f\| \leq (\lambda + \bar{\varepsilon}) \|f\|. \tag{27}$$

The inequalities (25) and (26) yield

$$\|Tf\| \geq \|T_a f\| - \|T_c f\| \geq \left(\frac{\lambda - \bar{\varepsilon}}{1 + \bar{\varepsilon}} - \bar{\varepsilon} \right) \|f\| . \tag{28}$$

By (27) and (28), we complete the proof by choosing

$$\bar{\varepsilon} < \varepsilon \quad \text{so that} \quad \frac{\lambda - \bar{\varepsilon}}{1 + \bar{\varepsilon}} - \bar{\varepsilon} > 0 \quad \text{and}$$

$$(\lambda + \bar{\varepsilon}) \left(\frac{\lambda - \bar{\varepsilon}}{1 + \bar{\varepsilon}} - \bar{\varepsilon} \right)^{-1} < 1 + \varepsilon .$$

REMARK. The proof of Theorem 1.1' may be obtained directly from our discussion as follows: Let $X \subset L^1$ be isomorphic to L^1 and $T:L^1 \to X$ a surjective isomorphism. So $T_a \neq 0$. Our proof of the Theorem of Section 2 showed that given $\bar{\varepsilon} > 0$ we can choose $S \in S^+$ with $\|T_a | S\| > \lambda - \bar{\varepsilon}$ and an atom $V:L^1 \to L^1$ with $\|(V-T_a) | S\| < \bar{\varepsilon}$, $\|V | S\| \, \|(V | S)^{-1}\| < 1 + \bar{\varepsilon}$.

Choosing A as in the above proof, we may choose $U:L^1 \to L_0^1(A)$ an into isometry. Set $Z = VUL^1$. Since VU maps disjoint functions to disjoint functions and is of course an isomorphism, it follows that Z is indeed isometric to L^1. Let $P:L^1 \to Z$ be a surjective contractive projection. Now it follows that $\|TU-VU\| < 2\bar{\varepsilon}$. We obtain that for $\bar{\varepsilon}$ sufficiently small, depending on λ and ε, $\tilde{T} = P|Y$ satisfies the conclusion of Theorem 1.1' where $Y = TUL^1$.

Theorem 1.2 follows from Alspach's result and our preceding discussion. Here is a sketch of the proof: Suppose $T:L^1 \to B$ is such that TZ is an isomorphism with Z isomorphic to L^1. By Theorem 1.1', $\varepsilon > 0$ given we may choose an $X \subset L^1$ isometric to L^1, a $Y \subset Z$ and a surjective $S:Y \to X$ with $\|Sy-y\| \leq \varepsilon\|y\|$ for all $y \in Y$. Let $x \in X$ and suppose $Sy = x$. It follows that setting $\delta = \|T^{-1}\|^{-1}$,

$$\begin{aligned} \|Tx\| &\geq \|Ty\| - \|T(y-x)\| \\ &\geq (\delta - \varepsilon) \|y\| \geq \frac{\delta - \varepsilon}{1 + \varepsilon} \|x\| ; \end{aligned}$$

hence $T|X$ is an isomorphism if $\varepsilon < \delta$.

Now suppose $B = L^1$ itself and let X be as above. Since X is isometric to L^1, it follows from our proof of Theorem 3.2 that there is a $\lambda > 0$ so that for all $\varepsilon > 0$, there is a subspace X' of X with X' isometric to L^1 and $\|T|X'\| \, \|(T|X')^{-1}\| < 1 + \varepsilon$, $\|T|X'\| \geq \lambda$. The final conclusion of Theorem 1.2 now follows from Alspach's result [1].

We conclude with a discussion of the Enflo-Starbird maximal function defined in Section 1 and a proof of the implication (c) $\Rightarrow$ (b) of Theorem 1.5. We first formulate matters in the setting of general finite-measure spaces $(\Omega, \mathcal{S}, \mu)$, $(X, \mathcal{F}, \nu)$ and an operator $T:L^1(\mu) \to L^1(\nu)$. Let $\mathcal{D}$ be the set of all finite-measurable partitions of $\mathcal{S}$ by members of $\mathcal{S}^+$, directed by refinement. That is; the members α of $\mathcal{D}$ are finite subsets of $\mathcal{S}^+$, say $\alpha = \{S_1,\dots,S_n\}$ with $\Omega = \Sigma S_i$ a.e. Let $\beta \in \mathcal{D}$; say that $\alpha \leq \beta$ provided for all $\bar{S} \in \beta$, there is an $S \in \alpha$ with $\bar{S} \leq S$ a.e. Let now T be fixed. Given $\alpha \in \mathcal{D}$, set $\lambda_\alpha = \max\{|TS| : S \in \alpha\}$. We then have the following result:

THEOREM 3.2. <u>The net</u> $(\lambda_\alpha)_{\alpha\in\mathcal{D}}$ <u>converges in</u> $L^1(\nu)$ <u>to a function</u> λ_T. $\lambda_T \neq 0$ <u>if and only if</u> T <u>has atomic part</u>.

We do not present the proof here. In the case when $L^1 = L^1(\mu)$, it can be proved that λ_T equals the Enflo-Starbird function defined in the first section. Here is a further interpretation of λ_T: Suppose one assumes Ω a complete metric space, $\mathcal{S}$ its Borel subsets, and $T = T_a + T_c$ as in Theorem 1.4(a); suppose that T_a is the strong ℓ^1-sum of the "special" sequence of atoms (A_j) with the A_j's represented by the measurable maps (a_j) and (σ_j) as in Theorem 1.4(b). Then one has $\lambda_T(x) = |a_1(x)|$ a.e.

We conclude with the proof of (c) $\Rightarrow$ (b) of Theorem 1.5. Let $s \in S^+$. There exist S^+ and S^- in $\mathcal{S}_S^+$ with $S^+ \cap S^- = \emptyset$, $S^+ \cup S^- = S$ and an invertible map $\tau : S^+ \to S^-$ with τ, τ^{-1} measurable and measure-preserving. Now suppose $T:L^1 \to L^1$ and $\delta > 0$ are such that $\|Tf\| \geq \delta$ whenever $f \in L^1$, $|f| = \chi_S$, $\int f d\mu = 0$. It follows from Theorem 3.2 that $\lambda_T = \max\{\lambda_{T|S^+}, \lambda_{T|\sim S}\}$. Consequently, if $\lambda_T = 0$ a.e., we have $\lambda_{T|S^+} = 0$ and $\lambda_{T|S^-} = 0$ a.e.; thus by Theorem 3.2, $\varepsilon > 0$ given, we may choose measurable partitions α_o of S^+ and β_o of S^- so that $\int \lambda_\alpha d\nu < \varepsilon$ and $\int \lambda_\beta d\nu < \varepsilon$ for any partitions α and β refining α_o and β_o respectively. Choose a partition $\alpha = \{S_1,\dots,S_m\}$ refining α_o so that $\beta = \{\tau S_1,\dots,\tau S_m\}$ refines β_o. (Simply let $\{S_1,\dots,S_m\}$ equal $\{S \cap \tau^{-1}D : S \in \alpha_o, D \in \beta_o\} \cap S^+$.) Let $r_1,\dots,r_m$ be the first m Rademacher functions. For each ω, the function $f_\omega = \sum_{i=1}^{m} r_i(\omega)(\chi_{S_i} - \chi_{\tau S_i})$ satisfies $|f_\omega| = \chi_S$ and $\int f_\omega d\mu = 0$. Hence $\|Tf_\omega\| \geq \delta$. Hence $\int \|Tf_\omega\| d\omega \geq \delta$. But

$$\begin{aligned}
\int \|Tf_\omega\| d\omega &= \int \int |\Sigma r_i(\omega) TS_i - \Sigma r_i(\omega) T(\tau S_i)| d\nu d\omega \\
&\leq \int \int |\Sigma r_i(\omega) TS_i| d\omega d\nu + \int \int |\Sigma r_i(\omega) T(\tau S_i)| d\omega d\nu \\
&\leq \int (\Sigma (TS_i)^2)^{1/2} d\nu + \int (\Sigma (T\tau S_i)^2)^{1/2} d\nu \\
&\leq (\int \max |TS_i|))^{1/2} (\Sigma \|TS_i\|)^{1/2} + (\int \max |T\tau S_i|)^{1/2} \Sigma(\|T\tau S_i\|)^{1/2} \\
&\leq (\int \lambda_\alpha d\nu)^{1/2} (\|T\| \frac{|S|}{2})^{1/2} + (\int \lambda_\beta d\nu)^{1/2} (\|T\| \frac{|S|}{2})^{1/2} \\
&\leq (2\varepsilon \|T\| |S|)^{1/2}.
\end{aligned}$$

Evidently, we have a contradiction if $(2\varepsilon\|T\| |S|)^{1/2} < \delta$.

BIBLIOGRAPHY

1. D.E. Alspach, "Small into isomorphisms on L^p spaces", to appear.

2. P. Enflo and T.W. Starbird, "Subspaces of L^1 containing L^1", Studia Math. 65 (1979), 203-225.

3. H. Fakhoury, "Représentations d'opérateurs à valeurs dans $L^1(X, \Sigma, \mu)$", Math. Annalen 240 (1979), 203-213.

4. N. Ghoussoub and H.P. Rosenthal, "Martingales, G_δ-embeddings and quotients of L_1", to appear, Math. Annalen.

5. N.J. Kalton, "The endomorphisms of L_p, $0 \leq p \leq 1$", to appear.

6. J. Lindenstrauss and A. Pełczyński, "Contributions to the theory of the classical Banach spaces", J. Funct. Anal. 8(1971), 225-249.

7. H.P. Rosenthal, "Some remarks concerning sign-embeddings", Seminaire D'analyse Fonctionelle Université Paris VII, 1981-82 (to appear).

8. H.L. Royden, Real Analysis, second edition, Macmillan, New York, 1968.

9. H.H. Schaefer, Banach Lattices and Positive Operators, Springer-Verlag, New York, 1974.

THE UNIVERSITY OF TEXAS AT AUSTIN

Contemporary Mathematics
Volume 26, 1984

INNER AND BARELY LINEAR TIME CHANGES OF ERGODIC $\mathbb{R}^k$ ACTIONS

Daniel J. Rudolph

I. INTRODUCTION.

Recent work of D. Nadler and J. Feldman [5], [8] has lifted the earlier work of Feldman [4], Ornstein, Weiss and the author [9] on continuous orbit equivalence of $\mathbb{R}^1$ to certain time changes of $\mathbb{R}^k$, $k \geq 2$. Our basic motivation here is to expand the applicability of these results to as wide a class of time changes as possible.

By an action of $\mathbb{R}^k$ we mean a measurable, measure preserving free ergodic action of $\mathbb{R}^k$, written φ_v, where $v \in \mathbb{R}^k$, on a Lebesgue probability space $(\Omega,\mathcal{F},\mu)$. Thus

$$\varphi_v \circ \varphi_{v'} = \varphi_{v+v'} , \varphi_v(\mu) = \mu,$$

and $\bar{\varphi}(\omega,v) = \varphi_v(\omega)$ is a measurable map from $\Omega \times \mathbb{R}^k$ to Ω, where we use the Lebesgue sets on $\mathbb{R}^k$, and Lebesgue measure m.

If ψ_v is another $\mathbb{R}^k$ action on Ω which has the same orbits and preserves a measure ν equivalent to μ, i.e., $\psi_v(\omega) = \varphi_{r(v,\omega)}(\omega)$, we say ψ_v is a "time change" of φ_v. We consider only "measurable" time changes where $r : \mathbb{R}^k \times \Omega \to \mathbb{R}^k$ is measurable. Clearly r must satisfy.

(1.1) $$r(v+v',\omega) = r(v,\omega) + r(v', \varphi_{r(v,\omega)}(\omega)).$$

and for a.e. ω, $r(\cdot,\omega)$ must be a bijection of $\mathbb{R}^k$ to itself. If its inverse is $r'(\cdot,\omega)$, then $\varphi_v = \psi_{r'(v,\cdot)}$. We call r the time change function and r' the inverse time change function.

We say ψ_v and φ_v, possibly on different spaces, are "orbit equivalent" if there is a time change ψ_v' of φ_v isomorphic to ψ_v.

0271-4132/84 $1.00 + $.25 per page

The work of D. Nadler and J. Feldman [5], [8] has focused on a special class of time changes, those for which $r(\cdot,\omega)$ and $r'(\cdot,\omega)$ are C^1 with all their first order partial derivatives bounded, the so-called "tempered time changes". This defines a restricted notion of orbit equivalence and for this stronger notion they have proven a number of striking results, lifting essentially all of the one dimensional theory to $\mathbb{R}^k$.

What we will do here is investigate first a certain trivial kind of time change, and then describe precisely what time changes can be trivially modified to give a tempered time change.

We will call a time change function r "inner" if it arises in the following simple way. Let $\sigma\colon \Omega \to \Omega$ be a nonsingular invertible map with $\sigma(\omega) = \varphi_{s(\omega)}(\omega)$. One would say "$\sigma$ preserves the orbit of φ", or "σ is in the full group of φ". Thus, $\sigma(\varphi_v(\omega)) = \varphi_{s(\varphi_v(\omega)) + v}(\omega)$. Setting $\bar{s}(v,\omega) = s(\varphi_v(\omega)) + v$, $\bar{s}(\cdot,\omega)$ takes $\mathbb{R}^k$ to itself and is measurable for a.e. ω. Letting $\sigma^{-1}(\omega) = \varphi_{s'(\omega)}(\omega)$, we have

$$s'(\varphi_{s(\omega)}(\omega)) = -s(\omega) .$$

Let

$$(1.2) \qquad \psi_v = \sigma^{-1}\varphi_v\sigma = \varphi_{s'(\varphi_{v+s(\omega)}(\omega))+v+s(\omega)} = \varphi_{r(v,\omega)}$$

where

$$(1.3) \qquad r(v,\omega) = v + s(\omega) + s'(\varphi_{v+s(\omega)}(\omega)) .$$

This map r is trivially a time change function, and (1.2) makes ψ_v isomorphic to φ_v.

It is useful to envision time changes in the following way. Think of φ_v as imposing coordinates on its orbits. Thus for any ω,ω' on the same orbit, let $v(\omega,\omega')$ be that vector with $\varphi_{v(\omega,\omega')}(\omega) = \omega'$. Similarly, ψ_v, a time change of φ_v, imposes coordinates, and let $v'(\omega,\omega')$ be that vector with $\psi_{v'(\omega,\omega')}(\omega) = \omega'$.

$$(1.4) \qquad \begin{aligned} v(\varphi_{\bar{v}}(\omega), \varphi_{\bar{v}'}(\omega')) &= v(\omega,\omega') - \bar{v} + \bar{v}', \\ r(v'(\omega,\omega'), \omega) &= v(\omega,\omega') \quad \text{and} \\ r'(v(\omega,\omega'),\omega) &= v'(\omega,\omega'), \end{aligned}$$

where $r'(v,\omega)$ is defined siminarly as $r(v,\omega)$ was defined above.

Measurability of the time change means that the map $(\omega,v) \to (\omega,v'(\omega,\varphi_v(\omega)))$ from $\Omega \times \mathbb{R}^n$ to itself is measurable. Furthermore r is inner iff

(1.5) $$v'(\omega',\omega) = v(\varphi_{s(\omega)}(\omega),\varphi_{s(\omega')}(\omega')) = v(\omega,\omega') - s(\omega) + s(\omega').$$

We could have started our discussion with solely the equivalence relation on Ω whose equivalence classes are orbits, with the topology of $\mathbb{R}^k$ and defined a 1-cocycle as a map $s : \Omega \to \mathbb{R}^k$ with $\varphi_{s(\omega)}(\omega)$ a bijection on orbits. A two cocycle would be a map $v : \Omega \times \Omega \to \mathbb{R}^k$, with $v(\omega_1,\omega_2) + v(\omega_2,\omega_3) = v(\omega_1,\omega_3)$ and $\varphi_{v(\omega,.)}$ a bijection on orbits for a.e. ω, and two 2-cocycles differ by a 2-coboundary iff $v'(\omega,\omega') = v(\omega,\omega') - s(\omega) + s(\omega')$, where s is a 1-cocycle.

From here one could build a cohomology theory, similar but much stronger than that of Zimmer [12]. Stronger in that the two 2-cocycles differ by a 2-coboundary only if they give rise to isomorphic $\mathbb{R}^k$ actions, where for Zimmer's cohomology, all finite measure preserving $\mathbb{R}^k$ actions give the same cohomology [10].

We now proceed to understand two things: when is a time change function inner, and when can one time change be followed by an inner one so that the combined time change is tempered, i.e., when can the two orbits equivalent actions be shown to be tempered equivalent.

The two arguments are very similar, one starting with a boundedness condition, the other as we shall see in Section III, a condition of sublinear growth.

II. Inner Time Changes.

We characterize here those time changes of φ_v which are inner. Although our main focus is on $\mathbb{R}^k$, $k > 1$, this characterization is nontrivial even for $k = 1$. In fact, it is possible to carry all our work here over to $\mathbb{Z}$ as one can see in Belinskaja [1], [2].

For any $v \in \mathbb{R}^k$ we call a set A a "d fixing set" for v if for any $\omega \in A$ and $\varphi_{nv}(\omega) \in A$, $|v'(\omega,\varphi_{nv}(\omega)) - v| < d$. In Section III we define an "a,b,v'-pinning set" for v. Notice that a d fixing set is a o,d,v-pinning set. What we will prove is the following result.

Theorem 2.1. A time change ψ_v of φ_v is inner iff for some basis $v_1 \dots v_k$ of $\mathbb{R}^k$, there is a d and a d-fixing set of positive measure for each v_i.

We first check the necessity of the condition. Suppose r is inner. For any $\varepsilon > 0$, pick a value d so large that there is a set $D \subset \Omega$, $\mu(D) > 1 - \frac{\varepsilon}{2}$, and for $\omega \in D$, $|s(\omega)| < \frac{d}{2}$. For $\omega,\omega' \in D$ and on the same orbit, $|v(\omega,\omega') - v'(\omega,\omega')| \le |s(\omega)| + |s(\omega')| < d$. This is much more than the condition of Theorem 2.1. We now, though, show that this condition is implied by that of Theorem 2.1. Since this argument is a special case of those for pinning sets we state it as a series of corollaries to arguments that appear in Section III.

Corollary 2.2. If there is a d-fixing set for v, of positive measure, then for any $\varepsilon > 0$, there is a d' and d'-fixing set A for v with $\mu(A) > 1 - \varepsilon$.

Proof. Follows from Lemma 3.6.

Corollary 2.3. Suppose for the time change ψ_v of φ_v there is a basis $v_1 \dots v_k$ each with a d-fixing set. Then for any $\varepsilon > 0$, there is a set A, $\mu(A) > 1 - \varepsilon$ and a d' so that for any ω, $\varphi_v(\omega) \in A$, $|v'(\omega,\varphi_v(\omega)) - v| < d'$.

Proof. Follow through the argument of Theorem 3.8 knowing a is already 0 and $v_i' = v_i'' = v$.

Thus the result we now want is the following.

Theorem 2.4. <u>Suppose there is a</u> d <u>and a set</u> A <u>of positive measure so that whenever</u> $\omega, \varphi_v(\omega) \in A$, <u>then</u> $|v'(\omega,\varphi_v(\omega)) - v| < d$. <u>Then</u> $r(v,\omega)$ <u>is inner</u>.

To prove this we must construct the map $\sigma = \varphi_{s(\omega)}(\omega)$ with $\sigma^{-1}\varphi_v\sigma = \psi_v$. The map s we will build componentwise, $s(\omega) = (s_1(\omega)\ldots s_k(\omega))$. Hence write $v(\omega,\omega') = (v_1(\omega,\omega')\ldots v_k(\omega,\omega'))$, $v'(\omega,\omega') = (v_1'(\omega,\omega')\ldots v_k'(\omega,\omega'))$. Now fix $j \in \{1,\ldots,k\}$. Let

$$b_j(\omega) = \sup_{\substack{\omega',\omega''\in A \\ \text{in the orbit} \\ \text{of } \omega}} (v_j(\omega',\omega'') - v_j'(\omega',\omega'')).$$

It is clear $b_j(\omega)$ is shift invariant, hence a constant b_j a.e. Furthermore

$$-b_j = \inf_{\substack{\omega',\omega''\in A \\ \text{in the orbit} \\ \text{of } \omega}} (v_j'(\omega',\omega'') - v_j(\omega',\omega'')).$$

Now for $\varepsilon > 0$ let $B_\varepsilon = \{\omega \in A$ there is an $\omega' \in A$, $|(v_j(\omega,\omega')-v_j'(\omega,\omega'))-b_j| < \varepsilon\}$. We know $\mu(B_\varepsilon) > 0$ and if $\varepsilon > \varepsilon'$, $B_{\varepsilon'} \subseteq B_\varepsilon$.

Lemma 2.5. <u>If</u> $\omega,\omega' \in B_\varepsilon$, <u>then</u> $|v_j(\omega,\omega') - v_j'(\omega,\omega')| < 2\varepsilon$.

Proof: Let $\bar\omega$ and $\bar\omega' \in A$ be those points putting ω and ω' in B_ε. Thus

$$\begin{aligned} b_j &\ge (v_j(\omega,\bar\omega') - v_j'(\omega,\bar\omega')) \\ &= (v_j(\omega,\omega') - v_j'(\omega,\omega')) \\ &\quad + (v_j(\omega',\bar\omega) - v_j'(\omega',\bar\omega')) \\ &\ge v_j(\omega,\omega') - v_j'(\omega,\omega') + b_j - \varepsilon. \end{aligned}$$

Thus $v_j(\omega,\omega') - v_j'(\omega,\omega') \le \varepsilon$. Also as

$$\begin{aligned} b_j &\ge (v_j(\omega',\bar\omega) - v_j'(\omega',\bar\omega)) \\ &= (v_j(\omega,\bar\omega) - v_j'(\omega,\bar\omega)) - (v_j(\omega,\omega') - v_j'(\omega,\omega')) \\ &\ge b_j - \varepsilon - (v_j(\omega,\omega') - v_j'(\omega,\omega')), \\ &\quad (v_j(\omega,\omega') - v_j'(\omega,\omega')) \ge -\varepsilon. \end{aligned}$$

What this has done is to improve our bound d to an ε. Now we make a first pass at a definition of σ_j. Let

$$s_\varepsilon^j(\omega) = \overline{\{v_j(\omega,\omega') - v_j'(\omega,\omega') \mid \omega' \in B_\varepsilon\}}.$$

Clearly if $\varepsilon > \varepsilon'$, $s_\varepsilon^j(\omega) \supseteq s_{\varepsilon'}^j(\omega)$. Furthermore

$$\begin{aligned} s_\varepsilon^j(\varphi_v(\omega)) &= \overline{\{v_j(\omega,\omega') - v_j(\omega,\varphi_v(\omega))} \\ &\quad \overline{- v_j(\omega,\omega') + v_j(\omega,\varphi_v(\omega)) \mid \omega \in B_\varepsilon\}} \\ &= s_\varepsilon^j(\omega) - v_j(\omega,\varphi_v(\omega)) + v_j(\omega,\varphi_v(\omega)). \end{aligned}$$

Lemma 2.6. <u>For a.e.</u> ω,

$$\text{dia } s_\varepsilon^j(\omega) < \varepsilon.$$

Proof. Suppose $v_1 = v_j(\omega,\omega') - v_j(\omega,\omega')$ and $v_2 = v_j(\omega,\omega'') - v_j(\omega,\omega'')$ are in $s_\varepsilon^j(\omega)$. Then $|v_1 - v_2| = |v_j(\omega'',\omega') - v_j(\omega'',\omega')| < \varepsilon$. ∎

Now let $s_j(\omega)$ be that single point in $\bigcap_{\varepsilon\to 0} s_\varepsilon^j(\omega)$. Thus

$$s(\varphi_v(\omega)) = s(\omega) - v(\omega,\varphi_v(\omega)) + v'(\omega,\varphi_v(\omega)),$$

i.e., for any ω,ω' on the same orbit,

$$v'(\omega,\omega') = v(\omega,\omega') - s(\omega) + s(\omega').$$

This is (1.5), and if we set $\sigma(\omega) = \varphi_{s(\omega)}(\omega)$, $\varphi_v\sigma = \sigma\psi_v$, and as $\sigma^{-1}(\omega) = \psi_{-s(\omega)}(\omega)$, σ is invertible.

As $s(\varphi_v(\omega)) = s(\omega) - v(\omega,\varphi_v(\omega)) + v'(\omega,\varphi_v(\omega))$, $\bar{s}(v,\omega) = s(\varphi_v(\omega))$ for fixed ω is continuous in v. The map σ is nonsingular as it carries μ to a ψ_v invariant measure which has the same generic points as ω, hence must be ν. This completes the proof of Theorem 2.4 and hence Theorem 2.1.

III. Barely Linear Time Changes.

What we now to investigate is precisely when an orbit equivalence can be shown to actually be tempered. It is worth noticing that these arguments are trivial for $\mathbb{R}$, as there it is well known that any time change of $\mathbb{R}$ can be made tempered. First we examine some properties of a tempered equivalence. Let $\psi_v = \varphi_{r(v,\cdot)}$ be tempered, i.e., outside of a shift invariant set of measure zero, $r(\cdot,\omega)$ and $r'(\cdot,\omega)$ are C^1 and there is a b with

$$m_{i,j}(\omega) = \frac{\partial r_i(\cdot,\omega)}{\partial v_j}, \quad m'_{i,j}(\omega) = \frac{\partial r'_i(\cdot,\omega)}{\partial v_j}$$

bounded by b.

Letting $M(\omega) = (M_{i,j}(\omega))$, we get $M^{-1}(\omega) = (m'_{i,j}(\omega))$, and

$$r(v,\omega) = \int_0^{|v|} M\left(\psi_{t\frac{v}{|v|}}(\omega)\right)\left(\frac{v}{|v|}\right) dt, \tag{3.1}$$

$$r'(v,\omega) = \int_0^{|v|} M^{-1}\left(\varphi_{t\frac{v}{|v|}}(\omega)\right)\left(\frac{v}{|v|}\right) dt,$$

and so for a unit vector v,

$$\frac{r(Tv,\omega)}{T} = \frac{1}{T}\int_0^T M(\psi_{tv}(\omega))(v)dt, \tag{3.2}$$

$$\frac{r'(Tv,\omega)}{T} = \frac{1}{T}\int_0^T M^{-1}(\varphi_{tv}(\omega))(v)dt,$$

which converge, as $T \to \infty$, for a.e. ω, by the ergodic theorem, to some $R(\omega,v)$ and $R'(\omega,v)$ which are ψ_{tv} and φ_{tv} invariant respectively. We now show they are in fact constant $R(v)$, $R'(v)$ a.e.

Lemma 3.1. <u>The maps</u> $R(\omega, v)$, $R'(\omega, v)$ <u>are constant</u> $R(v)$, $R'(v)$ <u>a.e.</u>, <u>and are linear maps in</u> v, $R' = R^{-1}$ <u>and</u> $R = \int_\Omega M(\omega)d\nu$, $R^{-1} = \int_\Omega M^{-1}(\omega)d\mu$.

Proof: Let $\omega' = \psi_{v'}(\omega)$. Now $r(Tv,\omega) - r(Tv,\omega') = r(v',\psi_{Tv}(\omega)) < kb|v'|$, so $\lim_{T\to\infty} \left(\frac{r(Tv,\omega)}{T} - \frac{r(Tv,\omega')}{T}\right) = 0$ and $R(\varphi_{v'}(\omega),v) = R(\omega,v)$ for all v'. Thus $R(\omega, v)$, and similarly $R'(\omega,v)$, are constants $R(v)$, $R'(v)$ a.e. From (3.2) then

$$R(v) = \lim_{T\to\infty}\left(\frac{1}{T^k}\int_{\substack{\text{unit}\\ \text{cube}}} M(\psi_{Tv}(\omega))(v)dm\right)$$

$$= \int_\Omega M(\omega)(v)d\nu,$$

$$R'(v) = \lim_{T\to\infty}\left(\frac{1}{T^k}\int_{\substack{\text{unit}\\ \text{cube}}} M'(\varphi_{Tv}(\omega))(v)dm\right)$$

$$= \int_\Omega M'(\omega)(v)d\mu.$$

Now consider, for a unit vector v,

$$\begin{aligned} |Tv-RR'(Tv)| &= |r(r'(Tv,\omega),\omega) - R(R'(Tv))| \\ &\le |r(r'(Tv,\omega),\omega) - r(R'(Tv),\omega)| \\ &+ |r(R'(Tv),\omega) - R(R'(Tv))| \\ &\le kb|r'(Tv,\omega) - R'(Tv)| + |r(R'(Tv),\omega) - R(R'(Tv))|. \end{aligned}$$

For a fixed v, and a.e. ω, once T is large enough, this is $\le T\varepsilon$, hence $|v-RR'v| < \varepsilon$, so $R' = R^{-1}$.

At this point we know from the ergodic theorem that for any $v \neq 0$, $\varepsilon > 0$, there is a T and a set A, $\mu(A) > 1 - \varepsilon$, and if $\omega \in A$, $t > T$, $|r(tv,\omega) - R(tv)| \le t|v|\varepsilon$. We now show that this is uniform in v.

Lemma 3.2. <u>Let ψ_v be a tempered time change of φ_v</u>, R <u>as above. Then given</u> $\varepsilon > 0$, <u>there is a</u> T <u>and a set</u> A, $\mu(A) > 1 - \varepsilon$, <u>so that for</u> $|v| > T$, $\omega \in A$, $|r(v,\omega) - R(v)| < \varepsilon|v|$, $|r'(v,\omega) - R^{-1}(v)| < \varepsilon|v|$.

Proof: Fix ε, and let $v_1 \dots v_\ell$ be unit vectors, $\min(\frac{\varepsilon}{3kb}, \frac{\varepsilon}{3\|R\|})$ - dense in the unit ball. Now let T be so large that for a set A, $\mu(A) > 1-\varepsilon$, if $t > T$, $\omega \in A$, then $|r(tv_i,\omega) - R(tv_i)| < \frac{\varepsilon}{3}t$. Now for any v, $|v| > T$, there exists v_i, t, $|v_i|v| - v| < |v| \min(\frac{\varepsilon}{3kb}, \frac{\varepsilon}{3\|R\|})$. Now $|r(v,\omega) - R(v)| \le |r(v,\omega) - r(v_i|v|,\omega)| + |r(v_i|v|,\omega) - R(v_i|v|)| + |R(v_i|v|) - R(v)| \le kb|v-v_i|v|| + \frac{\varepsilon}{3}|v| + \|R\|(v_i|v|- v) \le \varepsilon|v|$. The other inequality follows symmetrically.

Thus if we allow v to range over $T\cdot$(unit cube) then for all but ε of the values ω, $\frac{v(\omega,\psi_v(\omega))}{T}$ is R (unit cube $\pm\ \varepsilon$), nearly a parallelapiped.

Now suppose ψ_v is equivalent to φ_v by a map which can be further modified by an inner equivalence to $\bar{\psi}_v$ so that the combination is tempered, i.e.,

$$(3.3) \qquad \psi_v = \varphi_{r(v,\cdot)} \quad \text{and} \quad \bar{\psi}_v = \varphi_{\tilde{r}(v,\cdot)} = \psi_{\bar{r}(v,\cdot)} = \psi_{r(\tilde{r}(v,\cdot)\cdot)}$$

where $\tilde{r}$ is inner and $\bar{r}$ is tempered. We will call the map r "barely linear" for reasons which shall arise later. What can we say about r in this case?

Lemma 3.3. <u>If the time change</u> r <u>is barely linear then for any</u> $\varepsilon > 0$ <u>there is a</u> $b > 0$ <u>and a set</u> A, $\mu(A) > 1 - \varepsilon$, <u>so that if</u> $\omega, \omega' \in A$ <u>are on the same orbit, then</u> $|v'(\omega,\omega') - R(v(\omega,\omega'))| \le \varepsilon|v(\omega,\omega')| + b$.

Proof. Let $\psi_{v'(\omega,\omega')}(\omega) = \omega'$, $\varphi_{v(\omega,\omega')}(\omega) = \omega'$ and $\bar{\psi}_{\bar{v}(\omega,\omega')}(\omega) = \omega'$. As $\bar{r}(v,\cdot)$ is tempered, there is a set A_1, $\mu(A_1) > 1 - \frac{\varepsilon}{3}$ and a $T > 0$ so that if $\omega \in A_1$, $|v| > T$ then $|v'(\omega,\psi_v(\omega)) - R(\bar{v}(\omega,\psi_v(\omega)))| < \frac{\varepsilon}{3}|\bar{v}(\omega,\psi_v(\omega))|$. Choose b_1 so large that there is a set A_2, $\mu(A_2) > 1 - \frac{\varepsilon}{3}$ and if $\omega \in A_2$, $|v| < T$, then $|v'(\omega,\psi_v(\omega)) - R(\bar{v}(\omega,\psi_v(\omega))| < b_1$. Now for any $\omega,\omega' \in A_1 \cap A_2$ and on the same orbit,

$$|v'(\omega,\omega') - R(\bar{v}(\omega,\omega'))| < \frac{\varepsilon}{3}|\bar{v}(\omega,\omega')| + b_1.$$

As $\tilde{r}$ is inner, there is a set A_3, $\mu(A_3) > 1 - \frac{\varepsilon}{3}$ and a b_2 so that if ω, $\omega' \varepsilon A_3$ are on the same orbit, $|v(\omega,\omega') - v(\omega,\omega')| < b_2$ from Corollary 2.2. Thus for ω, $\omega' \varepsilon A = A_1 \cap A_2 \cap A_3$,

$|v'(\omega,\omega') - R(v(\omega,\omega'))| \leq \varepsilon|v(\omega,\omega')| + b_1 + b_2\|R\|$. Letting $b = b_1 + b_2\|R\|$ we have it.

■

It is this condition we wish to show is sufficient for r to be barely linear. We will in fact start with a condition on the surface much weaker, work back to this and proceed to prove the converse of Lemma 3.3, much as we did in Section II for inner time changes.

For a vector $v \varepsilon R^k$ we say a set A is an "a,b,v' - pinning set for v" if for any ω, $\varphi_{nv}(\omega) \varepsilon A$, $|v'(\omega,\varphi_{nv}(\omega)) - nv'| < an + b$.

Our result will be as follows.

Theorem 3.4. <u>If for a basis</u> $v_1 \ldots v_k$, <u>there is an</u> a,b,v_i<u>-pinning set for some</u> a,b <u>for each</u> v_i, <u>then</u> r <u>is barely linear</u>.

We now begin to construct a proof of this.

First a technical lemma we will need at several points.

Lemma 3.5. <u>Let</u> ψ_v <u>be a measurable time change of</u> φ_v. <u>For any</u> $T > 0$, $\varepsilon > 0$, <u>there is a</u> $B > 0$ <u>and a set</u> $A \subset \Omega$, $\mu(A) > 1 - \varepsilon$, <u>so that for any</u> v, $|v| < T$, <u>and any</u> ω, $\varphi_v(\omega)$ <u>both in</u> A, <u>we have</u>

$$|v'(\omega,\varphi_v(\omega))| < B.$$

Proof: Let $\bar{T} = \frac{8Tk}{\varepsilon}$. As the time change is measurable, there is a $B > 0$ and a set $V \subset \Omega \times [0,\bar{T}]^k$, $(\mu \times m)(V) > (1 - (\frac{\varepsilon}{4})^2)(\bar{T}^k)$ so that for $(\omega,v) \varepsilon V$, $|v'(\omega,\varphi_v(\omega))| < B/2$. Let $D = \{\omega | m(\{v| (\omega,v) \in V\}) > \bar{T}^k(1-\frac{\varepsilon}{4})\}$. We know $\mu(D) > 1 - \varepsilon/4$.

Now use the strong Rochlin lemma (7) to find a cross section F, with sectional measure $\bar{\mu}$, so that for $v \in [0,\bar{T}]^k$, $\varphi_v(F) \cap F = \phi$, the map $F \times [0,\bar{T}]^k \to \Omega$ given by $(\omega,v) \to \varphi_v(\omega)$ is measurable, taking $\mu \times \frac{m}{\bar{T}^k} \to \bar{\mu}$. Furthermore $\mu(\bigcup_{v\in[0,\bar{T}]^k} \varphi_v(F)) > 1 - \frac{\varepsilon}{4}$, and $\bar{\mu}(F \cap D) > 1 - \frac{\varepsilon}{4}$. Now let $A = \{\varphi_v(\omega) | \omega \in F \cap D, \ (\omega,v) \in V$ and $v \in [T,\bar{T}-T]^k\}$. It is a computation that $\mu(A) > 1 - \varepsilon$.

If ω, $\varphi_v(\omega) \in A$ and $|v| < T$, then $\omega = \varphi_{\bar{v}}(\omega')$, $\varphi_v(\omega) = \varphi_{\bar{v}+v}(\omega')$ where $(\omega,\bar{v})$ and $(\omega,\bar{v}+v)$ are both in V. Thus $|v'(\omega,\varphi_v(\omega))| = |v'(\omega',\varphi_{\bar{v}}(\omega')) - v'(\omega', \varphi_{\bar{v}+v}(\omega'))| < B$. ∎

Lemma 3.6. <u>If there is an</u> a,b,v'-<u>pinning set for</u> v, <u>then for any</u> $\varepsilon > 0$, <u>there is a</u> b' <u>and an</u> a,b',v'-<u>pinning set</u> A' <u>for</u> v <u>with</u> $\mu(A') > 1 - \varepsilon$.

Proof: Notice we have not assumed ${}_\infty\varphi_v$ is ergodic. Let A be the a,b,v'-pinning set for v. The set $\bar{A} = \bigcup_{i=1}^{\infty} \varphi_{iv}(A)$ is φ_v invariant, and hence has measure one or zero in a.e. ergodic component of φ_v. Pick an N so large that for a φ_v invariant set $\bar{A}_1 \subset \bar{A}$, $\mu(\bar{A}_1) > 0$, setting

$$A_1 = \bigcup_{i=1}^{N} \varphi_{iv}(A), \text{ if } \omega \in \bar{A}_1, \text{ then } \lim_{n\to\infty} \frac{1}{n} \sum_{i=1}^{n} \chi_{A_1}(\varphi_{-iv}(\omega)) > 1 - \frac{\varepsilon}{3} \text{ and let}$$

$A_2 = \bar{A}_1 \cap A_1$.

Now select a suitably dense set of vectors $v_1,\ldots,v_s$, that $\bar{A}_2 = \bigcup_{i=1}^{s} \varphi_{v_i}(\bar{A}_1)$ has measure $> 1 - \frac{\varepsilon}{3}$. Let $A_3 = \bigcup_{i=1}^{s} (\varphi_{v_i}(A_2) \cap \bigcup_{j=1}^{i-1} \varphi_{v_j}(\bar{A}_1)) = \bigcup_{i=1}^{s} A_3^i$.

Now $\bar{A}_2$ is φ_v invariant, $A_3 \subset \bar{A}_2$ and any φ_v orbit either lies outside $\bar{A}_2$, or intersects A_3 in a set of density at least $1 - \frac{\varepsilon}{4}$. Thus $\mu(A_3) > 1 - \frac{2\varepsilon}{3}$. Furthermore, if ω, $\varphi_{jv}(\omega) \in A_3$, they must be in the same A_3^i, i.e., $\varphi_{-v_i}(\omega)$, $\varphi_{jv}(\varphi_{-v_i}(\omega)) \in A_2$.

Now use Lemma 3.5 to select b_1 so large that there is a set A_4, $\mu(A_4) > 1 - \frac{\varepsilon}{3NS}$, and if $\omega,\varphi_{\tilde{v}}(\omega) \in A_4$, $|\tilde{v}| < \sup_i |v_i| + N|\bar{v}|$, then $|v'(\omega,\varphi_{\tilde{v}}(\omega))| < b_1$. Let $A' = A_3 \cap (\bigcap_{i=1}^{s} \bigcap_{j=1}^{N} \varphi_{jv+v_i}(A_4))$. Now if $\omega,\varphi_{nv}(\omega) \in A'$, then for some v_j and $n_1, n_2 < N$, $\varphi_{-v_j-n_1v}(\omega)$, $\varphi_{-v_j-(n_2+n)v}(\omega) \in A \cap A_4$. Letting $\omega' = \varphi_{-v_j-n_1v}(\omega)$, we get $|v'(\omega', \varphi_{(n_1-n_2+n)v}(\omega)) - (n_1-n_2+n)v'| <$

$a(n_1-n_2 + n) + b$. Thus $|v'(\omega,\varphi_{nv}(\omega)) - nv'| < an + (b+2b_1 + 2N|v'|)$. Let $b' = b + 2b_1 + 2N|v'|$. ∎

Lemma 3.7. <u>If there is an</u> a,b,v'-<u>pinning set for</u> v, <u>then for any</u> $\varepsilon > 0$, <u>there are</u> b', v'' <u>and an</u> ε,b',v''-<u>pinning set for</u> v <u>with measure</u> $> 1 - \varepsilon$.

Proof: From the last lemma, let A be an a,b,v'-pinning set for v with $\mu(A) > 1 - \varepsilon$. Let U be the transformation φ_v induced on A, $U(\omega) = \varphi_{n(\omega)v}(\omega)$, $n(\omega)$ the return time to A. Define the function $v_1(\omega) = v'(\omega,U(\omega))$ on A. We know $|v_1(\omega)| < (a + |v'|)n(\omega) + b$, and as $n(\omega) \in L_1(A)$, $v_1(\omega) \in L_1(A)$. Thus $\frac{1}{n}\sum_{i=1}^{n} v_1(U^i(\omega))$ converges pointwise a.e. to a U invariant function $v_2(\omega)$. Now choose N so large that for a set $A_1 \subset A$, $\mu(A_1) > \mu(A) - \frac{\varepsilon}{4}$, if $\omega \in A_1$, $n > N$, $|\frac{1}{n}\sum_{i=1}^{n} v_1(U^i(\omega)) - v_2(\omega)| = |\frac{1}{n}v'(\omega,U^n(\omega)) - v_2(\omega))| < \varepsilon$. Now choose b so large that there is a set $A_2 \subset A_1$, $\mu(A_2) > \mu(A_1) - \frac{\varepsilon}{4}$, if $\omega \in A_2$, $n \le N$, $|v_i'(\omega,U^n(\omega))| \le b$. Thus for any ω, $\varphi_{nv}(\omega) \in A_2$, $|v'(\omega,\varphi_{nv}(\omega)) - nv_2(\omega)| < \varepsilon n + b$. All we need is $v_2(\omega)$ constant v'' a.e. on A.

We argue this by a variant of Lemma 3.1. Let $a = \text{dia}(v_2(A))$. There are then subsets $S_1,S_2 \subset A$ of positive measure so that for any $\omega_1 \in S_1$, $\omega_2 \in S_2$, $|v_2(\omega_1) - v_2(\omega_2)| > \frac{a}{2}$.

As S_1, S_2 have positive measure, for some $\bar{v}$, $\varphi_{\bar{v}}(S_1) \cap S_2 = S_3$ has positive measure. Choose $\bar{b}$ so large that for a subset $S_4 \subset S_3$, $\mu(S_4) > \frac{\mu(S_3)}{2}$, if $\omega \in S_3$, $|v'(\omega,\varphi_v(\omega))| < \bar{b}$. Now as S_3 has positive measure, there are arbitrarily large values n, $\varphi_{nv}(S_4) \cap S_4$ has positive measure.

If $\omega \in \varphi_{nv}(S_4) \cap S_4$, then

$$|v'(\omega,\varphi_{nv}(\omega))-nv_2(\omega)| < \varepsilon n + b,$$

$$|v'(\varphi_{\bar{v}}(\omega),\varphi_{nv}(\varphi_{\bar{v}}(\omega)))-nv_2(\varphi_{\bar{v}}(\omega))| < \varepsilon n + b,$$

but

$$|v'(\omega,\varphi_{\bar{v}}(\omega))| < \bar{b}$$

and

$$|v'(\varphi_{nv}(\omega),\varphi_{nv}(\varphi_{\bar{v}}(\omega))| < \bar{b}$$

so

$$|v_2(\omega)-v_2(\varphi_v(\omega))| < 2\varepsilon + \frac{2b+2\bar{b}}{n} .$$

But this is $> \frac{a}{2}$. As ε could have been chosen as small as we like, and n as large, $a = 0$. This completes the proof.

■

Theorem 3.8. <u>Suppose for a time change</u> ψ_v <u>of</u> φ_v, <u>there is a basis</u> $v_1 \ldots v_k$ <u>each with an</u> a,b,v_i'<u>-pinning set for some</u> a,b. <u>It follows that for any</u> ε <u>there is a linear map</u> $R; R^k \circlearrowleft$ <u>and a set</u> $A, \mu(A) > 1 - \varepsilon$, <u>and a</u> $b > 0$ <u>so that for</u> $\omega, \varphi_v(\omega) \in A$,

(3.4) $$|v'(\omega,\varphi_v(\omega)) - R(v)| < \varepsilon|v| + b.$$

Proof: For each v_i, use Lemma 3.7 to find a b' and v_i'' so that for each v_i we have an $\frac{\varepsilon}{2k}$, b', v_i''-pinning set, of measure $> 1 - (\frac{\varepsilon}{3})^{2^k} \cdot \frac{1}{k}$, making sure $\varepsilon < \frac{3}{4} \cdot \frac{1}{2^k}$. Let $R(v_i)$ be the corresponding v_i'', and A_0 be the intersection of these pinning sets. Thus $\mu(A_0) > 1 - (\frac{\varepsilon}{3})^{2^k}$.

We now build inductively a string of sets. Choose N_1 so large that for a set $A_1 \subset A_0$, $\mu(A_1) > 1 - (\frac{\varepsilon}{3})^{2^{k-1}}$, if $\omega \in A_1$, $n > N_1$, then at least $(1 - (\frac{\varepsilon}{3})^{2^{k-1}})$ of the points $\omega, \varphi_{v_1}(\omega), \ldots, \varphi_{nv_1}(\omega)$ are in A_1, as are this fraction of $\omega, \varphi_{-v_1}(\omega), \ldots, \varphi_{-nv_1}(\omega)$.

Now if $A_j \subset A_{j-1} \ldots \subset A_0$ have been defined, $\mu(A_j) > 1 - (\frac{\varepsilon}{3})^{2^{k-j}}$, let N_{j+1} be chosen so large that for a set $A_{j+1} \subset A_j, \mu(A_{j+1}) > 1 - (\frac{\varepsilon}{3})^{2^{k-j-1}}$, if $\omega \in A_{j+1}$, then at least $(1 - (\frac{\varepsilon}{3})^{2^{k-j-1}})n$ of the points

$\omega, \varphi_{v_{j+1}}(\omega) \ldots \varphi_{nv_{j+1}}(\omega)$ are in A_j, as are this many of

$\omega, \varphi_{-v_{j+1}}(\omega) \ldots \varphi_{-nv_{j+1}}(\omega)$ if $n > N_{j+1}$.

Let $N = \sup(N_j)$ and notice $\mu(A_k) > 1 - \frac{\varepsilon}{3}$. Now choose b_1 so large that $b_1 > Nk \sup|v_i|$ and for a subset $A' \subset A_k$, $\mu(A') > 1 - \frac{2\varepsilon}{3}$, so that if $\omega \in A'$ then the density

$$\frac{\int_{|v| < b_1} \chi_{A_k}(\psi_v(\omega))dm}{\int_{|v| \leq b_1} 1dm} > 1 - \frac{2\varepsilon}{3} .$$

Using Lemma 3.5, we can find a b_2 so large that for a subset $\bar{A} \subset \Omega$, $\mu(\bar{A}) > 1 - (\frac{\varepsilon}{3})^2$, if $\omega, \varphi_v(\omega) \in \bar{A}$, $|v| < b_1$ then $|v'(\omega,\varphi_v(\omega))| < b_2$. It follows that for a set $A'' \subset \bar{A}$, $\mu(A'') > 1 - \frac{\varepsilon}{3}$, if $\omega \in A''$ then $V(\omega) = \{v \mid |v| < b$, and $\varphi_v(\omega) \in A\}$ has density at least $1 - \frac{\varepsilon}{3}$ in $\{v \mid |v| < b_1\}$. Letting $A = A' \cap A''$, $\mu(A) > 1 - \varepsilon$.

Thus for $\omega \in A$,

$$\frac{\int_{|v|<b_1} \chi_{A_k \cap A}(\varphi_v(\omega))\,dm}{\int_{|v|<b_1} 1\,dm} > 1 - \varepsilon > \frac{5}{8}. \tag{3.5}$$

Now we are ready to compute $|v'(\omega,\varphi_v(\omega)) - R(v)|$ for $\omega,\varphi_v(\omega) \in A$. Write $v = \Sigma_{i=1}^k n_i v_i + v_0$ where $n_i = 0$ or $n_i > N$ and $|v_0| < Nk \sup|v_i| < b_1$.

Now writing $\Sigma n_i v_i = \bar{v}$, consider

$$[A_k \cap \bar{A} \cap (\bigcup_{|v|<b_i} \varphi_v(\omega))] \cap \varphi_{-v}[A_k \cap \bar{A} \cap (\bigcup_{|v|<b_1} \varphi_v(\omega))].$$

From (3.5), this is nonempty. Let $\omega_1, \varphi_v(\omega_1) \in A_k \cap \bar{A} \cap (\bigcup_{|v|<b_1} \varphi_v(\omega))$.

Thus $|v'(\omega,\omega_1)|$, $|v'(\varphi_v(\omega), \varphi_v(\omega_1))| < b_2$, and $|R(v) - R(\bar{v})| < 2\|R\|b_1$. It remains to compute $|v'(\omega_1,\varphi_{\bar{v}}(\omega_1)) - R(\bar{v})|$. The idea here is that if we had $\omega_1, \varphi_{n_1v_1}(\omega_1), \varphi_{n_1v_1+n_2v_2}(\omega_1),\ldots,\varphi_{\bar{v}}(\omega_1)$ all in A_0, the fact that it is a pinning set for all v_i would give the result. They may not be, so we want to replace the path $\omega_1 \to \varphi_{n_1e_1}(\omega_1) \to \varphi_{n_1e_1+n_2e_2}(\omega_1) \ldots \to \varphi_{\bar{v}}(\omega_1)$ by a new one, all of whose vertices are in A_0, using the inductive sequence $A_0, A_1,\ldots,A_k$ we built earlier.

For notational convenience assume all the $n_i > N$, as whenever $n_i = 0$ we can simply omit that step of the following inductive construction of the new path $\omega_1 = \bar{\omega}_1 \to \bar{\omega}_2 \ldots \to \bar{\omega}_{2k-2} = \varphi_{\bar{v}}(\omega_1)$.

Let $\bar{v}_i = \Sigma_{j=i}^k n_i v_i$, and assume $\bar{\omega}_{2k-1-i} = \varphi_{v_i}(\bar{\omega}_i)$, which is true for $i = 1$, and that $\bar{\omega}_i = \bar{\omega}_{2k-1-i} \in A_{k-i+1}$, which again is true for $i = 1$, and that for $i \geq j > 1$, $\bar{\omega}_i = \varphi_{m_{i-1}v_{i-1}}(\bar{\omega}_{i-1})$, $\bar{\omega}_{2k-1-i} = \varphi_{-m'_{i-1}v_{i-1}}(\bar{\omega}_{2k-i})$, where $m_{i-1} + m'_{i-1} = n_{i-1}$, and all three have the same sign.

Now to construct $\bar{\omega}_{i+1}$ and $\bar{\omega}_{2k-2-i}$. As $\bar{\omega}_i, \bar{\omega}_{2k-1-i} \in A_{k-i+1}$, among the points $\bar{\omega}_i, \varphi_{v_i}(\bar{\omega}_i) \ldots \varphi_{n_i v_i}(\bar{\omega}_i)$ and the points $\bar{\omega}_{2k-1-i}, \varphi_{-v_i}(\bar{\omega}_{2k-1-i}) \ldots \varphi_{-n_i v_i}(\bar{\omega}_{2k-1-i})$ are subsets of density $> \frac{1}{2}$ in A_{k-i}. Hence there is a value j between 0 and n_i with both $\varphi_{jv_i}(\bar{\omega}_i)$ and $\varphi_{(n_i-j)v_i}(\bar{\omega}_{k-i+1})$ in A_{k-i}. Let $m_i = j$, $m_i' = (n_i-j)$, and we have it.

The induction ends when $i = k-1$. As all $A_i \subset A_0$, all $\bar{\omega}_i \in A_0$. It follows that $|v'(\omega_1,\varphi_{\bar{v}}(\omega_1))-R(\bar{v})| \le 2kb' + \frac{\varepsilon}{2k}(k|\bar{v}|) < 2kb' + \frac{\varepsilon}{2}|\bar{v}|$, and finally $|v'(\omega,\varphi_v(\omega))-R(v)| < \frac{\varepsilon}{2}|v| + 2b_2 + 2\|R\|b_1 + 2kb_1$. Letting $b = 2b_2 + 2\|R\|b_1 + 2kb_1$ we are done. ∎

The conclusion of this lemma differs from that of Lemma 3.3 only in that we do not have the map R independent of ε and invertible. We now show this.

Lemma 3.9. <u>Let</u> ψ_v <u>be a time change as in Lemma</u> 3.8. <u>There is, then, an invertible matrix</u> R <u>so that for any</u> $\varepsilon > 0$ <u>there is a</u> $b > 0$ <u>and a set</u> A, $\mu(A) > 1-\varepsilon$ <u>and if</u> $\omega, \varphi_v(\omega) \in A$, $|v'(\omega,\varphi_v(\omega))-R(v)| < \varepsilon|v| + b$.

Proof: Select $\varepsilon_i \to 0$ and $\Sigma\varepsilon_i < 1$ and let R_i, b_i and A_i be the matrices constants and sets from Theorem 3.8 corresponding to ε_i. As $\Sigma\varepsilon_i < 1$, $\bar{A} = \cap A_i$ has positive measure. If $\omega, \varphi_v(\omega) \in \bar{A}$, for all i we have $|v'(\omega,\varphi_v(\omega)) - R_i(v)| < \varepsilon_i|v| + b_i$. Thus for $j \ge i$,

$|R_j(v) - R_i(v)| < \varepsilon_i|v| + b_i$. As $\mu(\bar{A}) > 0$, for any unit vector v, there are arbitrarily large n so that $\mu(\varphi_{nv}(\bar{A}) \cap \bar{A}) > 0$. Thus for all unit vectors v, $|R_j(v) - R_i(v)| < \varepsilon_i + b_i/n$, for arbitrarily large n. Thus

$\|R_j-R_i\| < \varepsilon_i$, and R_i converges to some R.

Now $\|R_i-R\| < \sum_{j=i}^{\infty} \varepsilon_j = \bar{\varepsilon}_i$, so $|v'(\omega,\varphi_v(\omega))-R(v)| < \bar{\varepsilon}_i|v| + b_i$ if $\omega,\varphi_v(\omega) \in A_i$. All we now need is R invertible. Suppose not. Let v_0 be a unit vector with $R(v_0) = 0$.

Using Lemma 3.5, find a $\bar{b}$ so large that for a subset $\bar{A}' \subset \bar{A}$ of positive measure, if $\omega, \varphi_v(\omega) \in \bar{A}'$, and $|v| < 1$, then $|v'(\omega,\varphi_v(\omega))| < \bar{b}$ and $B(\omega) = \{v'(\omega,\varphi_v(\omega)) \,|\, |v| < 1, \varphi_v(\omega) \in \bar{A}'\}$ has Lebesgue measure at least $\frac{1}{\bar{b}}$.

Using the ergodic Theorem pick an $\omega \in \bar{A}'$ and $0 < n_1 < n_2 < \cdots$ so that all $\varphi_{n_i v_0}(\omega) \in \bar{A}'$, and the n_i have positive density d in $\mathbf{Z}$. We know the sets $v'(\omega,\varphi_{n_k v_0}(\omega)) + B(\varphi_{n_k v_0}(\omega))$ are disjoint. Thus for all k,

$\overline{B}_k = \bigcup_{i=1}^{k} (v'(\omega,\varphi_{n_i v_0}(\omega)) + B(\varphi_{n_i v_0}(\omega)))$ has Lebesgue measure at least $\frac{k}{\overline{b}}$.

But as $|v'(\omega,\varphi_{n_i v_0+v}(\omega))| < \overline{\varepsilon}_j n_i + b_i + \overline{b}$ for all i and j, $\overline{B}_k$ has Lebesgue measure at most $\overline{\varepsilon}_j n_k + b_j + \overline{b}$. For k sufficiently large, $\frac{k}{n_k} > \frac{d}{2}$. Thus for k large enough $\overline{\varepsilon}_j n_k + b_j + b > m(\overline{B}_k) > \frac{dn_k}{2\overline{b}}$. Picking $\overline{\varepsilon}_j < \frac{d}{4\overline{b}}$, $n_k > \frac{4\overline{b}(b_j+\overline{b})}{d}$, we have a conflict. Thus R is invertible. ∎

Now we are ready to prove Theorem 3.4. Let the time change ψ_v of φ_v satisfy Theorem 3.8, and hence Lemma 3.9. Fix an $1 > \varepsilon > 0$, for the remainder of the argument, making sure $\sqrt{\varepsilon} < \frac{|R(v)|}{|v|}$ for all v.

Build the set A and matrix R of Lemma 3.9, so that $\mu(A) > 1 - \frac{\varepsilon}{6}$ and if (ω,ω') are on the same orbit, $|v'(\omega,\omega') - R(v(\omega,\omega'))| < \frac{\varepsilon}{6}|v(\omega,\omega')|+b$. We now apply the strong Rochlin Theorem [7] for free R^k actions. Using it, find a set F so that any point ω in $T = \bigcup_{v \,\varepsilon\, \text{unit cube}} (\varphi_{(6bv)/\varepsilon}(F))$ has a unique representation $(\overline{\omega}, v)$ where $\overline{\omega} \,\varepsilon\, F$, $v \,\varepsilon$ (unit cube) $\frac{6b}{\varepsilon}$, $\omega = \varphi_v(\overline{\omega})$, and $\mu(T) > 1 - \varepsilon$. Furthermore F can be regarded as a nonatomic Lebesgue probability space $(F,\mathcal{S},\eta)$, where $\mathcal{F}/\overline{F} = \mathcal{S} \times$ (Lebesgue sets) and μ is $\frac{\eta \times \text{(Lebesgue measure)}}{(\frac{6b}{\varepsilon})^k}$. Last and most important, $\eta(F \cap A) > 1 - \frac{\varepsilon}{6}$.

Set $\overline{A} = F \cap A$. The above facts about F make the following construction measurable. Notice that if $\omega,\omega' \,\varepsilon\, \overline{A}$ are on the same orbit,

$$(3.6)\qquad \begin{aligned} |v'(\omega,\omega') - R(v(\omega,\omega'))| &< \frac{\varepsilon}{6}\,|v(\omega,\omega')| + b \\ &< \frac{\varepsilon}{3}\,|v(\omega,\omega')| \end{aligned}$$

as, if $\omega \neq \omega'$, $v(\omega,\omega') < \frac{6b}{\varepsilon}$.

We are now ready to build $\overline{v}$. Fix a point ω and for each $\omega'' \,\varepsilon\, \overline{A}$ in the orbit of ω, take the vector $R(v(\omega,\omega'')) - v'(\omega,\omega'')$ and center there a closed k-dimensional cube oriented parallel to the coordinate axes with side lengths $\frac{2\varepsilon}{3}|v(\omega,\omega'')|$. Call this cube $C(\omega,\omega'')$.

Lemma 3.10. <u>For every</u> ω, <u>except those on orbits not intersecting</u> $\bar{A}$,

$$\bigcap_{\substack{\omega'' \in A \\ \text{on the orbit} \\ \text{of } \omega}} C(\omega,\omega'')$$

is a nonempty k-dimensional rectangle.

Proof: Let two of the cubes be $C(\omega,\omega_1'')$, $C(\omega,\omega_2'')$, centered at $R(v(\omega,\omega_1''))- v'(\omega,\omega_1'')$ and $R(v(\omega,\omega_2'')) - v'(\omega,\omega_2'')$. These two vectors differ by

$$|R(v(\omega_1'',\omega_2''))- v'(\omega_1'',\omega_2'')| < \frac{\varepsilon}{3} |v(\omega_1'',\omega_2'')| < \frac{\varepsilon}{3} |v(\omega,\omega_1'')| + \frac{\varepsilon}{3} |v(\omega,\omega_2'')|.$$

Thus any two cubes intersect. It is an easy geometrical exercise that if any two of a collection of cubes with parallel edges intersect nontrivially, then the whole collection has a nonempty intersection.

∎

Call this rectangle $P(\omega)$. $P(\omega)$ will play much the same role as $S_\varepsilon^j(\omega)$ did for inner time changes. Notice if $\omega' \in \bar{A}$, $P(\omega') = \{w'\}$. Let $P(\omega) = \{v | p_j(\omega) \le v_j \le p_j'(\omega)\}$ define $p_j(\omega)$ and $p_j'(\omega)$.

Lemma 3.11. <u>If</u> ω_1,ω_2 <u>are on the same orbit, then for all</u> $j = 1 \ldots k$, $|p_j(\omega_1) - (p_j(\omega_2) - (v_j'(\omega_1,\omega_2) - R_j(v(\omega_1,\omega_2))))|$ <u>and</u> $|p_j'(\omega_1) - (p_j'(\omega_2) - (v_j'(\omega_1,\omega_2) - R_j(v(\omega_1,\omega_2))))| < \frac{\varepsilon}{3} |v(\omega_1,\omega_2)|$.

Proof: Fix j, and now for $\varepsilon'' > 0$ find $\omega'' \in \bar{A}$ so that the jth edge of $C(\omega_1,\omega'')$ has j^{th} coordinate $c_j(\omega_1,\omega'')$, and $c_j(\omega_1,\omega'') < p_j(\omega_1) < c_j(\omega_1,\omega'') + \varepsilon''$.

Now $c_j(\omega_1,\omega'') = c_j(\omega_2,\omega'') + (R_j(v(\omega_1,\omega_2)) - v_j'(\omega_1,\omega_2)) + \frac{\varepsilon}{3}(|v(\omega_1,\omega'')| - - |v(\omega_2,\omega'')|)$. Thus $p_j(\omega_2) - (v_j'(\omega_1,\omega_2) - R_j(v(\omega_1,\omega_2))) \le p_j(\omega_1) + \frac{\varepsilon}{3} |v(\omega_1,\omega_2)| + \varepsilon''$. Replacing ω_1 by ω_2 we get $p_j(\omega_2) - (v_j'(\omega_1,\omega_2) - R_j(v(\omega_1,\omega_2))) \ge p_j(\omega_1) - \frac{\varepsilon}{3}|v(\omega_1,\omega_2)| - \varepsilon''$.

Now let $\varepsilon'' \to 0$. These together give the result for p_j, and analogously p_j'.

∎

For each ω, let $p(\omega)$ be the midpoint of $P(\omega)$. Clearly the map $\omega \to \psi_{p(\omega)}(\omega)$ is continuous on orbits.

Corollary 3.12. For any ω_1,ω_2 on the same orbit,

$$|p(\omega_1) - (p(\omega_2) - (v'(\omega_1,\omega_2) - R(v(\omega_1,\omega_2))))| < \frac{2\varepsilon}{3}|v(\omega_1,\omega_2)|.$$

Proof: Follows directly from Lemma 3.11.

If all we were interested in was a Lipschitz condition on $\bar{v}$, we could now just set $\bar{v}(\omega_1,\omega_2) = v'(\omega_1,\omega_2) + p(\omega_1) - p(\omega_2)$. We in fact want a C^∞ condition, so we now smooth $p(\omega)$ in a very simple manner. Let $\kappa\colon \mathbb{R}^k \to \mathbb{R}^+$ be a C^∞ integrating kernel of norm 1, with bounded support. Now let

$$\bar{p}(\omega) = \int_{R^k} \kappa(v)\cdot(p(\varphi_v(\omega)) - v'(\omega,\varphi_v(\omega)) + R(v))dv.$$

Corollary 3.13. $|\bar{p}(\omega_1) - (\bar{p}(\omega_2) - (v'(\omega_1,\omega_2) - R(v(\omega_1,\omega_2))))| < \frac{2\varepsilon}{3}|v(\omega_1,\omega_2)|.$

Proof: Since $v'(\omega_1,\omega_2) + v'(\omega_1,\varphi_v(\omega_2)) - v'(\omega_1,\varphi_v(\omega_1)) = v'(\varphi_v(\omega_1),\varphi_v(\omega_2))$, we have

$$\begin{aligned}
&|\bar{p}(\omega_1) - (\bar{p}(\omega_2) - (v'(\omega_1,\omega_2) - R(v(\omega_1,\omega_2))))| \\
&= \Big|\int_{R^k} \kappa(v)[p(\varphi_v(\omega_1)) - v'(\omega_1,\varphi_v(\omega_1)) + R(v) \\
&\quad - p(\varphi_v(\omega_2)) + v'(\omega_2,\varphi_v(\omega_2)) - R(v) + v'(\omega_1,\omega_2) - R(v(\omega_1,\omega_2))]dv\Big| \\
&= \Big|\int_{R^k} \kappa(v)[p(\varphi_v(\omega_1)) - p(\varphi_v(\omega_2)) \\
&\quad + v'(\varphi_v(\omega_1),\varphi_v(\omega_2)) - R(v(\varphi_v(\omega_1),\varphi_v(\omega_2)))]dv\Big| \\
&\le \int_{R^k} \kappa(v)\frac{2\varepsilon}{3}|v(\omega_1,\omega_2)|dv.
\end{aligned}$$

This follows from Corollary 3.12 and $v(\omega_1,\omega_2) = v(\varphi_v(\omega_1),\varphi_v(\omega_2))$. ∎

We now define $\bar{v}$ by $\bar{v}(\omega_1,\omega_2) = v'(\omega_1,\omega_2) + \bar{p}(\omega_1) - \bar{p}(\omega_2)$.

Lemma 3.14. For a.e. ω, the map $\bar{r}'(\cdot,\omega);\ \mathbb{R}^k \to R^k$ given by $\bar{r}'(v,\omega) = \bar{v}(\omega,\varphi_v(\omega))$ is C^∞ and if its matrix of first partials at $\vec{0}$ is given by $\bar{R}(\omega)$, then $\|\bar{R}(\omega) - R\| < \frac{2\varepsilon}{3}$.

Proof: $\bar{r}'(v',\omega) = \bar{v}(\omega,\varphi_{v'}(\omega)) = v'(\omega,\varphi_v(\omega)) + \bar{p}(\omega) - \bar{p}(\varphi_{v'}(\omega)) = v'(\omega,\varphi_v(\omega) + \int_{R^k}\kappa(v)\cdot(p(\varphi_v(\omega)) - v'(\omega,\varphi_v(\omega)) + R(v))dv - \int_{R^k}\kappa(v)(p(\varphi_{v+v'}(\omega)) - v'(\varphi_{v'}(\omega),\varphi_{v+v'}(\omega)) + R(v))dv = \int_{R^k}\kappa(v)(p(\varphi_v(\omega)) - p(\varphi_{v+v'}(\omega)) - v'(\varphi_v(\omega),\varphi_{v+v'}(\omega)))dv = \int_{R^k}(\kappa(v) - \kappa(v-v'))(p(\varphi_v(\omega)) - v'(\omega,\varphi_v(\omega)))dv.$

Now $f(v) = p(\varphi_v(\omega)) - v'(\omega,\varphi_v(\omega))$ is uniformly continuous, as

$$|f(v_1) - f(v_2)| = |p(\varphi_{v_1}(\omega)) - p(\varphi_{v_2}(\omega)) - v'(\varphi_{v_1}(\omega),\varphi_{v_2}(\omega))| \le (\|R\| + \tfrac{2\varepsilon}{3})|v_2 - v_1|.$$

Thus, as κ is C^∞, $\bar{r}'(\cdot,\omega)$ is C^∞. From Corollary 3.13, for a basis vector

e_i, $\left|\frac{\bar{r}'(\delta e_i,\omega)}{\delta} - R(e_i)\right| < \frac{2\varepsilon}{3}$.

$$\left|\frac{\partial F_j'(\omega)}{\partial e_i} - R_{ij}\right| = |R_{i,j}(\omega) - R_{ij}| \le \frac{2\varepsilon}{3} .$$

Notice now that $\frac{|\bar{R}(\omega)(v)|}{|v|} > (\sqrt{\varepsilon} - \frac{2\varepsilon}{3}) > \frac{\sqrt{\varepsilon}}{3}$, and so $\bar{R}(\omega)$ is always invertible, and

$$\frac{|R^{-1}(v) - \bar{R}^{-1}(\omega)(v)|}{|v|} < \frac{2\varepsilon}{3}\left(\frac{3}{\sqrt{\varepsilon}}\right)\|R\| < 2\|R\|\sqrt{\varepsilon} .$$

Lemma 3.15. <u>For a.e. ω, the map $\bar{r}'(\cdot,\omega)$ is 1-1 onto R^k and obviously from the above remarks about $\bar{R}^{-1}$, its inverse $\bar{r}(\cdot,\omega)$ is C^∞ and its matrix of first partials at $\vec{0}$ is $\bar{R}^{-1}(\omega)$.</u>

Proof: For a.e. ω and v we have $|\bar{v}(\omega,\varphi_v(\omega)) - R(v)| < \frac{2\varepsilon}{3}|v|$ and $R(v) > \varepsilon|v|$. Hence $|\bar{v}(\omega,\varphi_v(\omega))| > \frac{\varepsilon}{3}|v|$. Fix ω. If $\bar{r}'(v_1,\omega) = \bar{r}'(v_2,\omega)$, then $0 = v(\varphi_{v_1}(\omega), \varphi_{v_2}(\omega)) > \frac{\varepsilon}{3}|v_2 - v_1|$, hence $r'(\ ,\omega)$ is 1-1. Furthermore the image of a ball of radius b must contain a ball of radius $\frac{\varepsilon b}{3}$, as its boundary is mapped at least this far away from 0. Thus $r(\cdot,\omega)$ is onto.

■

Now let $\bar{\psi}_v(\omega) = \varphi_{\bar{r}(v,\omega)}(\omega)$. We already know that $\bar{r}(\cdot,\omega)$ and $\bar{r}'(\cdot,\omega)$ are C^∞ for a.e. ω and their matrices of first partials satisfy $|\bar{R}(\omega)(v) - R(v)| < \frac{2}{3}\varepsilon|v|$ and $|\bar{R}^{-1}(\omega)(v) - R^{-1}(v)| < 2\|R\|\sqrt{\varepsilon}\,|v|$.

Lemma 3.16. <u>The map $\bar{\psi}_v$ is an inner time change of ψ_v.</u>

Proof: As $\bar{v}(\omega,\omega') = v'(\omega,\omega') + \bar{p}(\omega) - \bar{p}(\omega')$, and $\bar{p}$ is defined and continuous on a.e. orbit all we really need is $\omega \to \psi_{-\bar{p}(\omega)}(\omega)$ nonsingular. Now $\bar{\psi}_v$ preserves the measure $\mu\det(R(\omega))$ as $\bar{r}$ is C^∞, which has support on the generic points of ψ_v, hence must be $(\psi_{-p(\omega)}(\omega))^{-1}(\nu)$.

■

This completes the proof of Theorem 3.4, with something extra.

Corollary 3.17. <u>If the time change $r(v,\omega)$ is barely linear, for any $\varepsilon > 0$ it can be modified by an inner time change to $\bar{r}(v,\omega)$ a C^∞ tempered time change so that the matrices of first partials of $\bar{r}(\cdot,\omega)$ and $\bar{r}'(\cdot,\omega)$ are uniformly within ε of constant.</u>

Proof: Choose the ε in the argument above to be $\min(\varepsilon, (\frac{\varepsilon^2}{2\|R\|}))$.

This completes our basic argument. This corollary explains the name "barely linear", as it says such can be made as close as we like to linear.

Corollary 3.18. <u>Suppose the time change</u> $\psi_v = \varphi_{r(v,\cdot)}$ <u>has for a basis</u> $v_1,\ldots,v_k$, $r_j(v_i,\cdot) \in L^1(\nu)$ <u>for</u> $j = 1\ldots k$. <u>Then</u> $r(v,\omega)$ <u>is barely linear</u>.

Proof: Let $v'_{ji} = \int r_j(v_i,\omega)d\nu$. Then for any ε there is an N and set A, $\mu(A) > 1 - \varepsilon$ so that if $n > N$, $\left|\sum_{\ell=1}^{n} r_j(v_i,\psi_{\ell v_i}(\omega)) - nv'_{j,i}\right| < n\varepsilon$, by the ergodic theorem (obvious if ψ_{v_k} is ergodic. If not, an argument as in Lemma 3.9 shows the limit of averages is constant anyway). Thus $|v'_j(\omega,\psi_{nv_i}(\omega)) - nv'_{j,i}| < n\varepsilon$ or $|v'(\omega,\psi_{nv_i}(\omega)) - nv'_i| < n\varepsilon$. Find a subset $\bar{A}$ of A, $\mu(\bar{A}) > 0$, so that no point of $\bar{A}$ returns to $\bar{A}$ in fewer than N steps. This is an exercise requiring only aperiodicity.

Now if ω, $\psi_{nv}(\omega) \in \bar{A}$, $|v'(\omega,\psi_{nv_i}(\omega)) - nv'_i| < n\varepsilon$ and $\bar{A}$ is an ε, 0, v'_i-pinning set for v_i. Thus $r'(v,\omega)$ is an almost tempered of ψ_v, hence so is its inverse. ■

Corollary 3.19. <u>If the time change</u> $r(v,\omega)$ <u>of</u> φ_v <u>is</u> C^1 <u>with all its partials</u>

$$R_{i,j}(\omega) = \frac{\partial r_i(\cdot,\omega)}{\partial e_j}(\vec{0})$$

<u>in</u> $L^1(\mu)$, <u>then</u> r <u>is barely linear</u>.

Proof: We can write $r_j(e_i,\omega) = \int_0^1 R_{j,i}(\psi_{te_i}(\omega))dt$. Thus $r_j(e_i,\omega) \in L^1(\mu) = L^1(\nu)$ and the result follows from Corollary 3.20. ■

We can also write down extensions of the Nadler, Feldman [5], [8] work to the barely linear case. We indicate a few.

Corollary 3.20. <u>A barely linear time change preserves entropy class</u>, 0, <u>positive finite, or infinite</u>.

We can make this more precise in the following way.

Corollary 3.21. <u>If the time change</u> ψ_v <u>is barely linear and</u> R <u>is the matrix of Lemma</u> 3.11, <u>then</u> $h(\psi_v) = \det(R)h(\varphi_v)$.

Proof: We know $h(\psi_v) = h(\bar{\varphi}_v)$. It is also clear that the Radon-Nikodym derivative $\Delta_{\bar{r}}$ of μ with respect to $\bar{\nu}$ at ω is

$$\det\left(\frac{\partial \bar{r}_i(\cdot,\omega)}{\partial e_j}(e_j)\right)$$

and for any ε, we can build $\bar{\psi}_v$ so that

$$\int_\Omega \det\left(\frac{\partial \bar{r}_i(\cdot,\omega)}{\partial e_j}\right) d\bar{\nu} - \det(R) < \varepsilon.$$

Thus by Nadler's result (7), $h(\bar{\psi}_v) = \int_\Omega \Delta_{\bar{r}}(\omega)d\bar{\nu}h(\varphi)$. Thus $h(\psi_v) - \det(R)h(\varphi) < \varepsilon h(\varphi)$ for all ε. Hence the result. ∎

Special versions of Corollary 3.21 exist for the time changes of Collaries 3.18 and 3.19. For 3.18, let $R(\omega)$ be the matrix mapping v_i to $r(v_i,\omega)$, and let $R = \int_\Omega R(\omega)d\nu$. For 3.19 let $R = \int_\Omega R(\omega)d\nu$ for the $R(\omega)$ given there. These are obviously the R's of Lemma 3.11.

As a final application, Katok [6] has shown that for any $\mathbb{R}^k$ action φ_v and basis $e_1 \ldots e_k$, there is a time change ψ_v with a special form. Perhaps the most intuitive way to describe it is that under the coordinates $v'(\omega,\omega')$ on orbits, orbits can be measurably paved by the translates of the cell $C = \{v \mid (v,e_i) \le |e_i|\}$. More importantly for our purposes, for any $\varepsilon > 0$, we can also have $|v(\omega,\varphi_v(\omega))-v| < \varepsilon|v|$. This Lipschitz condition is more than enough to make the time change barely linear. Hence, Katok's representation can be obtained by a C^∞ tempered time change. This fact could have been obtained within Katok's argument, and he was in fact aware of it. This extra finesse is now, though, a simple application of our work here.

REFERENCES

[1] Belinskaja, R. M., "Partitionings of a Lebesgue space into trajectories which may be defined by ergodic automorphisms", (Russian), Funkcional Analiz Prilozen 2 (1968) No. 3: 4-16.

[2] Belinskaja, R. M., "Generalized degrees of automorphisms and entropy", Sib. Mat. Zh. 11 (1970) No. 4: 739-749.

[3] Dye, H., "On groups of measure-preserving transformations I", Am. J. of Math. 81 (1959): 119-159.

[4] Feldman, J., "New K-automorphisms and a problem of Kakutani", Israel J. of Math. 24 (1976) No. 1: 16-37.

[5] Feldman, J., and Nadler, D., "Reparametrization of N-flows of zero entropy", Trans. of AMS, 256 (1979), 289-304.

[6] Katok, A., The special representation theorem for multidimensional group actions, Dynamical Systems, I, Warsaw, Asterisque 49 (1977), 117-140.

[7] Lind, D., "Locally compact measure preserving flows", Adv. in Math. 15 (1975), 175-193.

[8] Nadler, D., "Abramov's formula for reparametrization of N-flows", preprint.

[9] Ornstein, Rudolph and Weiss, Equivalence of Measure Preserving Transformations, Memoirs of the AMS, 37 No. 262, (1982).

[10] Rudolph, D., "Smooth orbit equivalence of ergodic $\mathbb{R}^d$ actions, $d \geq 2$", Trans. of AMS, 253 (1979), 291-302.

[11] Zimmer. R., "Algebraic topology of ergodic Lie group actions and measurable foliations", preprint.

University of Maryland
College Park, MD 20742

Contemporary Mathematics
Volume **26**, 1984

PROCESSES EVOLVING FROM THE INDEFINITE PAST

M. J. Sharpe*

Spurred in part by Kakutani's famous paper [10] connecting planar Brownian motion with the Dirichlet problem and the notion of logarithmic capacity, Hunt [9] and a host of later workers developed Markov process theory with an eye toward general potential theoretic principles. It was an entirely natural development then that Dynkin [4] formalized Markov process theory in terms of a single process $(X_t)_{t\geq 0}$ run under a family of (P^x) of probability laws with $P^x\{X_0 = x\} = 1$, all evolution being controlled by a single transition semigroup (P_t). The theory is richest when some kind of duality hypothesis is imposed on (P_t). By weak duality of (P_t) and another transition semigroup $(\hat{P}_t)$ relative to a σ-finite measure m on the state space E of (P_t) and $(\hat{P}_t)$ is meant

$$\int P_t f \cdot g \, dm = \int f \cdot \hat{P}_t g \, dm \tag{1}$$

for every pair f,g of positive Borel functions on E. Strong duality of (P_t), $(\hat{P}_t)$ relative to m requires that, in addition to (1), the corresponding resolvents $U^q (\equiv \int_0^\infty e^{-qt} P_t \, dt)$ and $\hat{U}^q$ are absolutely continuous relative to m. That is, if f is a positive Borel function with $\int f \, dm = 0$ then $U^q f$ and $\hat{U}^q f$ both vanish identically. Strong duality is a rather classical area (see Chapter VI of [2]), long known [13] to have close connections with time reversion. Under strong duality hypotheses, it can be shown that there is a nice potential kernel density (or Green's function) $u^q(x,y)$ such that $U^q(x,dy) = u^q(x,y)m(dy)$, $\hat{U}^q(y,dx) = m(dx)u^q(x,y)$, $x \to u^q(x,y)$ being q-excessive for (P_t) and $y \to u^q(x,y)$ being q-excessive for $(\hat{P}_t)$. Reserving the prefix "co" for objects over the dual semigroup $(\hat{P}_t)$, the latter property may be stated as $y \to u^q(x,y)$ is q-co-excessive. In the special case of Brownian motion on $\mathbf{R}^d$ $(d \geq 3)$ with m Lebesgue measure, $(\hat{P}_t)$ is the same as (P_t) and $u^0(x,y)$ is a constant multiple of $|y - x|^{2-d}$.

*Research supported in part by NSF Grant MCS 79-23922.

During the last decade, the setup $(X_t, P^x, P_t, \ldots)$ has lost its status as the universal setting for Markov processes. Influenced by some new points of view of Dynkin during this period, current workers have recognized the value of considering a very general situation which envisions a single Markov process $(Z_t)_{-\infty<t<\infty}$ with random birth time α, random death time β, run under a single σ-finite measure P. I intend to describe here some results about dual Markov processes which are described most naturally in terms of such a two-sided *stationary* process Z. The new results here were obtained in collaboration with Getoor, and I shall refer to [8] for most of the details. See [1] for some related work. In order to make this exposition more comprehensible to non-experts, I shall describe only a limited version of what is obtained in [8]. In particular, I shall assume below that

(2) (X, P_t) *and* $(\hat{X}, \hat{P}_t)$ *are Markov processes in strong duality relative to* m;

(3) X *and* $\hat{X}$ *are standard processes.*

Standardness of X (on its lifetime interval $]0,\zeta[$) is defined to mean that if T_n are stopping times with $T_n \uparrow T$ then $X_T = \lim_n X_{T_n}$ a.s. on $\{T < \zeta\}$. [It is shown in [8], without any duality hypotheses whatsoever, that for standardness, it suffices that the above conditions hold only in case the T_n are the hitting times of a decreasing sequence of finely closed Borel sets in E. That is, standardness has potential theoretic meaning.] It follows from standardness of X that if $B \subset E$ is Borel then a.s., the closure $\{X_- \in B\}^-$ of $\{X_- \in B\} \equiv \{t\colon 0 < t < \zeta(\omega),\ X_{t-}(\omega) \in B\}$ is contained in the closure $\{X \in B\}^-$ of $\{X \in B\} \equiv \{t\colon 0 < t < \zeta(\omega),\ X_t(\omega) \in B\}$.

Under (2) and (3) one may construct a stationary process $(Z_t)_{-\infty<t<\infty}$ under a σ-finite measure P, shifts θ_t, birth time α and death time β, such that for every $s \in \mathbf{R}$, $t \to Z_{t+s}$ restricted to $\{\alpha < t < \beta\}$ is a copy of $t \to X_t$ under P^m and $t \to Z_{s-t}$ restricted to $\{\alpha < t < \beta\}$ is a copy of $t \to \hat{X}_{t-}$ under $\hat{P}^m$, the two halves of Z being conditionally independent given Z_s, on $\{\alpha < s < \beta\}$. The details of this construction were given by Mitro [11]. In case m is a finite invariant measure for both (P_t) and $(\hat{P}_t)$, P is simply a finite measure, $\alpha \equiv -\infty$ and $\beta \equiv \infty$, the stationary process Z being then the usual two-sided stationary process formed from the one-sided stationary process (X, P^m).

Let $\hat{Z}_t$ denote $Z_{(-t)-}$ so that $\hat{Z}_t$ corresponds to $\hat{X}$ just as Z corresponds to X. Here is a quick indication of the power of duality. Using $\{X_- \in B\}^- \subset \{X \in B\}^-$ it follows that $\{Z_- \in B\}^-$. Because the same inclusion holds for $\hat{Z}$ this gives $\{Z_- \in B\}^- = \{Z \in B\}^-$ P a.s., hence $\{X_- \in B\}^- =$

$\{X \in B\}^-$ P^m a.s. This is one of the main simplifications brought about by the standardness hypothesis.

Suppose now that one is given a raw additive functional (RAF) A of X. That is, A is increasing, right continuous, measurable (though it may anticipate the future), satisfies $A_{t+s} = A_t + A_s \circ \theta_t$, and is constant on $[\zeta,\infty]$. The prime examples are $A_t = \int_0^t f(X_s)\,ds$ (f bounded positive Borel on E), or A_t the local time for X at x_0 (supposing it exists), or, what turns out to be closely connected with capacities, $A_t = 1_{[L,\infty]}(t)1_{\{0<L<\zeta\}}$, where $L = L_B \equiv \sup\{t: X_t \in B\}$, with B Borel in E. It was shown by Revuz [14] that to any reasonable RAF A there corresponds a measure ν_A on E defined for f positive Borel by

$$(4) \qquad \nu_A(f) = \lim_{t\downarrow 0} t^{-1}E^m \int_0^t f(X_{s-})\,dA_s .$$

It was shown in [14] that if ν_A is σ-finite, then

$$(5) \qquad E^x \int_0^\infty f(X_{t-})\,dA_t = \int u(x,y)f(y)\nu_A(dy) ,$$

where $u(x,y) \equiv u^0(x,y)$ is the potential kernel density discussed earlier. (A weak duality version of (5) is given in [8].) In particular, in the third example given above, let π_B denote ν_A so that, according to (5)

$$(6) \qquad E^x\{f(X_{L_B-});\ 0 < L_B < \zeta\} = \int u(x,y)f(y)\pi_B(dy) .$$

As was shown in [7] (see also [3]) this establishes that π_B is the equilibrium measure for the set B, and its total mass is called the capacity $c(B)$ of B. In the case of a Brownian motion on $\mathbb{R}^d$ ($d \geq 3$), with B relatively compact, $c(B)$ is precisely the Newtonian capacity of B, and (6) gives Kakutani's result [10] that B has zero Newtonian capacity if and only if the process a.s. never hits B.

The Revuz measure ν_A for a RAF A of X has a very nice interpretation in terms of Z. First of all, we may construct a homogeneous random measure (HRM) κ over Z which lifts A over X. The procedure is discussed in detail in [12] and [8], but suffice it here to say that the lifting of $\int_0^t f(X_t)\,dt$ over X is the HRM $\kappa(dt) = f(Z_t)\,dt$, and the lifting of $1_{[L_B,\infty]}(t)1_{\{0<L_B<\zeta\}}$ is $\kappa(dt)$ = unit mass at $\lambda_B \equiv \sup\ \{t: Z_t \in B\}$, provided $\alpha < \lambda_B < \beta$. One obtains then the simple identification

$$\nu_A(f) = \int dP \int_0^1 f(Z_{s-})\kappa(ds) \tag{7}$$

where κ is the lifting of A to Z. In particular, because of the last example above, we see that if we suppose $P\{\lambda_B = \beta\} = 0$, then

$$\pi_B(f) = \int dP\ f(Z(\lambda_B-))1_{\{0<\lambda_B<1\}} . \tag{8}$$

Of course, stationary of P shows that if the last term $1_{\{0<\lambda_B<1\}}$ is replaced by $1_{\{\lambda_B \in]a,b]\}}$ then the left side of (8) should be multiplied by (b-a). That is, (8) takes the differential form

$$P\{Z(\lambda_B-) \in dx,\ \lambda_B \in dt\} = \pi_B(dx)dt , \tag{9}$$

which gives a very direct meaning to the equilibrium measure π_B.

Let us now set $\tau_B = \inf\{t: \hat{Z}_t \in B\}$, and suppose $P\{\tau_B = \alpha\} = P\{\lambda_B = \beta\} = 0$. Arguing on $\hat{Z}$ as we did for Z, there exists a measure $\hat{\pi}_B$ on E so that the dual of (9) holds: namely,

$$P\{Z(\tau_B) \in dx,\ \tau_B \in dt\} = \hat{\pi}_B(dx)dt . \tag{10}$$

Under more restrictive hypotheses, Hunt [9] proved the rather puzzling fact that $\hat{c}(B)$ $(\equiv \hat{\pi}_B(E)) = c(B)$. It turns out that by making use of Z we are able to prove an identity which makes Hunt's equality transparent. Using the strong Markov property of Z at the stopping time τ_B together with (10) we have

$$P\{\lambda_B = \tau_B \in]0,1]\} = P\{P^{Z(\lambda_B)}[L_B = 0];\ \tau_B \in]0,1]\}$$

$$= \int \hat{\pi}_B(dx)P^x[L_B = 0].$$

On the other hand, (6) gives $P^x[L_B > 0] = \int u(x,y)\pi_B(dy)$, so that

$$\int \hat{\pi}_B(dx)P^x[L_B > 0] = \iint \hat{\pi}_B(dx)u(x,y)\pi_B(dy) .$$

Adding the last two equalities gives

$$\hat{c}(B) = \iint \hat{\pi}_B(dx)u(x,y)\pi(dy) + P\{\lambda_B = \tau_B \in]0,1]\} . \tag{11}$$

When we compare (11) with its dual, we see that the double integral terms, which represent a type of mutual energy of $\hat{\pi}_B$ and π_B, are identical. The last term in (11) is precisely the P measure of those paths which touch B exactly once, and the contact occurs in the time period $]0,1]$. As hits by Z_t are the same as those by Z_{t-} (using standardness here) this is enough to show $\hat{c}(B) = c(B)$. It should be observed that the term $P\{\lambda_B = \tau_B \in]0,1]\}$ will vanish unless the process Z is sufficiently nonsymmetric to admit very non-regular sets B which may be hit only once.

REFERENCES

1. B. Atkinson and J. Mitro. Applications of Revuz and Palm type measures for additive functionals in weak duality. To appear in Seminar on Stochastic Processes, 1982. Birkhauser, Boston, 1983.

2. R. M. Blumenthal and R. K. Getoor. *Markov Processes and Potential Theory*. Academic Press, New York, 1968.

3. K. L. Chung. Probability approach to the equilibrium problem in potential theory. Ann. Inst. Fourier, 23 (1973), 313-322.

4. E. B. Dynkin. *Foundations of the Theory of Markov Processes* (English Translation), Prentice-Hall, Englewood Cliffs, New Jersey, 1961.

5. E. B. Dynkin. Green's and Dirichlet spaces to a symmetric Markov transition function. To appear.

6. E. B. Dynkin. Green's and Dirichlet spaces associated with fine Markov processes. To appear.

7. R. K. Getoor and M. J. Sharpe. Last exit times and additive functionals. Ann. Prob., 1 (1973), 550-569.

8. R. K. Getoor and M. J. Sharpe. Markov processes in weak duality. to appear.

9. G. A. Hunt. Markov processes and potentials I, II, III. Ill. J. Math., 1, 44-93 and 316-369 (1957), 2, 151-213 (1958).

10. S. Kakutani. Two dimensional Brownian motion and harmonic functions. Proc. Imp. Acad. Tokyo, 20 (1944), 706-714.

11. J. Mitro. Dual Markov processes: construction of a useful auxiliary process. Z. Wahrs. verw. Geb., 48 (1979), 97-114.

13. M. Nagasawa. Time reversions of Markov processes. Nagoya Math. J., 24 (1964), 177-204.

14. D. Revuz. Mesures associées aux fonctionelles additives de Markov I. Trans. Am. Math. Soc., 148 (1970), 501-531.

Department of Mathematics
University of California, San Diego
La Jolla, CA 92093

Contemporary Mathematics
Volume 26, 1984

AN INFINITESIMAL CHARACTERIZATION OF GELFAND PAIRS

by

Erik G.F. Thomas

SUMMARY. We prove the following theorem: Let G be a connected Lie group, $H \subset G$ a compact subgroup. Then (G,H) is a Gelfand pair if (and only if) the algebra of G-invariant differential operators on $G/_H$ is commutative.

INTRODUCTION. Let G be a locally compact topological group, H a compact subgroup and $X = G/_H$. The pair (G,H) is said to be a Gelfand pair if the convolution algebra $L^1_{H,H}(G)$ consisting of L^1 functions which are both left and right H-periodic, is commutative. This is for instance the case for the classical spherical, Euclidean and Lobachevski spaces X with their respective groups of motions, H being the stability subgroup of a point $p \in X$. The dimension of such a space being >1 one obviously has that for every x and y in X there exists $g \in G$ with $gx = y$ and $gy = x$; equivalently g and g^{-1} lie in the same double cosets modulo H, and so all functions in $L^1_{H,H}$ are symmetric i.e. $f(g) = f(g^{-1}) = \check{f}(g)$. Since $f \to \check{f}$ changes the order of a convolution product, $L^1_{H,H}$ is commutative. It has been shown more generally by Gelfand [3] that all Riemannian symmetric spaces have the above commutativity property (cf. Helgason [5]). A trivial example of a Gelfand pair is obviously G, $\{e\}$ with G abelian. A great deal of the abelian harmonic analysis: Fourier transform, inversion, Plancherel formula, can be generalized to Gelfand pairs (cf. Godement [4]).

The above definition of Gelfand pairs makes sense only if H is a compact subgroup, $L^1_{H,H}$ being reduced to 0 if H is merely closed. It is of some interest to give characterizations of Gelfand pairs which retain their meaning in the case where H is a closed subgroup. The main theorem of this paper is one such characterization (also see proposition 3 below). It is well known that if G is a Lie group and (G,H) is a Gelfand pair the algebra of G-invariant differential operators on $X = G/_H$ is commutative. The purpose of this paper is to prove the converse, namely the following theorem:

THEOREM: Let G be a connected Lie group, $H \subset G$ a compact subgroup. Then (G,H) is a Gelfand pair if and only if the algebra of G-invariant differential

0271-4132/84 $1.00 + $.25 per page

operators on $G/_H$ is commutative.(1)

We make essential use of a fact observed by Godement [4], namely that with respect to an appropriate topology every distribution on G having compact support can be approximated by distributions having their support in the point e (proposition 8 below). This seems to be the key to the links between the global and the infinitessimal. Perhaps some of the following propositions will facilitate use of this method in other cases.

1. NOTATIONS. We use the standard notations from the theory of distributions [1], [6], in particular $\mathcal{D}'$ and $\mathcal{E}'$ denote the spaces of distributions and compactly supported distributions on the Lie group G. The space of continuous functions with compact support on G is denoted C^c. Left Haar measure on G is denoted dg, the normalized left Haar measure on the compact subgroup H is denoted either by dh or by δ_H. The Dirac measure at the point $x \in G$ is denoted by δ_x. Given any space S of functions or distributions on G we denote $S_{\cdot,H}$ the subspace consisting of right H-periodic elements of S (i.e. those which satisfy $f(gh) = f(g)$ $\forall h \in H$ or $T * \delta_h = T$ $\forall h \in H$). $S_{H,H}$ similarly denotes the subspace of elements in S which are both left and right H-periodic. Locally integrable functions f are identified with distributions fdg. This is compatible with the two definitions of periodicity above because the modular function of G reduces to 1 on H. We denote $\overset{o}{\mathcal{E}}'$ the set of distributions on G concentrated on {e}, e the neutral element of G.

Let us recall that the space $\mathcal{D}'(G/_H)$ is naturally isomorphic to $\mathcal{D}'_{\cdot,H}$. If $\pi : G \to G/_H$ is the quotient map and $\varphi^o(\pi g) = \int_H \varphi(gh)dh$, it is known that the map $\varphi \to \varphi^o$ is a continuous linear map from $\mathcal{D}(G)$ onto $\mathcal{D}(G/_H)$. The transpose $\mathcal{D}'(G/_H) \to \mathcal{D}'(G)$ has as its image $\mathcal{D}'_{\cdot,H}$, and it is an isomorphism of $\mathcal{D}'(G/_H)$ onto $\mathcal{D}'_{\cdot,H}$. This isomorphism is just the extension to distributions of the map $f \to f \circ \pi$ which maps $L^1_{loc}(G/_H)$ onto $L^1_{loc}(G)_{\cdot,H}$ and $\mathcal{E}(G/_H)$ onto $\mathcal{E}_{\cdot,H}$.

2. INVARIANT OPERATORS.

PROPOSITION 1. Let $S \in \mathcal{E}'_{H,H}$, and let A_S be the linear operator in $\mathcal{D}'_{\cdot,H}$ defined by $A_S(T) = T * S$. Then A_S is continuous and commutes with left translations. Conversely every continuous linear operators A in $\mathcal{D}'_{\cdot,H}$ which commute with left translations is of the form A_S for a unique $S \in \mathcal{E}'_{H,H}$. Moreover $A_{S_1 * S_2} = A_{S_2} A_{S_1}$.

The first assertion is obvious. We prove the converse. If $A(T) = T * S$

(1) This theorem was proved by G. van Dijk in the case of semi-simple Lie groups (personal communication).

for some $S \in \mathcal{E}'_{H,H}$ we have $S = A(\delta_H)$. Conversely, let $S = A(\delta_H)$. Then S belongs to $\mathcal{D}'_{.,H}$. Since A commutes with left translations, $\delta_h * S = A(\delta_h * \delta_H) = A(\delta_H) = S$, and so $S \in \mathcal{D}'_{H,H}$. We now prove that the support of S is compact. In fact, L_g denoting left translation, for any sequence (g_n) tending to infinity in G, and any sequence of numbers (c_n), $c_n L_{g_n} \delta_H$ tends to zero in $\mathcal{D}'(G)$, $< L_g \delta_H, \varphi >$ being zero for n sufficiently large. Hence $c_n L_{g_n} S = A(c_n L_{g_n} \delta_H)$ tends to zero for all such sequences (c_n) and (g_n), and this is impossible unless S has compact support ([6], VI §3). In fact, if g_n tends to infinity in the support of $\check{S}$, e belongs to the support of $L_{g_n} S$ for all n. If K is a compact neighborhood of e, there exist functions $\varphi_n \in \mathcal{D}_K$ such that $< L_{g_n} S, \varphi_n > = 1$ for all n. There exists a fundamental sequence of continuous norms (p_n) on $\mathcal{D}_K$, with $p_n \leq p_{n+1}$. Let $c_n = np_n(\varphi_n)$. Since the sequence $c_n L_{g_n} S$ tends to zero, its restriction to $\mathcal{D}_K$ is equicontinuous by the theorem of Banach Steinhauss. Thus there exist M and m such that $|< c_n L_{g_n} S, \varphi >| \leq M p_m(\varphi)$ $\forall \varphi \in \mathcal{D}_K$ $\forall n$. For $\varphi = \varphi_n$ this gives $n \leq M$ for all $n \geq m$, a contradiction. Thus we have $S \in \mathcal{E}'_{H,H}$. The formula $A(T) = T * S$ is correct for $T = \delta_g * \delta_H$, hence for $T = \nu * \delta_H$, ν being a finite linear combination of point masses, and the latter being dense in $\mathcal{D}'$, the formula is correct for all $T \in \mathcal{D}'_{.,H}$. The last assertion is just associativity of the convolution product, two of the three factors having compact support.

PROPOSITION 2. <u>Consider the following convolution algebras</u>: $\mathcal{D}_{H,H}$, $C^c_{H,H}$, $L^1_{H,H}$ <u>and</u> $\mathcal{E}'_{H,H}$. <u>Then if one of these is commutative, so are the others.</u>

If $\mathcal{S}$ is one of the spaces $\mathcal{D}$, C^c, L^1 or $\mathcal{E}'$ the operator $P_{H,H}$ defined by $P_{H,H}(T) = \delta_H * T * \delta_H$ is continuous from $\mathcal{S}$ onto $\mathcal{S}'_{H,H}$. It follows that $\mathcal{D}_{H,H}$ is dense in $C^c_{H,H}$, $L^1_{H,H}$ and $\mathcal{E}'_{H,H}$, and so the result follows from standard approximation arguments.

If one (hence everyone) of the convolution algebras mentioned in proposition 2 is commutative, the pair (G,H) is said to be a Gelfand pair.

Combining proposition 1, 2 and the isomorphism $\mathcal{D}'_{.,H} \simeq \mathcal{D}'(G/_H)$ we obtain:

PROPOSITION 3. <u>Let</u> G <u>be a Lie group</u>, H <u>a compact subgroup</u>. <u>Then</u> (G,H) <u>is a Gelfand pair if and only if the algebra of continuous linear operators</u> $A: \mathcal{D}'(G/_H) \to \mathcal{D}'(G/_H)$ <u>which commute with the action of</u> G, <u>is a commutative algebra</u>.

(REMARK: analogous properties, applicable to locally compact groups, hold with $\mathcal{D}'(G/_H)$ replaced by spaces of measures or functions, e.g. $L^2(G/_H)$).

In particular we see that if (G,H) is a Gelfand pair the algebra of invariant differential operators $D: \mathcal{D}'(G/_H) \to \mathcal{D}'(G/_H)$ is commutative.

In order to obtain the converse (when G is connected) we first identify the subalgebra of $\mathcal{E}'_{H,H}$ corresponding to the invariant differential operators on $G/_H$. Let $\mathcal{A}$ be the set of distributions S of the form $S = \delta_H * E * \delta_H$ where $E \in \overset{o}{\mathcal{E}}'$, i.e. E having support in $\{e\}$.

PROPOSITION 4. Let $S \in \mathcal{E}'_{H,H}$. Then the operator $A_S : \mathcal{D}'_{.,H} \to \mathcal{D}'_{.,H}$ corresponds to a differential operator on $G/_H$ if and only if S belongs to $\mathcal{A}$.

We first note that E is not uniquely determined by $S = \delta_H * E * \delta_H$. Since $\delta_H * \delta_h = \delta_h * \delta_H = \delta_H$ for all $h \in H$, S is not altered as we replace E by $\delta_h * E * \delta_{h^{-1}}$ or by $I(E) = \int_H \delta_h * E * \delta_{h^{-1}}\, dh$. Note that $I(E)$ again belongs to $\overset{o}{\mathcal{E}}'$. Moreover $\delta_h * I(E) * \delta_{h^{-1}} = I(E)$, briefly: $I(E)$ is invariant under H. If we now replace E by $I(E)$, i.e. if we assume E to be invariant under H, we have $\delta_h * E = E * \delta_h$, whence $\delta_H * E = E * \delta_H$. Thus $S = \delta_H * E = E * \delta_H$. In particular it is clear why $\mathcal{A}$ is a subalgebra of $\mathcal{E}'_{H,H}$. Moreover, if $T \in \mathcal{D}'_{.,H}$, we have $T * S = T * \delta_H * E = T * E$. Thus the operators A_S in $\mathcal{D}'_{.,H}$, with $S \in \mathcal{A}$ are the same as the operators in $\mathcal{D}'_{.,H}$ of the type $T \to T * E$ with $E \in \overset{o}{\mathcal{E}}'$, invariant under H. Now clearly the operators in $\mathcal{D}'(G)$, $T \to T * E$, with $E \in \overset{o}{\mathcal{E}}'$ invariant under H, are the left invariant differential operators on G, which commute with right translations in H. But it is known that the restriction of these to $\mathcal{D}'_{.,H}$ are the operators corresponding to G-invariant differential operators on $G/_H$ (cf. Helgason [5] X§2). Thus the proof is complete.

3. USE OF ANALYTIC FUNCTIONS. We now consider the space $C^{\omega}(G)$ of real-analytic functions $f: G \to \mathbb{C}$.

PROPOSITION 5. $C^{\omega}(G)$ is a dense subspace of $\mathcal{E}(G)$.

A proof of this well known result does not seem to be readily available in the literature (the result being obvious for matrix groups). We indicate two proofs. The first refers to Gårding's proof [2] of Nelson's theorem on the density of analytic vectors for representation of G. The second avoids the technicalities of the heat equation on G, but instead makes use of Nelson's theorem to construct an approximation of δ composed of analytic functions.

1. Let $u(x,t)$ be the solution of the heat equation with initial data $f \in \mathcal{D}(G)$ as obtained in [2]. Then Gårding's arguments and Sobolev inequalities imply that $\lim_{t \downarrow 0} u(.,t) = f$ in the space $\mathcal{E}(G)$. Now u being analytic with respect to x, and $\mathcal{D}$ being a dense subspace of $\mathcal{E}(G)$, this gives the required result.

2. We apply Nelson's theorem to the space $\mathcal{B} = L^2(G) \cap C_o(G)$ of continuous square integrable functions tending to zero at infinity, equipped with the norm $\|f\|_{\mathcal{B}} = \|f\|_2 + \|f\|_\infty$, in which G acts continuously by left translation. It is clear that analytic vectors in this space are analytic functions. Starting with a sequence $\varphi_n \in C_c$ such that

a) $\lim_{n\to\infty} \int |\varphi_n|^2 dg = 1$ and b) $\lim_{n\to\infty} \int_{V^c} |\varphi_n|^2 dg = 0$ for every open

neighborhood V of e, we choose analytic vectors $f_n \in \mathcal{B}$ such that $\|f_n - \varphi_n\|_{\mathcal{B}} \le 1/n$. Then the sequence (f_n) has the same properties a) and b), and by multiplying $|f_n|^2$ by appropriate positive numbers we obtain a sequence (F_n) of non negative analytic functions on G, such that $\int F_n dg = 1$ and $\lim_{n\to\infty} \int_{V^c} F_n dg = 0$ for every neighborhood V of e. Now

for every left invariant differential operator D, and function $\varphi \in \mathcal{D}$, we have $D(F_n * \varphi) = F_n * D\varphi \to D\varphi$ uniformly, in particular $F_n * \varphi$ converges to φ in the space $\mathcal{E}(G)$, which, $F_n * \varphi$ being analytic (cf. proposition 6), gives the required result.

PROPOSITION 6. <u>Let</u> $f \in C^\omega(G)$ <u>and</u> $T \in \mathcal{E}'(G)$. <u>Then</u> $f * T$ <u>and</u> $T * f$ <u>belong to</u> $C^\omega(G)$.

PROOF. It will be sufficient to consider the case of $f * T$. First let $T = \mu$ be a measure with compact support. Then μ is a linear combination of a finite number of positive measures with compact support contained in the domain of an analytic chart. For measures with this property the function $x \to \int f(xy^{-1}) d\mu(y)$ is analytic by direct integration of the power series. The modular function being also analytic the result follows in this case. The general case will follow with the help of the next proposition (Schwartz structure theorem):

PROPOSITION 7. <u>If</u> T <u>is a distribution with compact support on</u> G <u>there exist a finite number of Radon measures with compact support</u> $\mu_1, \mu_2, \dots, \mu_n$ <u>and left invariant differential operators</u> $D_1, \dots, D_n$ <u>such that</u>

$$T = \sum_{i=1}^{n} D_i \mu_i .$$

This proved, we shall have $f * T = \sum_{i=1}^{n} f * D_i \mu_i = \sum_{i=1}^{n} D_i (f * \mu_i)$, which is obviously analytic.

PROOF OF PROPOSITION 7: The topology of $\mathcal{E}(G)$ is defined by the seminorms $P_{D,K}(f) = \sup_{x\in K} |Df(x)| = \|Df\|_{\infty,K}$, where D is an arbitrary differential operator in $\mathcal{E}(G)$, and K a compact subset of G. We obtain a fundamental system of seminorms however, by taking the left invariant differential operators

$X^{(\alpha)} = X_1^{\alpha_1} X_2^{\alpha_2} \dots X_n^{\alpha_n}$ where $(X_i)_{i=1,\dots,n}$ is a basis of the Lie algebra of left invariant vector fields on G (cf. [7] 1,1,2). Thus if T is a distribution with compact support there exist left invariant differential operators D_i (of the form $X^{(\alpha)}$), a compact set K (which may be taken to be an arbitrary neighborhood of the support of T) and a number M, such that $|T(\varphi)| \le M \sum_{i=1}^{n} \|D_i\varphi\|_{K,\infty}$, for all $\varphi \in \mathcal{E}(G)$. It follows by a standard application of the Hahn-Banach theorem that there exist Radon measures $\mu_1, \mu_2, \dots, \mu_n$, with support in K, such that $T(\varphi) = \sum_{i=1}^{n} \mu_i(D_i\varphi)$, i.e. $T = \sum_{i=1}^{n} {}^tD_i\,\mu_i$. The transposed operators ${}^tD_i$ being also left invariant the proof is complete.

A consequence of proposition 5 is that the topology $\sigma(\mathcal{E}'(G), C^{\omega}(G))$, the weakest for which the maps $T \to T(f)$, $f \in C^{\omega}(G)$, are continuous, is Hausdorff.

We now come to a simple but remarkable result, due to Godement [4], which was used by him to establish the connection between the integral and differential characterizations of spherical functions:

PROPOSITION 8. <u>Let</u> G <u>be a connected Lie group. Then the space</u> $\overset{\circ}{\mathcal{E}}'$ <u>of distributions on</u> G, <u>with support in</u> $\{e\}$, <u>is dense in</u> $\mathcal{E}'$ <u>when this space is equipped with the topology</u> $\sigma(\mathcal{E}', C^{\omega})$.

This is of course an immediate consequence of the Hahn-Banach theorem: If $f \in C^{\omega}(G)$ and $T(f) = 0$ for all $T \in \overset{\circ}{\mathcal{E}}'$, we have $Df(e) = 0$ for all left (or right) invariant differential operators D, and so $f = 0$.

(Note that in the above applications of the Hahn-Banach theorem, the spaces being separable, only countable choice is involved).

PROPOSITION 9. <u>Let</u> $S \in \mathcal{E}'(G)$. <u>Then the operators</u> $T \to T * S$ <u>and</u> $T \to S * T$ <u>are continuous in</u> $\mathcal{E}'$ <u>if this space is equipped with the topology</u> $\sigma(\mathcal{E}', C^{\omega})$.

We have for instance $T * S(f) = T(f * S')$ with some $S' \in \mathcal{E}'$, and $f * S'$ is again analytic by proposition 6.

4. PROOF OF THE MAIN THEOREM . By proposition 9 we know that the operator $P_{H,H}: \mathcal{E}' \to \mathcal{E}'$, defined by $P_{H,H}(T) = \delta_H * T * \delta_H$, is continuous when $\mathcal{E}'$ is equipped with the topology $\sigma(\mathcal{E}', C^{\omega})$. By proposition 4 we know that the image $\mathcal{A} = P_{H,H}(\overset{\circ}{\mathcal{E}}')$ is the subalgebra of $\mathcal{E}'_{H,H}$ corresponding to the invariant differential operators on $G/_H$. Thus, by proposition 8, $\mathcal{A}$ is dense in $\mathcal{E}'_{H,H}$, equipped with the topology induced by $\sigma(\mathcal{E}', C^{\omega})$. Therefore, if $\mathcal{A}$ is commutative, it follows from proposition 9 that $\mathcal{E}'_{H,H}$ is commutative, hence (G,H) is a Gelfand pair.

References

[1] Bruhat, F., Distributions sur un groupe locallement compact. Bull. Soc. Math. France 89 (1961), 43-75.

[2] Gårding, L., Vecteurs analytiques dans les representations des groupes de Lie. Bull. Soc. Math. France 88 (1960), 73-93.

[3] Gelfand, I.M., Spherical functions on symmetric spaces. Doklady Akad. Naak. S.S.S.R. 70 (1950), 5-8. Amer. Math. Soc. Transl.(2)37(1964)39-44.

[4] Godement, R., Introduction aux travaux de A. Selberg. Seminaire Bourbaki 1957.

[5] Helgason, S., Differential Geometry and Symmetric Spaces. Academic Press, New York 1962.

[6] Schwartz, L., Théorie des distributions Hermann, Paris 1950, 1951.

[7] Varadarajan, V.S., Lie groups, Lie algebras, and their representations. Prentice-Hall, Inc. 1974.

MATHEMATICS DEPARTMENT
UNIVERSITY OF GRONINGEN
POSTBUS 800
9700 AV GRONINGEN
NETHERLANDS

Contemporary Mathematics
Volume **26**, 1984

MULTIPLE POINTS OF BROWNIAN MOTION

by

Nils Tongring

For Brownian motion X taking values in Euclidean space $\mathbb{R}^d$, $d \geq 2$, the probability that a sample path ever revisits $X(t_0)$ for $t > t_0$ is zero. It is still possible that some points on the path are visited more than once. Such self-intersections are called multiple points.

Consider first the case when the dimension $d = 2$. Let L^k designate the set of all points of a path that are at least k-fold self-intersections, k an integer ≥ 1; for example, L^1 is the path itself. One might think that the sets L^k become increasingly sparse as k increases, and that for k large enough L^k is empty. But as Brownian travelers know this is false. Dvortesky, Erdös, and Kakutani [DEK] proved that L^k is a.s. dense in a sample path, for any k, and similarly for $L^{\aleph_0} = \bigcap_{k=1}^{\infty} L^k$. There are even points of multiplicity c.

The problem sketched in this note[1] is to determine how large a fixed set must be to have a non-empty intersection with the random set L^k.

Assume that the Brownian motion on the probability space Ω is in a representation where the elements ω of Ω are continuous functions of t, the sample paths of the process. The results below will be understood in terms of the measure Pr, without always including the proviso "for almost all ω in Ω". For example, paths in $\mathbb{R}^3$ can be pictured with triple points, but the set of such paths has probability zero, and there are therefore no triple points.

---

[1] Details will appear elsewhere.

0271-4132/84 $1.00 + $.25 per page

DEFINITION 1. A point $z_0 \in \mathbb{R}^d$, $d \geq 2$, is a k-tuple point, or a point of multiplicity k of the path ω in Ω if there exist k times $t_1 < t_2 < \ldots < t_k$ such that $X(t_i,\omega) = z_0$ for $i = 1,2,\ldots,k$. The set of all k-tuple points of the path ω is denoted $L^k(\omega)$. There is a natural chain of inclusions

$$L(\omega) = L^1(\omega) \supseteq L^2(\omega) \supseteq \ldots \supseteq L^k(\omega) \supseteq \ldots \supseteq L^{\aleph_0}(\omega) = \bigcap_{k=1}^{\infty} L^k(\omega) .$$

In $\mathbb{R}^3$, $L^3 = \phi$; there are only double points. In $\mathbb{R}^4$, $L^2 = \phi$: there are no multiple points of order greater than one.

$\mathbb{R}^2$.

If I is a fixed line in the plane, are there k-tuple points on I? With the line given by $x_1 = a$, and the Brownian motion $X = (X_1,X_2)$ starting from the origin, introduce the random variable

$$g_k = \inf\{ \sum_{i=1}^{k} (a-X_1(t_i))^2 + \sum_{i=2}^{k} (X_2(t_i) - X_2(t_{i-1}))^2\}^{1/2} ,$$

where the infimum is taken over all sequences $t_1,t_2,\ldots,t_k$ satisfying $0 < t_i \leq 1$, $\frac{1}{2} \leq t_i - t_{i-1} \leq 1$ $(i = 2,3,\ldots,k)$. If

$$\Pr(g_k < \varepsilon) > c_1 > 0 \qquad (1)$$

for every $\varepsilon > 0$, where c_1 is a constant independent of ε, k-tuple points exist with probability greater than zero. The idea of the proof is the same as that for the DEK Theorem. Find a number n of disjoint disks, each of radius ρ depending inversely on n, for which an estimate of the probability that at least one of the disks is entered k times by the path is greater than a positive constant. The disks are chosen with their centers on the line I. By a proper choice of the form of ρ, inequality (1) can be verified. Letting ρ approach zero gives then the existence of k-tuple points. A 0-1 argument leads to

PROPOSITION 1. Paths of Brownian motion in $\mathbb{R}^2$ have k-tuple points on any fixed line.

The type of probability estimates used in the proof require the separation of the times $\{t_i\}$ given above. There are of course other kinds of multiple points; for example, for any $t_0 > 0$ and any times a,b with $a < t_0 < b$, there exist times t_1,t_2 with $a < t_1 < t_0 < t_2 < b$ such that $X(t_1,\omega) = X(t_2,\omega)$.

If the argument of Proposition 1 is tried on the standard one-dimensional Cantor set K embedded in $\mathbb{R}^2$, k-tuple points in K again appear.

The line and Cantor set are simple examples covered by the following theorem, which connects the existence of a non-empty intersection of a fixed set A with the random set L^k to a capacity of A.

First recall the definition of the generalized transfinite diameter of a compact set. If Φ is a continuous function, monotone increasing, with $\lim_{s\to 0+}\Phi(s) = \infty$, let

$$D_N^{\Phi}(A) = \inf \binom{N}{2}^{-1} \sum_{1\le i\le j\le N} \Phi(|x_i - x_j|) ,$$

where the infimum is taken over all sets of N points $\{x_1,x_2,\ldots,x_N\}$ in A. D_N^{Φ} is monotone increasing in N. Let $V^{\Phi} = \lim_{N\to\infty} D_N^{\Phi}$, and introduce the capacity C^{Φ} of A:

$$C^{\Phi}(A) = \Phi^{-1}(V^{\Phi}(A)) ,$$

where Φ^{-1} is the inverse function of Φ ($C^{\Phi}(A) > 0$ means that $V^{\Phi}(A)$ is bounded). Now take for the function Φ

$$\Phi(s) = (\log 1/s)^{2k-1}$$

and denote the associated capacity function C^{Φ} by $C^{(k)}$.

THEOREM 1. _If_ A _is a compact set in_ $\mathbb{R}^2$ _and_ $C^{(k)}(A) > 0$, _then there are k-tuple points in_ A.

Notice that $C^{(1)}(A)$ is the usual (logarithmic) capacity, and that in this case the theorem is well-known: if a compact set has positive logarithmic capacity, it is hit by almost all paths.

DEFINITION 2. For A any Borel set in $\mathbb{R}^2$,

$$\begin{aligned} T_A^k(\omega) = \inf\{t > 0:\ & \text{there exists times} \\ & t_1 < t_2 < \ldots < t_k = t \\ & \text{with } X(t_i,\omega) = X(t,\omega) \in A \\ & \text{for } i = 1,2,\ldots,k\} \end{aligned}$$

is the _k-th hitting time_ of the set A. It is not clear that T_A^k is an optional time for any Borel set, but at least for A closed it is immediate. $T_A^1 = T_A$ is the usual hitting time of A,

and

$$T_A^1 \le T_A^2 \le \ldots$$

Theorem 1 then gives a sufficient condition for a compact set A to have $T_A^k < \infty$.

COROLLARY. If A is a set in $\mathbb{R}^2$ with (Hausdorff) dim $A > 0$, then A has k-tuple points of all orders.

This follows immediately from the observation that Hausdorff dimension is the same as capacity dimension, and if the capacity of A with respect to the function $\Phi(s) = \frac{1}{s^\alpha}$ is positive for some $\alpha > 0$, $C^{(k)}(A) > 0$ for any k.

A converse to Theorem 1 is not known, nor is the theorem itself necessarily the best possible.

$\mathbb{R}^3$.

The problem of the existence of multiple points in a fixed set A takes a slightly different form for Brownian motion in $\mathbb{R}^3$. First, paths have a.s. no points of multiplicity higher than two; and second, "large" sets may not be hit at all, since the process is transient. So the question in $\mathbb{R}^3$ is whether given a set A there is a collection of paths with positive probability having double points in A.

If A is open, the original proof for the existence of double points in $\mathbb{R}^3$ can be used to show that A has double points with positive probability.

If A is a set with Lebesgue measure greater than zero, it contains double points. The standard proof does not seem applicable here, but the method given by Hawkes [H] for the existence of multiple points of a symmetric Lévy process can be modified to give this result.

What if A is a line? Consider a plane normal to A and the Brownian motion projected onto this plane. If A contains double points, there is a point in the plane which be hit by the projected Brownian motion with probability greater than zero, which is impossible. Therefore a fixed line contains no double points.

Does the plane have double points? Consider the problem of the occurence of double points in a fixed compact set A as a function of its Hausdorff dimension. First of course A must be hit, so its (Newtonian) capacity must be positive; i.e., it must have infinite Hausdorff measure with respect to the function $h(s) = s$, and therefore its Hausdorff dimension must be at least one. As many authors have observed, for pairs of small sets A,B in $\mathbb{R}^d$ which we would consider independent, it often happens that their Hausdorff dimensions are related by

$$\dim A \cap B = \dim A + \dim B - d$$

Now in $\mathbb{R}^3$, $\dim L^2 = 1$, so that

$$\dim A \cap L^2 = \dim A - 2,$$

and there are double points in A if $\dim A > 2$, and no double points in A if $\dim A < 2$. $\dim A = 2$ would sum to be an ambiguous case. These heuristic considerations turn out to be valid.

PROPOSITION 2. *If $A \subset \mathbb{R}^3$ is a compact set with $0 \le \Lambda^2 A < \infty$, where $\Lambda^2 A$ denotes the Hausdorff measure with respect to the function $h(s) = s^2$, then there are no double points in A.*

Therefore a compact piece of the plane has no double points, nor the plane itself, even if "crumpled up".

An example where double points appear in a compact set of Hausdorff dimension two is given by the following observation.

PROPOSITION 3. *If A is a compact set in $\mathbb{R}^3$ with positive 2-capacity (that is, with respect to the function $\Phi(s) = \frac{1}{s^2}$) double points occur in A, with probability greater than zero.*

PROOF. Positive 2-capacity is equivalent to requiring that A carry a probability measure μ such that

$$\int_A \int_A 1/\|z_1 - z_2\|^2 \, d\mu(z_1) d\mu(z_2) < \infty .$$

For a number $\varepsilon > 0$ let K_ε be the function on the reals given by

$$K_\varepsilon(u) = \begin{cases} 1, & \text{if } |u| < \varepsilon \\ 0, & \text{if } |u| \ge \varepsilon \end{cases}$$

Consider the multiple integral

$$I_\varepsilon = \int_1^2 ds \int_3^4 dt \int_A \varepsilon^{-6} K_\varepsilon(|X(s)-z|) K_\varepsilon(|X(t)-z|) d\mu(z)$$

If $\mathbb{E}(I_\varepsilon)$ is bounded below and $\mathbb{E}(I_\varepsilon^2)$ bounded above by positive constants, both uniformly in ε, the Paley-Zygmund inequality [2] implies that

$$\Pr(I_\varepsilon > \delta) > \delta$$

[2] The suggestion for the use of a Paley-Zygmund argument is due to Robert Kaufman.

for some positive constant δ, and for all ε sufficiently small. There are then double points with probability at least δ, in particular, ones with occurence times t_1 in $[1,2]$, t_2 in $[3,4]$.

The estimates for $\mathbb{E}(I_\varepsilon)$ and $\mathbb{E}(I_\varepsilon^2)$ are elementary, and can be given in some detail.

Consider first

$$\mathbb{E}(I_\varepsilon) = \int_1^2 ds \int_3^4 dt \int_A \varepsilon^{-6} \Pr\{|X(s)-z| < \varepsilon, |X(t)-z| < \varepsilon\} d\mu(z).$$

If the events $\{|X(s)-z| < \varepsilon\}$, $\{X(t)-z| < \varepsilon\}$ are denoted by C,D, resp.,

$$\Pr\{|X(s)-z| < \varepsilon,\ |X(t)-z| < \varepsilon\} = \Pr(C \cap D)$$

$$= \Pr C \bullet \Pr_c D \geq \Pr c \bullet \Pr D$$

The last inequality is not clear for z near the origin, but since A is compact, a change of coordinates insures that the distance from A to the origin is large.

The form of the distribution function for Brownian motion provides the estimates $\Pr C \geq c_2 \varepsilon^3$, $\Pr D \geq c_2 \varepsilon^3$, c_2 a positive constant, for all s on $[1,2]$ and t in $[3,4]$. Since A is compact,

$$\mathbb{E}(I_\varepsilon) \geq c_3 > 0 \quad \text{for some}$$

positive constant c_2, and for any ε, $0 < \varepsilon < 1$.

Next consider

$$\mathbb{E}(I_\varepsilon^2) =$$

$$\int_A \int_A \left[\int_1^2 ds_1 \int_1^2 ds_2 \int_3^4 dt_1 \int_3^4 dt_2 \varepsilon^{-12} \Pr\{|X(s_1)-z_1| < \varepsilon, \right.$$

$$|X(s_2)-z_2| < \varepsilon,\ |X(t_1)-z_1| < \varepsilon,$$

$$\left. |X(t_2)-z_2| < \varepsilon\} \right] d\mu(z_1) d\mu(z_2).$$

If we can show that the term in brackets is of the order $\varepsilon^{12}/|z_1-z_2|^2$, then

$$\mathbb{E}(I_\varepsilon^2) = \int_A \int_A \frac{0(1)}{|z_1-z_2|^2} \quad d\mu(z_1) d\mu(z_2) = 0(1),$$

and the Paley-Zygmund inequality follows. To make this estimate, the behavior of the double integral over A for $|z_1-z_2|$ small must be examined, and the estimate $0(1)\varepsilon^{12}/|z_1-z_2|^2$ established separately for various parts of the integrals over the variables s_1, s_2, t_1, t_2.

REFERENCES

[DEK] Dvoretzky, A., Erdös, P., and Kakutani, S., Multiple points of paths of Brownian motion in the plane, Bull. Res. Conn. Israel 3 (1954), 364-371.

[H] Hawkes, J., Multiple points for symmetric Lévy processes, Math. Camb. Phil. Soc. 83 (1978), 83-90.

DEPARTMENT OF MATHEMATICS
YALE UNIVERSITY
NEW HAVEN, CT 06520

Contemporary Mathematics
Volume 26, 1984

MEASURABLE DYNAMICS

Benjamin Weiss

INTRODUCTION

The study of dynamical systems in physics has led to the development of the mathematical disciplines ergodic theory and topological dynamics. In the first one abstracts out of the classical setting a measure space and flows or transformations that preserve the measure, or at least preserve the measure class. The second concentrates mainly on general compact spaces and continuous flows or transformations. More recently, smooth dynamics, in which the differential structure is restored and interacts with the measure theory, is witnessing a renaissance and has expanded the interactions between both of the above with other parts of geometry and analysis. This paper is a step in the opposite direction, namely it is an attempt to see what can be done with the techniques of ergodic theory in the setting of a general measurable structure $(X,\mathcal{B})$, $\mathcal{B}$ is a σ-algebra of subsets of X, and an automorphism T of this structure. The hope is that one would gain new insights by studying the dynamics of such a Borel automorphism without singling out any particular measure class. I restrict attention throughout to actions of Z, i.e. iterates of a single automorphism, on a <u>standard</u> Borel space (see below for precise definitions). It is quite likely that much of what is done here can be carried out for analytic Borel spaces but general measurable spaces, even if countably generated can be quite pathological. For example, one can construct nondenumerable countably generated measurable spaces $(X,\mathcal{B})$ such that <u>every</u> automorphism T of X onto itself differs from the identify only on a countable set. (see [AS §48]).

Beginning with an automorphism T of a standard Borel space $(X,\mathcal{B})$ the proof of an abstract version of Poincare's recurrence theorem leads one to introduce the σ-ideal $\mathcal{W}(T) \subset \mathcal{B}$ which is generated by the <u>wandering</u> sets W, i.e. those sets $W \in \mathcal{B}$ such that $W \cap T^n W = \emptyset$ for all $n \neq 0$. It is then natural to study properties modulo $\mathcal{W}(T)$ instead of modulo sets of measure zero for a fixed measure class. Prior to embarking upon such a study we first establish a basic result which gives a characterization of $\mathcal{W}(T)$ as those sets which are of measure zero with respect to any ergodic conservative quasi-invariant measure υ. If we call a set $A \in \mathcal{B}$ <u>completely positive</u> if $\upsilon(A) > 0$

for all ergodic conservative quasi-invariant υ then this characterization of $\mathcal{W}(T)$ is equivalent to the statement that A is completely positive if and only if $X \setminus \bigcup_0^\infty T^n A$ belongs to $\mathcal{W}(T)$. I first proved this result with S. Shelah in [SW], but I shall describe a different proof in §1 which depends upon a very basic fact that I learned, following the conference honoring S. Kakutani, from A. Ramsay. The basic fact is that for any Borel automorphism of a standard Borel space there is a topology on X which makes X a complete separable metric space with $\mathcal{B}$ as its Borel sets, and with respect to which T is a homomorphism. Still in §1, we get that if T is non dissipative, $X \notin \mathcal{W}(T)$, and aperiodic then there is a wealth of ergodic quasi-invariant measures, in fact one from each orbit equivalence class of type III or II_∞. As far as type II_1 goes there may be no invariant measures (see §4). In this way we recover some results of K. Schmidt and W. Krieger (see [K]) in a more general setting.

Next define the spectrum $\sigma(T)$ of a Borel automorphism as follows: $\lambda \in \sigma(T)$ if there exists a measurable function $f: X \to C$ that satisfies

$$f(Tx) = \lambda f(x), \quad \text{all} \quad x \in X$$

and is non-trivial in the sense that

$$\{x: f(x) \neq 0\} \notin \mathcal{W}(T).$$

This last condition implies that $\sigma(T) \subset T^1 = \{z \in C : |z| = 1\}$. If T is non-dissipative and aperiodic then it turns out that always $\sigma(T) = T^1$. The second section is devoted to proving this fact while in the third we give a generalization in which the rotations of T^1 are placed by an arbitrary homeomorphism S of a compact metric space Y. More precisely adjoin $\{0\}$ to Y as a discrete point and set $S0 = 0$, $\tilde{Y} = Y \cup \{0\}$, and call a function $f: X \to \tilde{Y}$ non trivial if $\{x: f(x) \neq 0\} \notin \mathcal{W}(T)$. Then we prove in §3 that there exists for any non dissipative aperiodic Borel automorphism T a non trivial function $f: X \to \tilde{Y}$ such that

$$f(Tx) = S\, f(x) \quad \text{all} \quad x \in X.$$

The results of §1-3 all point in the direction of a basic similarity between all non dissipative, aperiodic Borel automorphisms. Under the obvious definition of isomorphism between Borel automorphisms the only distinguishing invariant that I know is the class of finite invariant measures. In §4 I prove a measurable version of the basic Rohlin lemma and also establish the hyperfiniteness of any Z-action at the Borel level. I have not yet been successful in using these results to show that every automorphism has a countable generator- and it may be that this is simply not true and allows us to distinguish between the shift on $[0,1]^Z$ and the shift on N^Z after removing the periodic points from each. At all events a major open problem here is to find new invariants which will distinguish between Borel automorphisms.

The earlier work of V. S. Varadarajan [V] analyzes the structure of Borel automorphism groups but only in so far as the class of finite invariant measures go. He proves a beautiful decomposition tehorem in which the space X is measurably decomposed into invariant Borel sets on each of which there is a unique invariant measure. This analysis indicates that the true complexity of Borel automorphisms lies in those that have no invariant measure. Here are some simple concrete examples of such transformations. Let E be a closed nowhere dense subset of the circle T^1 with positive Lebesque measure, and $E^* = \bigcup_{-\infty}^{\infty} E + n\alpha$ where α is irrational. Then $T^1 \setminus E^* = X$ invariant under R_α, rotation by α, and it's easy to see that R_α is non dissipative and aperiodic and of course X supports no finite measure invariant under R_α. Varying α we obtain a family of Borel automorphisms, and I don't know if two of those can be isomorphic or not.

Another way that one can imagine going is in the direction of admitting finitely additive measures. In general if λ is a finitely additive measure on $(X,\mathcal{B})$ then the null sets of λ form only an ideal and not a σ-ideal. However there are f.a.m. whose null sets do form a σ-ideal, for example if S is a topologically transitive homeomorphism of a compact metric space Y then one can construct an S-invariant f.a.m. whose null sets are precisely the sets of first category in Y. I have no idea as to how rich a class measures of this type form and whether or not they can shed further light on Borel automorphisms. There has been some recent work on isomorphisms modulo sets of first category, in particular M. Keane has shown that the n-shifts $(n \geq 2)$ are mutually isomorphic after omitting sets of first category.

I wish to take this opportunity of acknowledging my debt to Prof. Shizuo Kakutani who both through his work and via our personal contacts helped to shape my mathematics. His direct influence is discernible in this paper in the basic role played by induced transformations, but indirectly the spirit of his work is visible throughout.

1. QUASI-INVARIANT MEASURES FOR BOREL AUTOMORPHISMS

By a <u>standard Borel space</u> $(X,\mathcal{B})$ we shall mean a Polish space X with the σ-algebra generated by the open sets in X. Recall that a topological space (X,τ) is <u>Polish</u> if there is a metric d on X such that (X,d) is complete and separable and the topology τ is induced by the metric d. For most of our discussions the topology may be safely suppressed and the reader can keep in mind any of the basic examples: the unit interval $[0,1]$, the Cantor set $\{0,1\}^N$, the irrationals N^N with the usual topologies. A <u>Borel automorphism</u> $T:X \to X$

is a one to one and onto mapping T such that $T(\mathcal{B}) = \mathcal{B}$, i.e. $T(B) \in \mathcal{B}$ if and only if $B \in \mathcal{B}$. The first step in the analysis of T is the periodic part. Since there is a countable family of sets $\{C_j\}_1^\infty \subset \mathcal{B}$ that separates points we can describe the periodic points of period n for T as follows:

$$\bigcap_{j=1}^{\infty} [(C_j \cap T^{-n}C_j) \cup (C_j^c \cap T^{-n}C_j^c)] \overset{\text{def}}{=\!=} \mathcal{P}_n(T) \tag{1}$$

where $C^c = X \setminus C$ denotes the complement of C. Since (1) is a Borel set we see that the periodic points of T:

$$\{x: \text{there exists } n \text{ with } T^n x = x\} \tag{2}$$

is a Borel set. It is well known that the restriction of $\mathcal{B}$ to a Borel set $B \in \mathcal{B}$, yields again a standard Borel space $(B, \mathcal{B}|B)$. Later on, following Lemma 4, we shall prove this and for the mean time observe that it suffices to analyze separately a purely periodic Borel automorphism T and an <u>aperiodic</u> one, where for all x and all $n \neq 0$, $T^n x \neq x$.

The basic fact about a purely periodic automorphism is that there exists a global cross section. A precise formulation is as follows:

PROPOSITION 1. *If for all* $X \in X$, $T^n x = x$, *but* $T^k x \neq x$ *for* $0 < k < n$ *then there exists a Borel set* A_0 *such that* $A_0 \cap T^k A_0 = \phi$ *for* $0 < k < n$ *and* $X = \bigcup_0^{n-1} T^k A_0$.

PROOF. It is convenient to use here a metric d which give the topology, and a countable family of Borel sets $\{S_{ij}\}_{i,j=1}^\infty$ such that for each i, $\bigcup_{j=1}^\infty S_{ij} = X$ and for all j $\operatorname{diam}(S_{ij}) < 1/i$. Begin with S_{11}, say, and form

$$A_1 = S_{11} \cap T^{-1}S_{11}^c \cap \dots T^{-(n-1)}S_{11}^c$$

i.e.

$$A_1 = \{x \in S_{11} : T^k x \notin S_{11},\ 0 < k < n\}.$$

Next set $X_1 = X \setminus (\bigcup_{k=0}^{n-1} T^k S_{11})$, observe that X_1 is invariant under T, and form

$$A_2 = (S_{12} \cap X_1) \cap T^{-1}(S_{12} \cap X_1)^c \cap \dots \cap T^{-(n-1)}(S_{12} \cap X_1)^c.$$ Set

$X_2 = X_1 \setminus (\bigcup_{k=0}^{n-1} T^k A_2)$, and continue with $S_{21} \cap X_2$, and so on. By hypothesis, for all $x \in X$, the $\min\limits_{0<k<n} d(x, T^k x) > 0$ and thus for all x, there is some S_{ij} such that $x \in S_{ij}$ but $T_k x \notin S_{ij}$ for $0 < k < n$. Thus setting $A_0 = \bigcup_1^\infty A_m$ we see that $X = \bigcup_{k=0}^{n-1} T^k A_0$, while by construction $A_0 \cap T^k A_0 = \phi$ for all $0 < k < n$. ∎

The proposition applies to $\mathcal{P}_n(T)\setminus \bigcup_{k<n} P_k(T)$, the points of pure period n, and thus the dynamical behavior of a periodic Borel automorphism is seen to depend purely on the cardinality of the sets of periodic points. Henceforth we shall always assume that our automorphisms are aperiodic. A last methodological point before leaving this case. In the proof of the proposition we used a metric only to get a countable family of sets S_{ij} with the property that for all x, there was some set S_{ij} such that $x \in S_{ij}$ but $T^k x \in S_{ij}$ for $0 < k < n$. We could have constructed such a sequence of sets just from the fact that $\mathcal{B}$ has a countable family that separates points so that we could have avoided the use of the metric.

Even for aperiodic automorphisms T there may exist global cross sections which trivialize the dynamical behavior of T. The basic concept is that of a wandering set, which is a set W that $W \cap T^n W = \phi$ for all $n \neq 0$. Denote by $\mathcal{W}$ the collection of all $B \in \mathcal{B}$ for which there exists some wandering set W with $B \subset \bigcup_{-\infty}^{\infty} T^n W$. Clearly $\mathcal{W}$ forms an ideal in $\mathcal{B}$ which is in fact a σ-ideal. To see this latter property assume that $\{W_i\}_1^{\infty}$ are wandering sets, form successively

$$\bar{W}_1 = W_1,\ \bar{W}_2 = W_2 \setminus \bigcup_{-\infty}^{\infty} T^n W_1, \ldots\ \bar{W}_k = W_k \setminus \bigcup_{i=1}^{k-1} \bigcup_{-\infty}^{\infty} T^n W_i, \ldots$$

and then check that $W = \bigcup_1^{\infty} \bar{W}_k$ is a wandering set and

$$\bigcup_{-\infty}^{\infty} T^n W = \bigcup_{i=1}^{\infty} T^n W_i .$$

Modulo $\mathcal{W}$, subinvariant sets are actually invariant. In other words if $A \supset T A$, then there is some $A_0 \subset A$ with $T A_0 = A_0$ and $A \setminus A_0 \in \mathcal{W}$. Indeed $A \setminus T A$ is a wandering set and since $TA \subset A$ we have that $A \setminus \bigcap_0^{\infty} T^n A \subset \bigcup_0^{\infty} T^j (A \setminus TA)$ whilst $\bigcap_0^{\infty} T^n A$ is clearly T-invariant. We shall frequently make use of this observation without further comment. The fundamental recurrence theorem of Poincare in this setting becomes (cf. [Ox] ch. 17):

POINCARE RECURRENCE. If $B \in \mathcal{B}$ then there exists $B_0 \subset B$ such that $B \setminus B_0 \in \mathcal{W}$ and for all $x \in B_0$, $T^n x \in B_0$ for infinitely many values of $n \geqq 1$.

PROOF. Set $W = \{x \in B\colon T^n x \notin B \text{ for all } n \geqq 1\}$, observe that W is a wandering set and put $B_0 = B \setminus \bigcup_{-\infty}^{0} T^n W$. Noticing now that any $x \in B$ that returns to B only finitely many times belongs to $T^{-n} W$ for some n we see that points in B_0 return to B, and in fact to B_0, infinitely many times as required. ∎

At this point we can define the useful notion of the induced transformation on a set $T_B(x) = T^{n(x)}(x)$, $T_B: B \to B$, where $n(x) = \min\{n \geq 1: T^n x \in B\}$. If x never returns to B then T_B is not defined for that x, and the B_0 in Poincare's recurrence theorem can be described as $\bigcap_{k=1}^{\infty} T_B^{-k}(B)$. In a more pendantic treatment we would define T_B by our T_{B_0}.

The main result in connection with $\mathcal{W}$ is an alternative description of it in terms of quasi-invariant conservative measures . Recall that a measure μ on $\mathcal{B}$ is quasi-invariant for T if $\mu(B) = 0$ if and only if $\mu(TB) = 0$. A measure μ in conservative if $\mu(W) = 0$ for every wandering set W. For example if μ is non atomic and ergodic, i.e. $B = TB$ implies that $\mu(B) = 0$ or $\mu(B^c) = 0$, then μ is conservative since if W is a wandering set and has positive μ measure then since μ is non atomic $W = W_1 \cup W_2$ with $W_1 \cap W_2 = \phi$ and each of positive μ measure and then $\bigcup_{\infty}^{\infty} T^n W_1$ is T-invariant but neither it nor its complement has zero measure.

Let's denote by $\mathcal{M}$ the class of non atomic quasi-invariant ergodic measures for T, and for $\mu \in \mathcal{M}$ let $\mathcal{N}(\mu)$ denote the σ-ideal of μ-null sets. Then we have the obvious inclusion

$$\mathcal{W} \subset \bigcap_{\mu \in \mathcal{M}} \mathcal{N}(\mu)$$

and our main result here is that equality actually holds. This was proven in [SW], but to keep this paper self contained another proof will be given here. It depends upon a remarkable representation theorem which I learned from Arlan Ramsay who attributed it (or parts of it) to George Mackey.

THEOREM 2. *If $(X,\mathcal{B})$ is a standard Borel space and $T:X \to X$ a Borel automorphism then there is a topology τ on X with the following properties:*

(i) *(X,τ) is a complete separable metric space,*

(ii) *the Borel sets of (X,τ) are precisely $\mathcal{B}$,*

(iii) *T is a homeomorphism of (X,τ).*

Note that the underlying set X is not changed at all, so that in fact this theorem says that one gains no generality in discarding the topology. Nontheless this is not the usual framework of topological dynamics where compactness, or local compactness play an essential role. For the proof of the theorem we need some lemmas.

Denote by τ_0 some Polish topology on X with $\mathcal{B}$ its Borel sets, so that $\tau_0 \subset \mathcal{B}$, and τ_0 generates $\mathcal{B}$.

LEMMA 3. *If $\{\tau_j\}_1$ are Polish topologies on X with $\tau_0 \subset \tau_j \subset \mathcal{B}$ for all $1 \geqq j < \infty$, then there exists a Polish topology $\tau_\infty \subset \mathcal{B}$ such that $\tau_\infty \supset \bigcup_1^\infty \tau_j$, with τ_∞ generated by finite intersections $\bigcap_1^n G_j$, $G_j \in \tau_j$.*

PROOF. Let φ be the diagonal imbedding of X into the Cartesian product $\prod_1^\infty (X,\tau_j)$ with the product topology which is of course Polish. Now the fact that all the τ_j's contain τ_0 implies that $\varphi(X)$, the diagonal, is a closed subset and we take for τ_∞ the restriction of the product topology to $\varphi(X)$, pulled back via φ to X. It is clear that τ_∞ has all the required properties. ∎

LEMMA 4. *If $B \in \mathcal{B}$ then there exists a Polish topology $\tau_0 \subset \bar{\tau} \subset \mathcal{B}$ such that $B \in \bar{\tau}$.*

PROOF. The collection of sets that satisfy the demands of the lemma will be denoted by $\mathcal{C}$. Clearly $\mathcal{C}$ contains τ_0. Suppose $C \in \mathcal{C}$ and τ_1 the topology containing C. We construct a topology $\tau_1 \subset \tau_2 \subset \mathcal{B}$ that also contains $C^c = F$ which is τ_1-closed. To this end let d_1 be a metric giving the topology τ_1 and consider the function $f:X \to R$ defined by:

$$f(x) = \begin{cases} 1/d(x,F), & x \in C \\ 0 & x \in F \end{cases}$$

Let $\varphi: X \to X \times R$ denote the map from X to the graph of f, i.e. $\varphi(x) = (x,f(x))$. Since C is open, $\varphi(X)$ is a closed subset of $X \times R$ and we take for τ_2 the restrictions of the product topology on $X \times R$ to $\varphi(X)$. Clearly $\varphi(C)$ is closed in this relative topology and hence $F \in \tau_2$.

Next suppose $C_1, C_2, \ldots \in \mathcal{C}$ and let τ_j be the corresponding topologies. Apply lemma 3 to obtain τ_∞ that includes all τ_j, hence in particular $\bigcup_1^\infty C_j$. Thus $\mathcal{C}$ is closed under countable unions and complementation and thus $\mathcal{C}$ is a σ-algebra which necessarily is all of $\mathcal{B}$ since $\mathcal{C} \supset \tau_0$. ∎

At this point we observe that we have established the fact the restricting the Borel sets to a fixed Borel set $(B, \mathcal{B}|B)$ yields again a standard Borel space since we can choose a Polish topology under which B is closed.

COROLLARY 5. *For any countable family $\{B_j\}_1^\infty \subset \mathcal{B}$ there is a Polish topology $\tau_0 \subset \bar{\tau} \subset \mathcal{B}$ such that $\{B_j\}_1^\infty \subset \bar{\tau}$.*

PROOF. Apply Lemma 4 to find τ_j's and then Lemma 3 to combine them to the desired $\bar{\tau}$.

PROOF OF THEOREM 2. Let $\{G_j^0\}_1^\infty$ be a basis for the topology τ_0. Apply the corollary to the collection $\{TG_j^0, T^{-1}G_j^0\}^\infty$ to obtain a topology $\mathcal{B} \supset \tau_1 \supset \tau_0$ in which $TG_j^0, T^{-1}G_j^0$ are all open. Next let $\{G_j^1\}_1^\infty$ be a basis for τ_1 and apply the corollary to (X,τ_1) and the collection $\{TG_j^1, T^{-1}G_j^1\}_{j=1}^\infty$ to obtain a Polish topology $\tau_2 \supset \tau_1$ in which $T G_j^1$, $T^{-1}G_j^1$ are all open. Continue in this fashion to obtain a sequence $\tau_0 \subset \tau_1 \subset \tau_2 \subset \ldots$ and bases $\{G_j^k\}_{j=1}^\infty$ for τ_k such that $T^{\pm}G_j^k \in \tau_{k+1}$. Apply now lemma 3 to find a $\tau_\infty \supset \overset{\infty}{\cup} \tau_k$ which is Polish and is generated by the G_j^k. By construction T is bi-continuous with respect to τ_∞ which is the τ required by the theorem. ∎

For a homeomorphism T of metric space (X,d) we can related the global recurrence properties of T to point wise recurrent properties. A point $x \in X$ is said to be recurrent if

$$\liminf_{n\to\infty} d(x,T^n x) = 0.$$

PROPOSITION 6. If T is a homeomorphism of a separable metric space (X,d) and T has no recurrent points then $\mathcal{W}(T) = \mathcal{B}$, i.e. there is a wandering set W with $X = \cup\, T^n W$.

PROOF. For each $k \geq 1$ let

$$F_k = \{x: d(x,T^n x) \geq 1/k, \text{ all } n \geq 1\ .$$

Since T is continuous the F_k are closed sets. By the separability of (X,d) there is a countable family of sets $\{W_j^k\}_{j=1}^\infty$, each of diameter less than $1/2k$ with $F_k = \bigcup_{j=1}^{\infty} W_j^k$.

Clearly each W_j^k is a wandering set and thus $F_k \in \mathcal{W}$ for all k. The hypothesis that T has no recurrent points is equivalent to the assertion that $X = \bigcup_1^\infty F_k$ which proves the proposition. ∎

The basic construction will now be carried out for a homeomorphism T of a complete separable metric space (X,d) that has a nonperiodic recurrent point. It is essentially the construction given in [KW], which as we recently learned, was preceded by a similar construction in [G] (cf. also [E]) who did not, however, focus on a recurrent point. The goal is to produce a closed set $C \subset X$, and an identification of C with the Cantor set $\{0,1\}^N = \Omega$ so that T_C corresponds to the odometric map $\theta:\Omega \to \Omega$ which is defined by $\theta(\omega) = \omega + 1$, where the addition is the 2-adic addition when Ω is identified with the 2-adic integers. In "plain" language we add the sequence $(100\ldots.\ )$ to $(\omega_1\omega_2\omega_3\ldots)$ with the addition bit by bit modulo 2 with carry to the right. This will enable us to reduce theorems about an arbitrary Borel automorphism to theorems about towers over the dyadic odometer. Suppose then that x_0 is a non periodic recurrent point for T. To get started use the continuity of T to find $a_1 > 0$ so that $T(B(a_1)) \cap B(a_1) = \phi$ where

$$B(a) = \{x : d(x_0,x) \le a\}$$

and set $F(0) = B(a_1)$, $F(1) = T(B(a_1))$, $n_1 = 1$. By the recurrence of x_0 we can find a positive n_2 so that $d(x_0, T^{n_2}x_0) < a_1$ and then by the continuity of T we can find an $a_2 > 0$ so that:

(i) $T^{n_2}(B(a_2)) \subset B(a_1)$

(ii) $T^i(B(a_2)) \cap B(a_2) = \phi$ for $0 < i \le n_2$,

(iii) $\operatorname{diam} T^i(B(a_2)) \le 1/2$ for $0 \le i \le n_1 + n_2$

We put now $F(00) = B(a_2)$, $F(10) = T(B(a_2))$, $F(01) = T^{n_2}(B(a_2))$, $F(11) = T^{n_2+1}(B(a_2))$, and observe that the transformation induced by T on $F(00) \cup F(01) \cup F(10) \cup F(11)$ behaves like the odometric mapping on the 4 cylinder sets determined by the first two coordinates in Ω.

Use the recurrence once again to find n_3 so that

$$d(x_0, T^{n_3}x_0) < a_2$$

and the continuity of T to find $a_3 > 0$ so that

(i) $T^{n_3}(B(a_3)) \subset B(a_2)$

(ii) $T^i(B(a_3)) \cap B(a_3) = \phi$ for $0 < i \le n_3$

(iii) $\operatorname{diam} T^i(B(a_3)) \le 1/3$, for $0 \le i \le n_1 + n_2 + n_3$.

Define:

$$F(000) = B(a_3) \qquad F(100) = T(B(a_3))$$

$$F(010) = T^{n_2}(B(a_3)) \qquad F(110) = T^{n_2+1}(B(a_3))$$

$$F(001) = T^{n_3}(B(a_3)) \qquad F(101) = T^{1+n_3}(B(a_3))$$

$$F(011) = T^{n_3+n_2}(B(a_3)) \qquad F(111) = T^{1+n_2+n_3}(B(a_3)).$$

The pattern should now be clear, namely having found n_k and $B(a_k)$ so that

(i) $T^{n_k}(B(a_k)) \subset B(a_{k-1})$

(ii) $T^i(B(a_k)) \cap B(a_k) = \phi$ for $0 < i \le n_k$

(iii) diam $T^i(B(a_k)) \le 1/k$ for $0 \le i \le n_1 + n_2 + \ldots + n_k$

for all $k \le m$ we continue the definition to stage $m + 1$ using the recurrence of x_0 and the continuity of T.

Note that by the construction, for any $\omega \in \Omega, F(\omega_1) \supset F(\omega_1\,\omega_2) \supset \ldots \supset F(\omega_1 \ldots \omega_k) \supset \ldots$ and the intersection consists of precisely one point since (X,d) was assumed complete, and the F's are all closed and have diameter $\to 0$ by (iii). Our desired set C is precisely

$$C = \bigcap_{k=1}^{\infty} \bigcup_{\omega_1\omega_2 \ldots \omega_k} F(\omega_1 \ldots \omega_k) = \bigcup_{\omega \in \Omega} \bigcap_{k=1}^{\infty} F(\omega_1 \ldots \omega_k).$$

The first representation shows that C is closed and the second that it is naturally homeomorphic to the cantor set Ω. Furthermore, with the exception of the countable set of points that correspond to $\omega \in \Omega$ with $\omega_k = 1$ for all sufficiently large k, each $x \in C$ returns to C infinitely often in accordance with the odometric map. We can even say precisely what the first return time function to C is, namely if we write

$$t(x) = \min \{t \geq 1 \colon T^t x \in C\}$$

then $t(x)$ is constant on the sets $C_1 = C \cap F(0)$, $C_2 = C \cap F(10), \ldots, C_k = C \cap F(1^{k-1}0), \ldots$ with the values:

$$t(x) = \begin{cases} 1 & x \in C_1 \\ n_2 - 1 & x \in C_2 \\ \vdots & \vdots \\ n_k - \sum_1^{k-1} n_j & x \in C_k \\ \vdots & \vdots \end{cases}$$

To establish this one proves by induction that (ii) may be strenghtened to

$$T^i(B(a_n)) \cap B(a_n) = \phi \quad \text{all} \quad 0 < i \le \sum_1^k n_j$$

This last fact means that we describe $\bigcup_0^{\infty} T^k C$ in a simple way, namely we can break it up into mutually disjoint sets

$\{T^i C_k\}$, $1 \le k < \infty$, $0 \le i < n_k - \sum_1^{k-1} n_j$, and this set is T-invariant (we ignore henceforth the complete orbits of the points corresponding to $\omega \in \Omega$ that are eventually constant). We summarize this construction as follows, where we denote by $\hat{\Omega} \subset \Omega$ those sequences $\omega \in \Omega$ that have infinitely many zeros and infinitely many ones. Naturally $\hat{\Omega}$ in θ-invariant and differs from Ω by a countable set.

THEOREM 7. If T is a homeomorphism of a Polish space with a non periodic recurrent point x_0 then there is a Borel set $C \subset X$ and a one to one onto mapping $\varphi: C \to \hat{\Omega}$ such that $\theta\varphi = \varphi T_C$, and furthermore there are integers $t_k \geqq 1$, so that the sets

$$\{T^i C_k\}\ ,\ 1 \le k < \infty,\ 0 \le i < t_k$$

are pairwise disjoint and exhaust $\bigcup_{-\infty}^{\infty} T^n C$, where

$$C_k = \varphi^{-1}(\{\omega : \omega_j = 1,\ 1 \le j \le k-1,\ \omega_k = 0\}).$$

An immediate corollary of the above theorem, Proposition 6 and Theorem 2 is:

COROLLARY 8. If T is a Borel automorphism of a standard Borel space $(X,\mathcal{B})$, aperiodic and non dissipative, i.e. $X \notin \mathcal{W}(T)$, then the conclusion of Theorem 6 holds.

The meaning of this corollary is that if T is any nondissipative aperiodic Borel automorphism then one can find a T-invariant subset F such that T restricted to F is isomorphic to an odometric tower. Now the odometer has a wealth of quasi-invariant measures, and in fact up to orbit equivalence any ergodic non singular transformation can be represented by some quasi-invariant measure on the dyadic odometer (cf. [W] for a discussion of the facts that we will be needing concerning orbit equivalence). Furthermore, any measure on the odometer can be extended to a measure on a tower over the odometer, and with the exception of a type II_1 transformation becoming II_∞ (i.e. a finite invariant measure becoming infinite) the orbit equivalence type doesn't change. In fact, T may not have any finite invariant measures. For an example we can remove from the circle $\mathbb{T}$ a set A, invariant under R_α, rotation by an irrational α, with full Lebesgue measure such that $\mathbb{T}\backslash A$ is non dissipative for R_α. The unique ergodicity of R_α then says that $(\mathbb{T}\backslash A, R_\alpha)$ has no finite invariant measures, while our discussion above shows that it has measures representing all other types: II_∞ and all III's. We formalize this as a theorem.

THEOREM 9. If T is an aperiodic, non dissipative Borel automorphism of a Borel automorphism of a Borel space $(X,\mathcal{B})$, and $(Y,\mathcal{C},\nu,S)$ is any non singular ergodic transformation, of type II_∞, or III, then there exists a quasi-invariant measure μ for T such that $(X,\mathcal{B},\mu,T)$ is orbit equivalent to $(Y,\mathcal{C},\nu,S)$.

Returning to $\mathcal{W}$, observe that if $A \not\in \mathcal{W}$ then we can restrict T to the invariant set $\bar{A} = \bigcup_{-\infty}^{\infty} T^n A$ and then by theorem 8 there is an ergodic quasi-invariant measure μ that charges A. If $\mu(A)$ were to vanish then so would $\mu(\bar{A})$ since μ is quasi-invariant and thus $A \not\in N(\mu)$ and we have proved.

COROLLARY 10. For an aperiodic Borel automorphism T is $\mathcal{W}(T)$ is the σ-ideal generated by the wandering sets, and $M(T)$ the family of non-atomic, quasi-invariant ergodic measures and $N(\mu)$ the null sets of μ then

$$\mathcal{W}(T) = \bigcap_{\mu \in M(T)} N(\mu)$$

§2. THE SPECTRUM OF AN AUTOMORPHISM

According to §1, a non dissipative Borel automorphism T has a wealth of quasi-invariant measures. In this section we shall show that such automorphisms also have a wealth of eigenfunctions, in fact for any eigenvalue we shall construct an invariant set $A \not\in \mathcal{W}(T)$ and a function

$$f\colon A \to \mathbb{T} \quad \text{such that} \quad f(Tx) = \lambda f(x) \quad \text{for all} \quad x \in A.$$

The emphasis of course is on the fact that $A \not\in W(T)$, since for wandering sets such f's may be trivially constructed with values $\{\lambda^n : n \in Z\}$. The general case will be carried out in several steps in the first of which we treat the case when T itself is the dyadic odometer in the second we pass to odometric towers and finally apply Corollary 1.8 to complete the argument.

STEP ONE.

Fix some $\lambda \in T^1$, $\lambda \neq 1$ but possibly rational. Our strategy is to construct a measure μ of product type on Ω, that is to say $\mu = \prod_1^\infty \mu_k$ where μ_k is the distrubution of the digits ω_j for $j \in J_k$, and $J_1, J_2, J_3 \dots$ is some countable partition of N into finite successive blocks of consecutive numbers. The odometric map, θ, is non singular and ergodic with respect to μ and thus if μ is non-atomic any set of full μ-measure is not in $\mathcal{W}(\theta)$. Together with the construction of μ we will also construct an f such that

$$f(\theta\omega) = \lambda f(\theta) \ \mu - \text{a.e.}$$

and this will accomplish what we want.

Begin by putting $f_1(\omega) = \lambda^{\omega_1}$ and setting $J_1 = \{1\}$, $\mu_1(0) = \mu_1(1) = 1/2$. We will denote elementary sets as follows:

$$C(J;\ a_1,\dots,a_k) = \{\omega : \omega_{j(i)} = a_i \quad 1 \le i < k\}$$

where $J = j(1) < \dots < j(k)$. For $\omega \in C(J_1;0)$ we have that $f_1(\theta\omega) = f_1(\omega)$. In the second stage, supposing that $J_2 = \{2,\dots,k_2\}$ we will put

$$f_2(\omega) = \lambda^{\omega_1+2\omega_2+\dots+2^{k_2-1}\omega_{k_2}}$$

Now for all $\omega \notin C(J_1 \cup J_2;\ 1,1,\dots,1)$ we will have as required that $f_2(\theta\omega) = \lambda f_2(\omega)$, however for all $\omega \notin C(J_1 \cup J_2;\ 0,0,\dots,0)$ $f_2(\omega) \ne f_1(\omega)$ (if λ is irrational, which is the usual case). On the various sub cylinders of $C(J_1;0)$ where $f_1(\omega)$ equalled one, $f_2(\omega)$ equals $\lambda^2, \lambda^4, \lambda^6$ etc. According to Kronecker's theorem there are i_2 and j_2 so that

$$|\lambda^{2i_2} - 1| < 1/10 \quad \text{and} \quad |\lambda^{2j_2} - 1| < 1/10.$$

Now that we do is choose k_2 large enough so that $i_2, j_2 < 2^{k_2-1}$ and then define μ_2 so that most of its mass is concentrated precisely on those subcylinders of $C(J_1;0)$ where $f_2(\omega)$ equals λ^{2i_2} and λ^{2j_2}. In this way $f_2(\omega)$ will differ only slightly from $f_1(\omega)$ on most of the space and thus the successive f_n's will converge. The reason for using 2 cylinder sets is to guarantee that the eventual $\mu = \prod_1^\infty \mu_n$ is non atomic. Here is an explicit definition of μ_2:

$$\mu_2(C(J_2;a_1,\dots a_{k_2-1})) = \begin{cases} 1/10(2^{k_2-1}-2) & \text{if } \sum_1^{k_2-1} 2^{m-1} a_m \notin \{i_2,j_2\} \\ 1/2 - 1/20 & \text{if } \sum_1^{k_2-1} 2^{m-1} a_m \in \{i_2,j_2\} \end{cases}$$

We'll write out one more step in complete detail. Apply Kronecker's theorem to find $i_3 < j_3$ with

$$|\lambda^{2^{k_2} i_3} - 1| < 1/10^2, |\lambda^{2^{k_2} j_3} - 1| < 1/10^2$$

Set $J_3 = \{k_2+1, k_2+2, \dots, k_3\}$ where k_3 is chosen so that $j_3 < 2^{k_3-k_2}$, define

$$f_3(\omega) = \lambda^{\sum_1^{k_3} 2^{m-1}\omega_m}$$

and

$$\mu_3(J_3;a_1,\dots,a_{k_3-k_2}) = \begin{cases} 1/10^2\,(2^{k_3-k_2}-2)) & \text{if } \sum_{m=1}^{k_3-k_2} 2^{m-1} a_m \notin \{i_3,j_3\} \\ 1/2 - 1/(2.10^2) & \text{if } \sum_{m=1}^{k_3-k_2} 2^{m-1} a_m \in \{i_3,j_3\}. \end{cases}$$

Our choices of i_3, j_3 and μ_3 gives:

$$(1)_3 \quad (\mu_1\times\mu_2\times\mu_3)(\{\omega : |f_2(\omega) - f_3(\omega)| \geq 1/100\}) \leq 1/10^2$$

$$(2)_3 \quad (\mu_1\times\mu_2\times\mu_3)\ (\{\omega : |f_3(\theta\omega) \neq \lambda f_3(\omega)\}) < 1/(10^2(2^{k_3-k_2}))$$

Continuing in this way we inductively define intervals J_n, measures, μ_n on 2^{J_n} and functions f_n that depend only upon ω_j for $j \in \bigcup_{m=1}^{n} J_m$ and satisfy

$$(1)_n \quad (\prod_1^n \mu_m)\ (\{\omega : |f_{n-1}(\omega) = f_n(\omega)| \geq 10^{-n+1}\}) \leq 10^{-n+1}$$

$$(2)_n \quad (\prod_1^n \mu_m)\ (\{\omega : f_n(\theta\omega) \neq \lambda f_n(\omega)\}) \leq 10^{-n+1}.$$

Applying the Borel-Cantelli lemma to the information in $(1)_n$ we get that f_n converges μ-a.e. to a function which we denote by f. The Borel-Cantelli lemma applied again to $(2)_n$ shows that $f(\theta\omega) = \lambda f(\omega)$ μ-a.e. from which it follows immediately that there is a θ-invariant set of full μ-measure on which $f(\theta\omega) = \lambda f(\omega)$ which is what we set out to construct.

STEP TWO.

For the odometric tower we are given a height function $h: \Omega \to N$ that is constant on the cylinder sets

$$C_1 = C(\{1\};0),\ C_2 = C(\{1,2\};1,0),\dots,C_n = C(\{1,\dots,n\};1,1,\dots,1,0),\dots$$

assuming the values h_m on C_m. These correspond to the sets $C_1, C_2, \dots$ of Theorem 1.7. The odometer tower $T = \theta^h$ is defined on $X = \Omega^h$ as follows:

$$\Omega^h = \{(\omega,\ell) : 0 \leq \ell < h(\omega)\}$$

$$\theta^h(\omega,\ell) = \begin{cases} (\omega,\ell+1) & \text{if } \ell+1 < h(\omega) \\ (\theta\omega,0) & \text{if } \ell+1 = h(\omega) \end{cases} .$$

Our strategy will be as before, to try to define inductively a measure μ on $X_0 = \{(\omega,0)\} : \omega \in \Omega\} \subset X$ and a function $f: X \to T^1$ so that $f(Tx) = \lambda f(x)$ a.e. with respect to the measure. The measure will be extended from X_0 to X in the usual way with sets in $X_1 = \{(\omega,\ell) \in X : \ell = 1\}$ getting a measure half of what they would have in X_0 etc, so that the map T is non-singular and the total measure remains finite.

After the first step in which we put $J_1 = \{1\}$, $\mu_1(0) = \mu_1(1) = 1/2$ and

$$f_1(\omega,\ell) = \lambda^{\omega_1+\ell}$$

as before we come up against an obstacle. We would like to define f_2 in such a way as to guarantee that for all (ω,ℓ) with $\omega \notin C(J_1 \cup J_2;1,1,\dots,1)$

$$f_2(T(\omega,\ell)) = f_2(\omega,\ell)$$

with $J_2 = 2,\dots,k_2$, as before, for some choice of k_2. However, on the successive subcylinders of $C(J_1;0)$ instead of assuming the values $\lambda^2,\lambda^4,\lambda^6,\dots$ etc., now f_2 assumes the values

$$\lambda^{h_1+h_2},\lambda^{2h_1+h_2+h_3},\lambda^{3h_1+2h_2+h_3},\lambda^{4h_1+2h_2+h_3+h_4},\dots$$

where the exponents are the even partial sums of the sequence

$$h_1,h_2,h_1,h_3,h_1,h_2,h_1,h_4,h_1,h_2,h_1,h_3,h_1,h_2,h_1,h_5,h_1,\dots$$

whose law of formation we will now describe explicitly. Define the function $r(n)$ in such a way that

$$\theta^{n-1}(\bar{\omega}) \in C_{r(n)}$$

where $\bar{\omega} = \{(\omega_i) : \omega_i = 0 \text{ all } i\}$. Thus $r(n)$ describes the sequence of sets of the partition $\{C_m\}$ into which the orbit of $\bar{\omega}$ under θ falls. Every second term is 1, every fourth term is 2 etc.

Needless to say Kronecker's theorem doesn't apply and we need another tool to pick out subcylinders where f_2 is close to f_1 where we can concentrate the mass of μ_2. We do this by discerning in the even partial sums of sequence above the structure of a <u>finite IP set</u>. Generally, an IP-set generated by $\{a_1 \dots a_m\}$ is the sequence of all partial sums of the a_i, explicitly:

$$\text{IP}\ \{a_1,\dots,a_m\} = \{\sum_{i=1} \varepsilon_i a_i : \varepsilon_i \in \{0,1\}\}$$

One can easily check that the even partial sums above are an IP-set generated by

$$a_1 = h_1 + h_2,\ a_2 = 2h_1 + h_2 + h_3, \ldots$$

$$a_m = 2^{m-1} h_1 + 2^{m-2} h_2 + \ldots + h_m + h_{m+1}, \ldots$$

We need an elementary lemma:

LEMMA 1. *If* $\lambda \in T^1$ *and* $A = IP\{a_1,\ldots,a_m\}$ *with all* $a_i > 0$, *then for some* $\alpha \in A$, $\alpha \neq 0$ *we have*

$$|\lambda^\alpha - 1| \leq 2\pi/m.$$

PROOF. Consider the sequence $\lambda^{a_1}, \lambda^{a_1+a_2}, \ldots \lambda^{a_1+a_2+\ldots+a_m}$ and divide T^1 into m equals arcs of length $2\pi/m$. Either the m elements of the sequence lie in distinct arcs, in which case the one that falls in the arc containing zero does the trick, or two elements lie in the same arc, say $\lambda^{a_1+\ldots+a_p}$ and $\lambda^{a_1+\ldots+a_q}$ with $p < q$. In the latter case, λ^α does the trick with $\alpha = a_{p+1}+\ldots+a_q$ which also lies in A. ∎

Note that by splitting $a_1,\ldots,a_m$ into two nearly equal sets we can guarantee two distinct such α's at the cost of replacing $2\pi/m$ by $4\pi/m-1$. Now to carry out step 2, we choose k_2 large enough so that we can use the lemma to find two sub-cylinders where f_2 is close to f_1 and concentrate μ_2 on those two subcylinders as before. At the next stage, instead of even partial sums we will have the partial sums of the same sequence $h_1, h_2, h_1, h_3, \ldots$ in blocks of 2^{k_2}. This once again is seen to have the structure of an IP-set, its generators are the $\{a_i\}$ just defined with $i \geqq k_2-1$.

Thus once again Lemma 1 can replace our previous use of Kronecker's theorem. Continuing in this fashion we complete the construction of a measure $\mu = \pi\mu_n$ and a function $f:X \to T^1$ as in step one.

Finally we apply Corollary 1.8 to imbed some odometric tower inside a given non-dissipative Borel automorphism and we have proved the following theorem:

THEOREM 2. *If* $(X,\mathcal{B},T)$ *is a non-dissipative Borel automorphism of a standard Borel space and* $\mathcal{B} \in T^1$ *then there exists a T-invariant set* $B \in \mathcal{B}$, $B \notin \mathcal{W}(T)$ *and a* $\mathcal{B}$*-measurable function* $f: B \to T^1$ *so that*

$$f(Tx) = \lambda f(x)$$

for all $x \in X$.

§3. A GENERALIZATION OF SPECTRUM

In this section we show that we can replace the circle T^1 and multiplication by λ, by an homeomorphism of a compact metric space. Here is the result we are after:

THEOREM 1. *If* $(X,\mathcal{B},T)$ *is a non-dissipative Borel automorphism and* S *is a homeomorphism of a compact metric space* Y, *then there is a T-invariant set* $B \in \mathcal{B}$, $B \notin \mathcal{W}(T)$, *and a* $\mathcal{B}$*-measurable function* $f : B \to Y$ *such that for all* $x \in B$

$$f(Tx) = Sf(x).$$

We begin once again with the case when T is the dyadic odometer. By restricting to a minimal subset of (Y,S) we may as well assume that (Y,S) is already minimal. Fix some $y(0) \in Y$ and begin by setting

$$f_1(\omega) = S^{\omega_1}(y(0))$$

and putting $\mu_1(C(\{1\};0)) = \frac{1}{2} = \mu_1(C(\{1\} ; 1))$. To simplify the writing we will put $y(m) = S^m(y(0))$. Now we need a basic lemma concerning recurrent points:

LEMMA 2. *If* $S : Y \to Y$ *is continuous and*

$$\liminf_{n\to\infty} d(S^n y_0, y_0) = 0$$

for some $y_0 \in Y$ *then for any positive integer* k,

$$\liminf_{n\to\infty} d(S^{nk} y_0, y_0) = 0.$$

PROOF. Suppose k and $\varepsilon > 0$ are given. Choose in succession positive integers $a_1, a_2, \ldots, a_k$ so that for any $\alpha \in IP\{a_1,\ldots,a_k\}$

(1) $$d(S^\alpha y_0, y_0) < \varepsilon,$$

Indeed supposing that $a_1,\ldots,a_i$ have already been chosen so that (1) holds for $\alpha \in IP\{a_1,\ldots,a_i\}$ we use the continuity of S to find δ so that if $d(y,y_0) < \delta$ then we still have

(2) $$d(S^\alpha y, y_0) < \varepsilon, \quad \alpha \in IP\{a_1,\ldots,a_i\}$$

Now by assumption we can find a_{i+1} so that

$$d(S^{a_{i+1}} y_0, y_0)$$

and then using (2) we see that (1) holds for all $\alpha \in IP\{a_1,\ldots,a_{i+1}\}$. Now consider $a_1, a_1+a_2, \ldots, a_1+a_2+\ldots+a_k$, either they are all incongruent modulo k, in which case one is congruent to 0 mod k and we are done -

namely we've found some α with $k|\alpha$ and $d(S^{\alpha}y_0,y_0) < \varepsilon$. If they are not all incongruent modulo k, two of them are congruent mod k and then their difference which is still in the IP-set does the trick. ■

Since (Y,S) is minimal our starting point $y(0)$ is necessarily a recurrent point and using the lemma we can find $i_2 < j_2$ with $d(y(2i_2),y(0))$ and $d(y(2j_2),y(0))$ as small as we please. Now since S in not an isometry we have to worry separately about $d(y(2i_2+1),y(1))$ and $d(y(2j_i+1),y(1))$ and we use the continuity of S to ensure that all 4 distances at issue here are less than 1/10. As before we choose now k_2 so that 2^{k_2-1} is greater than i_2 and j_2 and define f_2 by

$$f_2(\omega_1,\omega_2,\ldots,\omega_{k_2}) = y(\omega_1+2\omega_2 +\ldots+ 2^{k_1-1}\omega_{k_2})$$

and μ_2 by the same formula as in §2.

In the next step we shall have to find values of m that are divisible by 2^{k_2} and are such that $y(m+i)$ is as close as we please to $y(i)$ for $0 \le i < 2^{k_2}$. Lemma 2 will provide us with such values and enables us to continue the construction in complete analogy with <u>step one</u> of §2. The continuity of S is used to control $d(y(m+i), y(i))$ for $i > 0$ in terms of $d(y(m),y(0))$ while the completeness of the space Y assures that in the limit of the f_n's that we define there is a value for f to assume. In this fashion we prove the following proposition:

PROPOSITION 3. <u>If S is a continuous mapping of a complete metric space Y with a recurrent point y_0, then there is a quasi-invariant measure μ on the dyadic odometer (Ω,θ) and a function $f: \Omega \to Y$ so that</u>

$$f(\theta\omega) = Sf(\omega) \qquad \mu\text{-a.e.}$$

In order to extend the proposition to odometric towers we have to really use the fact that Y is compact. While I don't have an actual counterexample it seems unlikely to me that the theorem will hold without some kind of compactness in Y. As we saw in §2, given an odometric tower with values $h_1,h_2,h_3,\ldots$ we have associated IP-set generated by $a_1 = h_1+h_2$, $a_2 = 2h_1+h_2+h_3,\ldots$ etc., and to carry out the construction of an "eigenfunction" f with values in (Y,S) we need a point $y_0 \in Y$ with the property that for any k and any $\varepsilon > 0$ there is some m and $\alpha \in IP\{a_k,a_{k+1},\ldots,a_m\}$ with $d(s^{\alpha}y_0,y_0) < \varepsilon$. For a given IP set the existence of such a recurrent point can be deduced from the existence of idempotents in the enveloping semi-group of S or from the combinatorial version of this fact expressed in Hindman's theorem cf. [F]. Since these results are perhaps not yet widely known I shall describe in some more detail how this is done. If $\{a_i\}_1^{\infty}$ is an infinite sequence of positive integers the IP-set generated by it is the set of all finite partial sums, i.e. $A = \mathrm{IP}\{a_1,\dots,a_n,\dots\} = \{\sum_{i\in I} a_i : I$ is a finite subset of $N\}$. For a given IP-set A, if $\{x_\alpha : \alpha \in A\}$ is a sequence of points in a metric space we say that x converges - IP to x, in symbols

$$\mathrm{IP} - \lim_{\alpha\in A} x_\alpha = x$$

if for every $\varepsilon > 0$, there is a k such that

$$d(x_\alpha, x) < \varepsilon \quad \text{for all} \quad \alpha \in \mathrm{IP}\{a_{k+1}, a_{k+2},\dots\} .$$

If $I_1, I_2,\dots$ are disjoint finite subsets of N, and

$$b_m = \sum_{j\in I_m} a_j$$

then the IP-set generated by the b_m's is said to be sub-IP set of $\mathrm{IP}\{a_1,\dots,a_m,\dots\}$. Hindman's theorem is equivalent to the following proposition which we shall assume:

PROPOSITION 4. _If_ $\{x_\alpha : \alpha \in \mathrm{IP}\{a_1,\dots,a_n,\dots\}\}$ _is an_ IP-_system of points in a compact metric space then there is a sub_ IP, $B \subset A$ _and_ $y \in Y$ _such that_

$$\mathrm{IP} - \lim_{\beta\in B} y_\beta = y .$$

We deduce from this the proposition that we shall need:

PROPOSITION 5. _If_ S _is a continuous mapping of a compact metric space_ Y, _and_ $A = \mathrm{IP} - \{a_1,\dots,a_n,\dots\}$ _is given there exists a_ $y_0 \in Y$ _and a sub_ - IP _set_ $B \subset A$ _such that_

$$(*) \qquad \mathrm{IP} - \lim_{\beta\in B} S^\beta y_0 = y_0 .$$

PROOF. We can start with any $y \in Y$ and apply Proposition 4 to ge a sub-IP set $A' \subset A$ with $\{S^{\alpha'} y : \alpha' \in A'\}$ converging in IP to some y_0. We shall construct for this y_0 a sub-IP set $B \subset A'$ which will satisfy (*). To ease notation we suppose $A' = A$. To begin with we prove that for any $\varepsilon > 0$ and any k there is some $\alpha \in A_k = \mathrm{IP}\{a_k, a_{k+1},\dots\}$ with

$$d(S^\alpha y_0, y_0) < \varepsilon .$$

To this end, use the definition of IP-convergence to find some $m \geq k$ with $d(S^\alpha y, y_0) < \varepsilon/2$ for all $\alpha \in A_m$. Next use the continuity of S^{a_m} to find some δ so that $d(u, y_0) < \delta$ implies

$$d(S^{a_m} u, S^{a_m} y_0) < \varepsilon/2.$$

Now use the IP-convergence again to find $\alpha_1 \in A_{m+1}$ with

$$d(S^{\alpha_1}y, y_0) < \delta \ .$$

We have then

$$d(S^{a_m}y_0, y_0) < d(S^{\alpha_1+a_m}y, S^{a_m}y_0) + d(S^{\alpha_1+a_m}y, y_0)$$

$$< \varepsilon/2 + \varepsilon/2 = \varepsilon$$

since $\alpha_1 + a_m \in A_m$.

The sub-IP set B is now built up inductively as follows:

Choose some $b_1 \in A_1$ so that $d(S^{b_1}y_0, y_0) < 1/10$. Next use that we have proved to find b_2 disjoint from b_1 so that $S^{b_2}y_0$ is so close to y_0 that not only is $d(S^{b_2}y_0, y_0) < 1/1000$ but $d(S^{b_2+b_1}y_0, y_0)$ is still less than $1/10$. This is possible since S^{b_1} is continuous. Continuing in this way we construct a sub-IP $B \subset A$ with $\text{IP} - \lim_{\beta \in B} S^{\beta}y_0 = y_0$ as required. ∎

In order to prove Theorem 1 we apply Proposition 5 to the IP-set associated with the odometric tower which is imbedded in the non-dissipative Borel automorphism. This gives us a point $y_0 \in Y$ for which the construction in step two of §2 can be imitated in a striaghtforward fashion and thus we prove Theorem 1. We should emphasize again that since S is not an isometry, in the m-th stage of the construction when we want f_m to be close to f_{m-1} we must worry "separately" about each of the previous levels. The main point is that the number of these levels is fixed and so we can control simultaneously all of the levels by means of the first and the continuity of the iterates of S.

§4. ROHLIN TOWERS AND HYPERFINITENESS

The results of the proceding sections all point in the direction of a basic similarity between all non-dissipative aperiodic Borel automorphisms. Indeed under the natural definition of isomorphism, namely that $(X_i, \mathcal{B}_i, T_i)$, $i = 1,2$ are isomorphic if and only if there is a Borel isomorphism φ between $(X_i, \mathcal{B}_i)$ equivariant with respect to T_1, T_2, we have not yet seen non-isomorphic Borel automorphisms. A elementary invariant is the class of finite invariant measures for $(X, \mathcal{B}, T)$ denoted by $M_0(T)$. If $(X_i, \mathcal{B}_i, T_i)$ are to be isomorphic then there must be a one-to-one correspondence between $M_0(T_1)$ and $M_0(T_2)$. Thus for example the two shift σ_2 and three shift σ_3 are not Borel isomorphic since there is an element μ in

$M_0(\sigma_3)$ with $h_\mu(\sigma_3) = \log 3$ while for all $\nu \in M_0(\sigma_2)$ we have $h_\nu(\sigma_2) \leq \log 2$. To eliminate this type of invariant one can restrict attention to those Borel automorphisms for which $M_0(T)$ is empty. For easy examples of such T, take any uniquely ergodic (X,T) and some quasi-invariant ergodic measure λ different from the unique invariant measure μ, and let $X_0 \subset X$ be a T-invariant Borel set with $\lambda(X_0) = 1$, $\mu(X_0) = 0$. Then (X_0,T) is a non dissipative Borel automorphism with $M_0(T) = \phi$. Taking for T irrational rotations gives a wide class of such automorphisms.

A more explicit construction of such an $X_0 \subset X$ when (X,T) is strictly ergodic is as follows. Let μ be the unique invariant measure and construct by the usual procedures a set $A \subset X$, of first category with $\mu(A) > 0$. Then $X_0 = X \setminus \bigcup_{-\infty}^{\infty} T^j A$ is a second of second category with $\mu(X_0) = 0$. Since T is minimal no set of the second category can belong to $\mathcal{W}(T)$ and thus (X_0,T) is non-dissipative with $M_0(T|X_0) = \phi$. The fact that no set of second category belongs to $\mathcal{W}(T)$ follows at once from the fact that every Borel set has the Baire property, i.e., differs from an open set by a set of first category. If now $C \in \mathcal{W}(T)$ is of second category then there would exist a wandering set W of second category. But $W \supset U \setminus E$ for some non empty open set U and E of first category. Since T is minimal (topological transitivity would do) there exists some $n_0 > 0$ with $U \cap T^{n_0} U \neq \phi$, but then since X is compact and hence complete also $W \cap T^{n_0} W \neq \phi$ contradicting the fact that W is a wandering set.

In seeking possible invariants to distinguish Borel automorphisms one is led to the questions of the existence of generating partitions. We say that a finite or countable partition $\mathcal{P}$ of X generates if the σ-algebra generated by the $T^i\mathcal{P}$'s is all of $\mathcal{B}$ modulo $\mathcal{W}(T)$. Recall that Rohlin proved that countable generators always exist for finite invariant measures, while Krengel has shown that modulo the null set of an ergodic measure of type II_∞ or III two-set generators exist. I haven't been able to decide whether or not countable generators always exist in the Borel-sense defined above. The general case is essentially equivalent to settling one concrete example - the shift based on [0,1], i.e. $X = [0,1]^Z$ with $(Tx)(n) = x(n+1)$. I have been successful in extending one of the basic tools for these questions to the Borel framework, namely Rohlin's lemma, and it is perhaps worth recording at least that. Since there is no definite measure in this picture I take the point of view in [LW] and pose the following question:

Given $A \subset X$ and $n \in N$ when does there exist a $B \subset X$ with $B, TB, \ldots, T^{n-1}B$ disjoint and $A \setminus \bigcup_{j=0}^{n-1} T^j B \in \mathcal{W}(T)$?

If there is any quasi-invariant ergodic measure μ with respect to which T^n is ergodic and $\mu(A) = 1$ then clearly such an A cannot exist. We are thus led to a definition: $B \in \mathcal{B}$ will be termed a completely positive set if $\mu(B) > 0$ for all $\mu \in M(T)$, and it is a natural condition to demand of A that $X \setminus A$ be a completely positive set. For prime n it turns out that no further obstruction can occur, and we have the following theorem:

THEOREM 1. *If T is a non-periodic Borel automorphism of a standard Borel space $(X,\mathcal{B})$ and $A \in \mathcal{B}$ has a completely positive complement then for any prime p, there exists a set $B \in \mathcal{B}$ with*

(i) $B, T,B, \ldots T^{p-1}B$ mutually disjoint, and

(ii) $A \setminus \bigcup_{0}^{p-1} T^jB \in \mathcal{W}(T)$.

Before proving the theorem we need some properties of completely positive sets. Basic for us in the fact that C is a completely positive set if and only if $X \setminus \bigcup_{j=0}^{\infty} T^nC$ belongs to $\mathcal{W}(T)$ or recalling our remark preceding the Poincare recurrence theorem, the smallest T-invariant set containing C is modulo $(\mathcal{W})$ all of X. This is nothing but a restatement of the fundamental Corollary 1.9. The notion of a completely positive set differs from that of a set of positive measure, for example if C is completely positive and $C = C_1 \cup C_2$ then it may be that neither C_1 nor C_2 is completely positive, but what is true is that X splits into two T-invariant sets $X_1 \cup X_2$ with $X_1 \cap X_2 = \phi$ and $C_1 \cap X_1$ completely positive in X_1 and $C_2 \cap X_2$ completely positive in X_2. The decomposition is of course not unique, one such may be obtained by taking for $X_1 = \bigcup_{0}^{\infty} T^jC_1$ and for $X_2 = X \setminus X_1$ (always modulo $\mathcal{W}(T)$ which will be systematically repressed in the sequel).

LEMMA 2. *If T is an aperiodic Borel automorphism then for any $k \geq 2$, there is a completely positive set C with $C, TC, \ldots, T^{k-1}C$ mutually disjoint.*

PROOF. We may assume that X is a compact metric space (no assumption on the continuity properties of T.) For each $\ell \in N$ let

$$A_\ell = \{x : d(x, T^ix) > 1/\ell \quad , \text{ all } \; 0 < i < k\}$$

Then $\bigcup_{1}^{\infty} A_\ell = X$ since T is assumed aperiodic. For each ℓ, we can partition the space into a finite number of sets each of diameter less than $1/\ell$, say $\{B_{\ell,j}\}_{j=1}^{J(\ell)}$ and thus we have for all $\ell \in N$ and $1 \leq j \leq J(\ell)$ that $B_{\ell,j} \cap A_\ell = C_{\ell,j}$ satisfies

$$T^i C_{\ell,j} \cap C_{\ell,j} = \phi \quad \text{all} \quad 0 < i < k$$

and $\bigcup C_{\ell,j} = X$. We take now in lexicographic order the $C_{\ell,j}$'s and put $C_1 = C_{1,1}$; then $C_2 = (X \setminus \bigcup_{-\infty}^{\infty} T^i C_1) \cap C_{1,2}$; $C_3 = (X \setminus \bigcup_{-\infty}^{\infty} T^i(C_1 \cup C_2)) \cap C_{1,3}, \ldots$ etc., is easy to check that $C = \bigcup_1^{\infty} C_j$ satisfies the requirement. ∎

A refinement of the above argument will show that we can require in addition that C lies inside a preassigned completely positive set.

LEMMA 3. <u>If</u> T <u>is an aperiodic Borel automorphism then for</u> $k \geq 2$ <u>and any completely positive set</u> C, <u>there is a completely positive</u> $C_0 \subset C$ <u>such that</u> $C_0 \cap T^i C_0 = \phi$ <u>all</u> $0 < i < k$.

PROOF. (a) Suppose first that $k = 2$, and set $D = T^{-1}(TC \setminus C)$. On the one hand $D \cap TD = \phi$, while if $X_1 = \bigcup_0^{\infty} T^j D$ then modulo $\mathcal{W}$ X_1 is T-invariant and $X \setminus X_1 \subset C$. If $X \setminus X_1$ is not in $\mathcal{W}$, use Lemma 2 to find there a completely positive E, and then $D \cup E$ satisfies $T(D \cup E) \cap D \cup E = \phi$ while $D \cup E$ is completely positive. Thus the lemma holds for $k = 2$.

(b) Suppose true for k. Then to see that the lemma holds for $2k$, given C completely positive, find $C_0 \subset C$ completely positive with $T^i C_0 \cap C_0 =$ for $0 < i < k$. Then apply the argument of (a) to T_{C_0} to get $C_1 \subset C_0$ which is completely positive for the induced transformation T_{C_0} and satisfies $C_1 \cap T_{C_0} C_1 = \phi$. One checks that C_1 is completely positive for T and for $0 < i < 2k$, $C_1 \cap T^i C_1 = \phi$. Finally note that if the lemma is true for k it is true for all $k' < k$, thus (a) and (b) give an inductive proof of the lemma. ∎

We are now in a position to imitate the proof of [LW]. To see the idea we shall begin with a special case.

SPECIAL CASE OF THEOREM. <u>Assume that</u> $X \setminus A$ <u>is completely positive for</u> T^n. In this special case we do not need the fact that n is prime. Denote by $D = X \setminus A$, and notice that TD is contained in $\bigcup_{j=1}^{\infty} T^{-jn} D$, hence for each $x \in D$ (modulo $\mathcal{W}$) there is some $j \geq 1$ such that $T^{jn+1} x \in D$. Let $f_1(x)$ denote the least such x. Continuing by induction we can define measurable functions $f_1(x) < f_2(x) < \ldots < f_{n-1}(x)$ on D such that

$$T^{f_i(x)n+i} x \in D, \text{ all } 1 \leq i \leq i.$$

Divide D into a countable number of sets according to the value of $f_{n-1}(x)$ and denote those sets (not in $\mathcal{W}$) by $D_1, D_2, \ldots$ so that

(i) $D = \bigcup_1^\infty D_j$

(ii) for each D_j there is some M_j with $f_{n-1}(x) = M_j$ for all $x \in D_j$.

Start with D_1 and set $X_1 = \bigcup^\infty T^j D_1$, which is T-invariant. Apply Lemma 3 to find a completely positive set (for $T|X_1$) $D_1^* \subset D_1$ such that

$$D_1^* \cap T^j D_1^* = \phi , \quad 0 < j < 2M_1 .$$

We represent X_1 as a tower over D_1^*, which means that we consider the space partitioned into the sets, D_1^*, $T D_1^* \setminus D_1^*$, $T^m D_1^* \setminus (\bigcup_0^{m-1} T^j D_1^*), \ldots$ and then divide the tower into pure columns relative to the partition of X_1 into $X_1 \cap A$ and $X_1 \setminus A$. This gives a countable number of columns (since modulo $\mathcal{W}$ the height is everywhere finite) corresponding to a countable partition of D_1^* into sets $D_{1,\ell}^*$ such that for all $x,y \in D_{1,\ell}^*$ and all i less than $r(\ell)$, which is the common return time to D_1^* for all points in $D_{1,\ell}^*$ and gives the height of the column above $D_{1,\ell}^*$, $T^i x$ and $T^i y$ either both belong to A or both belong to $X_1 \setminus A$. The definition of the set B will be given separately for each such column.

Matters have been arranged so that $r(\ell) \geq 2M_1$ for all ℓ, and therefore we will be able to divide the column above $D_{1,\ell}^*$ into blocks of length divisible by n by omitting levels that belong to D. This is done by removing the levels at height: $0, f_1(x), f_2(x), \ldots, f_{i-1}(x)$ where $r(\ell) \equiv i \pmod n$. For each remaining block one assigns to B every n-th level. Thus $T^j B \cap B = \phi$ for $0 < j < n$ and only points of $D \cap X_1$ are not covered by $\bigcup_0^{n-1} T^j B$.

Next one considers $D_2 \setminus X_1$, sets $X_2 = \bigcup_0^\infty T^j (D_2 \setminus X_1)$ and continues as described above. After a countable number of steps all of X will be accounted for and we have established.

PROPOSITION 4. *If $X \setminus A$ is completely positive for T^n then there is a set B with $T^j B \cap B = \phi$ for $0 < j < n$ and $\bigcup_0^{n-1} T^j B \supset A$.*

To complete the proof of Theorem 1 we have to analyze what can happen when a set C is completely positive for T but not for T^n. Here the fact that n is prime plays a decisive role in simplifying the analysis. The situation is described by the following lemma:

LEMMA 5. *If* C *is completely positive for* T, *and* p *is prime then* X *can be partitioned into two T-invariant sets* X_1, X_2 *such that*:

(i) $C \cap X_1$ *is completely positive for* $T^p \, X_1$;

(ii) *there is a set* $D \subset X_2$ *such that* $T^j D \cap D =$ *for* $0 < j < p$ *and* $\bigcup_{0}^{p-1} T^j D = X_2$ (modulo $\mathcal{W}$).

PROOF. Let us put $C^* = \bigcup_{j=0}^{\infty} T^{jp} C$. Then C^* is invariant under T^p modulo $\mathcal{W}(T^p)$ or, what is the same, modulo $\mathcal{W}(T)$. One can either use Corollary 1.9 to prove $\mathcal{W}(T) = \mathcal{W}(T^p)$ or see this fact by a direct argument. Now since C is completely positive for T we have that $X = \bigcup_{j=0}^{p-1} T^j C^*$. If $TC^* = C^*$ then C is already completely positive for T^p and we put $X = X_1$. Otherwise $TC^* \setminus C^*$ is non trivial and we consider $C_1^* = C^* \setminus T^{-1} C^*$, which is again T^p-invariant and satisfies $TC_1^* \cap C_1^* = \phi$. We put $X_2 = \bigcup_{j=0}^{p-1} T^j C_1^*$ and $X_1 = X \setminus X_2$. It is easy to check that $C^* \setminus X_2$ is T-invariant and thus $C \cap X_1$ is completely positive for $T^p | X_1$. It remains to find a set D as required by (ii). For this we observe that the primality of p implies that there cannot be a T^p-invariant set $E \subset X_2$ with $E \cup TE \cup \ldots \cup T^{m-1} E = X_2$, $E \cup T^i E = \phi$ for $0 < i < m$ for any $p < m$. Thus if $p > 2$, $T^2 C_1^* \setminus (C_1^* \cup TC_1^*)$ must be non trivial, and we put $C_2^* = C_1^* \setminus (T^{-2} C_1^* \cup T^{-1} C_1^*)$ and consider the tower over C_2^*. This must coincide with X_2 by our previous remark and continuing in this way we eventually get the required D. The model to keep in mind here is the cyclic permutation on p-elements, Z_p where C_1^* is some subset $E \subset Z_p$ and we are going after a singleton in the algebra generated by E and its complement. ∎

PROOF OF THE THEOREM 1. Apply Lemma 5 to decompose the space into two pieces X_1 and X_2. On X_1 use Proposition 4 while on X_2 the result is manifestly true, putting together the two bases we get a B as required. ∎

A question that is closely related to the validity of a Rohlin lemma is the notion of hyperfinitess. Here one focuses on the equivalence relation R_T defined by T, namely xR_Ty if and only if there is some $n \in Z$ such that $y = T^n x$. As a subset of $X \times X$ with the product Borel structure $\mathcal{B} \times \mathcal{B}$ the equivalence relation R_T,

$$R_T = \{(x,y) : y = T^n x \quad \text{for some} \quad n \in Z\}$$

is $\mathcal{B} \times \mathcal{B}$ measurable. Such an equivalence relation R, is said to be *hyperfinite* if there is an increasing sequence of *finite* sub-equivalence relations, $R_1 \subset R_2 \subset \ldots$ with $\bigcup_{1}^{\infty} R_n = R$. Here of course by a *finite* equivalence relation we mean one in which each equivalence class is a finite set. It is not hard to see that if R is hyperfinite then we can also find $R_1 \subset R_2 \subset \dots$, with $\bigcup_1^\infty R_n = R$ and each R_n bounded uniformly, i.e. there is some M_n such that all equivalence classes contain at most M_n elements. By an analysis somewhat analogous to our analysis of the periodic points of T one sees that finite equivalence relations $R \subset X \times X$ are unions of functions and thus have a simple structure. It is an easy consequence of some of the preliminary lemmas above that any Borel automorphism is hyperfinite.

THEOREM 6. Any Borel automorphism is hyperfinite.

PROOF. Let $B_1 \supset B_2 \supset \dots$ be a decreasing sequence of completely positive sets with $T^i B_n \cap B_n = \phi$ for $0 < i < 10^n$, for $n = 1,2,\dots$. The existence of such a sequence follows at once from Lemma 3. We shall carry out the construction of $R_1 \subset R_2 \subset \dots$ with $\bigcup R_n = R_T$ neglecting what happens on some sets in $\mathcal{W}(T)$. There will be a countable number of such sets, and collecting them together they are all contained in a single T-invariant set in $\mathcal{W}(T)$, say U. Since R_T restricted to U is manifestly hyperfinite, we can conclude that all of R_T is hyperfinite, and thus no harm is caused by neglecting sets in $\mathcal{W}(T)$.

To define R_1, represent X (modulo $\mathcal{W}$) as a tower over B_1, and partition B_1 according to height of the tower, say $B_1 = \bigcup_{10}^\infty B_1^\ell$ with $x \in B_1^\ell$ if

$$\ell = \min \{m \geq 1 : T^m x \in B_1\}.$$

For each $x \in B_1^\ell$ we put $\{T^i x : 0 \leq i \leq \ell\}$ in a single equivalence class. This clearly defines a measurable finite sub-equivalence relation which we denote by R_1. In general R_n is defined by the same recipe after representing the whole space (modulo $\mathcal{W}$) as a tower over B_n. The fact that R $R_1 \subset R_2 \subset \dots$ is a consequence of the fact that $B_1 \supset B_2 \supset \dots$. It only remains to check that $\bigcup_1 R_n = R_T$. To see this consider the set of those $x \in X$ such that the pair $(x, T^{-1}x) \notin \bigcup_1^\infty R_n$. Clearly this set consists precisely of those x that are in $\bigcap_1 B_n$. By the defining properties of the B_n, this latter intersection is a wandering set, so that omitting once more a set from $\mathcal{W}(T)$ we have a T-invariant set X_0 such that for all $x \in X_0$, $(x_0, T^{-1}x_0) \in \bigcup_1^\infty R_n$ which implies that $\bigcup_1^\infty R_n$ coincides with R_T on X_0. As we've remarked already this completes the proof. ■

I don't know if this last theorem is valid for Borel actions of a general amenable group. It was established in [CFW] that for any given non singular measure the equivalence relation derived from the action of an amenable group is hyperfinite, but I don't know if the procedure described there can be carried out in a Borel fashion without using the measure.

REFERENCES

[AS] R. Aumann and L. Shapley, Values of Non-Atomic Games, Princeton Univ. Press 1974.

[CFW] A. Connes, J. Feldman and B. Weiss, An amenable equivalence relation is generated by a single transformation, Ergodic Theory and Dynamical Systems, v. 1 (1981), pp. 431-450.

[E] E. G. Effros, Transformation groups and C^*-algebras, Ann. of Math. v. 81 (1965), pp. 38-55.

[F] H. Furstenberg, Recurrence in Ergodic Theory and Combinatorial Number Theory, Princeton Univ. Press, 1981.

[G] J. Glimm, Locally compact transformation groups, Trans. Amer. Math. Soc. v. 101 (1961), pp. 124-138.

[KW] Y. Katznelson and B. Weiss, The construction of quasi-invariant measures, Israel J. of Math. v. 12 (1972), pp. 1-4.

[K] W. Krieger, On Borel automorphisms and their quasi-invariant measures, Math. Z. v. 151 (1976), pp. 19-24.

[LW] E. Lehrer, and B. Weiss, An ε-free version of Rohlin's lemma, Ergodic Theory and Dynamical Systems, v. 2 (1982), pp. 45-48.

[OX] J. Oxtoby, Measure and Category, Springer, New York, 1971.

[SW] S. Shelah and B. Weiss, Measurable recurrence and quasi-invariant Israel J. of Math. v. 43 (1982), pp. 154-160.

[V] V. S. Varadarajan, Groups of automorphisms of Borel spaces, Trans. Amer. Math. Soc. v. 109 (1963), pp. 191-220.

[W] B. Weiss, Orbit equivalence of non-singular actions, in Theorie Ergodique, Monog. 29, L'Eenseignement Math. 1981, pp. 77-107.

Department of Mathematics
The Hebrew University of Jerusalem
Jerusalem, Israel

Contemporary Mathematics
Volume 26, 1984

A NOTE ON MIKUSIŃSKI'S PROOF OF THE TITCHMARSH CONVOLUTION THEOREM

Kôsaku Yosida* and Shigetake Matsuura**

THE TITCHMARSH CONVOLUTION THEOREM

This theorem was discovered in 1926 by E. C. Titchmarsh [4]. It reads:

THEOREM. Let $C[0,\infty)$ denote the totality of complex-valued continuous functions $f = f(t)$ defined on the interval $[0,\infty)$. If the convolution $f\cdot g$ of two functions f and g of $C[0,\infty)$ satisfies

$$f\cdot g(t) = \int_0^t f(t-\tau)g(\tau)d\tau \equiv 0,$$

then either $f(t) \equiv 0$ or $g(t) \equiv 0$ must hold.

The original proof of Titchmarsh is not easy since it involves deep theorems on analytic functions. An elementary proof was given in 1951 by J. Mikusiński [2] jointly with C. Ryll-Nardzewski by making use of the Lerch moment theorem [1]. The Lerch theorem was derived in [2] from a theorem of E. Phragmén [3].

The purpose of the present note is to propose a modification of the Mikusiński proof so that we obtain a plain proof, by making use of the Liouville theorem in analytic function theory combined with the Weierstrass polynomial approximation theorem.***

II. PROOF OF THE THEOREM

The first step (the special case $f = g$). We start with

$$\int_0^t f(t-\tau)f(\tau)d\tau \equiv 0. \tag{1}$$

Let $T > 0$ and β be any real number. Put

$$\hat{f}(t) = e^{-i\beta t}f(t), \quad \text{where } i = \sqrt{-1}. \tag{2}$$

---

* Department of Mathematics, the University of Tokyo.
** Research Institute of Mathematical Sciences, Kyoto University.
*** After the present note was finished, Professor K. Masuda communicated to us another proof which makes use of the Liouville theorem combined with the Fourier transform.

Then, for any real number α, we obtain

$$(3)\qquad (\int_{-T}^{T} e^{\alpha u}\hat{f}(T-u)du)^2 = \int_{-T}^{T}\int_{-T}^{T} e^{\alpha(u+v)}\hat{f}(T-u)\hat{f}(T-v)dudv$$

$$= \int_{-T}^{T}dv\int_{u=-v}^{T} e^{\alpha(u+v)}\hat{f}(T-u)\hat{f}(T-v)du$$

$$+ \int_{-T}^{T}dv\int_{-T}^{u=-v} e^{\alpha(u+v)}\hat{f}(T-u)\hat{f}(T-v)du \quad = I_{\alpha,\beta,1} + I_{\alpha,\beta,2}\ .$$

By the substitution

$$u = T - (t-\tau), \quad v = T - \tau$$

with

$$2T \geqq u+v \geqq 0, \quad T \geqq u \geqq -T, \quad T \geqq v \geqq -T$$

and

$$\partial(u,v)/\partial(t,\tau) = 1,$$

we have, by (1) and (2),

$$(4)\qquad I_{\alpha,\beta,1} = \int_0^{2T}dt\int_0^t e^{\alpha(2T-t)}\hat{f}(t-\tau)\hat{f}(\tau)d\tau$$

$$= \int_0^{2T} e^{\alpha(2T-t)-i\beta t}(\int_0^t f(t-\tau)f(\tau)d\tau)dt = 0.$$

If $\alpha > 0$ then $e^{\alpha(u+v)} \leqq 1$ in the integration domain of $I_{\alpha,\beta,2}$. Hence

$$(5)\qquad |I_{\alpha,\beta,2}| \leqq \int_{-T}^{T}dv\int_{-T}^{u=-v}|\hat{f}(T-u)\hat{f}(T-v)|du$$

$$\leqq \int_{-T}^{T}\int_{-T}^{T}|f(T-u)f(T-v)|dudv = M^2 \quad \text{with} \quad 0 \leqq M < \infty.$$

Therefore, by (3), (4) and (5),

$$(6)\qquad |\int_{-T}^{T} e^{(\alpha+i\beta)u}f(T-u)du| \leqq M, \quad \text{whenever} \quad \alpha \geqq 0.$$

Hence the Laplace integral

$$(7)\qquad E_T(z) = \int_0^T e^{zu}f(T-u)du$$

satisfies, for $\alpha = \text{Re}\, z \geqq 0$,

$$(8)\qquad |E_T(z)| \leqq N, \quad \text{where} \quad N = M + \int_{-T}^{0}|f(T-u)|du.$$

It is easy to see that the entire analytic function

$$E_T(z) = \int_0^T e^{zu}f(T-u)du$$

satisfies

$$(9)\qquad |E_T(z)| = |\int_0^T e^{zu}f(T-u)du| \leqq N_1 \quad \text{for all} \quad z, \quad \text{where}$$

$$N_1 = \max(N,K) \quad \text{with} \quad K = \max_{0\leqq u\leqq T} T \times |f(T-u)|.$$

Hence, by the Liouville theorem, we have $F_T(z) =$ a constant, and so, by differentiating $F_T(z)$ at $z = 0$, we obtain

$$\int_0^T u^n f(T-u)du = 0 \qquad (n = 0,1,2,\ldots).$$

Therefore, by the Weierstrass polynomial approximation theorem, we have

$$\int_0^T q(u)f(T-u)du = 0$$

for every continuous function $q(u)$. Thus $f(T-u)$ must vanish for $0 \leqq u \leqq T$, i.e., $f(t)$ must vanish for $0 \leqq t \leqq T$. This proves the Theorem for the special case $f = g$.

The second step (the general case $f \neq g$). Let us start with

(10) $$f\cdot g(t) = \int_0^t f(t-\tau)g(\tau)d\tau = 0 \quad \text{for all} \quad t \geqq 0.$$

Following after [2], we put

(11) $$f_1(t) = tf(t) \quad \text{and} \quad g_1(t) = tg(t).$$

Then $f_1\cdot g(t) + f\cdot g_1(t) \equiv 0$, because, by (1),

$$\int_0^t (t-\tau)f(t-\tau)g(\tau)d\tau + \int_0^t f(t-\tau)\tau g(\tau)d\tau = t\int_0^t (f(t-\tau)g(\tau)d\tau.$$

Hence, again by (10),

$$\begin{aligned} 0 &\equiv (f\cdot g_1)\cdot(f_1\cdot g + f\cdot g_1)(t) \\ &\equiv (f\cdot g)\cdot(f_1\cdot g_1)(t) + (f\cdot g_1)\cdot(f\cdot g_1)(t) = (f\cdot g_1)\cdot(f\cdot g_1)(t). \end{aligned}$$

That is, we have obtained

(12) $$\int_0^t f(t-\tau)\tau g(\tau)d\tau \equiv 0.$$

Iterating the same argument, we have

(12)' $$\int_0^t f(t-\tau)\tau^n g(\tau)d\tau \equiv 0 \qquad (n = 0,1,2,\ldots).$$

Thus, by the Weierstrass polynomial approximation theorem, we see that $\int_0^t f(t-\tau)g(\tau)q(\tau)d\tau \equiv 0$ for every continuous function $q(t)$. This proves that

$$f(t-\tau)g(\tau) \equiv 0 \quad \text{for} \quad 0 \leqq \tau \leqq t < \infty,$$

that is, either $f(t) \equiv 0$ or $g(t) \equiv 0$ must hold.

REFERENCES

[1] E. Lerch, Sur un point de la théorie des fonctions génératrices d'Abel. Mathematica 27 (1903), 339-352.

[2] J. Mikusiński, Operational Calculus, Pergamon Press (1967).

[3] E. Phragmén, Sur une extension d'une théoréme classique de la théorie des fonctions, Acta Mathematica 28 (1904), 331-369.

[4] E. C. Titchmarsh, Introduction to the Theory of Fourier Integrals, (1937), p. 327- .

Contemporary Mathematics
Volume 26, 1984

ERGODIC ACTIONS OF ARITHMETIC GROUPS AND THE KAKUTANI-MARKOV FIXED POINT THEOREM[1]

Robert J. Zimmer[2]

Our aim in this lecture is two-fold. First we describe some recent results concerning ergodic actions of certain discrete subgroups of Lie groups. Second, we will give some indication of the role that the classical Kakutani-Markov fixed point theorem (proven independently by these mathematicians over 40 years ago) plays in the proof of these results. The groups in question will be lattice subgroups of $SL(n,R)$, $n \geq 3$ (or other higher rank semisimple groups). While in recent years some significant progress has been made concerning the measure theoretic behavior of the actions of these groups, the results we describe here can be viewed as a beginning of "smooth ergodic theory" for these groups. Thus, we study actions of these groups on manifolds and the smooth structure of the manifold and not only the measure theoretic structure will play a basic role. We will see a rigidity phenomenon that has no analogue for actions of the integers or real line, and from a more algebraic point of view, our results can be considered as the first rigidity results for homomorphisms of the groups in question into diffeomorphism groups.

Although our results are valid for more general semisimple groups, for simplicity we will restrict our attention here to the groups $G_n = SL(n,R)$. We shall always assume $n \geq 3$. We let $\Gamma \subset G_n$ be a lattice subgroup, i.e. Γ is discrete and G_n/Γ has a finite G_n-invariant measure. A basic example is $\Gamma = SL(n,Z)$. In this case G_n/Γ is of finite (G_n-invariant) volume but is not compact. Cocompact lattices (i.e. G_n/Γ compact) also exist for every n. We can now pose a very general question.

QUESTION. <u>What are the smooth, volume preserving, ergodic actions of Γ on compact manifolds?</u>

EXAMPLES.

a) Let H be a Lie group and $\Lambda \subset H$ a subgroup such that H/Λ is compact with a finite H-invariant measure. Let $\pi: G_n \to H$ be a homomorphism. Then via π, Γ acts in a smooth volume preserving fashion on H/Λ. This will often

---

[1] Research partially supported by NSF Grant MCS 8004026

[2] Sloan Foundation Fellow

be ergodic. (For example, if H is simple and π is non-trivial, this will always be ergodic.)

b) $SL(n,Z)$ acts on $T^n = R^n/Z^n$ by automorphisms. This action is volume preserving and ergodic. More generally, we may have Γ acting by automorphisms of a nilmanifold N/Λ. (Here N is a nilpotent Lie group, Λ a lattice subgroup.)

c) We may have homomorphisms of Γ into a compact Lie group K with dense image. Then Γ acts ergodically and isometrically on the homogeneous spaces of K.

All these examples are of course of an algebraic nature. One way to attempt to generate new examples of actions would be to try to perturb a given action. Our first main result shows that for the isometric actions, sufficiently smooth perturbations do not yield new actions, at least up to topological conjugacy.

THEOREM 1. *Let* M *be a compact Riemannian manifold,* $\dim M = m$. *Set* $r = m^2 + m + 1$. *Assume* Γ *acts by smooth isometries of* M. *Let* $\Gamma_0 \subset \Gamma$ *be a finite generating set*. *Then any volume preserving action of* Γ *on* M *which*

i) *for elements of* Γ_0 *is a sufficiently small* C^r *perturbation of the original action;*

and ii) *is ergodic;*

actually leaves a C^0*-Riemannian metric invariant. In particular, there is a* Γ*-invariant topological distance function and the action is topologically conjugate to an action in example* (c) *above.*

Being algebraic in nature, the examples presented above carry dimension restrictions. For instance, in example (c) we have the following consequence of the work of Margulis [2] and Raghunathan [4]: If K is a compact Lie group and $\Gamma \to K$ is a dense range homomorphism, then $\dim K \geq \dim G_n (= n^2-1)$. We now put forward:

CONJECTURE: *Suppose* $n > \dim M > 0$. *Then there is no smooth, volume preserving, ergodic action of* Γ *on* M.

Our second main result is in the direction of this conjecture. It asserts that the conjecture is true if we add one further hypothesis, namely "near C^r-isometry" on the generators of the group. More precisely, let ξ be a smooth metric on a compact manifold M. For each integer $r \geq 0$, there is a natural C^r-topology on the space of all smooth metrics. Let $\mathcal{O}$ be a neighborhood of ξ in the C^r topology. If S is a set of diffeomorphisms of M, we say that ξ is $(\mathcal{O},S)$-invariant if $f^*\xi \in \mathcal{O}$ for all $f \in S$. We then have:

THEOREM 2. *Let* M *be a compact manifold,* $\dim M = m$. *Let* $r = m^2 + m + 1$.

Assume $n > m > 0$. *Let* $\Gamma_0 \subset \Gamma$ *be a finite generating set. Then for every smooth metric* ξ *on* M, *there is a* C^r *neighborhood* $\mathcal{O}$ *of* ξ *such that there are no smooth* $\mathrm{vol}(\xi)$*-preserving ergodic actions of* Γ *on* M *for which* ξ *is* $(\mathcal{O},\Gamma_0)$*-invariant*.

To put the condition of $(\mathcal{O},\Gamma_0)$-invariance into sharper focus, we remark that if ψ_t is a smooth flow on M, then for any $\xi,\mathcal{O}$ we have for t_0 sufficiently small that ξ is $(\mathcal{O},\psi_{t_0})$-invariant. Thus this condition does not present an obstruction to a single diffeomorphism being mixing or even Bernoulli. However, for Γ we have the following general result from which both Theorems 1 and 2 can be deduced.

THEOREM 3. *Let* M *be a compact manifold*, $m = \dim M$, $r = m^2 + m + 1$. *Let* $\Gamma_0 \subset \Gamma$ *be a finite generating set. Then for every smooth metric* ξ *on* M *there is a* C^r *neighborhood* $\mathcal{O}$ *of* ξ *such that any smooth*, $\mathrm{vol}(\xi)$*-preserving, ergodic action of* Γ *on* M, *with* ξ $(\mathcal{O},\Gamma_0)$*-invariant, leaves a* C^0*-Riemannian metric on* M *invariant*.

To deduce Theorem 1 from Theorem 3 is straightforward. To deduce Theorem 2, we use (among other things) the results of Margulis and Raghunathan mentioned above and the well known fact that for a manifold of dimension m, the group of isometries can have dimension at most $m(m + 1)/2$.

We now turn to a discussion of the role of the Kakutani-Markov fixed point theorem in the proof of these results. We should emphasize that our goal here is limited. The proofs of the theorems are rather involved, even in outline, and we do not purport to present an overview of the proof. However, the Kakutani-Markov theorem is one of the ingredients in the proof and we will try to indicate how it arises.

For simplicity, we shall assume we are in the context of Theorem 2, so that $\dim M = m < n$. Our goal is to construct a Γ-invariant C^0-Riemannian metric on M. Γ of course acts on the space of all C^0 Riemannian metrics on M, and so we are looking for a fixed point in this space. Let us recall the Kakutani-Markov theorem, a basic theorem asserting the existence of a fixed point under certain conditions.

KAKUTANI-MARKOV THEOREM [1]: *A solvable group acting by affine automorphisms of a compact convex set has a fixed point*.

We are of course very far from being able to apply this result to our situation. First of all, Γ is not solvable, and secondly, the space of C^0 metrics, while it is convex, is not compact. Non-compactness arises in fact from two directions: The space of inner products on a single tangent space is not compact, and generally speaking, a space of continuous functions (even if the values are constrained to lie in a compact set) will not be compact. The second non-compactness problem can be avoided at the expense of taking measurable functions rather than only continuous ones (using a suitable weak-*-topology.) To deal with the first non-compactness problem we will replace inner products by measures.

More precisely, suppose V is a finite dimensional real vector space. Let $F\ell(V)$ be the (compact) space of (full) flags of V. An inner product ξ on V canonically determines a probability measure on $F\ell(V)$. (Namely, the orthogonal group $O(\xi)$ is compact and acts transitively on $F\ell(V)$. Thus there is a unique $O(\xi)$-invariant probability measure on $F\ell(V)$.) If M is a manifold, we let $F\ell(M) \to M$ be the flag bundle. Thus the fiber at each point $x \in M$ is $F\ell(TM_x)$ where TM_x is the tangent space at x. Any Riemannian metric ξ on M thus gives us in a canonical way a "measurable field of measures in $F\ell(M)$", where the latter means that for each $x \in M$ we have a probability measure $\mu(x)$ on the flags at x, and $x \to \mu(x)$ is measurable. Γ acts on the measurable fields of measures in $F\ell(M)$, and the existence of a Γ-invariant (measurable) Riemannian metric on M clearly implies the existence of a Γ-invariant measurable field of measures in $F\ell(M)$. In general a measure on $F\ell(V)$ need not come from an inner product on V, and correspondingly for an arbitrary group action the existence of an invariant measurable field of measures will not imply the existence of an invariant Riemannian metric (even if we allow the Riemannian metric to only be measurable in $x \in M$.) However, for Γ we have the following basic fact.

THEOREM 4. <u>If Γ acts by volume preserving diffeomorphisms of a compact manifold M, there is a Γ-invariant measurable Riemannian metric on M if and only if there is a Γ-invariant measurable field of measures in $F\ell(M)$.</u>

The space of all measurable fields of measures in $F\ell(M)$ is a compact convex set in a suitable weak-*-topology. Thus, Theorem 4 suggests that we can use the Kakutani-Markov theorem if we can deal with the following two (major) problems:

1) How can we use the Kakutani-Markov theorem to obtain a fixed point in the space of measurable fields of measures in $F\ell(M)$ when Γ is not solvable?
2) How do we pass from the existence of a measurable Γ-invariant Riemannian metric to the existence of a C^0 Γ-invariant Riemannian metric?

Both of these questions are highly non-trivial and the solutions in both cases use in very strong ways special properties of the group Γ (and its ambient Lie group G_n) that we are considering. We next describe the solution to problem (1).

Since Γ is not solvable, one is tempted to apply the Kakutani-Markov theorem to solvable subgroups of Γ. However, Γ does not possess a natural solvable subgroup Γ_0 such that the quotient Γ/Γ_0 carries a manageable structure. On the other hand, in the ambient Lie group G_n, the subgroup of upper triangular matrices, say B_n, will be solvable and G_n/B_n can be naturally identified with the compact space $F\ell(R^n)$ of full flags in R^n. Thus, we shall induce the Γ-actions to G_n-actions in a standard manner and apply the Kakutani-Markov theorem to the solvable subgroup B_n. We recall that if X is any Γ-space then there is a naturally associated G-space, $X \times_\Gamma G_n$ defined as follows. Let Γ act on $X \times G_n$ by $\gamma(x,g) = (\gamma x,\gamma g)$, and let G_n act on $X \times G_n$ by $h(x,g) = (x,gh^{-1})$. Then the Γ action and the G_n action commute, so we have an induced G_n action on $X \times_\Gamma G_n = (X \times G_n)/\Gamma$. In particular, since Γ acts on M and $F\ell(M)$, we obtain a G_n-map of G_n spaces $F\ell(M) \times_\Gamma G_n \to M \times_\Gamma G_n$. This is also a fiber bundle with the base of finite volume (since both M and G_n/Γ have finite volume) and the fiber of this bundle is still $F\ell(R^m)$ where $m = \dim M$. One can verify that there is a Γ-invariant measurable field of measures in $F\ell(M)$ if and only if there is a G_n-invariant measurable field of measures in $F\ell(M) \times_\Gamma G_n$. Of course, G_n is not solvable either, but by our discussion above we can apply the Kakutani-Markov theorem to obtain a B_n-invariant measurable field of measures in $F\ell(M) \times_\Gamma G_n$. Since the fiber of the bundle $F\ell(M) \times_\Gamma G_n \to M \times_\Gamma G_n$ is $F\ell(R^m)$ and we are at the moment only discussing measurable sections, let us identify each fiber with $F\ell(R^m)$. Then the B_n-invariant field of measures is a measurable map $\mu: M \times_\Gamma G_n \to \mathrm{Meas}(F\ell(R^m))$.

Quite generally, we recall that if Y is a G-space and there is a B-fixed point in Y where $B \subset G$ then there is an associated G-map $G/B \to Y$. This map is constant if and only if the B-fixed point is a G-fixed point. In our situation, making some standard identifications, the map μ gives us, for each $x \in M \times_\Gamma G_n$, a map $\mu_x: G_n/B_n \to \mathrm{Meas}(F\ell(R^m))$, i.e. $\mu_x: F\ell(R^n) \to \mathrm{Meas}(F\ell(R^m))$. If each μ_x were constant, the B_n-invariant field would be G_n-invariant, and problem (1) would be solved.

At this point we recall (only in vague terms) the work of Mostow [3] and Margulis [2]. One of its consequences is that a map $F\ell(R^n) \to \mathrm{Meas}(F\ell(R^m))$ satisfying certain algebraic relations related to a discrete subgroup implies the existence of a related group homomorphism $SL(n,R) \to SL(m,R)$. In our situation we have a family μ_x of such maps indexed by a G_n-space of finite volume, and B_n invariance implies certain algebraic relations between various μ_x. The results of Margulis can be generalized to apply to this situation as well (i.e. to obtain "superrigidity for cocycles" [5]) and we can deduce that if μ_x are not degenerate then there is a non-trivial homomorphism $G_n \to G_m$. In this manner, we are able to deduce that either the required G_n-invariant measurable field of measures exists or there is a non-trivial homomorphism $G_n \to G_m$. In the present situation $m < n$, so the latter is impossible. This completes our description of the solution of problem (1

Problem (2) is not concerned with the Kakutani-Markov theorem dire and we will say only a few words about it. Theorem 4 and the superrigidi arguments sketched above are valid under the more general situation of an action on an arbitrary vector bundle (not just the tangent bundle) and wh base space is of finite (invariant) measure , not necessarily compact. W apply the entire apparatus not just to the action on the tangent bundle b the action on certain jet bundles defined on the ergodic components of th Γ-action on the frame bundle of M. This enables us to derive results in suitable Sobolev space and to apply the Sobolev embedding theorem to dedu the required continuity. The arguments are involved, entailing in partic some geometric constructions on the frame bundle and some unitary represe theory of Γ . Details can be found in [6].

We conclude by mentioning one other point that arises in connectio problem (2) and which leads to a natural question. In constructing the S space alluded to in the preceding paragraph we need L^2 bounds on a measu invariant metric relative to an arbitrary smooth metric. More precisely i is a measurable Riemannian metric on M and σ is a smooth metric let us that ξ is L^p if there exist $f,h \in L^p(M)$ such that $\|x\|_\xi \leq f(p)\|x\|_\sigma$ $\|x\|_\sigma \leq h(p)\|x\|_\xi$ for all $x \in TM_p$. (Since M is compact, this is indepe of σ.) Using the condition of "near isometry" on the generators of the and Kazhdan's property concerning unitary representations of Γ , we are a to derive sufficient conditions that a Γ-invariant measurable Riemannian metric be L^p. It would probably be generally useful to develop alternate conditions ensuring that a measurable invariant Riemannian metric be L^p.

References

1. S. Kakutani, Two fixed point theorems concerning bicompact convex sets Proc. Imp. Acad. Tokyo 14 (1938), 242-245.

2. G. A. Margulis, Discrete groups of motions of manifolds of nonpositive curvature, A.M.S. Translations, 109 (1977), 33-45.

3. G. D. Mostow, Strong rigidity of locally symmetric spaces, Annals of M Studies, no. 78 (1973).

4. M. Raghunathan, On the congruence subgroup problem, Publ. Math. I.H.E. 46 (1976), 107-161.

5. R. J. Zimmer, Strong rigidity for ergodic actions of semisimple groups Annals of Math., 112 (1980), 511-529.

6. R. J. Zimmer, Volume preserving actions of lattices in semisimple grou compact manifolds, forthcoming.

UNIVERSITY OF CHICAGO
Chicago, Illinois

ABCDEFGHIJ-AMS-8987654